# Fabrication of Carbon and Related Materials/Metal Hybrids and Composites

# Fabrication of Carbon and Related Materials/Metal Hybrids and Composites

Editors

**Walid M. Daoush**
**Fawad Inam**
**Mostafa Baboli**
**Maha M. Khayyat**

MDPI • Basel • Beijing • Wuhan • Barcelona • Belgrade • Manchester • Tokyo • Cluj • Tianjin

*Editors*

Walid M. Daoush
Production Technology
Helwan University
Cairo
Egypt

Fawad Inam
Engineering and Computing
University of East London
London
United Kingdom

Mostafa Baboli
Chemical Engineering
Sohar University
Sohar
Oman

Maha M. Khayyat
Nanotechnology
Semiconductors Center,
Materials Science Research
Institute
King Abdulaziz City of
Science and Technology,
Riyadh
Saudi Arabia

*Editorial Office*
MDPI
St. Alban-Anlage 66
4052 Basel, Switzerland

This is a reprint of articles from the Special Issue published online in the open access journal *Crystals* (ISSN 2073-4352) (available at: www.mdpi.com/journal/crystals/special_issues/Carbon_Fabrication).

For citation purposes, cite each article independently as indicated on the article page online and as indicated below:

LastName, A.A.; LastName, B.B.; LastName, C.C. Article Title. *Journal Name* **Year**, *Volume Number*, Page Range.

**ISBN 978-3-0365-4434-2 (Hbk)**
**ISBN 978-3-0365-4433-5 (PDF)**

© 2022 by the authors. Articles in this book are Open Access and distributed under the Creative Commons Attribution (CC BY) license, which allows users to download, copy and build upon published articles, as long as the author and publisher are properly credited, which ensures maximum dissemination and a wider impact of our publications.

The book as a whole is distributed by MDPI under the terms and conditions of the Creative Commons license CC BY-NC-ND.

# Contents

**About the Editors** . . . . . . . . . . . . . . . . . . . . . . . . . . . . . . . . . . . . . . . . . . . . . . . . . . . . vii

**Preface to "Fabrication of Carbon and Related Materials/Metal Hybrids and Composites"** . . ix

**Walid M. Daoush, Turki S. Albogmy, Moath A. Khamis and Fawad Inam**
Syntheses and Step-by-Step Morphological Analysis of Nano-Copper-Decorated Carbon Long Fibers for Aerospace Structural Applications
Reprinted from: *Crystals* **2020**, *10*, 1090, doi:10.3390/cryst10121090 . . . . . . . . . . . . . . . . . . 1

**Norah Hamad Almousa, Maha R. Alotaibi, Mohammad Alsohybani, Dominik Radziszewski, Saeed M. AlNoman and Bandar M. Alotaibi et al.**
Paraffin Wax [As a Phase Changing Material (PCM)] Based Composites Containing Multi-Walled Carbon Nanotubes for Thermal Energy Storage (TES) Development
Reprinted from: *Crystals* **2021**, *11*, 951, doi:10.3390/cryst11080951 . . . . . . . . . . . . . . . . . . 17

**Abu Bony Amin, Syed Muhammad Shakil and Muhammad Sana Ullah**
A Theoretical Modeling of Adaptive Mixed CNT Bundles for High-Speed VLSI Interconnect Design
Reprinted from: *Crystals* **2022**, *12*, 186, doi:10.3390/cryst12020186 . . . . . . . . . . . . . . . . . . 31

**Murni Handayani, Nurin Nafi'ah, Adityo Nugroho, Amaliya Rasyida, Agus Budi Prasetyo and Eni Febriana et al.**
The Development of Graphene/Silica Hybrid Composites: A Review for Their Applications and Challenges
Reprinted from: *Crystals* **2021**, *11*, 1337, doi:10.3390/cryst11111337 . . . . . . . . . . . . . . . . . . 45

**A. T. Hamed, E. S. Mosa, Amir Mahdy, Ismail G. El-Batanony and Omayma A. Elkady**
Preparation and Evaluation of Cu-Zn-GNSs Nanocomposite Manufactured by Powder Metallurgy
Reprinted from: *Crystals* **2021**, *11*, 1449, doi:10.3390/cryst11121449 . . . . . . . . . . . . . . . . . . 73

**Oleg Lisovski, Sergei Piskunov, Dmitry Bocharov, Yuri F. Zhukovskii, Janis Kleperis and Ainars Knoks et al.**
$CO_2$ and $CH_2$ Adsorption on Copper-Decorated Graphene: Predictions from First Principle Calculations
Reprinted from: *Crystals* **2022**, *12*, 194, doi:10.3390/cryst12020194 . . . . . . . . . . . . . . . . . . 87

**Omayma A. Elkady, Hossam M. Yehia, Aya A. Ibrahim, Abdelhalim M. Elhabak, Elsayed. M. Elsayed and Amir A. Mahdy**
Direct Observation of Induced Graphene and SiC Strengthening in Al–Ni Alloy via the Hot Pressing Technique
Reprinted from: *Crystals* **2021**, *11*, 1142, doi:10.3390/cryst11091142 . . . . . . . . . . . . . . . . . . 99

**Salem Mohammed Aldosari and Sameer Rahatekar**
Extrusion Dwell Time and Its Effect on the Mechanical and Thermal Properties of Pitch/LLDPE Blend Fibres
Reprinted from: *Crystals* **2021**, *11*, 1520, doi:10.3390/cryst11121520 . . . . . . . . . . . . . . . . . . 119

**Salem Mohammed Aldosari, Muhammad A. Khan and Sameer Rahatekar**
Influence of High-Concentration LLDPE on the Manufacturing Process and Morphology of Pitch/LLDPE Fibres
Reprinted from: *Crystals* **2021**, *11*, 1099, doi:10.3390/cryst11091099 . . . . . . . . . . . . . . . . . . 133

**Sameh H. Ismail, Ahmed Hamdy, Tamer Ahmed Ismail, Heba H. Mahboub, Walaa H. Mahmoud and Walid M. Daoush**
Synthesis and Characterization of Antibacterial Carbopol/ZnO Hybrid Nanoparticles Gel
Reprinted from: *Crystals* **2021**, *11*, 1092, doi:10.3390/cryst11091092 . . . . . . . . . . . . . . . . . **149**

**Madiha Batool, Shazia Khurshid, Walid M. Daoush, Sabir Ali Siddique and Tariq Nadeem**
Green Synthesis and Biomedical Applications of ZnO Nanoparticles: Role of PEGylated-ZnO Nanoparticles as Doxorubicin Drug Carrier against MDA-MB-231(TNBC) Cells Line
Reprinted from: *Crystals* **2021**, *11*, 344, doi:10.3390/cryst11040344 . . . . . . . . . . . . . . . . . **173**

**Maha M. Khayyat**
Silica Microspheres for Economical Advanced Solar Applications
Reprinted from: *Crystals* **2021**, *11*, 1409, doi:10.3390/cryst11111409 . . . . . . . . . . . . . . . . . **193**

**Maha M. Khayyat**
Crystalline Silicon Spalling as a Direct Application of Temperature Effect on Semiconductors' Indentation
Reprinted from: *Crystals* **2021**, *11*, 1020, doi:10.3390/cryst11091020 . . . . . . . . . . . . . . . . . **201**

**Saad A. Aljlil**
Development of Red Clay Ultrafiltration Membranes for Oil-Water Separation
Reprinted from: *Crystals* **2021**, *11*, 248, doi:10.3390/cryst11030248 . . . . . . . . . . . . . . . . . **213**

**Fadel S. Hamid, Omayma A. Elkady, A. R. S. Essa, A. El-Nikhaily, Ayman Elsayed and Ashraf K. Eessaa**
Analysis of Microstructure and Mechanical Properties of Bi-Modal Nanoparticle-Reinforced Cu-Matrix
Reprinted from: *Crystals* **2021**, *11*, 1081, doi:10.3390/cryst11091081 . . . . . . . . . . . . . . . . . **229**

**Mohamed Ali Hassan, Hossam M. Yehia, Ahmed S. A. Mohamed, Ahmed Essa El-Nikhaily and Omayma A. Elkady**
Effect of Copper Addition on the AlCoCrFeNi High Entropy Alloys Properties via the Electroless Plating and Powder Metallurgy Technique
Reprinted from: *Crystals* **2021**, *11*, 540, doi:10.3390/cryst11050540 . . . . . . . . . . . . . . . . . **243**

**Abdulla I. Almazrouee, Khaled J. Al-Fadhalah and Saleh N. Alhajeri**
A New Approach to Direct Friction Stir Processing for Fabricating Surface Composites
Reprinted from: *Crystals* **2021**, *11*, 638, doi:10.3390/cryst11060638 . . . . . . . . . . . . . . . . . **263**

# About the Editors

**Walid M. Daoush**

Dr. Daoush is a professor of materials science at the department of production technology of Helwan University. He moved from CMRDI in Cairo to Helwan University in 2010 as an associate professor of materials science. He started his work as a research assistant at CMRDI. Since he joined CMRDI in 1996, his work has focused on the fabrications and characterizations of materials. In 2004, he obtained a Ph.D. from University of Ain Shams in Cairo. Recently, his efforts have been applied to a wide range of materials and he is a coauthor of more than 60 published articles in different fields such as metal-supported CNTs, biomaterials, tungsten heavy alloys and metallic foams. He had Research Assistant Professorships at (KAIST) from 2007 to 2008, while serving as a Post-Doc at (NTUST) in 2011. In addition, he has been a visiting scientist in several international institutes, including Japan Fine Ceramic Center (JFCC) and Materials Research Institute of Slovak republic, and recently as a Fulbright visiting scholar in the college of engineering at San Diego State University. Dr. Walid is listed in the membership board of KRF, Who's Who in the world 2009, American Chemical society and Materials Research Society. He received the prize of young scientists of Materials Science from the Egyptian academy of Science and Technology in 2010. Dr. Daoush has cosupervised more than 15 theses and published over 50 articles, one patent, and chapters in text books. He collaborated with several national and international organizations for start-up projects in Egypt and different countries in Austrian Institute of Technology (AIT), Materials Research Institute of Slovak Republic, TRC of the Egyptian Armed Forces, STDF in Egypt, the University Carlos III de Madrid and recently to the college of engineering at San Diego State University. Since 2016, he has been working as an adjunct professor at the college of science of the Imam Mohammad ibn Saud Islamic University in Saudi Arabia.

**Fawad Inam**

Based at the University of East London (UK), Professor Dr Fawad Inam has contributed to achieving more than GBP 2.5 million in research and enterprise income from government and commercial platforms. Followed by over 228k individuals worldwide in publicly accessible general engineering outreach social media channels, he has advised over 55 universities and institutes from all the four nations of the UK. As a British Chartered Engineer (CEng), fellow of the Institute of Mechanical Engineers (IMechE), and principal fellow of the Higher Education Academy (HEA, UK), he has formed/strengthened international partnerships, delivered engineering courses and influenced institutional higher education practice in other countries such as China, South Korea, Singapore, Malaysia, India, Pakistan, Saudi Arabia, and the Netherlands, to list a few. He is the recipient of the 'Materials World Award' conferred by the Institute of Materials, Minerals and Mining (IoM3, UK) in 2008. Prof Inam's research and enterprise expertise is primarily focused around improving the performance and functionalities of engineering products using materials science and engineering, in particular nano-influenced smart materials.

**Mostafa Baboli**

Dr. Mostafa Ghasemi Baboli is currently working as an Associate Professor in the Chemical Engineering department of the Sohar University, Oman. Before joining Sohar University, he worked at Universtiti Teknologi PETRONAS (2016–2020) and The National University of Malaysia (UKM) in 2013–2016. He also consults for water and wastewater treatment, green technology and renewable energy projects. He has published more than 80 manuscripts in high-level international journals (mostly Elsevier journals), with a total impact factor of more than 200.

**Maha M. Khayyat**

Dr. Maha M. Khayyat is currently a research professor of physics at KACST. Dr Khayyat received her PhD from Cavendish laboratory, University of Cambridge (Oct. 2001–July 2004), studying the mechanical, optical and electrical properties of crystalline and amorphous semiconductor materials (Si, Ge, and GaAs) using indentation techniques at both the micro- and nanoscale. Then, she worked as a collaborative researcher at Cavendish lab, where she was introduced to the topic of photovoltaic cells at Microelecronics Research Center (MRC) and Nanoscience Center of University of Cambridge. She worked within a joint project KACST/ IBM, undertaking research at T. J. Watson Research Center, USA. She worked on solar cells based on Si nanowires and then started conducting original research by improving nanowire technology and spalling at liquid nitrogen temperature. After this, she spent her sabbatical year at MIT, Nanoengineering group, funded by the Center of Clean Water of Energy of KFUPM. She aimed to develop surface structures that trap photons in thin films of Si. Then, she worked at Poitiers University, Poitiers, France, studying the properties of GaN nanomembraned for LED applications.

# Preface to "Fabrication of Carbon and Related Materials/Metal Hybrids and Composites"

This Special Issue covers various aspects of materials science and engineering in the field of the fabrication of hybrid and composite materials. It mainly focuses on the novel developments in and new processing methodologies for the fabrication and modification of carbon and related hybrid and composite materials for different applications. It consists of sixteen research papers and one review article describing diverse research topics. The contributions of this Special Issue may be classified into four groups: the first concerns fabrications of new composite and hybrid materials of the different allotropies of carbon, such as carbon fibers, carbon nanotubes and graphene; the second group considers the synthesis and properties of new polymeric composite and hybrid materials with various applications due to their electrical, mechanical and antimicrobial activities; the last group is strongly related to a discussion on alloys and metal matrix composites and their welding properties. Some of the articles generally contain the results of calculations and are compared with experimental observations. The results presented in the articles collected in this Special Issue clearly demonstrate that the carbon-based hybrid and composite materials are clearly very important in the field of advanced functional materials.

We would like to thank all authors whose contributions are included in this Special Issue for their excellent work, and for their inspiring and interesting articles.

Walid M. Daoush, Fawad Inam, Mostafa Baboli, and Maha M. Khayyat
*Editors*

Article

# Syntheses and Step-by-Step Morphological Analysis of Nano-Copper-Decorated Carbon Long Fibers for Aerospace Structural Applications

Walid M. Daoush [1,2,*], Turki S. Albogmy [1], Moath A. Khamis [1] and Fawad Inam [3]

1. Department of Chemistry, College of Science, Imam Mohammad ibn Saud Islamic University (IMSIU), Al Riyadh 11623, Saudi Arabia; torkealbgme@gmail.com (T.S.A.); imoath06@gmail.com (M.A.K.)
2. Department of Production Technology, Faculty of Technology and Education, Helwan University, Cairo 11281, Egypt
3. Department of Engineering and Construction, University of East London, London E16 2RD, UK; f.inam@uel.ac.uk
* Correspondence: wmdaoush@imamu.edu.sa

Received: 24 September 2020; Accepted: 26 November 2020; Published: 28 November 2020

**Abstract:** Carbon long fiber/copper composites were prepared using electroless and electroplating methods with copper metal for potential aerospace applications. Carbon fibers were heat-treated at 450 °C followed by acid treatment before the metallization processes. Three different methods of metallization processes were applied: electroless silver deposition, electroless copper deposition and electroplating copper deposition. The metallized carbon fibers were subjected to copper deposition via two different routes. The first method was the electroless deposition technique in an alkaline tartrate bath using formaldehyde as a reducing agent of the copper ions from the copper sulphate solution. The second method was conducted by copper electroplating on the chemically treated carbon fibers. The produced carbon fiber/copper composites were extensively investigated by Field-Emission Scanning Electron Microscopy (FE-SEM) supported with an Energy Dispersive X-Ray Analysis (EDAX) unit to analyze the size, surface morphology, and chemical composition of the produced carbon long fiber/copper composites. The results show that the carbon fiber/copper composites prepared using the electroplating method had a coated type surface morphology with good adhesion between the copper coated layer and the surface of the carbon fibers. However, the carbon fiber/copper composites prepared using the electroless deposition had a decorated type morphology. Moreover, it was observed that the metallized carbon fibers using the silver method enhanced the electroless copper coating process with respect to the electroless copper coating process without silver metallization. The electrical conductivity of the carbon fiber/copper composites was improved by metallization of the carbon fibers using silver, as well as by the electrodeposition method.

**Keywords:** carbon long fibers; copper composites; electroless copper deposition; electroless silver deposition; copper electroplating; contact electrical resistivity

## 1. Introduction

Advanced materials and carbon-fiber composites are used extensively throughout revolutionary aircrafts such as Boeing 787 Dreamliner and Airbus A350 family [1]. The stiffness, lightness, and toughness of the carbon fiber allowed technologists to create a very-low-drag delta wing body and fuselage. These advanced carbon fiber composites lead to not only lighter aircraft but also lower fatigue sensitivity, which means they require less maintenance. The Boeing 787's heavy maintenance interval was increased from 6 to 13 years [1]. Carbon-fiber-reinforced plastics (CFRPs) constitutes more than 50 vol.% of these aerospace mobile structures, as represented in Figure 1. CFRPs are micro-composites formed from a lightweight polymer binder (e.g., epoxy) with laid carbon fiber to manufacture structures having extraordinarily high stiffness and strength-to-weight ratios.

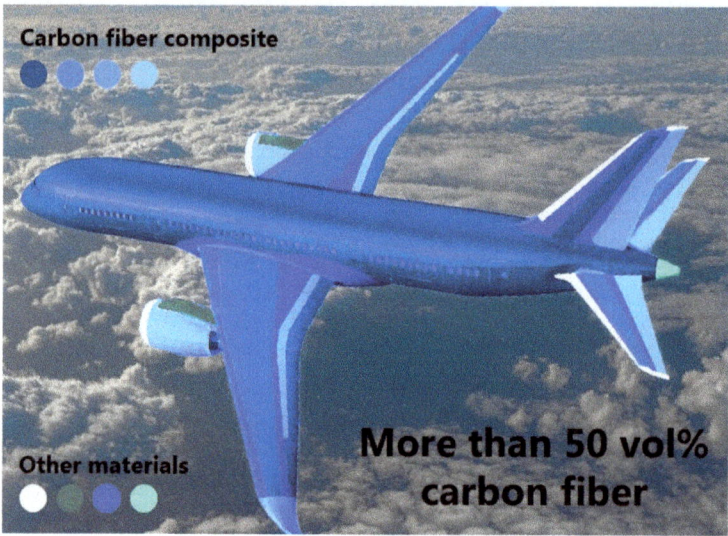

**Figure 1.** Materials analysis for a typical carbon-fiber-reinforced plastic (CFRP) aircraft.

This revolutionary technology (Figure 1) relies on the superlative combination properties of CFRPs primarily contributed by carbon fibers [2–6]. Carbon fibers, containing more than 92% by weight of carbon, have high strength, low density (1.8 g/cm$^3$, light weight, high breaking strength (2–7 GP), high tensile modulus (200–500 GPa), and a low thermal expansion coefficient (0.1–1.1 × 10$^{-6}$ K$^{-1}$) [3,7]. They are also characterized by high resistance to acids, alkalis, and organic solvents. Carbon fibers have a low coefficient of thermal expansion and a good electrical conductivity, as well as low x-ray absorption and nonmagnetic properties [8–10]. The as-produced carbon fibers usually have relatively smooth surfaces, low surface energy, low chemical reactivity, and lack of chemically active functional groups, which have a significant effect on their mechanical properties and restrict their extensive applications [11,12].

Wing boxes, made of CFRP, are able to support the load imposed during flight and support the whole aircraft aerodynamically while also strategically minimizing their overall contribution to the weight of an aircraft. However, CFRPs, unlike like their aluminum counterparts, do not conduct electricity. This makes them susceptible to lightning strike damage and to mitigate such a drawback, an electrically conductive expanded copper foil layer is usually laid on the outer surface of the composite structure layup [13]. If a lightning bolt strikes an unprotected composite structure, up to 200,000 amps of electricity seeks the path of least resistance and may vaporize metal control cables and weld hinges on control surfaces or explode fuel vapors within fuel tanks if the current arcs through gaps around fasteners [1]. High electrical conductivity is, therefore, required here to dissipate the high current and heat generated by a lightning strike. However, expanded copper foil (ECF) layer possess issues of its own. Temperature and atmospheric pressure variations (for instance, 50 °C to −50 °C and 100 to 25 kPa, respectively) during the ground-to-air flight cycle can lead to the expansion and reduction of the protective layer, which can damage the relatively less resilient epoxy matrix of CFRPs, reducing the overall effectiveness of the composite substitution.

Here, we synthesized nano-copper-influenced carbon fibers as an alternate technology for the ECF laid on top of the CFRP wing box. Carbon fibers, generally speaking, have poor wetting behavior with metals such as copper and aluminum [12,14–17].

It is necessary to modify the surfaces of carbon fibers to resolve this key issue. One of the widely researched solutions is to coat the carbon fibers with metal layers. This method also reduces their susceptibility to interfering with the matrix and avoids the interaction of carbon fibers with several metals such as iron [18]. Metal-coated carbon fibers can also be used as a reinforcement phase in different metal matrix composites for different applications such as electric contact materials and electric brushes [16,19], as well as for the fabrication of fiber composites used automotive and aerospace sectors and other electrical equipment [20]. Materials with high electrical and thermal conductivities in combination with a low coefficient of thermal expansions are currently required for electrical and electronic applications. Carbon fiber/Cu composites possess the properties of copper, i.e., the excellent electrical and thermal conductivities, and the properties of carbon fiber, i.e., small coefficient of thermal expansion. These composites can be used in electrical and electronic applications. The electrical conductivity of carbon fiber/Cu composite materials is very important, particularly if these materials are used for electrical and electronic applications. The materials for this application should possess high electrical and thermal conductivities. Carbon fiber/Cu composites have successfully solved this problem. In this type of material, carbon is utilized because of its good sliding and antifriction properties, whilst copper is used because of its high electrical and thermal conductivities [16,19]. Various other studies have been conducted to improve the wetting of carbon fibers to metals. An $SnCl_2/PdCl_2$ solution is used as activating solution by depositing Pd nanoparticles on the surface of the carbon fibers before coating using the electroless deposition technique [21,22]. Electroless deposition can take place after the surface has been activated by Pd particles via the autocatalytic reaction to deposit metal nanoparticles on the surface of carbon fibers [23].

In another report, deposition of silver nanoparticles or films using electroless silver deposition was used for obtaining surface activity and improving their electrical conductivity and physical properties. Activation by silver aerosols and copper electrolyte deposition was also considered. After annealing, silver-activated carbon fibers were effectively placed in a solution for electroplating copper, to obtain a uniform copper coating on their surface [24]. Electroplating Cu was utilized to increase the thickness of the interlayer and forming a coating layer with a good adhesion with the surfaces of the carbon fibers. The current work utilizes a superlative combination of properties offered by carbon fiber (PAN: Polyacrylonitrile type) with high electrical conductivities of copper to synthesize a nano-copper-decorated carbon fiber nanocomposite via the coating route. To make these composites suitable for powder technology processing, surface treatment of the carbon fibers was essential via thermal de-binding, acid treatments, and/or a tin/silver metallization process before encapsulating the carbon fibers into the copper matrix using two coating methods, (electroless or electrodeposition) to produce a continuous conductive coating with uniform thickness. The contact electrical resistivity of the produced carbon fiber/Cu nanocomposites using either electroless or electrodeposition techniques was measured as well.

## 2. Materials and Methods

### 2.1. Starting Materials

High-purity PAN-type long carbon fibers were provided by Mitsubishi Chemical Carbon Fiber and Composites Ltd. (Sacramento, USA). Each bundle was composed of around 1000 fibers, which were bonded together with an organic binder (sizing agent). Table 1 lists the physical and mechanical properties of the carbon fibers used in this study. Copper sulfate pentahydrate was purchased from Winlab Ltd., Leicester, UK. Silver nitrate and stannous chloride dihydrate were purchased from BDH Chemicals Ltd., East Yorkshire, UK. Potassium sodium tartrate and potassium dichromate were provided by Merck Ltd., Darmstadt, Germany. Formaldehyde and sodium hydroxide were purchased from Panreac AppliChem, Barcelona, Spain. Acetone, hydrochloric acid, ammonia solution, nitric acid, and sulfuric acid were provided by Riedel De-Haen, Seelze, Germany.

**Table 1.** Physical and mechanical properties of PAN-type carbon fibers.

| Property | Value |
|---:|:---:|
| Fiber diameter, μm | 7 |
| Density, g·cm$^{-3}$ | 1.78 |
| Tensile strength, GPa | 3.0 ± 0.2 |
| Tensile modulus, GPa | 221 ± 4 |
| Ultimate elongation, % | 1.4 |
| Specific heat, J·kg$^{-1}$·K$^{-1}$ | 711 |
| Thermal conductivity, W·m$^{-1}$·K$^{-1}$ | 8 |
| Electrical resistivity, Ω·cm | 2.2 ± 0.5 × 10$^{-3}$ |

## 2.2. Methods

### 2.2.1. Pretreatment of Carbon Fibers

Carbon fibers were treated as described in Figure 2. In brief, the bundles of as-received carbon long fibers were cut into strands of around 6 cm in length and then heat-treated at 450 °C for 30 min in an open oven (Figure 3) to remove any sizing, binding, and degreasing agents. They were then washed in acetone for 15 min followed by washing with distilled water to remove any organic remained contaminants. The obtained carbon long fibers were then acid-treated by concentrated nitric acid, and then stirred in a freshly prepared chromic acid solution for 15 min (by dissolving 5 g of potassium dichromate in 5 mL distilled water in the form of a paste, before adding 100 mL of 98% concentrated sulfuric acid dropwise). This was then followed by washing thoroughly (at least three times in deionized water) to remove any inorganic impurities on the surface of the carbon fibers. The obtained carbon fibers were dried for 15 min at 110 °C.

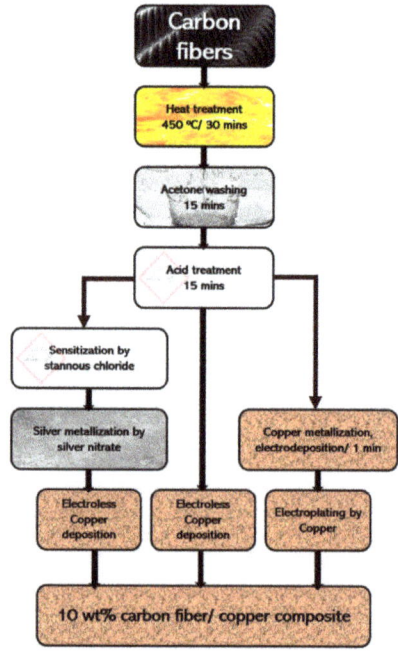

**Figure 2.** Schematic flowchart of the metallization process of carbon long fibers.

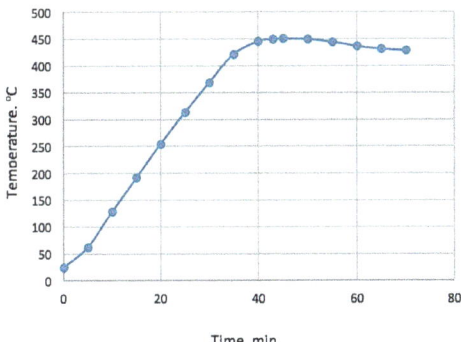

**Figure 3.** Heat treatment cycle of the carbon long fibers in the oven at 450 °C for 30 min.

### 2.2.2. Metallization of Carbon Fibers Using the Tin/Silver Process

As elaborated in Figure 2, a batch of heat-treated and etched carbon fibers were further activated. The activation process included two steps: sensitization and silver deposition. About 0.01 g of the treated carbon long fibers were treated using tin sensitization solution. The sensitizing agent was prepared by dissolving 0.1 g of $SnCl_2 \cdot 2H_2O$ in 10 mL of distilled water, and the pH of the solution was adjusted to around 1.8 using hydrochloric acid. The solution was stirred using a magnetic stirrer for 30 min. The sensitized carbon fibers were then washed with distilled water to remove any residuals of the sensitizing agent. The obtained sensitized carbon fibers were surface-activated using the silver deposition method. About 0.2 g of silver nitrate was dissolved in 100 mL of distilled water, and the pH of the solution was adjusted by ammonia to around 10.7. The sensitized carbon fibers were then added and the solution was stirred using magnetic stirring for 15 min. Then, 20 mL of formaldehyde was added to the solution. The reduction reaction was completed within 15 min, and then the activated carbon fibers were washed with distilled water, filtered, and dried at 110 °C for 30 min.

### 2.2.3. Metallization of Carbon Fibers Using the Copper Deposition Processes

Two processes were used to metallize the chemically treated carbon fibers with copper (Figure 2). The first method (electroless copper deposition) was conducted on the acid-treated carbon fibers in an alkaline tartrate bath. The solutions included in the process were 70 g/L copper sulfate as a source of copper, 170 g/L potassium sodium tartrate as a chelating agent of the copper ions, and 100 mL/L formaldehyde as a reducing agent of the copper ions to copper in the metallic state. The pH of the solution was adjusted to around 13, and the temperature was maintained at room temperature (~24 °C). After completion of the copper deposition reaction, the metallized carbon fibers underwent washing with distilled water, filtration, and drying at 110 °C for 30 min.

The second metallization process was the electrodeposition of copper nanoparticles on the surface of the chemically treated carbon fibers. The acid-treated carbon long fibers (around 6 cm in length) were stretched on a cathode frame made from plastic. The current and time were adjusted to get the required copper deposition on the carbon long fibers. In brief, 120 g of copper sulfate pentahydrate and 90 mL of sulfuric acid were dissolved in distilled water (via 30 min of magnetic stirring) to prepare 1 L of electrolyte solution. The copper electrodeposition on the surface of the carbon long fibers was achieved by passing 8 µA current for 1 min.

### 2.2.4. Syntheses of Carbon Fiber/Copper Composites

The acid-treated and tin/silver-metallized carbon fibers were coated using an electroless copper chemical reduction method to prepare 10 wt.% carbon fiber/copper composite samples (Figure 2). About 0.35 g of copper sulfate pentahydrate was dissolved in 10 mL of distilled water, and the solution was stirred using a magnetic stirrer. About 1.7 g of potassium sodium tartrate was added as chelating agent to prevent precipitation of copper as copper hydroxide at high pH of the alkaline solution. Then, 0.5 g of sodium hydroxide was added to adjust the pH to 12.5. A calculated amount of ~0.01 g equivalent to 10 wt.% carbon long fibers was added to the solution, and continuous magnetic stirring at 500 rpm was used to disperse the fibers in the solution. Then, 10 mL of formaldehyde solution (38%) was added as a reducing agent of the copper ion in the copper sulfate to copper metal. The reaction on the surface of the treated carbon fiber was completed within 30 min. The solution was filtered, and the obtained 10 wt.% carbon fiber/copper composite was dried at 110 °C.

On the other hand, the metallized carbon long fibers using copper electroplating were connected to the negative electrode of the electroplating cell of the same composition as mentioned in Section 2.2.3, where two copper plates were connected to the positive electrode. Direct current of density 12 µA/cm² was passed through the electroplating cell at time intervals of 5 min.

2.2.5. Characterization of Carbon Fiber/Copper Composites

The as-received, heat-treated, and chemically treated carbon fibers and the activated, metallized, and copper-coated carbon fibers underwent investigations using a field-emission scanning electron microscope (FE-SEM, model JEOL JSM-7600F). The powders were sputter coated by a platinum JFC 1600 auto fine coater. The compositional analysis of the samples was determined using the Energy Dispersive X-Ray Analysis (EDAX) unit connected with the FE-SEM.

The electrical resistivity of the prepared carbon fiber/Cu nanocomposites was measured using the four-probe method with an Omega multimeter device. A fixed direct current (DC) current of 1 A was passed through the test sample via two crocodile clips. For each test, the multimeter was zeroed with no current passing the specimen, and then the measurement was carried out. The resistivity ($\rho$) was calculated according to the following equation:

$$R = (\rho\ L)/A, \tag{1}$$

where R is the resistance in $\Omega$, L is the measured length in cm, A is the cross-section area in cm², and $\rho$ is the resistivity in $\mu\Omega \cdot cm$. The dimensions of the fibers were calculated according to the data listed in Table 1. Each bundle was composed of around 1000 fibers of diameter ~7 µm and length ~6 cm.

## 3. Results and Discussion

Figure 4 shows the SEM images of as-received carbon fibers. It was observed that the fibers had diameters of approximately 6.7 µm, as roughly confirmed in Table 1.

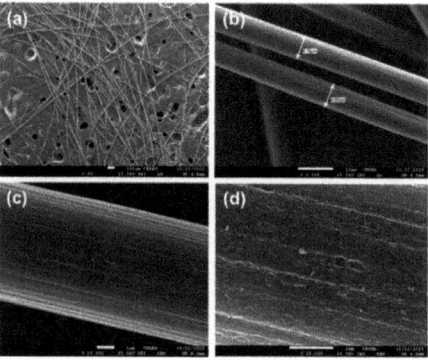

**Figure 4.** SEM images with different magnifications of the as-received carbon fibers. (**a–d**) Low to high resolutions.

Carbon long fibers were heat-treated to remove some of the volatile organic materials such as the sizing agents which were added to the fibers during the fabrication process. Figure 5 shows the SEM images with different magnifications and the EDAX compositional analysis of the carbon fibers after heat treatment. Comparing the surface morphology of the untreated carbon fibers (Figure 4) with the surface morphology of the heat-treated carbon fibers (Figure 5), it was observed that the layers of the sizing agent were partially removed as the carbon fiber surface in evident. Some other impurities composed of Ba, K, Cr, and oxygen were detected through the EDAX analysis of heat-treated carbon fibers due to the presence of remnants from the sizing agent. Figure 6 shows SEM images with different magnifications of the treated carbon fibers after washing and treating them with chromic acid. The results reveal that the diameter of the carbon fibers decreased from around 6.7 μm to around 5.1 μm due to heat and chemical treatments. The decrease in the diameter was due to the removal of the binding, sizing, and degreasing agents which adhered to the surface of carbon fibers as a consequence of heat and chemical treatments of the carbon fibers with acetone and chromic acid. A uniform morphological roughening of the carbon fiber was observed (Figure 6). This process was conducted to increase the bond strength of the applied deposits by increasing the surface roughness of the carbon fiber substrate. However, a very rough surface such as that of etched fibers is not recommended since it affects the smoothness and uniformity of the final deposits. Suitably rough surfaces create a network profile to which the subsequent deposit can be physically anchored and produce a uniform coating thickness.

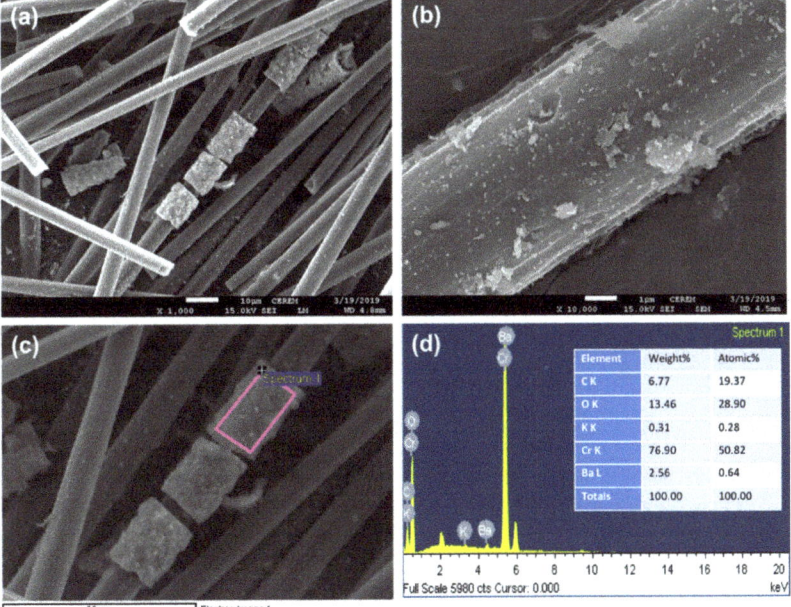

**Figure 5.** Heat-treated carbon fibers at 450 °C: (**a**) SEM image at low resolution; (**b**) SEM image at high resolution; (**c**) Selected area for EDAX analysis; and (**d**) EDAX elemental analysis.

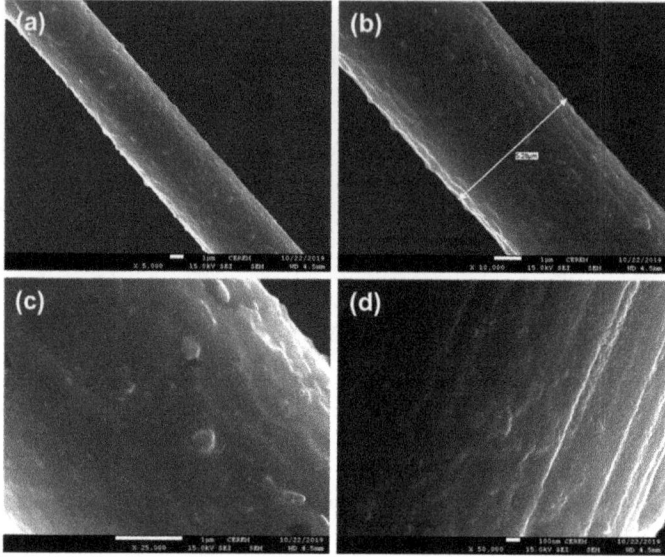

**Figure 6.** SEM images with different magnifications of the carbon fibers surface-treated with chromic acid. (**a**–**d**) Low to high resolutions.

The chemically treated carbon fibers were metallized using three different techniques, namely, electroless tin/silver deposition, electroless copper deposition, and copper electrodeposition.

*3.1. Metallization of Carbon Fibers Using Electroless Tin/Silver Deposition*

Some of the surface-etched carbon fibers were further sensitized and activated to impart a uniform conducting film on the fiber surfaces, ensuring uniform adhesion of subsequent metallization and further promoting better coating and plating.

Tin/silver sensitization and activation were carried out to deposit nanosized silver particles onto carbon fibers prior to the electroless copper coating operations. Tin(II) ions were adsorbed onto the carbon fiber surfaces and silver(I) ions were reduced to metallic silver nanoparticles. Silver nanoparticles were deposited onto the surface of the carbon fibers in the second step as shown in the chemical reaction below.

$$Sn^{++}_{(aq)} + 2\,Ag^{+}_{(aq)} \rightarrow Sn^{4+}_{(aq)} + 2Ag^{0}_{(s)} \qquad (2)$$

Tin(IV) ions were produced from the oxidation of tin(II) ions by silver(I) ions. Figure 7a,b show the SEM image of the silver-activated carbon fibers. Deposited silver nanoparticles in the range of around 40–130 nm could be observed. These deposited silver nanoparticles decorated and adhered to the surface of the carbon fibers. Figure 7c,d show the EDAX compositional area analysis of the silver-activated carbon fibers. It was observed from the results that silver particles were composed mainly of 2.04 wt.% Ag and 0.61 wt.% tin, as remnants from the sensitization process.

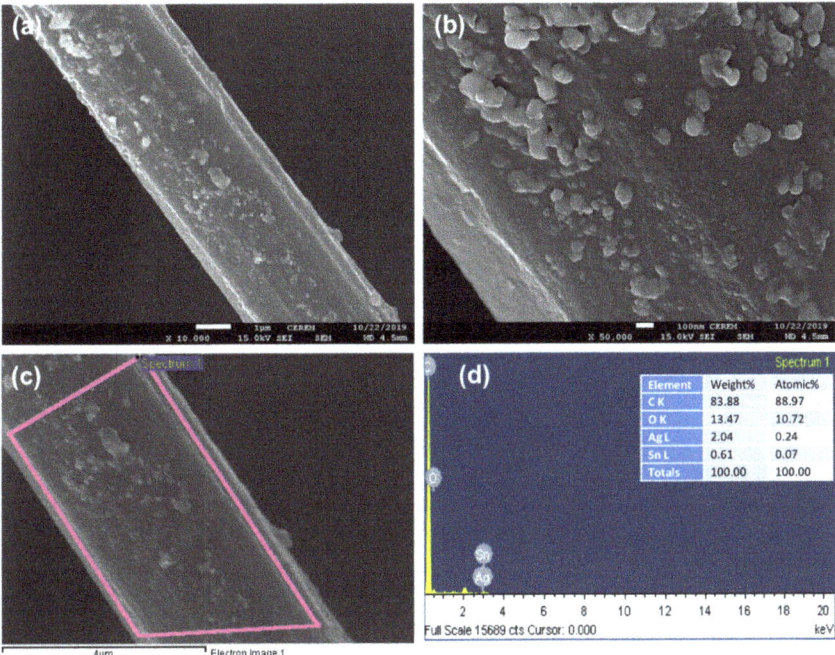

**Figure 7.** Carbon fibers silver-activated using the silver deposition method: (**a**) SEM image at low resolution; (**b**) SEM image at high resolution; (**c**) Selected area for EDAX analysis; and (**d**) EDAX elemental analysis.

*3.2. Metallization of Carbon Fibers Using Electroless Copper Deposition*

The chromic-acid-treated carbon fibers were subjected to autocatalytic electroless copper deposition on its surface in the alkaline tartrate copper sulfate solution. Figure 8a,b show the SEM images with different magnifications of the deposited copper nanoparticles on the acid-treated carbon fibers. It was observed from the results that the deposited copper nanoparticles had particle size range between 85 and 165 nm. The copper nanoparticles had polygonal particle shapes, and some agglomerated particles were also observed. The deposited copper nanoparticles adhered to the surface of the carbon fibers, imparting a decorative type copper layer. A good uniform deposition can be observed. Figure 7c,d show the EDAX compositional analysis of the metallized carbon fiber by the deposited copper nanoparticles. It was revealed that the copper nanoparticles were composed mainly of copper metal. The appearance of the small oxygen peak may be due to the oxidation of the surface of some deposited copper nanoparticles which were present in the aqueous solution during the process or due to the technical limitations of the EDAX technology with respect to the detection of oxygen.

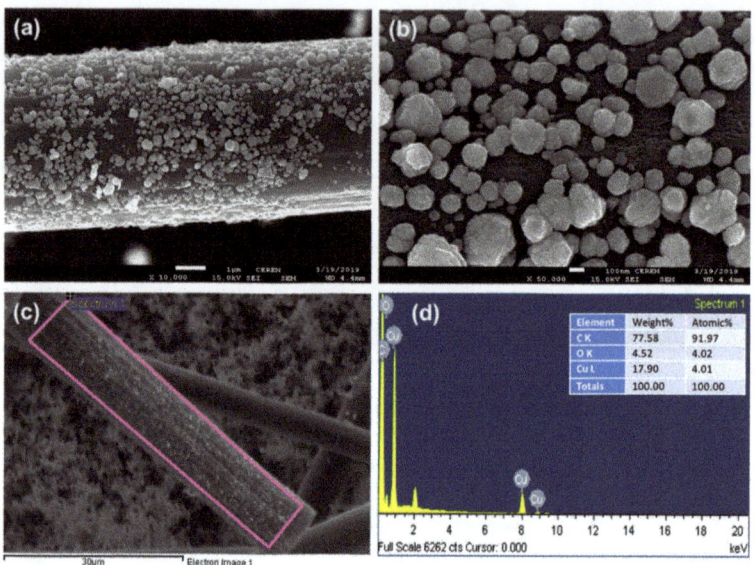

**Figure 8.** Carbon fibers metallized using the electroless copper deposition method: (**a**) SEM image at low resolution; (**b**) SEM image at high resolution; (**c**) Selected area for EDAX analysis; and (**d**) EDAX elemental analysis.

The alkaline tartrate electroless copper solution used formaldehyde as a reducing agent of the copper ions to the copper metal. The half-cell reaction for the electroless copper deposition is shown below.

$$Cu^{++} + 2e^- \rightarrow Cu^0 \quad E^0 = +0.340 \text{ V} \tag{3}$$

The rate of copper deposition was affected by the variation of the pH of the solution. Electroless copper solutions, using formaldehyde as a reducing agent, employed a high pH above 12. It was reported that the $E^0$ of formaldehyde depends on the pH of the solution [22], as shown below.

$$HCOOH + 2H^+ + 2e^- \rightarrow HCHO + H_2O \text{ (pH = 0, } E^0 = +0.05 \text{ V)} \tag{4}$$

$$HOO^- + 2H_2 + 2e^- \rightarrow HCHO + 3OH^- \text{ (pH = 14, } E^0 = -1.07 \text{ V)} \tag{5}$$

As copper salts (copper sulfate pentahydrate) are insoluble at pH above 4, the use of alkaline media necessitates the use of a complexing or chelating component, such as tartrate salts and ethylenediaminetetraacetic acid (EDTA) [22]. The full electroless copper deposition process at pH ~12 can be written according to the redox reaction below.

$$Cu^{2+} + 2HCHO + 4OH^- \rightarrow Cu^0 + H_2 + 2H_2O + 2HCO_2 \tag{6}$$

*3.3. Metallization of Carbon Fibers Using Copper Electrodeposition*

Electrodeposition is a common way of depositing of metals and its alloys on the surface of conductive materials. Copper is used in the electroplating process with either cyanide or sulfate baths. However, cyanide solutions are hazardous and should be avoided for industrial practices. An alternative approach is to use other acid baths to precipitate Cu. Chloride, oxalate, nitrate, thiosulfate, glycolate, lactate, and acetate have been reported; however, sulfate baths are the most commonly used [23].

A conventional acid electrochemical copper cell was used to electroplate carbon long fibers. This electrochemical cell consisted of copper sulfate and a sulfuric acid solution as the electrolyte. Two high-purity copper plates were used as anodes. A cathode of carbon fibers was inserted between the anodes in the solution. Copper ions of the copper sulfate were dissolved in the electroplating solution. The remaining $SO_4^{2-}$ anion played no part in the reactions and, therefore, does not appear in the equations [23]. The complete chemical reaction of the electroplating process of the carbon fibers confirmed the transfer of copper ions from the anode to the cathode passing through the electrolyte and depositing copper on the conductive carbon long fibers fixated on the cathode. It was assumed that the total copper ion concentration in the electrolyte remained unchanged.

Figure 9 shows SEM images with different magnifications of the deposited copper nanoparticles on the surface of the carbon long fibers upon passing a current of 8 µA for 1 min through the electrodeposition cell. It was observed that polygonal copper nanoparticles ranging between 85 and 135 nm in size were homogeneously deposited on the surface of the treated carbon fibers, giving a decorative morphology texture (Figure 9b).

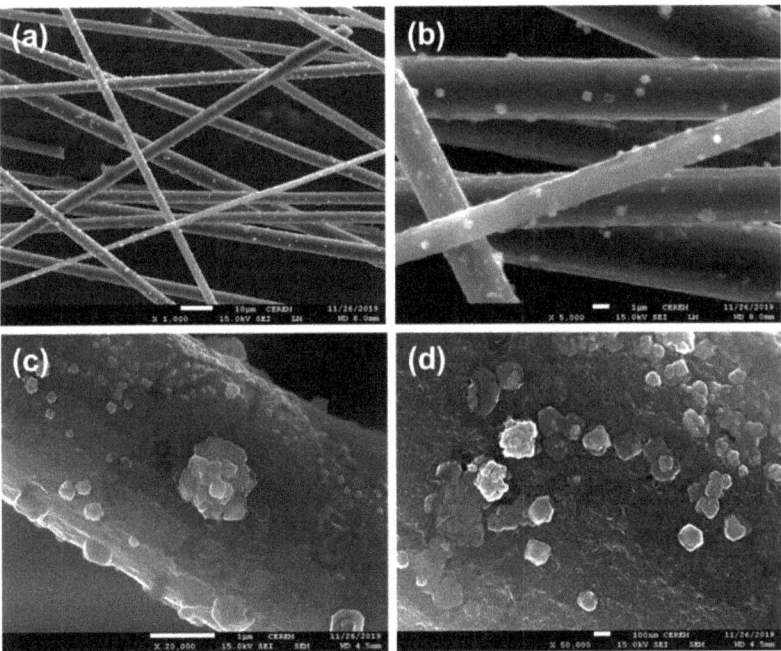

**Figure 9.** SEM images with different magnifications of the carbon long fibers copper-coated using the electroplating copper coated method upon passing 8 µA current in the electroplating cell for 1 min. (**a**–**d**) Low to high resolutions.

### 3.4. Syntheses of Carbon Fiber/Copper Composites

The acid-treated, tin/silver-metallized, and copper-metallized carbon fibers were used to fabricate the 10 wt.% carbon fiber/copper composite via electroless coating and copper electroplating.

Figure 10a,b show SEM images with different magnifications of 10 wt.% carbon fibers coated via electroless deposition of copper on silver-metallized carbon fibers. The surface of the fibers was completely covered with multilayers composed of polygonal copper nanoparticles ranging in size from 50–100 nm. Figure 10c,d show the EDAX semiquantitative analysis of the produced 10 wt.% carbon long fiber/copper composite via electroless deposition in the alkaline potassium sodium tartrate bath using formaldehyde as a reducing agent. A compositional analysis of the copper-coated silver-metallized carbon fibers showed mainly copper and carbon. In addition, some silver and tin remained in the copper-coated layers from the earlier tin/silver metallization process.

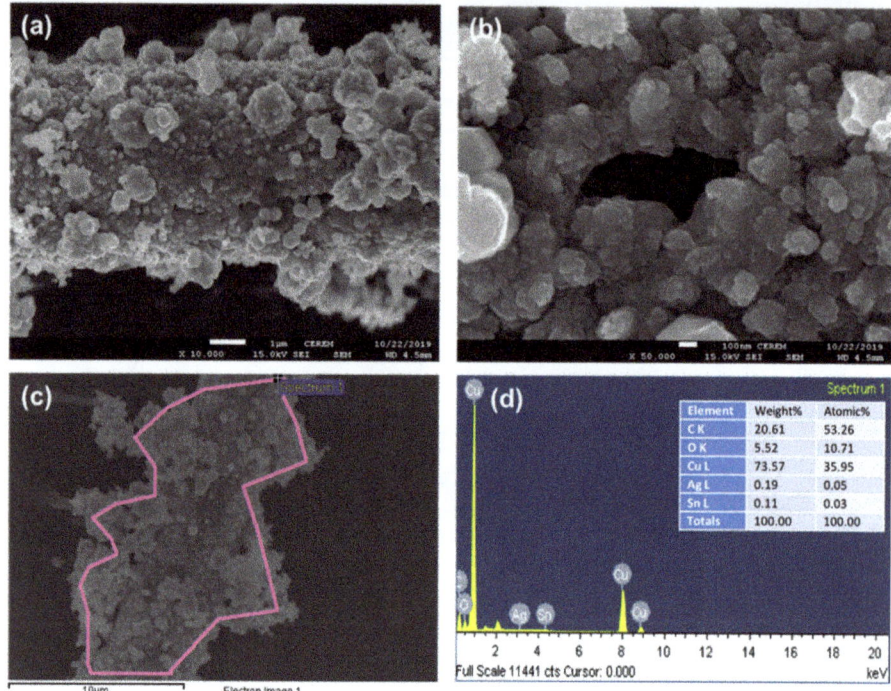

**Figure 10.** Carbon fibers activated via silver deposition followed by electroless copper deposition: (**a**) SEM image at low resolution; (**b**) SEM image at high resolution; (**c**) Selected area for EDAX analysis; and (**d**) EDAX elemental analysis.

Figure 11 shows SEM images with different magnifications for carbon fiber/copper composites prepared via the electroplating method. The morphology (Figure 11) was achieved using 12 µA/cm$^2$ current density for 5 min. Carbon fibers were completely coated and covered by a dense copper layer. Fine copper particles were deposited on the surface of the carbon fibers. These particles appeared to be growing laterally, simultaneously forming a network and eventually becoming a layer of bulky copper clusters of particles. It was also noticed that the Cu deposits on the carbon fibers adhered very well to the carbon fibers, and the degree of tightness was higher than the Cu-coated carbon fibers fabricated via the electroless deposition method due to the absence of pores in the deposited layers. However, Figure 12 shows that there were some regions where there was no coating. Such regions could create some interfacial regions for the application discussed in this work. The EDAX compositional analysis of the prepared carbon fiber/copper composites via the electrodeposition method with 12 µA/cm$^2$ current density is also shown in Figure 12. It was observed that the electrodeposited layer was composed mainly of copper on the surface of the carbon fibers.

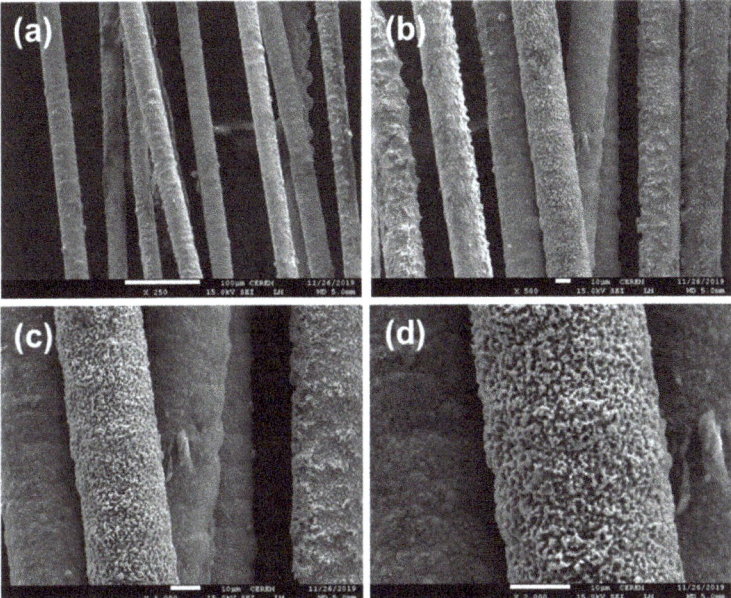

**Figure 11.** SEM images with different magnifications of the carbon long fibers copper-coated using the electroplating copper coated method with a current density of 12 µA/cm$^2$ for 5 min in the electroplating cell. (**a–d**) Low to high resolutions.

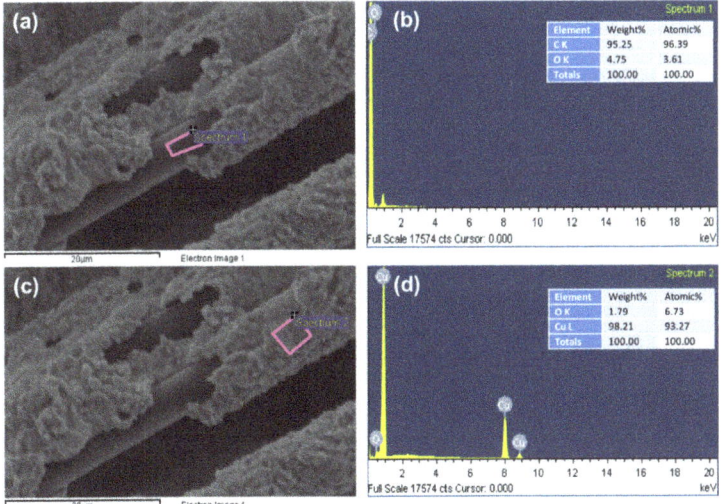

**Figure 12.** EDAX compositional area analysis of the carbon long fibers copper-coated using the electroplating copper-coated method with a current density of 12 µA/cm$^2$: (**a**) Selected uncoated fiber region for EDAX analysis; (**b**) EDAX elemental analysis for uncoated fiber region; (**c**) Selected coated region for EDAX analysis; and (**d**) EDAX elemental analysis for coated region.

### 3.5. Electrical Resistivity of Carbon Fiber/Copper Composites

The contact electrical resistivity for the fiber/deposit interface following electroless and electrolytic deposition for each Cu deposit condition was measured. The correlation provided qualitative analysis on the bonding and adhesion of the deposit to the carbon fibers. For example, a higher void content at the interface would result higher contact resistivity and lower adhesion. When measuring the contact electrical resistivity of the carbon fiber/deposit interfaces of the investigated Cu-coated fibers, as shown in Figure 13, it was found that the contact electrical resistivity of the carbon fiber/copper composites fabricated via electrodeposition was lower than that of the carbon fiber/copper composites fabricated via silver metallization and that of the non-metallized carbon fiber/copper composites fabricated via electroless deposition. This was probably due to the high purity of the deposited copper when using the electrodeposition method than the electroless one, as shown from Figures 10d and 12b,d. Copper and carbon fibers have no mutual wettability and solubility. Accordingly, carbon fiber is mechanically bonded to copper matrix solely due to the fiber's roughness. Thus, the interfacial adhesion between carbon fibers and the copper matrix remains weak. When carbon fibers are oxidized with $CrO_3$, greater surface roughness is produced, resulting in better adhesion properties between the carbon fibers and Cu and leading to improved electrical conductivity. When the oxidized carbon is surface-activated and metallized with Sn/Ag solution, Ag acts as the active center for Cu deposition, consequently improving the carbon fiber/copper matrix interface and decreasing the percentage porosity, thereby resulting in a higher electrical conductivity. In addition, the resistivity of all the produced carbon fibers/copper composites was lower than that of the uncoated fibers (Figure 13). This means that the conductivity of carbon fibers was improved by the contribution of copper in the composite. Since copper is a face-centered cubic (FCC) crystalline material, it theoretically contributes a free electron per atom to the conduction bond where it is available for conducting electrical current. At any time, a certain number of free electrons can be at any given distance outside of the copper surface. Carbon fibers, similar to copper, also possess free electrons, which are available for conduction. In the case of two clean surfaces (i.e., carbon fiber and copper) placed together closely in an intimate contact, the free electrons are able to exchange positions without interference [25].

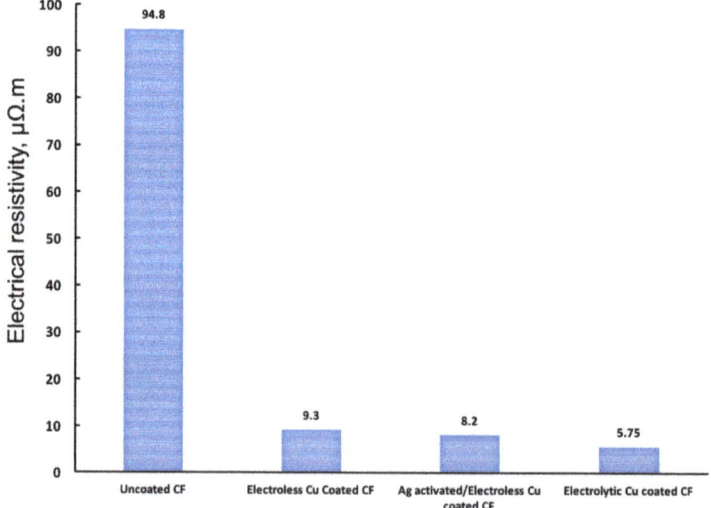

**Figure 13.** Electrical resistivity of the prepared carbon fiber/copper composites.

## 4. Conclusions

The present work provided an economical strategy highlighting the step-by-step synthesis and morphological analysis of nano-copper-decorated carbon fibers for aerospace structural applications. Carbon long fibers (PAN type) were successfully coated with copper using electroless or electrodeposition methods. The loading of copper coating was about 90 vol.%. Electroless silver deposition, electroless copper deposition, and copper electroplating were used to produce and analyze the conductive area coatings. The 10 wt.% carbon fiber/copper nanocomposites were subjected to heat and acid treatments followed by stannous chloride sensitization and the silver deposition method and/or electroless copper deposition and copper electroplating method, thereby producing continuous and selective area coatings of copper. The coating protocol was optimized to achieve a high degree of coating uniformity and conductivity as all samples were characterized for use in different conditions. Silver and copper nanoparticles ranging in size from 40–170 nm were deposited/decorated and adhered to the surfaces of treated carbon fibers. The morphology and thickness of conductive clusters of copper could be manipulated by varying the electrochemical variables as demonstrated in this study. It was concluded from the measurement of the electrical conductivity that the adhesion of the electrodeposited Cu coating on carbon fibers was the highest, followed by silver-metallized/electroless copper-coated carbon fibers and, lastly, electroless copper-coated carbon fibers. Such an optimization of variables could alter the performance of lightning protection measures for the outer layers of aerospace structures (e.g., wing box), which will be the subject of future research.

**Author Contributions:** Conceptualization, W.M.D. and F.I.; methodology, W.M.D., T.S.A. and M.A.K.; validation, W.M.D., T.S.A. and F.I.; formal analysis, W.M.D., T.S.A. and M.A.K.; investigation, W.M.D., T.S.A., M.A.K. and F.I.; resources, W.M.D., T.S.A., M.A.K. and F.I.; data curation, W.M.D., T.S.A. and M.A.K.; writing—original draft preparation, W.M.D., T.S.A., M.A.K. and F.I.; writing—review and editing, W.M.D. and F.I.; visualization, T.S.A., M.A.K. and W.M.D.; supervision, W.M.D. and F.I.; project administration, W.M.D. and F.I. All authors have read and agreed to the published version of the manuscript.

**Funding:** This research received no external funding.

**Acknowledgments:** The authors acknowledge all technicians of the SEM unit at the King Saud University for their cooperation and help during the investigations of our samples. Also, the authors would like to thank Nasr M. Khattab at Central Metallurgical R&D Institute for his guidance during the work.

**Conflicts of Interest:** The authors declare no conflict of interest.

## References

1. Mathijsen, D. What does the future hold for composites in transportation markets? *Reinf. Plast.* **2017**, *61*, 41–46. [CrossRef]
2. Hegde, S.; Shenoy, B.; Chethan, K. Review on carbon fiber reinforced polymer (CFRP) and their mechanical performance. *Mater. Today Proc.* **2019**, *19*, 658–662. [CrossRef]
3. Hiken, A. The Evolution of the Composite Fuselage: A Manufacturing Perspective, Aerospace Engineering, George Dekoulis, IntechOpen. 2018. Available online: https://www.intechopen.com/books/aerospace-engineering/the-evolution-of-the-composite-fuselage-a-manufacturing-perspective (accessed on 2 August 2020).
4. Gupta, G.; Kumar, A.; Tyagi, R.; Kumar, S. Application and future of composite materials: A review. *Int. J. Innov. Res. Sci. Technol.* **2016**, *5*, 6907–6911.
5. Hughes, H. The carbon–fiber epoxy interface—A review. *Compos. Sci. Technol.* **1991**, *41*, 13–45. [CrossRef]
6. Naito, K.; Yang, M.; Kagawa, Y. Tensile properties of high strength polyacrylonitrile (PAN)-based and high modulus pitch-based hybrid carbon fibers–reinforced epoxy matrix composite. *J. Mater. Sci.* **2012**, *47*, 2743–2751. [CrossRef]
7. Shirvanimoghaddam, K.; Hamim, S.; Akbari, K.; Fakhrhoseini, S.M.; Khayyam, H.; Pakseresht, A.H.; Ghasali, E.; Zabet, M.; Munir, K.S.; Jia, S.; et al. Carbon fiber reinforced metal matrix composites: Fabrication processes and properties. *Compos. Part A* **2017**, *92*, 70–96. [CrossRef]
8. Inam, F.; Doris, W.; Wong, Y.; Kuwata, M.; Peijs, T. Multiscale Hybrid Micro-Nanocomposites Based on Carbon Nanotubes and Carbon Fibers. *J. Nanomater.* **2010**, *2010*, 453420. [CrossRef]
9. Barton, B.; Behr, M.; Patton, J.; Hukkanen, E.J.; Landes, B.G.; Wang, W.; Horstman, N.; Rix, J.E.; Keane, D.; Weigand, S.; et al. High–modulus low–cost carbon fibers from polyethylene enabled by boron catalyzed graphitization. *Small* **2017**, *13*, 1701926–1701932. [CrossRef]
10. Jabbour, L.; Chaussy, D.; Eyraud, B.; Beneventi, D. Highly conductive graphite/ carbon fiber/cellulose composite papers. *J. Compos. Sci. Technol.* **2012**, *72*, 616–623. [CrossRef]

11. Calderon, N.; Voytovych, R.; Narciso, J.; Eustathopoulos, N. Wetting dynamics versus interfacial reactivity of AlSi alloys on carbon. *J. Mater. Sci.* **2010**, *45*, 2150–2156. [CrossRef]
12. Morgon, P. *Carbon Fibers and Their Composites*; CRC Press: Boca Raton, FL, USA, 2005; pp. 130–162.
13. Lightning Strike Protection for Composite Structures. Available online: https://www.compositesworld.com/articles/preview/8fb77671-c21b-4a68-8fd9-5dc16c7862c0 (accessed on 9 June 2020).
14. Li, S.; Qi, L.; Zhang, T.; Zhou, J.; Li, H. Interfacial microstructure and tensile properties of carbon fiber reinforced Mg–Al–RE matrix composites. *J. Alloy. Compd.* **2016**, *663*, 686–692. [CrossRef]
15. Kachold, F.; Singer, R. Mechanical properties of carbon fiber-reinforced aluminum manufactured by high-pressure die casting. *J. Mater. Eng. Perform.* **2016**, *25*, 3128–3133. [CrossRef]
16. Li, W.; Liu, L.; Zhong, C.; Shen, B.; Hu, W. Effect of carbon fiber surface treatment on Cu electrodeposition: The electrochemical behavior and the morphology of Cu deposits. *J. Alloy. Compd.* **2011**, *209*, 3532–3536. [CrossRef]
17. Abraham, S.; Pai, C.; Satyanarayana, G.; Vaidyan, V.K. Copper coating on carbon-fibers and their composites with aluminum matrix. *J. Mater. Sci.* **1992**, *27*, 3479–3486. [CrossRef]
18. Zhang, J.; Liu, S.; Lu, Y.; Yin, X.; Zhang, Y.; Li, T. Liquid rolling of woven carbon fibers reinforced Al5083-matrix composites. *J. Mater. Process. Technol.* **2016**, *95*, 89–96. [CrossRef]
19. Fitzer, E.; Figueiredo, J.L.; Bernardo, C.A.; Baker, R.T.K.; Hüttinger, K.J. Carbon fibers-present state and future expectation; pitch and mesophase fibers; structure and properties of carbon fibers. In *Carbon Fibers Filaments and Composites*, 1st ed.; Springer: New York, NY, USA, 1989; pp. 3–146.
20. Wang, X.; Fu, X.; Chung, D.D.L. Electromechanical study of carbon fiber composites. *J. Mater. Res.* **1998**, *13*, 3081–3092. [CrossRef]
21. Di, L.; Liu, B.; Song, J.; Shan, D.; Yang, D.A. Effect of chemical etching on the Cu/Ni metallization of poly (ether ether ketone)/carbon ber composites. *Appl. Surf. Sci.* **2011**, *257*, 4272–4277. [CrossRef]
22. Balaraju, J.N.; Rajam, K.S. Surface morphology and structure of electroless ternary NiWP deposits with various W and P contents. *J. Alloy. Compd.* **2009**, *486*, 468–473. [CrossRef]
23. Lee, C.K. Structure, electrochemical and wear-corrosion properties of electroless nickel-phosphorus deposition on CFRP composites. *Mater. Chem. Phys.* **2009**, *114*, 125–133. [CrossRef]
24. Byeo, J.; Kim, J. Fabrication of a Pure, Uniform Electroless Silver Film Using Ultrafine Silver Aerosol Particles. *Langmuir* **2010**, *26*, 11928–11933.
25. Daoush, W.M.; Alkhuraiji, T.S.; Khamis, M.A.; Albogmy, T.S. Microstructure and electrical properties of carbon short fiber rein- forced copper composites fabricated by electroless deposition followed by powder metallurgy process. *Carbon Lett.* **2020**, *30*, 247–258. [CrossRef]

**Publisher's Note:** MDPI stays neutral with regard to jurisdictional claims in published maps and institutional affiliations.

© 2020 by the authors. Licensee MDPI, Basel, Switzerland. This article is an open access article distributed under the terms and conditions of the Creative Commons Attribution (CC BY) license (http://creativecommons.org/licenses/by/4.0/).

Article

# Paraffin Wax [As a Phase Changing Material (PCM)] Based Composites Containing Multi-Walled Carbon Nanotubes for Thermal Energy Storage (TES) Development

Norah Hamad Almousa [1], Maha R. Alotaibi [1], Mohammad Alsohybani [1], Dominik Radziszewski [2], Saeed M. AlNoman [3], Bandar M. Alotaibi [1] and Maha M. Khayyat [1,*]

[1] King Abdul Aziz City for Science and Technology, Riyadh 11442, Saudi Arabia; nalmousa@kacst.edu.sa (N.H.A.); mralotaibi@kacst.edu.sa (M.R.A.); sohybani@kacst.edu.sa (M.A.); bmalotaibi@kacst.edu.sa (B.M.A.)
[2] New Energy Transfer Ltd., 02-913 Warsaw, Poland; d.radziszewski@newenergytransfer.com
[3] Saudi Electricity Company, Riyadh 11416, Saudi Arabia; SMNoman@se.com.sa
* Correspondence: mkhayyat@kacst.edu.sa

**Abstract:** Thermal energy storage (TES) technologies are considered as enabling and supporting technologies for more sustainable and reliable energy generation methods such as solar thermal and concentrated solar power. A thorough investigation of the TES system using paraffin wax (PW) as a phase changing material (PCM) should be considered. One of the possible approaches for improving the overall performance of the TES system is to enhance the thermal properties of the energy storage materials of PW. The current study investigated some of the properties of PW doped with nano-additives, namely, multi-walled carbon nanotubes (MWCNs), forming a nanocomposite PCM. The paraffin/MWCNT composite PCMs were tailor-made for enhanced and efficient TES applications. The thermal storage efficiency of the current TES bed system was approximately 71%, which is significant. Scanning electron spectroscopy (SEM) with energy dispersive X-ray (EDX) characterization showed the physical incorporation of MWCNTs with PW, which was achieved by strong interfaces without microcracks. In addition, the FTIR (Fourier transform infrared) and TGA (thermogravimetric analysis) experimental results of this composite PCM showed good chemical compatibility and thermal stability. This was elucidated based on the observed similar thermal mass loss profiles as well as the identical chemical bond peaks for all of the tested samples (PW, CNT, and PW/CNT composites).

**Keywords:** TES; PCM; paraffin wax; multi-walled CNTs; SEM; EDX; TGA; FTIR

## 1. Introduction

Intermittent thermal energy, particularly from renewable resources such as solar energy, has entailed the need to develop reliable thermal energy storage (TES) technologies, mainly for heating and cooling applications. Furthermore, the balance of thermal energy supply and demand should be investigated thoroughly, such as the economic deployment of TES technologies, which could be achieved and be successful. In this regard, solar heating and cooling have been considered the most substantial applications requiring TES systems. Although most of the traditional thermal energy systems require short-term thermal storage solutions (i.e., water thermal storage), some significant applications require robust TES technologies, especially for long-term storage requirements in the industrial and commercial sectors.

There are three types of thermal energy storage technologies: sensible storage, latent or, more often, phase change storage, and thermochemical storage [1]. First, sensible thermal storage is based on the capability of storage materials to store thermal energy while varying its temperature without changing its state (i.e., solid or liquid). The most well-known sensible materials include rock, sand, and water. Meanwhile, latent thermal storage is associated with storing thermal energy by changing the storage material phase or state from solid to liquid or vice versa. Some common examples include molten salt, ice/water, and paraffin wax (PW). The third type (thermochemical storage) is different, as it stores thermal energy through a chemical process called adsorption, which differs from the previous two types associated with physical thermal storage. For example, silica gel and zeolite are considered thermochemical storage materials.

The key to the effective and widespread use of solar energy for low-temperature thermal applications is efficient and cost-effective heat storage. Latent heat storage is mostly desirable because of its capability to deliver a high-level energy storage intensity and its features to store heat at a constant temperature, which are related to the phase transition temperature of the heat storage element. For example, molten salt has been used in concentrated solar power systems to store solar heat. Additionally, ice storage has been widely used for chilled water applications. However, storage capacity and temperature range are the two main factors that determine the suitability of phase change materials for specific applications. Therefore, paraffin wax (PW) has been introduced as a promising PCM, especially for free cooling applications [2–5].

Carbon nanotubes (CNTs) are considered a high thermal conductivity additive due to their huge, homogeneous micropores, matchless physicochemical properties, particular surface area, minimal density, and thermal conductivity at high levels [2000–6000 W/(m K)] [4]. CNTs are being employed as an addition to improve the heat transfer properties of other chemical adsorbents [2,4]. Considering that CNTs are a one-dimensional nano-addition substance, they have a distinguished nano-scale effect, high thermal conductivity, and low mass loss [4]. Corresponding to related studies, CNTs have a thermal conductivity greater than 3000 W/(m K) [5,6]. Based on the previous facts, the mixing of CNTs with paraffin wax (PW) to enhance the thermal conductivity of PCMs has been conducted in numerous studies [7–10].

Wang et al. [10] integrated pristine multi-walled CNTs (MWCNTs) into paraffin and revealed that the thermal conductivity improvement ratios in the composite containing 2.0 wt.% were 35% and 40%, respectively, in solid and liquid states. In alternative research, Wang et al. [11] handled CNTs with a mechano-chemical treatment. Handled CNTs have been effectively spread regarding the palmitic acid solution caused by their chemical functional structure, mainly the hydroxide radical in the shell of said CNTs. Wang et al. [11,12] synthesized separate dual types of MWCNTs and compared the impact of the two variations on the thermal behavior of PW and palmitic acid. The authors mentioned that by adding low amounts of grafted MWCNTs, this modification could improve the PCMs' thermal conductivity at a heat above 60 °C compared to premier MWCNTs. Nevertheless, these modified MWCNT-containing nanocomposites have lower thermal conductivity than premier MWCNTs at lower temperatures.

Teng et al. [13] reported the addition of MWCNTs and graphite to paraffin to improve the properties of PCMs. This investigational outcome showed that the addition of MWCNTs is more efficient than graphite in changing the thermal storage of paraffin.

Ye et al. [14] used $Na_2CO_3$, MgO as a PCM, and MWCNTs as supportive substances. The obtained results showed that the thermal conductivity of MWCNTs improves as the mass fraction increases, and likewise raises with the increase in test temperature. Xu and Li [15] manufactured stable paraffin composites modified by diatomite and MWCNTs to produce cement-based thermal energy storage composites. They reported that the thermal conductivity and thermal storage rate increased significantly compared to the blank paraffin. Li et al. [16] combined CNTs with 1-octadecanol to improve the thermal conductivity of the PCM. Tang et al. [17] applied multi-walled CNTs plus n-octadecylamine (f-MWCNTs) to enhance the MWCNT dispersibility of PCMs in paraffin.

Xiao et al. [18] developed oxidized and grafted carbon nanotubes as a filler to improve the thermal energy storage of palmitic acid (PA). Wherever the grafted type was implanted on the 3-propyltrimethoxysilane based on oxidized type, the composite PCMs' thermal conductivity with different types of carbon nanotube raised by 34–40% compared to that of ordinary palmitic acid. However, it is worth noting that the latent heat of grafted CNTs doped with palmitic acid is higher than that of ordinary palmitic acid, whereas oxidized CNTs doped with palmitic have lower latent heat than palmitic acid. Tao et al. [3]. Employed single-walled carbon nanotubes, multi-walled carbon nanotubes, graphene, and C60 as fillers to enhance the properties of high-temperature salt PCM. the results showed that the capability of enhancing of thermal conductivity is in the following decreasing order: single-walled carbon nanotubes, multi-walled carbon nanotubes, and graphene, while the composite PCM with C60 has the lowest thermal conductivity. It was concluded that that the columnar framework encourages efficient heat conduction pathway connections. Therefore, single-walled carbon nanotubes and multi-walled carbon nanotubes exhibit additional advantages for thermal conductivity improvement. The thermal conductivity of composite PCMs with the load up of 1.5 wt.% of single-walled carbon nanotubes and multi-walled carbon nanotubes increased by nearly 57% and 50%, respectively. Moreover, new sorts of carbon nanotubes were investigated, such as grafted types CNTs. While the combination of high conductive additives into PCM revealed a substantial increase in thermal conductivity, latent heat storage/discharge, and overall performance improvements. However, there are still unanswered problems, such as whether PCM packed in the pores of supporting materials is required to build an effective heat transfer network. As a result, choosing the best heat transfer additive for PCM composites is crucial. This needs a complete research study to examine the performance of various additives and to define the elements impacting PCM composites, filling the knowledge gap in previous studies, as previously described. Our current research focuses on the use of paraffin wax and multi-walled carbon nanotube (MWCNT) composites for thermal energy storage applications.

In this study, paraffin wax was doped with nano additives of Multi-Walled Carbon Nanotubes (MWCNs), to forming a nanocomposite PCM. The properties of nanocomposite PCM have been investigated such that the thermal energy storage features and capabilities of the new materials can be greatly enhanced and improved. First, an experimental study of the Thermal Energy Storage (TES) system is described. Characterization of Paraffin-Wax doped with Multi-Walled Carbon Nanotubes Composites is then discussed. Furthermore, SEM, EDX, FTIR, and TGA are among the characterization techniques used.

## 2. Materials

### 2.1. TES System Evaluation and Experimental Set-Up

A thermal energy storage (TES) system mainly consists of the following parts (see Figure 1): a source of heat, a storage unit, and load resistance. To improve the overall performance and to increase the efficiency of the system, several approaches can be considered.

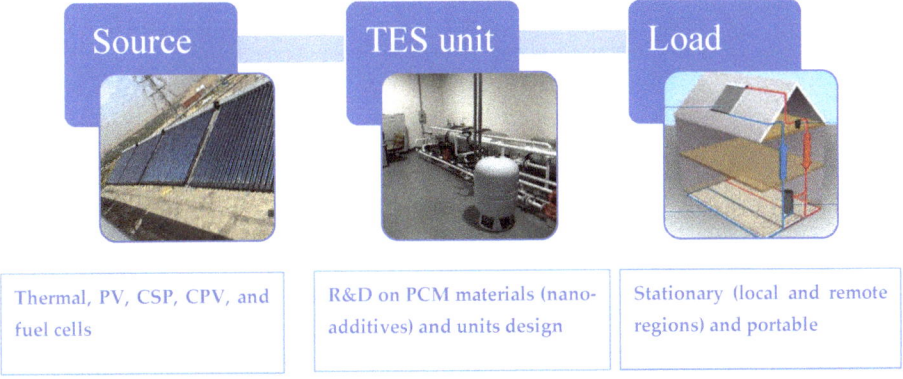

**Figure 1.** A thermal energy storage system mainly consists of a source of thermal energy (such as solar tubes, as shown) and a TES unit to store the thermal energy for some period of time to be used by a load for domestic or industrial use.

The course of the melting and solidifying process of the PCM was investigated during the experiments. The material was tested as follows: Temperature sensors were installed on the supply and return of the storage tank water system. The flow meter measured the water flow continuously, and the measurement results were saved every 10 s. The mass of the material used to fill the bed, which was used during the test, was 822.3 kg. It should be noted that the calculations did not take into account heat losses due to the environment or the heat accumulated in the structural elements of the tank itself, in which the phase change material was located.

### 2.2. Materials: Paraffin Wax (PW) and Multi-Walled Carbon Nanotubes (MWCNTs)

A phase change material with the properties described below was used during the tests. The phase change temperature of the PW was 54 °C with the other related properties tabulated, as shown in Table 1. Paraffin wax from the unit bed at our laboratory was provided by NET (New Energy Transfer Company, Poland). The MWCNTs had the following parameters: 9.5 nm in diameter, 1.5 µm in length, and a carbon purity of approximately 90%.

**Table 1.** Physical properties of the paraffin wax (PW) [5].

|    | Thermal-Related Parameters | Value | Unit |
|----|---|---|---|
| 1  | Melting temperature | 54.32 | °C |
| 2  | Latent heat of fusion | 184.48 | kJ/kg |
| 3  | Density of PW (liquid phase) | 775.00 | kg/m$^3$ |
| 4  | Density of PW | 833.60 | kg/m$^3$ |
| 5  | Specific heat of PW (liquid phase) | 2.44 | kJ/kgK |
| 6  | Thermal conductivity | 0.15 | W/mK |
| 7  | Viscosity | $6.3 \times 10^{-3}$ | P.S (Pascal Second) |
| 8  | Kinematic viscosity | $8.31 \times 10^{-5}$ | m$^2$/s |
| 9  | Prandtl number | 1001.23 | - |
| 10 | Thermal expansion coefficient | $7.14 \times 10^{-3}$ | 1/°C |

### 2.3. Preparation of PW Using the Melting Method

Paraffin and Paraffin/MWCNT compounds were prepared separately by adding CNTs and MWCNTs, respectively, to paraffin using a mechanical dispersion method. The first batch of samples was prepared with a different load of nanoparticles corresponding to 0.5% (PW1-CNT), 0.75% (PW-2CNT), and 1% (PW-3CNT) by mass. Paraffin wax was weighted and melted in a beaker on a hotplate stirrer at 70 °C. Magnetic stirring was applied for 1 h after adding different fractions of nanoparticles, followed by sonication for 3 h. Finally, the liquid compounds were cooled at room temperature to obtain a solid (Figure 2). Element analysis for PW from the TES unit and the pristine one from the Techno Pharm Chem company are showed in Appendix A.

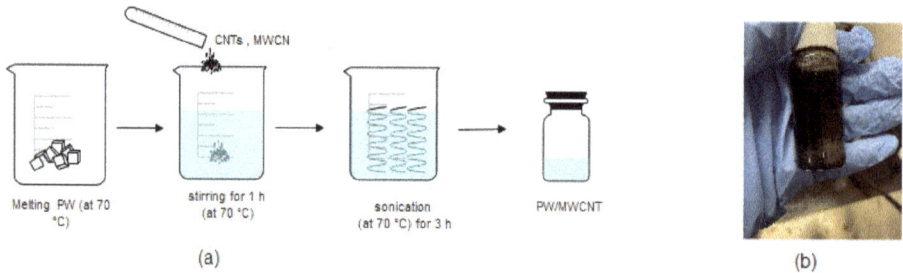

**Figure 2.** (a) Sample preparation using the mechanical dispersion method. (b) Final product following sample preparation of PCM with MWCNT.

## 2.4. Thoretical Background of TES System Evaluation

In this section, we review the necessary equations and theoretical foundations to evaluate the thermal energy storage (TES) and the temperature of the feed water, which was calculated at the start using Equation (1):

$$T_{w,aw} = \frac{T_{w1} + T_{w2}}{2} \quad (1)$$

By linearly interpolating the literature data, the density and specific heat of the water were calculated using Equation (2):

$$f(T) = f(T_0) + \frac{f(T_1) - f(T_0)}{T_1 - T_0} \times (T_{w,aw} - T_0) \quad (2)$$

Equation (3) was used to calculate the heat flux transferred from the water to the PCM material throughout the storage loading process as well as the amount of heat according to the dependency:

$$Q = \rho_w \times V_w \times C_w \times (T_{w,1} - T_{w,2}) \quad (3)$$

The heat accumulation during the charging and discharging of the paraffin bed were determined by Equation (4). Finally, we determined the theoretical heat using Equation (5):

$$\eta = \frac{Q}{Q_t} \times 100\% \quad (4)$$

$$Q_t = m_{PCM} \times C_{PCM,S} \times (T_m - T_1) + m_{PCM} \times L + m_{PCM} \times C_{PCM,L} \times (T_2 - T_m) \quad (5)$$

For more information on the abbreviations and symbols for equations, see Appendix B.

## 3. Results and Discussion

### 3.1. TES System Evaluation

Based on the obtained results, the heat transferred from the feed water to the PCM material (i.e., PW) when the heat storage was being charged and the heat received by the water from the phase modification material through the heat storage discharge were calculated. Therefore, the feedwater temperature was calculated at the beginning, according to Equation (1).

In order to conduct further analysis of the test results, the density and specific heat of the supply water needed to be determined. However, these values change as a function of temperature, as shown in Table 2. Since the literature data only provide the thermophysical properties of water at selected temperatures, analysis of the obtained test results and the density and specific heat of the water were calculated by linearly interpolating the literature data using Equation (2).

**Table 2.** The density and specific heat of water as a function of temperature.

| Temperature (°C) | Density (kg/m³) | Specific Heat (J/kg. K) |
|---|---|---|
| 0.01 | 999.90 | 4.212 |
| 10.00 | 999.70 | 4.191 |
| 20.00 | 998.20 | 4.183 |
| 30.00 | 995.70 | 4.174 |
| 40.00 | 992.20 | 4.174 |
| 50.00 | 988.10 | 4.174 |
| 60.00 | 983.20 | 4.179 |
| 70.00 | 977.80 | 4.187 |
| 80.00 | 971.80 | 4.195 |
| 90.00 | 965.30 | 4.208 |
| 100.00 | 958.40 | 4.220 |

In Equation (2), $T$ is the actual temperature for which the density or specific heat of the water is determined. $T_0$ and $T_1$ are the closest temperatures from the literature data below and above the actual temperature, respectively, while $f(T), f(T_0)$, and $f(T_1)$ are the specific density or heat values for the given temperatures $T$, $T_0$, and $T_1$. The average water temperature at the inlet and outlet from the reservoir, calculated according to Equation (1), was used to calculate the density and specific heat of the water.

The heat flux transferred from the water to the PCM material during the storage loading process as well as the amount of heat according to the dependency was then calculated using Equation (3).

The heat that could theoretically be accumulated in the PCM material (i.e., PW) and the efficiency Equation (4) of heat accumulation during charging and discharging of the paraffin bed were determined. The thermal efficiency during the loading and unloading of the heat storage was also determined. The theoretical heat formula (Equation (5)) consists of three parts: sensible heat accumulated in a solid substance in the range from the initial temperature ($T_1$) to the material phase change temperature ($T_m$), which is 54 °C, and the phase change heat and sensible heat accumulated in the liquid from the melting point to the final temperature ($T_2$) of water flowing from the exchanger.

The heat stored in the material while the paraffin bed was charging was lower than the theoretical heat. This is because some of the PCM material in the tank may not melt, and the properties of PCM materials change over time. PCM material partially degrades during the following cycles, and its heat storage capacity partially decreases. The theoretical thermal capacity of the paraffin bed, which resulted from the scope of the calculations, was 57.39 kWh according to the characteristics of the PCM material used. The heat absorbed by the deposit in the calculations was 35.04 kWh. Thus, the ratio of the heat stored in paraffin to the theoretical heat capacity of the bed equaled 61.07%. The results are presented in Table 3.

**Table 3.** Process of heat storage loading.

| Time (s) | Water Temperature at the Inlet to the Tank (°C) | Water Temperature at the Outlet to the Tank (°C) | Water Flow (kg/s) |
|---|---|---|---|
| 10 | 42.6 ± 0.4 | 34.6 ± 0.4 | 1.90 |
| 660 | 46.5 ± 0.4 | 44.8 ± 0.4 | 2.85 |
| 1340 | 53.0 ± 0.4 | 51.2 ± 0.4 | 2.86 |
| 2000 | 57.4 ± 0.4 | 55.8 ± 0.4 | 2.86 |
| 2660 | 62.7 ± 0.4 | 61.1 ± 0.4 | 2.86 |
| 3340 | 66.7 ± 0.4 | 66.0 ± 0.4 | 2.87 |
| 4000 | 67.7 ± 0.4 | 66.1 ± 0.4 | 2.87 |
| 4660 | 68.1 ± 0.4 | 66.8 ± 0.4 | 2.88 |
| 5340 | 67.2 ± 0.4 | 66.0 ± 0.4 | 2.87 |
| 6000 | 68.1 ± 0.4 | 66.3 ± 0.4 | 2.86 |
| 6660 | 67.9 ± 0.4 | 66.3 ± 0.4 | 2.87 |
| Total heat (kWh) | | | 35.04 |
| Theoretical heat (kWh) | | | 57.39 |
| Efficiency (%) | | | 61.07 |
| Average heat flux (kW/m$^2$) | | | 19.18 |

As shown in Table 3, the total heat removed from the PCM material was 39.41 kWh, while the heat that could theoretically be removed was 57.39 kWh. Thus, the heat storage discharging efficiency was 70.91%. The discharge efficiency of heat storage with PCM materials ought to be in the range of 45–78%. This value is in the upper range of the discharge efficiency for the paraffin bed, which proves the good performance of the bed.

Figure 3 shows the temperature difference between the water supply and the return from the bed. A sudden drop and increase in the value of the water return temperature from the ex-changer is related to the incorrect reading of the measuring system. The average difference in temperature was 1.93 K. This is an acceptable disparity to ensure greater test accuracy. This can be achieved, for example, by reducing the water flow through the exchanger.

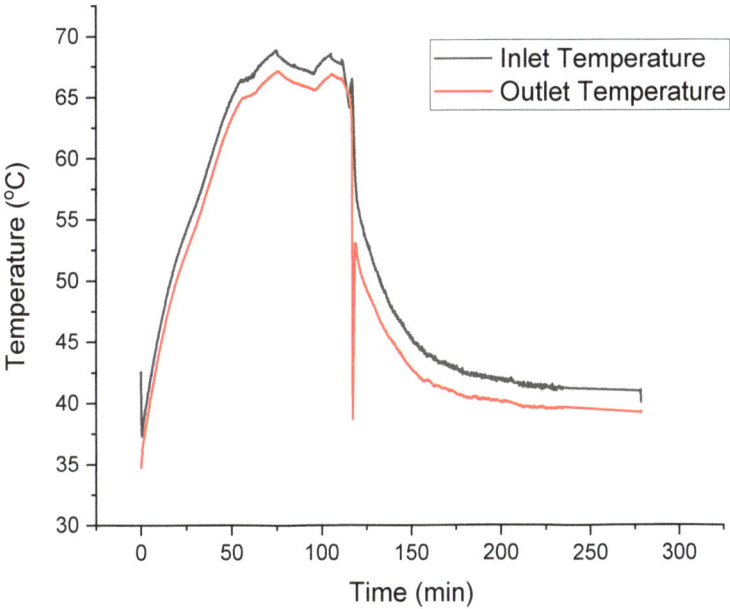

**Figure 3.** Temperature difference between the supply and return from the paraffin tank.

*3.2. Characterizations of Paraffin Wax Containing Multi-Walled Carbon Nanotube Composites*

Carbon nanotubes are categorized into single-wall carbon nanotubes (SWCNT) and multi-walled carbon nanotubes (MWCNT), where SWCNTs have been made of monolayer graphene. The dangling bonds are swiftly incorporated on the boundary while winding the graphene layer into a cylinder, resulting in the axis of CNTs becoming randomly dispersed [5]. When the graphite surface area is lined up lengthwise across the SWCNT axis, a two-dimensional geometry such as a graphene surface with a single layer is produced [6]. On the contrary, multi-walled carbon nanotubes have outstanding characteristics (thermal, electrical, and mechanical), which afford a wide range usage potential opportunities [7]. It has been determined that the thermal conductivity of single multi-walled carbon nanotubes near 37,000 W/(m K) at a temperature of 100 K with the macroscopic thermal conductivity are able to achieve 6000 W/(m K) [8]. The thermal conductivity of single multi-walled carbon nanotubes at room temperature can be comparable to an isotopically pure diamond and can even achieve a greater value [9].

3.2.1. SEM and EDX Characterizations

SEM images of carbon nanotube and carbon nanotube–paraffin nanocomposites were used to describe their micromorphology using Scanning Electron Microscope manufacturing by JEOL Ltd., model: JSM-7100F, Tokyo, Japan.

The multi-walled carbon nanotubes were made of a black powder with a laminar architecture and loosely packed particles, as shown in Figure 4, displaying PW-based MWCNTs nanocomposites. Figure 4 confirms that the MWCNT layers were distributed in the PW in various paths and spots. The MWCNTs created a framework that supports heat transfer. Moreover, the layers of the MWCNTs were totally and regularly covered by paraffin, where the MWCNTs and PW were strongly integrated as the content of the MWCNTs increased, without any microcracks or loose interfaces. These observations have been confirmed by the EDX Elemental Analysis (see Figure 5), of the spatial distribution of various emelments.

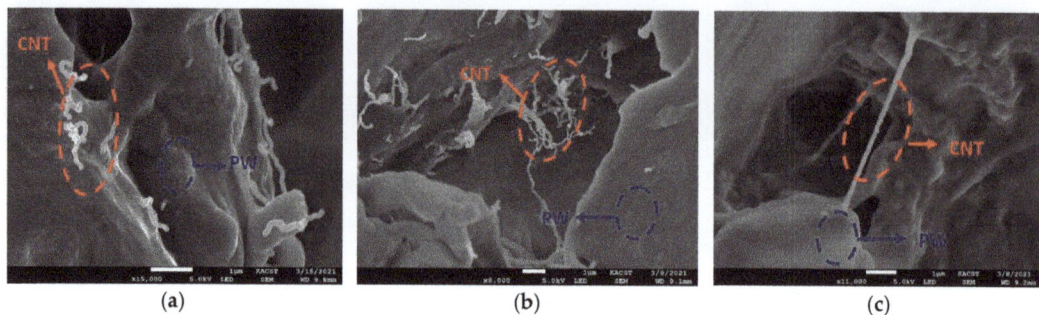

**Figure 4.** SEM images of the paraffin wax/MWCNT composites: PW-1CNT (**a**), PW-2CNT (**b**), and PW-3CNT (**c**). Scale bars of Figures 4 and 5 are the same (see Figure 5).

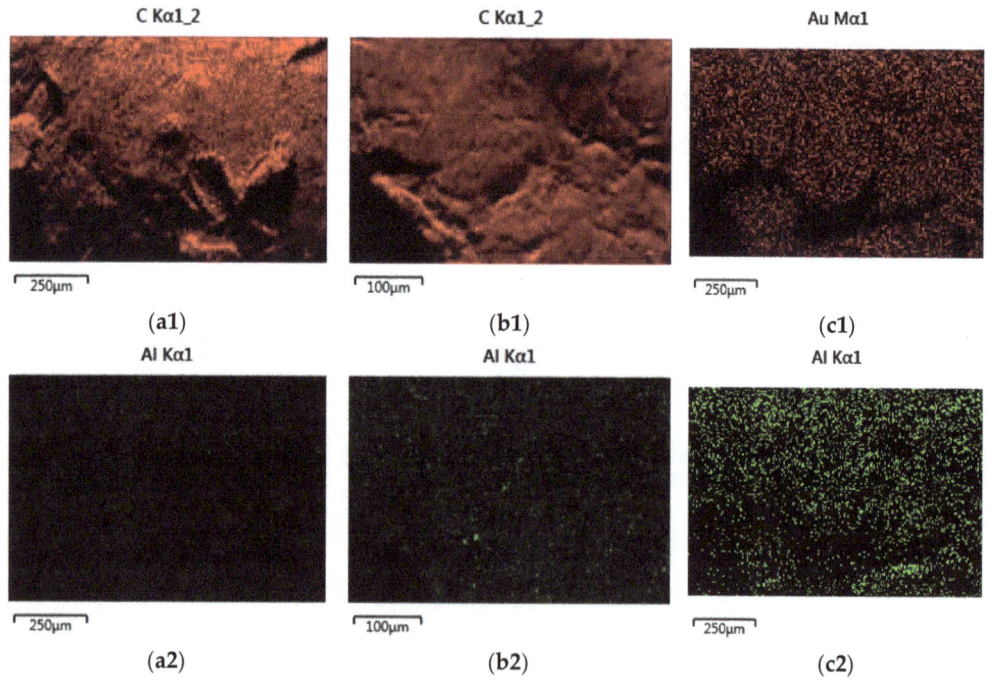

**Figure 5.** EDX images of the paraffin wax/MWCNT composites: (**a1,a2**) PW-1 CNT, (**b1,b2**) PW-2 CNT, and (**c1,c2**) PW-3 CNT.

It was anticipated that as the mass fraction of multi-walled carbon nanotubes rises, the thermal conductivity slowly rises, while the latent heat drops, which would indicate that the increase in the PCM's thermal conductivity using MWCNTs complements the decreased latent heat of the PCM composite. Consequently, further thermal measurements are required, with a fitting mass fraction of multi-walled carbon nanotubes needing to be measured, based on the application.

3.2.2. FTIR Analysis

The FTIR spectra of the paraffin, carbon materials, and composites are shown in Figure 6.

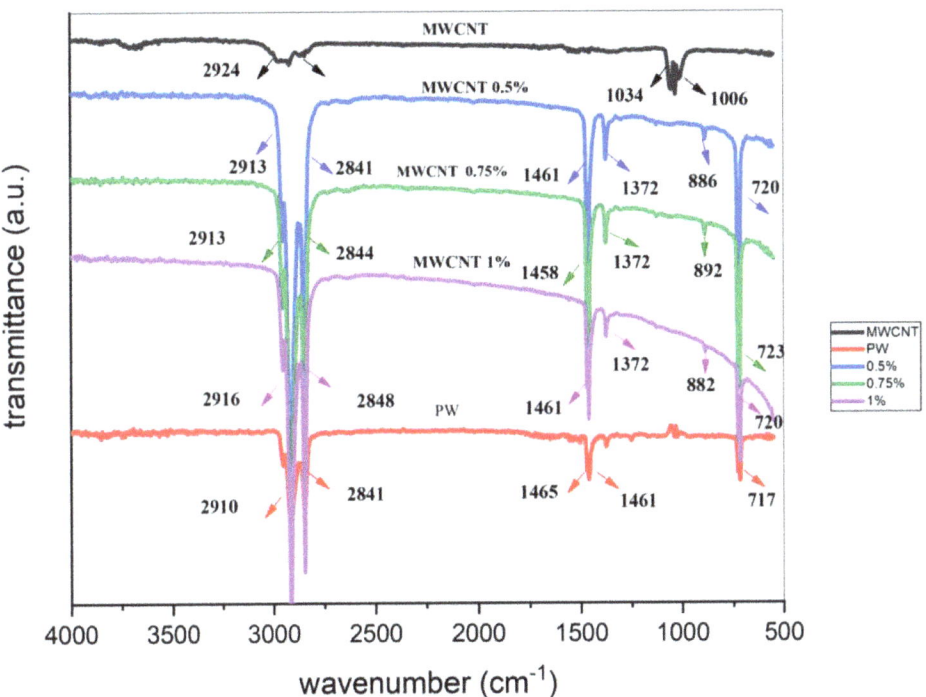

**Figure 6.** FTIR spectra of paraffin wax, CNTs, PW-1CNT (0.5% CNT), PW-2CNT (0.75% CNT), and PW-3CNT (1% CNT).

The PW-1CNT, PW-2CNT, PW-3CNT, and PW spectra displayed four remarkable peaks, corresponding to the presence of PW. The bonds detected at around 2800 cm$^{-1}$ are designated to the C-H stretching vibration of the $CH_3/CH_2$ groups. The peak near 1500 cm$^{-1}$ is assigned to the bending vibration of C-H. The 717 cm$^{-1}$ absorption band is credited to the in-plane deformation rocking vibration of the PW molecule [19,20]. The spectra corresponding to the CNT showed a band of absorption around 3000 cm$^{-1}$, which can be accredited to the stretching of the C=C bonds. Bonds corresponding to the stretching of the C-C of the CNT structure were observed at around 1000 cm$^{-1}$ [19,21–23].

The addition of CNT to the PW did not form any new peaks, and the PW-1CNT, PW-2CNT, and PW-3CNT only displayed a combination of peaks corresponding to the PW and CNT.

### 3.2.3. TGA Analysis

Thermogravimetric analyses (TGAs) were performed in alumina crucibles under nitrogen (flow rate, 100 mL/min) using a thermobalance Discovery SDT 650 from TA Instruments with a heating rate 10 °C/min.

Thermogravimetric analysis method: Figure 7 shows the measured TGA curves. The clear mass loss that is observed is due to paraffin degradation. Furthermore, the nearly overlapping TGA curves in Figure 7 not only reveal the close paraffin impregnation amount but also the CNTs' exceptionally stable thermal performance. In addition, the mass loss that is seen in the thermal behaviour curves and as calculated in Table 4 confirm that the CNT disperses well in the paraffin wax. In other words, CNT is known to be chemical inert and shows resistance to thermal degradation of the thermal decomposition of paraffin wax nanocomposites after the addition of CNT particles, as shown in Table 4 and Figure 7. It is shown that the thermal decomposition of paraffin wax nanocomposites gradually moves to a slightly higher temperature with increasing amounts of MWCNT particles. Even though the CNT interaction with PW is weak due to the fact that CNT did not functionalize, the CNT can still interact with the surrounding matrix and can improve the overall thermal stability of the PW/CNT nanocomposite. The residual weight at 600 °C, after considering the CNT weight loss and considering that the PW temperature was completely decomposed at this point, as well as the fact that the CNT showed a little degradation, was close to the initial loadings 0.5, 0.75 and 1 wt.% [24–26]. Also as seen in Figure 7, occurrences of rapid degradation of all nanocomposites are shown at around 210 °C due to the degradation of PW, but each single composite experienced different weight loss.

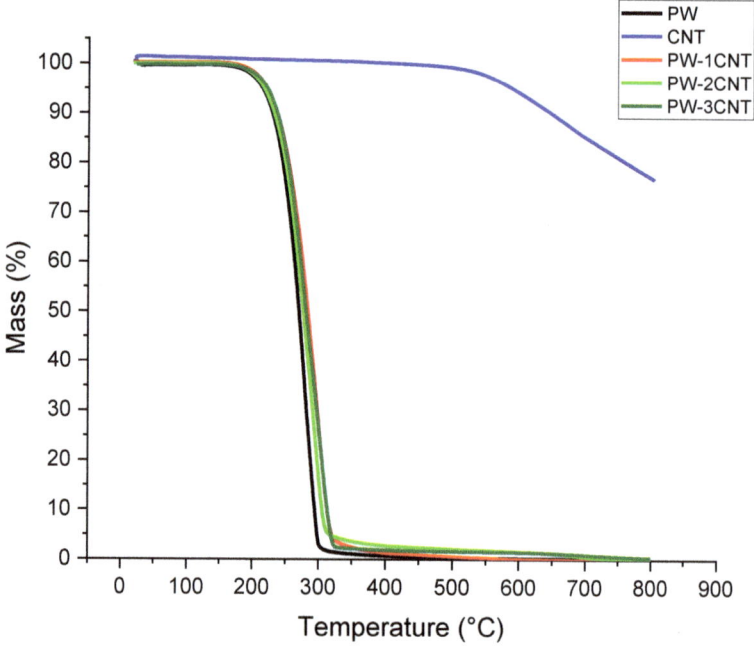

**Figure 7.** TGA thermogram of the paraffin wax/MWCNT and composites (PW-1CNT, PW-2CNT, and PW-3CNT).

**Table 4.** Mass loss of the PW, CNT, and composites 1, 2, and 3.

| Sample Code | Onset Temperature °C | Mass Loss at 600 °C | Mass Loss % at 700 °C |
|---|---|---|---|
| CNT | 533.56 | 94.237 | 85.234 |
| PW (2) | 243.12 | 0.281 | 0.195 |
| Composite 1 | 249.26 | 0.171 | −0.043 |
| Composite 2 | 244.56 | 1.603 | 0.639 |
| Composite 3 | 239.97 | 1.493 | 0.98 |

Weighing Accuracy for thermobalance Discovery SDT 650 = ±0.5%; Standard uncertainty (mK) of electronic thermometer = ±1.0.

It was reported by Kuziel et al. [27] that in MWCNT (0.5 wt.%)–paraffin nanocomposites, the thermal conductivity is enhanced by 37%, in addition to increasing by 6.3% to the enthalpy phase change ($\Delta Hm$), compared to the paraffin, which has brilliant cycling stability. Individual and fibrous ultra-long MWCNTs have superior properties due to the faster nucleation of larger crystallites by MWCNTs via short- and long-range templating as well as the intrinsic properties of individual and fibrous ultra-long MWCNTs [28–30].

## 4. Conclusions

Phase change materials (PCMs) are considered efficient for storing thermal energy due to their high latent temperature and slight temperature variation during the phase change process. Based on a literature review of PCM type, nanoparticle type, and fraction, as shown in Table 5, we started studying PW (as a PCM) as a material for storing thermal energy, as it has several advantages, including latent fusion, chemical stability, negligible supercooling, no phase separation, and low cost. The thermal storage performance of a TES bed system was approximately 71%, which can be considered relatively high. In order to enhance the thermal conductivity of the PW, the dispersion of high thermal conductivity materials such as MWCNTs was employed. The prepared composites of PW and MWCNTs of various weights were characterized by using various techniques, namely, SEM, EDX, FTIR, and TGA. First, the SEM and EDX results showed significant improvement in the molecular structure of the PW/MWCNT composites. When PW and MWCNTs were mixed, the MWCNT layers were distributed evenly and were integrated with paraffin layers through strong interfaces without microcracks. The FTIR results showed that adding CNT to PW did not form any new peaks, and the prepared composites only displayed a combination of peaks corresponding to the PW and CNT, such that a physical combination can be intuitively expected. TGA analysis elucidated that the addition of MWCNT to paraffin enhanced its thermal properties toward better thermal conductivity. At 700 °C, composites 1, 2, and 3 showed a mass loss of −0.043, 0.639, and 0.98, respectively. Moreover, it can be deduced that the relatively coinciding mass loss profiles (PW, CNT, and PW/CNT composites) could be due to the fact that a uniform distribution of the CNT layers within PW was accomplished.

Table 5. Summary of the previous work on PCM type, nanoparticle type, and fraction.

| Authors (Year) | PCM | Type of Additives | Fraction of Additives | Comments |
|---|---|---|---|---|
| Wang et al. (2009) [10] | Paraffin | MWCNTs | 0.2, 0.5, 1 and 2 | The composite containing 2.0 wt.% had a higher thermal conductivity of 35% and 40%, respectively, in solid and liquid states. |
| Wang et al. (2011) [12] | Paraffin Palmitic acid | G8-CNT G18-CNT | - | The thermal conductivity of paraffin and palmitic acid were improved by adding a small amount of G8-CNT. The results showed that the thermal conductivity of CNTs is clearly affected by the length of the grafted chain. |
| Teng et al. (2013) [13] | Paraffin | MWCNTs Graphite | 1, 2 and 3 | MWCNTs were more effective in enhancing paraffin performance in all experimental parameters compared to graphite. |
| Ye et al. (2014) [14] | $Na_2CO_3$/MgO | MWCNTs | 0.1%, 0.2%, 0.3% and 0.5% | As the weight percentage of the MWCNTs increased, the thermal conductivity of the composite PCMs increased by approximately 96% (the highest) for 0.5% of MWCNTs. |

Table 5. Cont.

| Authors (Year) | PCM | Type of Additives | Fraction of Additives | Comments |
| --- | --- | --- | --- | --- |
| Xu and Li (2014) [15] | paraffin (R27, Rubitherm) | MWCNTs Diatomite materials | 0.26% - | The thermal conductivity of PCM-DP600-CNTs was substantially improved, with an improvement level of up to 42.45%. |
| Li et al. (2014) [16] | Paraffin | CNTs CNTs-SA | - | The thermal conductivity of MicroPCMs/CNTs-SA with 4% CNTs increased by 79.2% when compared to MicroPCMs. |
| Tang et al. (2014) [17] | Paraffin | (f-MWCNT) | 1, 5 and 10% | With 10 wt.% of f-MWCNTs, the thermal conductivity and the heat transfer of the paraffin/f-MWCNTs composite PCMs increased by 86.7%. |
| Xiao et al. (2015) [18] | Palmitic acid | CNTs, oxidized CNTs and grafted CNTs | 1/100 | CNTs, O-CNTs, and G-CNTs improved the palmitic acid thermal conductivity, but G-CNT composites had the highest latent heat. |
| Tao et al. (2015) [3] | Salt | SWCNT, MWCNT, graphene and C60 | 0.1%, 0.5%, 1.0%, 1.5%, 2.5% | SWCNTs and MWCNTs exhibited significant enhancement in the PCMs' thermal conductivity with mass fractions near 1.5% by 57% and 50%, respectively. |

The main suggestions for future work drawn in the light of the obtained results are as follows:

1. Studying the effect of PW/MWCNT composites (of various lengths and diameters) on the thermal conductivity and diffusivity of PW nanocomposites.
2. Comparing the effect of MWCNTs on heat storage/release rates to pristine PW.
3. Investigating the durability of these composites considering the number of possible circular (heating/cooling) applications and any possible degradation of PCM over time.

**Author Contributions:** Conceptualization, experimental work, and writing—original draft, N.H.A.; experimental work and analysis, M.R.A.; design, methodology, and interpretation of results, M.A.; editing, D.R.; editing and interpretation of results discussion, S.M.A.; conceptualization, design, interpretation of results, writing—original draft and editing, M.M.K.; reviewing. B.M.A. All authors have read and agreed to the published version of the manuscript.

**Funding:** This research was funded by the King Abduaziz City for Science and Technology.

**Institutional Review Board Statement:** Not applicable.

**Informed Consent Statement:** Not applicable.

**Data Availability Statement:** The data used to support the findings of this study are included within the article. However, the corresponding author can provide the data used in this study upon request.

**Acknowledgments:** The authors would like to thank the King Abdulaziz City for Science and Technology and the Saudi Electricity Company for supporting this work. The authors would also like to thank SASO (Saudi Standards of Metrology and Quality) for providing access to their tools to conduct the DSC measurements. The authors are also grateful to Ali A. Algarni, Mohammed S. Alotaibi, Mohammed A. Alhajji, and Sultan A. Alburidi for the FT-IR, SEM, and TGA analysis.

**Conflicts of Interest:** The authors declare no conflict of interest.

## Appendix A. Element Analysis for PW from the TES Unit and the Pristine One from the Techno Pharm Chem Company

| | PW (TES Unit Bed) |
|---|---|
| Element | Conc (mg/kg) |
| Li | 0.022 ± 0.001 |
| B | 1.83 ± 0.011 |
| Na | 545 ± 0.2 |
| Mg | 7.88 ± 0.2 |
| Al | 31 ± 0.15 |
| Si | 85 ± 0.2 |
| Ca | 191 ± 0.5 |
| Sc | 0.02 ± 0.003 |
| Ti | 1 ± 0.005 |
| V | 0.058 ± 0.001 |
| Cr | 11 ± 0.05 |
| Mn | 0.8 ± 0.04 |
| Fe | 21 ± 0.1 |
| Co | 0.065 ± 0.005 |
| Ni | 0.33 ± 0.005 |
| Cu | 0.66 ± 0.007 |
| Zn | 2 ± 0.05 |
| Se | 0.22 ± 0.005 |
| Sr | 0.3 ± 0.015 |
| Mo | 0.29 ± 0.005 |
| Ag | 0.048 ± 0.001 |
| Sn | 107 ± 0.2 |
| Sb | 0.044 ± 0.001 |
| Ba | 0.29 ± 0.005 |
| W | 0.11 ± 0.004 |
| Pb | 0.11 ± 0.003 |
| Ge | UDL |
| Sb | UDL |
| Cs | UDL |

## Appendix B. Nomenclature of Quantities Used in Equations

$T_{w1}$—water inlet temperature
$T_{w2}$—water outlet temperature
$T_{w,aw}$—average temperature of the water supply and return
$T_m$—phase change temperature of a phase change material
$F(T)$—the actual density or specific heat of the water
$F(T_0)$—the density or specific heat of water at a stationary point
$F(T_1)$—the density or specific heat of water at another known stationary point
$\rho_w$—average water density at constant pressure
$V_w$—water volume flow
$C_w$—specific heat at constant pressure
$Q$—real heat capacity
$Qt$—theoretical heat capacity
$m_{PCM}$—mass of the phase change material
$C_{PCM,S}$—the sensible heat of a phase change material (solid)
$C_{PCM,L}$—the sensible heat of a phase change material (liquid)
$L$—the latent heat of a phase change material

## References

1. Karaipekli, A.; Biçer, A.; Sari, A.; Tyagi, V.V. Thermal characteristics of expanded perlite/paraffin composite phase change material with enhanced thermal conductivity using carbon nanotubes. *Energy Convers. Manag.* **2017**, *134*, 373–381. [CrossRef]
2. Yan, T.; Li, T.; Wang, R.; Jia, R. Experimental investigation on the ammonia adsorption and heat transfer characteristics of the packed multi-walled carbon nanotubes. *Appl. Therm. Eng.* **2015**, *77*, 20–29. [CrossRef]
3. Tao, Y.; Lin, C.; He, Y. Preparation and thermal properties characterization of carbonate salt/carbon nanomaterial composite phase change material. *Energy Convers. Manag.* **2015**, *97*, 103–110. [CrossRef]

4. Kim, P.; Shi, L.; Majumdar, A.; McEuen, P.L. Thermal transport measurements of individual multiwalled nanotubes. *Phys. Rev. Lett.* **2001**, *87*, 215502. [CrossRef]
5. Wu, S.; Yan, T.; Kuai, Z.; Pan, W. Thermal conductivity enhancement on phase change materials for thermal energy storage: A review. *Energy Storage Mater.* **2020**, *25*, 251–295. [CrossRef]
6. Hersam, M.C. Progress towards monodisperse single-walled carbon nanotubes. *Nat. Nanotechnol.* **2008**, *3*, 387–394. [CrossRef]
7. Iijima, S. Helical microtubules of graphitic carbon. *Nat. Cell Biol.* **1991**, *354*, 56–58. [CrossRef]
8. Warzoha, R.; Fleischer, A.S. Effect of carbon nanotube interfacial geometry on thermal transport in solid–liquid phase change materials. *Appl. Energy* **2015**, *154*, 271–276. [CrossRef]
9. Berber, S.; Kwon, Y.-K.; Tomanek, D. Unusually high thermal conductivity of carbon nanotubes. *Phys. Rev. Lett.* **2000**, *84*, 4613–4616. [CrossRef]
10. Wang, J.; Xie, H.; Xin, Z. Thermal properties of paraffin based composites containing multi-walled carbon nanotubes. *Thermochim. Acta* **2009**, *488*, 39–42. [CrossRef]
11. Wang, J.; Xie, H.; Xin, Z.; Li, Y.; Chen, L. Enhancing thermal conductivity of palmitic acid based phase change materials with carbon nanotubes as fillers. *Sol. Energy* **2010**, *84*, 339–344. [CrossRef]
12. Wang, J.; Xie, H.; Xin, Z. Preparation and thermal properties of grafted CNTs composites. *J. Mater. Sci. Technol.* **2011**, *27*, 233–238. [CrossRef]
13. Teng, T.-P.; Cheng, C.-M.; Cheng, C.-P. Performance assessment of heat storage by phase change materials containing MWCNTs and graphite. *Appl. Therm. Eng.* **2013**, *50*, 637–644. [CrossRef]
14. Ye, F.; Ge, Z.; Ding, Y.; Yang, J. Multi-walled carbon nanotubes added to $Na_2CO_3$/MgO composites for thermal energy storage. *Particuology* **2014**, *15*, 56–60. [CrossRef]
15. Xu, B.; Li, Z. Paraffin/diatomite/multi-wall carbon nanotubes composite phase change material tailor-made for thermal energy storage cement-based composites. *Energy* **2014**, *72*, 371–380. [CrossRef]
16. Li, M.; Chen, M.; Wu, Z. Enhancement in thermal property and mechanical property of phase change microcapsule with modified carbon nanotube. *Appl. Energy* **2014**, *127*, 166–171. [CrossRef]
17. Tang, Q.; Sun, J.; Yu, S.; Wang, G. Improving thermal conductivity and decreasing supercooling of paraffin phase change materials by n-octadecylamine-functionalized multi-walled carbon nanotubes. *RSC Adv.* **2014**, *4*, 36584–36590. [CrossRef]
18. Xiao, D.; Qu, Y.; Hu, S.; Han, H.; Li, Y.; Zhai, J.; Jiang, Y.; Yang, H. Study on the phase change thermal storage performance of palmitic acid/carbon nanotubes composites. *Compos. Part A Appl. Sci. Manuf.* **2015**, *77*, 50–55. [CrossRef]
19. Varshney, D.; Ahmadi, M.; Guinel, M.J.-F.; Weiner, B.R.; Morell, G. Single-step route to diamond-nanotube composite. *Nanoscale Res. Lett.* **2012**, *7*, 535. [CrossRef] [PubMed]
20. Nie, C.; Tong, X.; Wu, S.; Gong, S.; Peng, D. Paraffin confined in carbon nanotubes as nano-encapsulated phase change materials: Experimental and molecular dynamics studies. *RSC Adv.* **2015**, *5*, 92812–92817. [CrossRef]
21. Maleki, A.; Hamesadeghi, U.; Daraei, H.; Hayati, B.; Najafi, F.; McKay, G.; Rezaee, R. Amine functionalized multi-walled carbon nanotubes: Single and binary systems for high capacity dye removal. *Chem. Eng. J.* **2017**, *313*, 826–835. [CrossRef]
22. Ovsiienko, I.; Len, T.; Matzui, L.; Tkachuk, V.; Berkutov, I.; Mirzoiev, I.; Prylutskyy, Y.; Tsierkezos, N.; Ritter, U. Magnetore-sistance of functionalized carbon nanotubes. *Materwiss. Werksttech.* **2016**, *47*, 254–262. [CrossRef]
23. Metwally, N.H.; Saad, G.R.; El-Wahab, E.A.A. Grafting of multiwalled carbon nanotubes with pyrazole derivatives: Characterization, antimicrobial activity and molecular docking study. *Int. J. Nanomed.* **2019**, *14*, 6645–6659. [CrossRef] [PubMed]
24. Han, L.; Jia, X.; Li, Z.; Yang, Z.; Wang, G.; Ning, G. Effective encapsulation of paraffin wax in carbon nanotube agglomerates for a new shape-stabilized phase change material with enhanced thermal-storage capacity and stability. *Ind. Eng. Chem. Res.* **2018**, *57*, 13026–13035. [CrossRef]
25. Elgafy, A.; Lafdi, K. Effect of carbon nanofiber additives on thermal behavior of phase change materials. *Carbon* **2005**, *43*, 3067–3074. [CrossRef]
26. Jeong, N.; Park, Y.C.; Yoo, J.H. Preparation of highly pure and crystalline carbon nanotubes and their infiltration by paraffin wax. *Carbon* **2013**, *63*, 240–252. [CrossRef]
27. Kuziel, A.W.; Dzido, G.; Turczyn, R.; Jedrysiak, R.; Kolanowska, A.; Tracz, A.; Zieba, W.; Cyganiuk, A.; Terzyk, A.; Boncel, S. Ultra-long carbon nanotube-paraffin composites of record thermal conductivity and high phase change enthalby among par-affin-based heat storage materials. *J. Energy Storage* **2021**, *36*, 102396. [CrossRef]
28. Dincer, I.; Rosen, M. *Thermal Energy Storage: Systems and Applications*, 2nd ed.; Wiley: New York, NY, USA, 2011.
29. Souayfane, F.; Fardoun, F.; Biwole, P. Phase change materials (PCM) for cooling applications in buildings: A review. *Energy Build.* **2016**, *129*, 396–431. [CrossRef]
30. Günther, E.; Hiebler, S.; Mehling, H. Determination of the heat storage capacity of PCM and PCM-objects as a function of temperature. In Proceedings of the ECOSTOCK, 10th International Conference on Thermal Energy Storage, Galloway, NJ, USA, 31 May–2 June 2006.

Article

# A Theoretical Modeling of Adaptive Mixed CNT Bundles for High-Speed VLSI Interconnect Design

Abu Bony Amin [1,*], Syed Muhammad Shakil [2] and Muhammad Sana Ullah [2]

[1] Department of Electrical and Computer Engineering, University of Massachusetts Amherst (UMass), 100 Natural Resources Road, Amherst, MA 01003, USA
[2] Department of Electrical and Computer Engineering, Florida Polytechnic University, 4700 Research Way, Lakeland, FL 33805, USA; sshakil2152@floridapoly.edu (S.M.S.); mullah@floridapoly.edu (M.S.U.)
* Correspondence: abonyamin5908@floridapoly.edu

**Abstract:** The aroused quest to reduce the delay at the interconnect level is the main urge of this paper, so as to come across a configuration of carbon nanotube (CNT) bundles, namely, squarely packed bundles of mixed CNTs. The demonstrated approach in this paper makes the mixed CNT bundle adaptable to adopt for high-speed very-large-scale integration (VLSI) interconnects with technology shrinkage. To reduce the delay of the proposed configuration of the mixed CNT bundle, the behavioral change of resistance (R), inductance (L), and capacitance (C) has been observed with respect to both the width of the bundle and the diameter of the CNTs in the bundle. Consequently, the performance of the modified bundle configuration is compared with a previously developed configuration, namely, squarely packed bundles of dimorphic MWCNTs in terms of propagation delay and crosstalk delay at local-, semiglobal-, and global-level interconnects. The proposed bundle configuration is, ultimately, enacted as the better one for 32-nm and 16-nm technology nodes, and is suitable for 7-nm nodes as well.

**Keywords:** mixed CNT bundle; crosstalk delay; interconnect; propagation delay; *RLC* model

## 1. Introduction

The overwhelming exploitation of interconnects to the device delay makes researchers weigh Carbon Nanotubes (CNTs) for the possession pertinent to long mean free path [1], electrical properties [1,2], thermal properties [2,3], electromigration, and current density [4]. Moreover, crosstalk delay is a potential stymie for CNTs due to capacitive coupling between adjacent bundles [5]. While it is brought up, in previous literature [1], that the performance will be meliorated with further technology scaling, CNTs can render much better performance based on the exploration of some features.

It is claimed in [5,6] that mutual inductance does not have a considerable impact on crosstalk-induced delay and glitches; instead, coupling capacitance with electrostatic and quantum capacitance makes the main contribution. It is also noticed in [7] that the graphitized electron beam-induced deposition (EBID) carbon has the capability to produce a low-resistance ohmic contact to multiple shells of MWCNT, in the context of making high-performance electrical interconnect structures for next-generation electronic circuits. Although the densely packed configuration of bundled CNTs is seizing attention for improving performance, a trade-off between propagation delay and crosstalk delay is conspicuous [8]. Having noticed, from earlier work [5], that SWCNT and DWCNT shows poorer performance than Cu-based interconnects, owing to higher coupling capacitances, we endeavored to avoid putting any SWCNTs and DWCNTs on the edge of the bundle in our configuration.

To improve the crosstalk delay along with propagation delay, some works [6,9–12] are conducted by introducing different bundle configurations and by combining both MWCNT and SWCNT in the bundle. Rai et al. claimed that the structure with MWCNTs surrounded by SWCNTs yields better performance by considering the tunnelling and intershell coupling between adjacent shells by depicting four different structures in [10]. The same group, subsequently, showed that the structure with SWCNTs and MWCNTs possessing equal halves vertically was the best one in terms of frequency noise amplitude by delineating the same four structures in another work [11]. However, it is demonstrated in [6,9] that, by varying the relative position of MWCNTs and SWCNTs in the bundle, CNTs with spatial distribution, putting the SWCNTs entirely wrapped up by the MWCNTs in the bundle, indulges lower crosstalk delay than those with random arrangements. A delay-efficient configuration of a mixed bundle is proposed in [12] as well, though the crosstalk delay performance of this configuration is not well-proved since this work opposes the fact, mentioned in [6,9], that SWCNTs are mounted over the boundary of the bundle.

The aim of this paper is to present an innovative diameter-controlled configuration to alleviate the propagation delay and crosstalk delay of size shrinking interconnects, which is feasible from the fabrication aspect. This configuration is presented here with a detailed theoretical and mathematical model analysis and comparison results to assure better enquiries of its performance and enlighten its advantages. To analyze the delay performance, the analytical delay model has been obtained using the parameters from previous works [13–16].

The rest of this paper is organized in the following manner. A modified configuration of a mixed CNT bundle is proposed and illustrated in Section 2. Section 3 is used to develop the mathematical models for *RLC* elements for isolated CNTs, and eventually for a mixed bundle based on the configuration introduced in Section 2. Section 4 is dedicated to the interest of simulating and analyzing the performance indicators, propagation delay, and crosstalk noise for different technology nodes, and depicting a graphical comparison with the previously well-developed research work [17,18]. Section 5 comes up with the conclusion by appending the future work potentialities and improvements.

## 2. Modified Mixed CNT Bundle Configuration

Our endeavor in this paper is to enhance the performance by altering the configuration, shown in Figure 1a. In this newly introduced configuration, illustrated in Figure 1b, the replacement of smaller MWCNTs takes place with a bunch of SWCNTs, which are wrapped up by the larger MWCNTs. This modified approach is a virtue of increasing the number of CNTs in the bundle so that we can fill up the unoccupied space with SWCNTs more efficiently and densely. The spacing between shells of MWCNTs and adjacent CNTs, equivalent to the van der Waals distance ($\delta \approx 0.34$ nm) between graphene layers in graphite, is maintained concurrently [19,20].

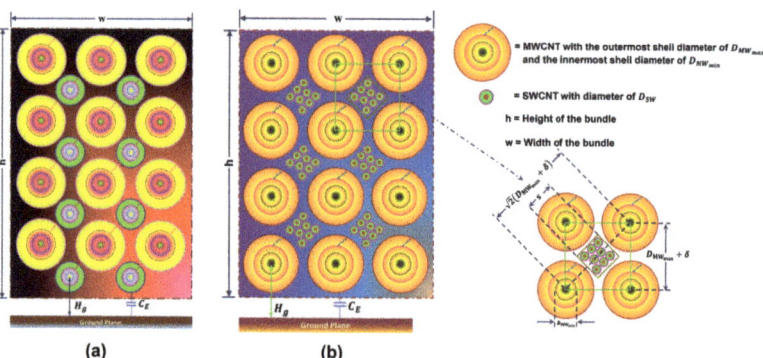

**Figure 1.** (**a**,**b**) In both architectures, the size of the larger MWCNTs are the same. The main modification happened in the introduced configuration by means of smaller MWCNTs replaced by a bunch of SWCNTs with the same predefined diameter, according to the space available based on the technology nodes.

The proposed configuration is inspired by the geometric pattern previously proposed in the works [17,21], to accommodate a greater number of CNTs in the bundle. In addition to every four larger MWCNTs forming a square by taking the vertices of the square in the center of those MWCNTs, another square forms in the center of the square. A certain number of SWCNTs is accommodated in this newly formed square, which will follow the hexagonally packed pattern. The number of larger MWCNTs and the number of SWCNTs in the bundle are calculated using Equation (1).

By considering one-third of the shells of the MWCNTs as metallic [22], the average number of conducting channels for a shell can be calculated by:

$$N_c(i) \approx \begin{cases} \alpha T D_i + \beta & \text{if } D_i > \frac{D_T}{T}; \\ \frac{2}{3} & \text{if } D_i \leq \frac{D_T}{T}; \end{cases} \quad (1)$$

where $\alpha = 2.04 \times 10^{-4}$ nm$^{-1}$K$^{-1}$, $\beta = 0.425$, $D_T = 1300$ nm·K, and $D_i$ is the diameter of the $i$th shell of the MWCNT.

To assure the simplicity of the calculation and to show the relation among all parameters, we are going to pursue further by considering a constant '$a$', which is the side of the square formed by the MWCNTs in Figure 1. Thus, the diameter of the outermost shell of the MWCNT is:

$$D_{MW_{max}} = a - \delta. \quad (2)$$

The number of shells of MWCNTs can be calculated using the following Formula (3), according to [9,12,23]:

$$n = \left\lceil \frac{D_{MW_{max}} - D_{MW_{min}}}{2\delta} \right\rceil. \quad (3)$$

According to the geometry of circle, we know that the diagonal of the bigger square from Figure 1 is $\sqrt{2}a$. Hence, we may calculate the side ($s$) of the smaller square from Figure 1 by:

$$s = \sqrt{2}a - (a + \delta) = \left(\sqrt{2} - 1\right)a - \delta. \quad (4)$$

To calculate the plausible number of accommodated SWCNTs in the smaller square of Figure 1:

$$N_{SW} = \left( N_{SW_H} N_{SW_V} - \left\lfloor \frac{N_{SW_V}}{2} \right\rfloor \right) \quad (5)$$

where

$$N_{SW_H} = \left\lceil \frac{s - D_{SW}}{D_{SW} + \delta} \right\rceil; \text{ and } N_{SW_V} = \left\lceil \frac{2(s - D_{SW})}{\sqrt{3}(D_{SW} + \delta)} \right\rceil.$$

The number of MWCNTs in Figure 1 is, thus:

$$N_{MW_h} = \left\lfloor \frac{w - D_{MW_{max}}}{D_{MW_{max}} + \delta} \right\rfloor \quad (6)$$

$$N_{MW_v} = \left\lfloor \frac{h - D_{MW_{max}}}{D_{MW_{max}} + \delta} \right\rfloor, \quad (7)$$

where $\lfloor X \rfloor$ and $\lceil X \rceil$ signifies that each element of $X$ has been rounded to the nearest integer less than or equal to that element, and to more than or equal to that element, respectively.

The number of smaller squares ($N_{Sq}$) in Figure 1 can be estimated by the following expression:

$$N_{Sq} = (N_{MW_h} - 1)(N_{MW_v} - 1). \quad (8)$$

The total number of SWCNTs in the bundle is given here:

$$N_{SW_\Sigma} = N_{Sq} N_{SW}. \tag{9}$$

## 3. Improved Mathematical Models

After obtaining the total number of CNTs (both MWCNT and SWCNT) from Section 2, we develop and extract the diameter-controlled $RLC$ elements for the mixed bundle of dimorphic CNTs at different technology nodes in this section. In Figures 2–4, the diameter of the MWCNTs is yielded as a function of '$a$', the diameter of SWCNTs is presumed to be the constant value of 1 nm, and the length of the interconnect and aspect ratio are considered as 100 µm and 2, respectively.

### 3.1. Resistance in Mixed CNT Bundle

The mathematical approach to determine the equivalent resistance of the mixed bundle is the extraction of the resistance components for isolated CNTs and, eventually, the total resistance of the bundle. To pursue the calculation, we will consider the equivalent single-conductor (ESC) model, where the resistance of the shells of the MWCNTs are in parallel, and adjacent CNTs are also in parallel [24]. According to [25], the quantum resistance ($R_q$) of SWCNTs can be estimated using the conductance $G = \left(\frac{2e^2}{\hbar}\right) MT$, where $e$ is the electron charge with the value of $1.62 \times 10^{-19}$ C, and $\hbar = 6.6262 \times 10^{-34}$ Js is the Planck constant:

$$R_q = \frac{\hbar}{4e^2} \approx 6.45 \text{ k}\Omega. \tag{10}$$

On the other hand, in the case of a single-wall nanotube length ($l$) exceeding the mean free path of electrons ($\lambda_{SWCNT}$), another resistance ($R_s$) comes up along with the former one, owing to scattering, which can be computed from the following expression:

$$R_s = \frac{\hbar}{4e^2} \left(\frac{1}{\lambda_{SWCNT}}\right). \tag{11}$$

Finally, the total resistance, emerging from the previous two components of the resistance, for an isolated SWCNT, is denoted by (12):

$$R_{SWCNT} = \begin{cases} R_c + R_q & \text{if } l \ll \lambda_{CNT}; \\ R_c + R_q + lR_s & \text{if } l \geq \lambda_{CNT}; \end{cases} \tag{12}$$

The lump resistance ($R_{lump}$), having the quantum or intrinsic resistance from Equation (10), caused by the quantum detainment of electrons in a nano-wire and imperfect metal–nanotube contact resistance ($R_c$), may vary from a few to several hundreds of kilo-ohms, based on the fabrication process [9,12,26]. The lump resistance for different isolated MWCNTs of the proposed bundle configuration using (13) has been acquired from [4,9,23,24,26,27]:

$$R_{lump} = \left[\sum_{i=1}^{n} \left(\frac{R_q}{N_c(i)} + R_c\right)^{-1}\right]^{-1}. \tag{13}$$

The per-unit length (p.u.l) scattering resistance ($R_s$) emerges for the length of the nano-wire surpassing the effective mean free path of the electron [24]. The scattering resistance ($R_s$) for different isolated MWCNTs of the proposed bundle configuration is estimated from Equation (14), based on [4,9,26]:

$$R_s = \sum_{i=1}^{n} \frac{R_q}{N_c(i) \lambda_i}. \tag{14}$$

The equivalent resistance of the bundle including both MWCNTs and SWCNTs can be reckoned by the following expression:

$$R_{bundle} = \left[\left(\frac{R_q}{N_{MW}} + l\frac{R_s}{N_{MW}}\right)^{-1} + \frac{R_{SW}}{N_{SW_\Sigma}}\right]^{-1}. \quad (15)$$

The characteristics of resistance (R) depend on both the width of the interconnect wire based on the technology node and the diameter of the used CNTs in the bundle. The simultaneous impact of both factors is observed in Figure 2. It is obvious that the lower resistance for the bundle is attainable by increasing the width of the bundle along with the diameter of the CNTs. Since it is taken into account that all the CNTs in the bundle are in parallel with each other, the increased number of CNTs can be obtained by increasing the width in a given space of the bundle, and can reduce the resistance significantly. Moreover, the increased diameter of the MWCNTs increases the number of shells, which are also in parallel to each other.

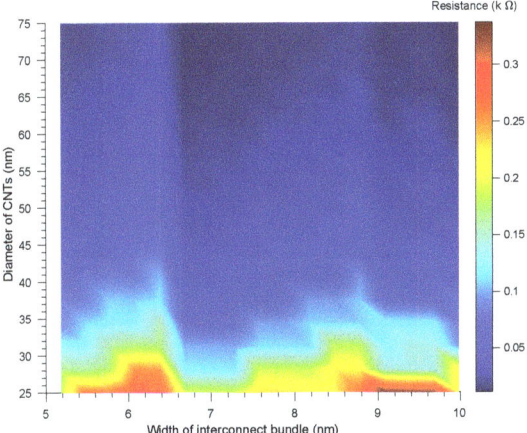

**Figure 2.** Synchronal variation of resistance of the squarely packed bundle of mixed CNTs from altering the width of the bundle and the diameter of the CNTs.

### 3.2. Inductance in Mixed CNT Bundle

To determine the overall inductance for our proposed configuration of the mixed bundle, we will first calculate the inductance for isolated SWCNT and then for the isolated MWCNT, and finally, the equivalent inductance for the entire bundle will be demonstrated, as given in Equation (20). The inductance of the SWCNT consists of two components, which are denoted as kinetic inductance ($L_k$) and magnetic inductance ($L_m$). Considering the ballistic conduction for a 1D conductor, the kinetic inductance ($L_k$) can be obtained by:

$$L_k = \frac{\hbar}{2e^2 v_F}, \quad (16)$$

where $v_F$ is the Fermi velocity of an electron with the value of approximately $8 \times 10^5$ ms$^{-1}$. Since $L_k$ is the function of some constant values, the approximate per-unit length (p.u.l.) value is 16 nH/μm [28]. On the other hand, the stored energy of carriers in a magnetic field engenders magnetic inductance ($L_m$) in SWCNTs [28] which is approximated by:

$$L_m = \frac{\mu_0}{2\pi} \ln\left(\frac{y}{d}\right), \quad (17)$$

with $\mu_0 = 4\pi \times 10^{-7}$ Hm$^{-1}$. Now, in the case of MWCNTs, the magnetic inductance for the $i$th shell can be approximated by the following expression:

$$L_{m_{MW}}(i) = \frac{\mu_0}{2\pi} \cosh\left(\frac{2h}{D_{MW}(i)}\right), \{i \in \mathbb{N} : 1 \leq i \leq n\}. \tag{18}$$

The magnetic and kinetic inductance of the isolated MWCNT in the proposed bundle is calculated using Equations (18) and (19), respectively:

$$L_{k_{MW}}(i) = \sum_{i=1}^{n} \frac{L_k}{2N_c(i)}. \tag{19}$$

Finally, the overall equivalent inductance of the bundle is estimated in (20), which indicates that the kinetic inductance component of SWCNT exists when the length of the interconnect wire exceeds the electron mean free path:

$$L_{\text{bundle}} = \begin{cases} \left(\frac{N_{MW}}{L_{m_{MW}} + L_{k_{MW}}} + \frac{N_{SW_\Sigma}}{L_{m_{SW}} + L_{k_{SW}}}\right)^{-1} & \text{if } l \leq \lambda_{CNT}; \\ \left(\frac{N_{MW}}{L_{m_{MW}} + L_{k_{MW}}} + \frac{N_{SW_\Sigma}}{L_{m_{SW}}}\right)^{-1} & \text{if } l > \lambda_{CNT}; \end{cases} \tag{20}$$

The inductance of the bundle also exhibits the same phenomena as the resistance does. The behavioral change of the inductance of the bundle, with the width and diameter of the CNTs in the bundle, is depicted in Figure 3. Based on the attained diameter, we estimate the delay of the bundle in Section 4.

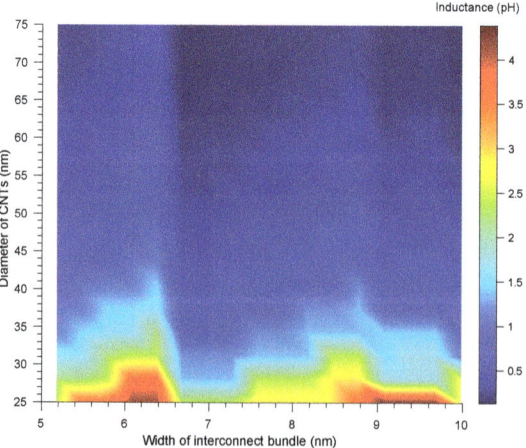

**Figure 3.** Concurrent extraction of inductance of the squarely packed bundle of mixed CNTs by varying the width of the bundle and the diameter of the CNTs.

### 3.3. Capacitance in Mixed CNT Bundle

The p.u.l. quantum capacitance for a CNT is estimated in Equation (21) by taking the analogy of the required energy to enclose an extra electron at an acquirable quantum state level beyond the Fermi energy level and effective capacitance. This capacitance comes into notification due to the quantum electrostatic energy stored in the nanotube while carrying the current [28]:

$$C_q = \frac{2e^2}{\hbar v_F} \approx 0.1 \text{ fF}/\mu\text{m}. \tag{21}$$

It has already been mentioned that, to estimate the inductance for a isolated SWCNT, that SWCNT must have four conducting channels, and these channels should form a parallel combination [28]. As a result, the equivalent effective quantum capacitance of an isolated SWCNT can be approximated here:

$$C_{SW} = 4C_q \approx 0.4 \text{ fF}/\mu\text{m}. \tag{22}$$

The electrostatic capacitance is expressed in the following expression by considering the SWCNT as a thin wire with the diameter $D_{SW}$ putting, at a distance of '$y$', away from the ground [28]:

$$C_e(SWCNT) = \frac{2\pi\epsilon_0\epsilon_r}{\cosh^{-1}\left(\frac{y}{D_{SW}}\right)}, \tag{23}$$

with absolute dielectric permittivity ($\epsilon_0$) = 8.854 × 10$^{-12}$ Fm$^{-1}$. Now, the capacitance for an isolated MWCNT is calculated using the recursive model. It is recommended in [6] that it is mandatory to determine the quantum capacitance of each shell before estimating the effective capacitance of a single MWCNT. The quantum capacitance is basically the estimation of the finite density of electronic states of quantum wire [24].

$$C_q = \frac{4e^2}{\hbar v_F} \sum_{i=1}^{n} N_c(i). \tag{24}$$

According to the ESC model of MWCNTs, it can be inferred from [27] that a shell-to-shell mutual capacitance between two adjacent shells of MWCNT is:

$$C_s(i+1,i) = \frac{2\pi\epsilon_0\epsilon_r}{\ln\left(\frac{D_i+2\delta}{D_i}\right)}, \{i \in \mathbb{N} : 1 \leq i \leq n\}, \tag{25}$$

where $D_i$ is the diameter of the $i$th shell of any isolated MWCNT. At first, in the case of the outermost shell, the equivalent capacitance ($C_{ESC}$), expressed in Equation (26), represents only the quantum capacitance of that shell. As much as we move toward the inner shell, the quantum capacitance of that particular shell makes a parallel combination with the equivalent capacitance ($C_{q-s}$), as shown in expression (28); a series combination of the capacitance of any shell and the mutual capacitance between that shell and previous shell obtained in (27), will continue until reaching the innermost shell.

$$C_{ESC}(1) = C_q(1) \tag{26}$$

$$C_{q-s}(i-1) = \left(\frac{1}{C_{ESC}(i-1)} + \frac{1}{C_s(i+1,i)}\right)^{-1} \text{ where, } \{i \in \mathbb{N} : 2 \leq i \leq n\} \tag{27}$$

$$C_{ESC}(i) = C_q(i) + C_{q-s}(i-1) \text{ where, } \{i \in \mathbb{N} : 2 \leq i \leq n\}. \tag{28}$$

The electrostatic capacitance demonstrates the potential difference between the ground and the CNT over the ground plane [24]. The p.u.l. electrostatic capacitance can be approximated in (29):

$$C_E = \frac{2\pi\epsilon_0\epsilon_r}{\ln\left(\frac{D_i+2\delta}{D_i}\right)}. \tag{29}$$

The conglomerate capacitance of the proposed mixed bundle is obtained in (30) by considering the overall effect of the SWCNTs and MWCNTs in the bundle. To estimate this, we considered the effect of electrostatic capacitance of MWCNTs over the ground plane on the effective capacitance in series:

$$C_{\text{bundle}} = \frac{N_{\text{MW}_\text{H}} C_\text{E} \left( N_{\text{MW}} C_{\text{ESC}_{\text{MW}}} + N_{\text{SW}_\Sigma} C_{\text{SW}} \right)}{N_{\text{MW}} C_{\text{ESC}_{\text{MW}}} + N_{\text{SW}_\Sigma} C_{\text{SW}} + N_{\text{MW}_\text{H}} C_\text{E}}. \tag{30}$$

Unlike the resistance and inductance, capacitance shows a descending behavior, with a lower diameter of CNTs in the bundle. We can also notice, from Figure 4, that capacitance decreases further in higher-technology nodes. The reason behind this phenomenon is that the capacitance components rising from parallel CNTs magnify the equivalent capacitance in the bundle.

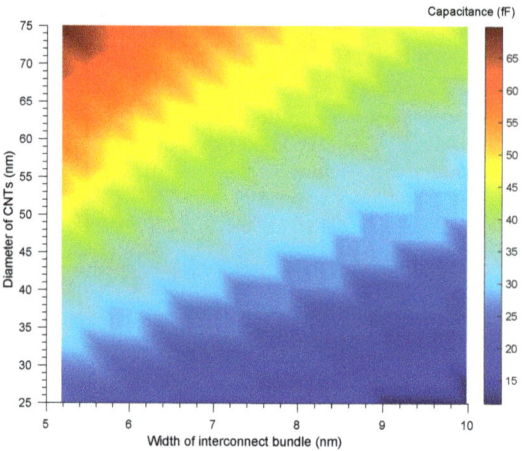

**Figure 4.** Contemporaneous denouement of capacitance of the squarely packed bundle of mixed CNTs by varying the width of the bundle and the diameter of CNTs.

## 4. Simulation Results

This section illustrates the performance comparison of a squarely packed bundle of dimorphic MWCNTs and that of mixed CNTs to exploit the feature of using mixed CNTs in the interconnect bundle. To observe the performance in terms of propagation delay, we simulate the Kahng's model, obtained from earlier work [13], using the extracted equivalent value of $RLC$ and the optimized dimensions of CNTs in Section 3. Having the $RLC$ value of our model, we validate the extracted conductance, inductance and capacitance rising in our configuration by comparing them with those in the previously discussed mixed CNT model in [18]. Subsequently, we assess the performance of the proposed configuration in terms of crosstalk delay, excerpted from [4], using the optimized size of particular CNTs and extracted $RLC$ in Section 3, and the number of CNTs in Section 2.

To validate the $RLC$ value of the proposed model, the conductance and inductance of the mixed CNT bundle is observed by varying the probability of metallic CNT (%) using Equation (1). We observe, from Figure 5, that the conductance of our mixed CNT bundle configuration varies proportionally with the percentage of metallic CNTs in the bundle. It can be deduced from expression (1) that the number of channels per shell increases with an increase in the percentage of metallic CNTs, which, in turn, reduces the resistance and increases the conductance of the bundle. As our proposed configuration is densely packed and geometrically organized, its capacity to hold a noticeably higher number of CNTs in the bundle makes the conductance of our configuration, as depicted in Figure 5, much higher than that of the mixed CNT bundle in [18]. For the same reason, our proposed configuration comes up with a lower inductance than the previous mixed CNT models in both Figures 5 and 6. Figure 6, meanwhile, demonstrates that the capacitance of the proposed model increases with increasing interconnect lengths, though our configuration still sustains better performance by generating lower capacitances. In this circumstance, we can infer that by increasing the percentage of metallic CNTs and delimiting the interconnect length, we can reduce the extracted *RLC* parameter and, eventually, the delay.

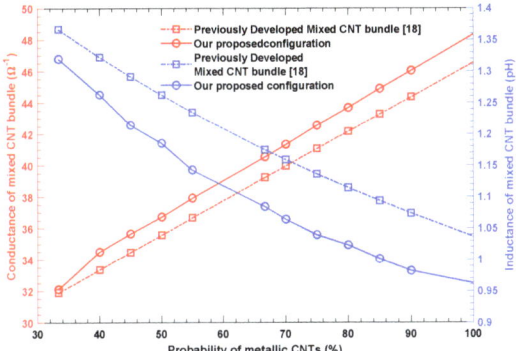

**Figure 5.** Impact of increasing metallic CNTs on the conductance and inductance and comparison of the results of the proposed configuration and the earlier mixed CNT bundle configuration. Overall conductance and inductance of the mixed CNT bundle is obtained, considering the interconnect length of 40 nm.

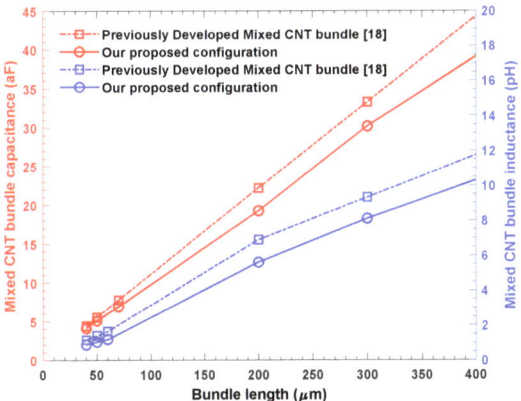

**Figure 6.** Findings of overall bundle inductance (*L*) and capacitance (*C*) for different interconnect lengths to observe the comparative parameter illustration of the proposed configuration and the previous model by considering that $\frac{2}{3}$ of the CNTs are metallic in the bundle.

It is demonstrable from Figures 7–9 that our proposed configuration yields a lower propagation delay than the preceding configuration in [17] does. It is conspicuous, from Table 1. that we can increase the number of CNTs without distorting the overall configuration of the squarely packed bundle using the proposed approach. As a consequence, the resistance and inductance decreases while the capacitance increases for any specific technology node. Finally, the overall impact decreases the propagation delay for the squarely packed bundle of mixed CNTs, which is represented in the Figures 7–9 for the local, semiglobal, and global levels, respectively. However, the preceding squarely packed bundle configuration does not seem suitable for 7-nm technology nodes, because of the unavailability of space to accommodate CNTs of various sizes. Hence, the simulation illustrations don not include the delay of a squarely packed bundle of dimorphic MWCNTs for 7-nm technology. In comparison with [18], the unique contribution of our proposed configuration is the feasibility for 32-nm, 16-nm, and 7-nm technology nodes, along with performance enhancements.

**Table 1.** Diameter and number of pertinent CNTs accommodated in the bundle during simulation to obtain propagation delay and crosstalk delay.

| Interconnect Length (μm) | Technology Node (nm) | Squarely Packed Dimorphic | | | | Squarely Packed Mixed | | | |
|---|---|---|---|---|---|---|---|---|---|
| | | $D^L_{MW_{max}}$ (nm) | $N^L_{MW}$ | $D^s_{MW_{max}}$ (nm) | $N^s_{MW}$ | $D_{MW_{max}}$ (nm) | $N_{MW}$ | $D_{SW}$ (nm) | $N_{SW_\Sigma}$ |
| Local (0–100) | 32 nm | 10 | 32 | 4.31 | 21 | 10 | 21 | 1 | 96 |
| | 16 nm | 8.5 | 8 | 3.72 | 3 | 8.5 | 8 | 1 | 15 |
| | 7 nm | - | - | - | - | 4.5 | 3 | 1 | 8 |
| Semiglobal (101–500) | 32 nm | 10 | 32 | 4.31 | 21 | 10 | 21 | 1 | 96 |
| | 16 nm | 8.5 | 8 | 3.72 | 3 | 8.5 | 8 | 1 | 15 |
| | 7 nm | - | - | - | - | 4.5 | 3 | 1 | 8 |
| Global (501–2500) | 32 nm | 10 | 48 | 4.31 | 33 | 10 | 33 | 1 | 160 |
| | 16 nm | 8.5 | 14 | 3.72 | 6 | 8.5 | 12 | 1 | 25 |
| | 7 nm | - | - | - | - | 4.5 | 12 | 1 | 5 |

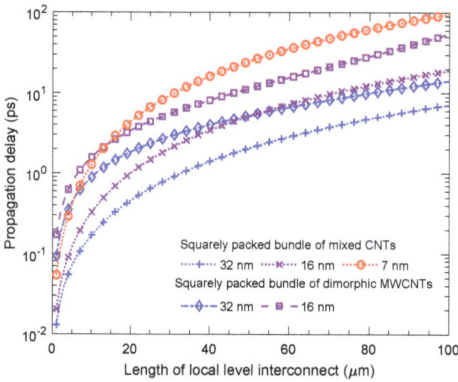

**Figure 7.** Comparison of delay performance of a squarely packed bundle of mixed CNTs and that of dimorphic MWCNTs for local-level interconnect lengths. The size and number of accommodated CNTs for different technology nodes are mentioned in Table 1.

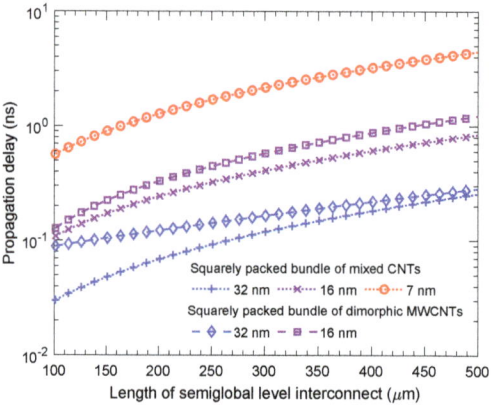

**Figure 8.** Illustration of comparative delay performance of a squarely packed bundle of mixed CNTs and that of dimorphic MWCNTs for semiglobal-level interconnect lengths. The size and number of accommodated CNTs for different technology nodes are mentioned in Table 1.

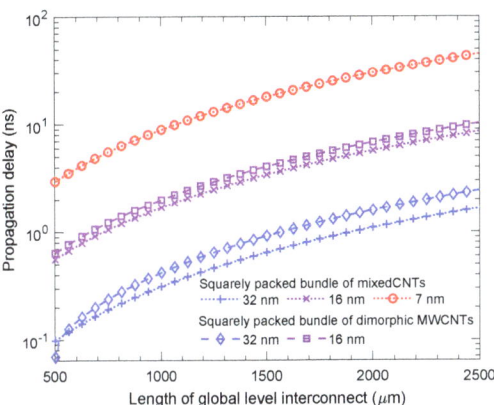

**Figure 9.** Demonstration of comparison between the propagation delay performance of a squarely packed bundle of mixed CNTs and that of dimorphic MWCNTs for global-level interconnect lengths. The size and number of accommodated CNTs for different technology nodes are mentioned in Table 1.

The crosstalk delay, basically, arises from the capacitance formed between the CNTs from different bundles, while it is considered that all CNTs in the bundles are in parallel [4]. The inter-bundle capacitance, the function of spacing between the the centers of two adjacent CNTs, the average diameter of the adjacent CNTs, and the relative permittivity based on the level of interconnect length, is estimated by Equation (31), where $D_{MW_{max}}$ is used as the diameter because we placed MWCNTs on the edge of the bundle to reduce the overall crosstalk impact, having been motivated by previous works [6,9]:

$$C_{cm_{ESC}} = \frac{\pi \epsilon_0 \epsilon_r}{\cosh^{-1}\left(\frac{S_p}{D_{MW_{max}}}\right)} N_{MW_h}. \tag{31}$$

Eventually, the crosstalk performance of our proposed configuration is depicted in Figures 10–12 for local-, semiglobal-, and global-level interconnects by comparing with the preceding configuration from [17]. It is noticeable from Figure 11 that the crosstalk performance of the proposed configuration is substantial for both the 32-nm and 16-nm technology nodes. On the other hand, Figure 11 illustrates significant crosstalk performance betterment in the 32-nm technology node, compared with that in the 16-nm technology node. In the case of the global-level interconnect, the amount of crosstalk delay performance enhancement of our proposed configuration for both 32-nm and 16-nm technology nodes is almost the same, as is illustrated in Figure 12. It is also demonstrable that our proposed configuration is appropriate for 7-nm technology nodes.

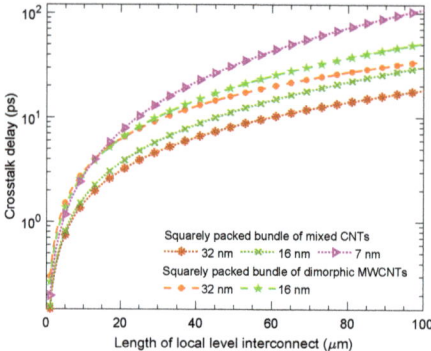

**Figure 10.** Comparative exhibition of the crosstalk delay of the proposed and previously developed bundle configurations for different technology nodes at local-level interconnect lengths. The dimension of the used CNTs are the same as those used in the simulation for obtaining propagation delays at the local level.

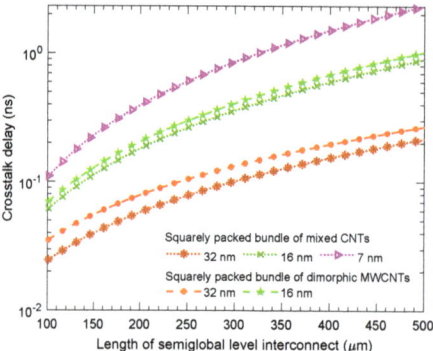

**Figure 11.** Comparative analysis of the crosstalk delay of the proposed and previously developed bundle configurations for different technology nodes at semiglobal-level interconnect lengths. The dimension of the used CNTs are the same as those used in the simulation for obtaining propagation delays at the semiglobal level.

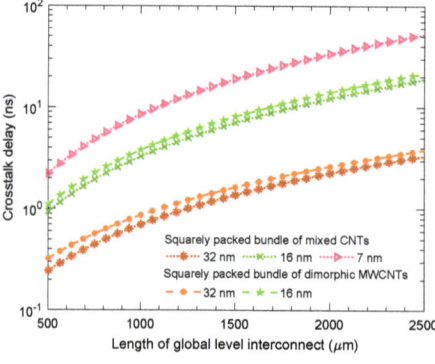

**Figure 12.** Comparative illustration of the crosstalk delay of the proposed and previously developed bundle configurations for different technology nodes at global-level interconnect lengths. The diameter and number of the used CNTs are the same as those used in the simulation for obtaining propagation delays at the global level.

## 5. Conclusions

A modified configuration of the squarely packed bundle of mixed CNTs is presented to assure the high speed VLSI interconnect with less area possession. By proposing this configuration, the simultaneous applicability of both MWCNTs and SWCNTs for scaled interconnects in future VLSI-integrated circuits is analyzed abstractly. The propagation delay and crosstalk delay performance are extracted and analyzed using an RLC model and a delay model. As a result, it exhibits the transcendence of squarely packed bundles of mixed CNTs for local-, semiglobal-, and global-level interconnects at 32-nm, 16-nm, and 7-nm technology nodes. In this approach, CNTs with only two different sizes are used. In the upcoming endeavor, our intention is to advance the work by adding the CNTs with various sizes to make the configuration more convenient in terms of fabrication process.

**Author Contributions:** Conceptualization, A.B.A.; methodology, A.B.A.; software, A.B.A.; validation, A.B.A.; formal analysis, A.B.A.; investigation, A.B.A.; resources, S.M.S.; data curation, A.B.A.; writing—original draft preparation, A.B.A.; writing—review and editing, A.B.A.; visualization, A.B.A.; supervision, M.S.U.; project administration, M.S.U.; funding acquisition, M.S.U. All authors have read and agreed to the published version of the manuscript.

**Funding:** This research received no external funding.

**Institutional Review Board Statement:** Not applicable.

**Informed Consent Statement:** Not applicable.

**Data Availability Statement:** Not applicable.

**Conflicts of Interest:** The authors declare no conflict of interest.

## References

1. Li, H.; Yin, W.Y.; Banerjee, K.; Mao, J.F. Circuit modeling and performance analysis of multi-walled carbon nanotube interconnects. *IEEE Trans. Electron Devices* **2008**, *55*, 1328–1337. [CrossRef]
2. Kabir, M.S. Controlled Growth of a Nanostructure on a Substrate, and Electron Emission Devices Based on the Same. U.S. Patent 7,977,761, 12 July 2011.
3. Liew, K.; Wong, C.; He, X.; Tan, M. Thermal stability of single and multi-walled carbon nanotubes. *Phys. Rev. B* **2005**, *71*, 075424. [CrossRef]
4. Kaushik, B.K.; Majumder, M.K. *Carbon Nanotube Based VLSI Interconnects: Analysis and Design*; Springer: Berlin/Heidelberg, Germany, 2015.
5. Pu, S.; Yin, W.; Mao, J.; Liu, Q.H. Crosstalk Prediction of Single- and Double-Walled Carbon-Nanotube (SWCNT/DWCNT) Bundle Interconnects. *IEEE Trans. Electron Devices* **2009**, *56*, 560–568. [CrossRef]
6. Subash, S.; Kolar, J.; Chowdhury, M.H. A new spatially rearranged bundle of mixed carbon nanotubes as VLSI interconnection. *IEEE Trans. Nanotechnol.* **2011**, *12*, 3–12. [CrossRef]
7. Kim, S.; Kulkarni, D.D.; Rykaczewski, K.; Henry, M.; Tsukruk, V.V.; Fedorov, A.G. Fabrication of an ultralow-resistance ohmic contact to MWCNT–metal interconnect using graphitic carbon by electron beam-induced deposition (EBID). *IEEE Trans. Nanotechnol.* **2012**, *11*, 1223–1230. [CrossRef]
8. Amin, A.B.; Ullah, M.S. Mathematical Framework of Tetramorphic MWCNT Configuration for VLSI Interconnect. *IEEE Trans. Nanotechnol.* **2020**, *19*, 749–759. [CrossRef]
9. Majumder, M.K.; Kaushik, B.K.; Manhas, S.K. Analysis of delay and dynamic crosstalk in bundled carbon nanotube interconnects. *IEEE Trans. Electromagn. Compat.* **2014**, *56*, 1666–1673. [CrossRef]
10. Rai, M.K.; Garg, H.; Kaushik, B. Temperature-dependent modeling and crosstalk analysis in mixed carbon nanotube bundle interconnects. *J. Electron. Mater.* **2017**, *46*, 5324–5337. [CrossRef]
11. Sharma, M.; Rai, M.K.; Khanna, R. Temperature-dependent crosstalk and frequency spectrum analyses in adjacent interconnects of a mixed CNT bundle. *J. Comput. Electron.* **2020**, *19*, 177–190. [CrossRef]
12. Sandha, K.S.; Thakur, A. Comparative Analysis of Mixed CNTs and MWCNTs as VLSI Interconnects for Deep Sub-micron Technology Nodes. *J. Electron. Mater.* **2019**, *48*, 2543–2554. [CrossRef]
13. Kahng, A.B.; Muddu, S. An analytical delay model for RLC interconnects. *IEEE Trans. Comput.-Aided Des. Integr. Circuits Syst.* **1997**, *16*, 1507–1514. [CrossRef]
14. Ullah, M.S.; Chowdhury, M.H. Analytical models of high-speed RLC interconnect delay for complex and real poles. *IEEE Trans. Very Large Scale Integr. (VLSI) Syst.* **2017**, *25*, 1831–1841. [CrossRef]

15. Sanaullah, M.; Chowdhury, M.H. Analysis of RLC interconnect delay model using second order approximation. In Proceedings of the 2014 IEEE International Symposium on Circuits and Systems (ISCAS), Melbourne, VIC, Australia, 1–5 June 2014; pp. 2756–2759. [CrossRef]
16. Sanaullah, M.; Chowdhury, M.H. A new real pole delay model for RLC interconnect using second order approximation. In Proceedings of the 2014 IEEE 57th International Midwest Symposium on Circuits and Systems (MWSCAS), College Station, TX, USA, 3–6 August 2014; pp. 238–241. [CrossRef]
17. Amin, A.B.; Ullah, M.S. Performance Analysis of Squarely Packed Dimorphic MWCNT Bundle for High Speed VLSI Interconnect. In Proceedings of the 2020 IEEE Canadian Conference on Electrical and Computer Engineering (CCECE), London, ON, Canada, 30 August–2 September 2020; pp. 1–6. [CrossRef]
18. Subash, S.; Chowdhury, M.H. Mixed carbon nanotube bundles for interconnect applications. *Int. J. Electron.* **2009**, *96*, 657–671. [CrossRef]
19. Li, H.; Lu, W.; Li, J.; Bai, X.; Gu, C. Multichannel ballistic transport in multiwall carbon nanotubes. *Phys. Rev. Lett.* **2005**, *95*, 086601. [CrossRef]
20. Naeemi, A.; Meindl, J.D. Compact physical models for multiwall carbon-nanotube interconnects. *IEEE Electron Device Lett.* **2006**, *27*, 338–340. [CrossRef]
21. Amin, A.B.; Ullah, M.S. Performance Analysis of Squarely Packed Polymorphic SWCNT Interconnect. In Proceedings of the 2019 IEEE 10th Annual Ubiquitous Computing, Electronics Mobile Communication Conference (UEMCON), New York, NY, USA, 10–12 October 2019; pp. 1199–1203. [CrossRef]
22. Naeemi, A.; Meindl, J.D. Physical modeling of temperature coefficient of resistance for single-and multi-wall carbon nanotube interconnects. *IEEE Electron Device Lett.* **2007**, *28*, 135–138. [CrossRef]
23. Das, D.; Rahaman, H. Analysis of crosstalk in single-and multiwall carbon nanotube interconnects and its impact on gate oxide reliability. *IEEE Trans. Nanotechnol.* **2011**, *10*, 1362–1370. [CrossRef]
24. Majumder, M.K.; Das, P.K.; Kaushik, B.K. Delay and crosstalk reliability issues in mixed MWCNT bundle interconnects. *Microelectron. Reliab.* **2014**, *54*, 2570–2577. [CrossRef]
25. Srivastava, N.; Li, H.; Kreupl, F.; Banerjee, K. On the applicability of single-walled carbon nanotubes as VLSI interconnects. *IEEE Trans. Nanotechnol.* **2009**, *8*, 542–559. [CrossRef]
26. Sahoo, M.; Ghosal, P.; Rahaman, H. Modeling and analysis of crosstalk induced effects in multiwalled carbon nanotube bundle interconnects: An ABCD parameter-based approach. *IEEE Trans. Nanotechnol.* **2015**, *14*, 259–274. [CrossRef]
27. Sarto, M.S.; Tamburrano, A. Single-conductor transmission-line model of multiwall carbon nanotubes. *IEEE Trans. Nanotechnol.* **2009**, *9*, 82–92. [CrossRef]
28. Banerjee, K.; Srivastava, N. Are carbon nanotubes the future of VLSI interconnections? In Proceedings of the 43rd annual Design Automation Conference, San Francisco, CA, USA, 24–28 July 2006; pp. 809–814.

*Review*

# The Development of Graphene/Silica Hybrid Composites: A Review for Their Applications and Challenges

Murni Handayani [1,*], Nurin Nafi'ah [2], Adityo Nugroho [2], Amaliya Rasyida [2], Agus Budi Prasetyo [1], Eni Febriana [1], Eko Sulistiyono [1] and Florentinus Firdiyono [1]

1. Research Center for Metallurgy and Materials-National Research and Innovation Agency (BRIN), Building 470, PUSPIPTEK Area, Tangerang Selatan 15314, Indonesia; agus080@lipi.go.id (A.B.P.); enif001@lipi.go.id (E.F.); ekos001@lipi.go.id (E.S.); flor001@lipi.go.id (F.F.)
2. Department of Materials and Metallurgical Engineering, Institut Teknologi Sepuluh Nopember, Surabaya 60111, Indonesia; nurin.18025@mhs.its.ac.id (N.N.); adityonugroho.18025@mhs.its.ac.id (A.N.); amaliya@mat-eng.its.ac.id (A.R.)
* Correspondence: murni.handayani@lipi.go.id or murni.handayani@brin.go.id

**Abstract:** Graphene and silica are two materials that have wide uses and applications because of their unique properties. Graphene/silica hybrid composite, which is a combination of the two, has the good properties of a combination of graphene and silica while reducing the detrimental properties of both, so that it has promising future prospects in various fields. It is very important to design a synthesis method for graphene/silica composite hybrid materials to adapt to its practical application. In this review, the synthesis strategies of graphene, silica, and hybrid graphene/silica composites such as hydrothermal, sol-gel, hydrolysis, and encapsulation methods along with their results are studied. The application of this composite is also discussed, which includes applications such as adsorbents, energy storage, biomedical fields, and catalysts. Furthermore, future research challenges and futures need to be developed so that hybrid graphene/silica composites can be obtained with promising new application prospects.

**Keywords:** graphene; silica; hybrid composites; nanocomposites; adsorbents; energy storages; biomedical fields; catalysts

**Citation:** Handayani, M.; Nafi'ah, N.; Nugroho, A.; Rasyida, A.; Prasetyo, A.B.; Febriana, E.; Sulistiyono, E.; Firdiyono, F. The Development of Graphene/Silica Hybrid Composites: A Review for Their Applications and Challenges. *Crystals* **2021**, *11*, 1337. https://doi.org/10.3390/cryst11111337

Academic Editor: Walid M. Daoush

Received: 20 September 2021
Accepted: 26 October 2021
Published: 1 November 2021

**Publisher's Note:** MDPI stays neutral with regard to jurisdictional claims in published maps and institutional affiliations.

**Copyright:** © 2021 by the authors. Licensee MDPI, Basel, Switzerland. This article is an open access article distributed under the terms and conditions of the Creative Commons Attribution (CC BY) license (https://creativecommons.org/licenses/by/4.0/).

## 1. Introductions

In the period 1982–1983, the major concept re-direction of the sol-gel process to create heterogeneous material was explained by Roy, Komarneni, and colleagues. During the process, the term 'nanocomposite' was often used [1]. Nanocomposite, solid phase material, has at least one dimension in the nanometre range on amorphous, semicrystalline, or crystalline or combinations thereof. Composition of nanocomposites are design based on multifunctional properties that refer to the inorganic or organic, or both [1,2].

To a great extent in the last two decades, the progression of many aspects built by nanocomposite has been developed. Nanocomposite properties are influenced by the structures. Therefore, the need for understanding and practicing material behavior across length scales from the atomic to the macroscopic scale is important for scientists and engineers. Commonly, their reinforced material is divided into particles, layered materials, and fibrous materials [2].

Similarly, nanocomposites are also divided by their matrix. There are polymer matrix nanocomposites, ceramic matrix nanocomposites, and metal matrix nanocomposites [3]. Extraordinary properties were obtained by combining various materials as previously mentioned. such as its based composition, reinforcement, or matrix on the design material process by the scientist and engineer. Essentially, a variety of applications of nanocomposites is achieved by versatile properties that could be enhanced from their manufacturing [4]. Recently, nanocomposites have widely been applied in environmental remedies, energy, medicine, sunscreens, biomaterials, etc. Nanocomposites are becoming an attractive field of materials which provide novel performance due to their remarkable properties [5]. Currently, nanocomposites are widely used in industry to replace the use of conventional fillers because most of the nanoscale fillers are able to improve the mechanical and thermal properties of nanocomposites [6].

Graphene, a semimetal thin material, is a single layer carbon atom with 2 dimensional sheets in a densely packed honeycomb lattice structure [5]. Since its discovery, it has continuously been making an impact on future material development, increasing research into advance synthesis methods from year to year, in order to confront industrial challenges from many angles, such as automotive, green energy, electronics, biomedical, catalyst and others [7]. The graphene-copper (II) phthalocyanine (CuPc) hybrid material has been used as an electrocatalyst for the electrochemical reduction of $CO_2$ [8]. Meanwhile, reduced Nickel/graphene oxide nanoparticles were applied as a catalyst to convert $CO_2$ to $CH_4$ by showing good activity in $CO_2$ methanation processes [9].

Starting from the academic field to isolated graphite in 2004, existing theoretical possibility in preparing tiny sheets of graphene, within 10 years becoming a high prospect project in growing economical and innovative research for society, aiming dynamics field and multitude of actors on commercial [7]. Graphene discovery was a sign of the new era for researchers and the material physics community to collect the "gold" from the "hidden gold mine". [10].

With numerous functions and a high chance of becoming a future material, graphene has been supported by its great properties. It has a large surface area (2630 m$^2$), high electrical conductivity ($10^6$ S cm$^{-1}$), high thermal conductivity (5000 W m$^{-1}$ K$^{-1}$), high mechanical strength (~40 Nm$^{-1}$), great optical transmittance (~97.7%), high modulus of elasticity (1 TPa), and high electron intrinsic mobility (250000 cm$^2$ V$^{-1}$ s$^{-1}$) [11–13]. Primarily, due to the potential value of graphene, analysis of this material is focused on its properties such as electrical, physical, mechanical, and optical [12]. For example, recognizing that properties needed for several characterization methods are likely optical microscopy, transmission electron microscopy (TEM), atomic force microscopy (AFM), angle-resolved photoemission spectroscopy (ARPES), Raman scattering and Rayleigh scattering [14]. In the case of graphene attractive properties, various applications have been listed including high-end composite materials, field effect transistors, electromechanical systems, strain sensors, electronics, supercapacitors, hydrogen storage and solar cells [13].

Silica is an inexhaustible resource on earth [15] that can be found both from natural resources [16] and inorganic compounds [17]. Agricultural waste such as rice husks, rice hulls, bagasse ash, semi-burned rice straw ash [18], bentonite, quartz sand, and diatomaceous earth [15] are the natural resources containing high silica. Meanwhile, silica obtained from inorganic compounds, namely TEOS (Tetraethyl Orthosilicate) [18], SSS (Sodium Silicate Solution), and TMOS (Tetramethyl Orthosilicate) can be synthesized from silane [17]. In recent years, the synthesis of silica from natural materials has been growing with various advantages. The natural evolution of plants over the years has been able to develop silica layers and produce highly reactive silica by a simple process [15].

Silica has wide applications in various fields including drug delivery systems, catalysis, biomedical, imaging, chromatography, sensors and as a filler composite material [18]. Porous silica ceramics as candidates for high temperature dielectrics and thermal shields with their dielectric and thermal properties as well as low density used in aerospace and engineering [19]. Naturally, silica has a tendency to produce the strength and hardness of ceramic materials [15]. In addition, silica aerogel is a promising material used in composite insulators with its porous, ultra-lightweight, and nanostructured properties. It also can be used as a material for water resistance, UV protection, fire resistance, and has excellent acoustic barrier properties [20]. Silica gel with a high specific surface area and a gas adsorption capacity due to the presence of micro and mesopores is used as an adsorbent such as for removal of heavy metals from wastewater and the adsorption of volatile organic compounds [21]. In other fields of application, silica nanoparticles are used as superhydrophobic materials by changing their hydrophilic properties to hydrophobic ones [22].

Based on the several advantages possessed by graphene and silica, both can be integrated to produce better new properties. The synthesis of graphene silica hybrid composites will produce properties that integrate the two and can avoid performance degradation caused by agglomeration of graphene [23]. Stack agglomeration that occurs in the graphene preparation process will greatly reduce the specific surface area, thereby reducing the performance of graphene. The combination of graphene and silica can effectively reduce agglomeration and produce advanced functional materials [24]. This material has complementary advantages and considerable application prospects in many fields. Currently, graphene and silica (G/SiO$_2$) hybrid composites are widely applied for electrodes, catalysts, hydrogen storage, batteries, displays, adsorbents, and sensors [23].

In this paper review, we report on the current focus on the development of graphene–silica hybrid composites and their applications in various fields. Furthermore, various synthesis methods for graphene, silica, and graphene–silica hybrid composites have been summarized according to sources. The functions and applications of graphene–silica hybrid composites in various fields are briefly discussed.

## 2. Synthesis Methods

### 2.1. Synthesis Method of Silica

Silica extraction from several sources with various methods aims to produce silica with a high purity and a high quality. Several researchers have conducted research to synthesize silica from its source with various methods and have focused on the environmental regulation of temperature and combustion time as well as chemical treatment so as to produce different structures and properties of silica amorphous to the crystalline phase [25].

#### 2.1.1. Sol-Gel Method

Sol-gel is the most frequently used method in silica synthesis and is suitable for the manufacturing industry. The reaction in this method is controlled by one of the acidic or basic conditions. By using this method, a uniform silica particle in size, high purity, ease of control and scalability was achieved [22]. Abbas, et al. [20] conducted an experiment on extracting silica aerogel from rice husks using the sol-gel method for application in environmentally friendly, lightweight, and heat-resistant cement composites. It uses sodium hydroxide solution to form hydrogel, is immersed in ethanol to make an alcogel and surface modification to form a hydrophobic gel, and is then dried to produce silica aerogel. The result showed a dominant mesoporous structure and a high surface area of 760 m$^2$/g. The addition of silica aerogel to the cement composite can reduce the density from 2102 kg/m$^3$ to 1133 kg/m$^3$ and lowers the thermal conductivity from 1.76 W/mK to 0.33 W/mK.

In the study by Ismail et al. [22], the extraction of silica nanoparticles from silica sand using is performed mechanically by grinding and the sol-gel method is used as a superhydrophobic material. To form a silica superhydrophobic, silica nanoparticles were mixed with stearic acid and ethanol. It was observed that the particle size increased with the increasing chain length of the alcohol. The particle size distribution was related to solvent polarity, which influenced the nucleation and growth of the silica particles. Methanol is the most polar alcohol solvent, which can increase the solubility of sodium silicate and therefore obtained a high concentration of very small silica nuclei. The superhydrophobic effect of silica nanoparticles can also be seen in their application, both as coatings and as mixed materials.

The sol-gel method was also used in the research of Sdiri et al. [21] to extract silica gel from siliceous sands. Silica sand was mixed with sodium carbonate to prepare sodium silicate, and hydrochloric acid was then added to obtain silica gel in micro and mesoporous pores with a high adsorption capacity. The silica content resulting from this method was 88.8–97.5%. The porosity of the silica gel reaches 57% and the specific surface area exceeds

340 m$^2$/G. Silica gel obtained from this method was in micro and mesoporous pores with a high adsorption capacity, which has potential to be applied to adsorbents, such as heavy metal removal from wastewater and the adsorption of volatile organic compounds.

2.1.2. Hydrothermal

The hydrothermal method is one of the processes used to synthesize silica with a low cost and a simple technique in the preparation of nanomaterials [26]. The synthesis of amorphous silica nanowires from commercial silicate glass was carried out by Zhu Y et al. [26] using the one-step hydrothermal method. The hydrothermal process was carried out at a temperature of 170 °C. The results showed that the silica in nanoscale was achieved in the form of an amorphous SiO$_2$ nanowire with a diameter of 20–100 nm and a length of several tens of micrometers.

In a study by Ortiz et al. [27], MCM-41 mesoporous silica was synthesized by a hydrothermal method using sodium silicate (Na$_2$SiO$_3$) as a source of silica, hexadecyltrimethylammonium bromide (CTAB) as a template agent, and ethyl acetate as a pH regulator. The synthesis was carried out at 80, 90, and 100 °C, resulting in an increase in temperature affecting the formation of MCM-41 silica negatively. Mesoporous silica MCM-41 is synthesized from these reaction conditions depicted in a well-ordered hexagonal array with spherical morphology and particle sizes of 200 to 500 nm. MCM-41 mesoporous silica could have promising applications in catalysis, drug delivery systems and the adsorption of organic molecules owing to its high specific surface area.

2.1.3. Leaching

The two main steps in extracting silica, especially from natural resources, are combustion and chemical treatment. Dissolution of materials in acid can be used to remove organic compounds in materials before combustion [28]. A method of chemical treatment with hydrochloric acid or sulfuric acid was used to prepare ultrafine size silica, high reactivity and high purity. The purpose of acid pretreatment is to increase purity by removing impurities and gives the silica a high surface area during its deposition [25]. Eko, et al. [29] applied sulfuric acid for the leaching process of low-grade silica from quartz sand. The result depicted sulfuric acid as being very effective to remove impurities of aluminum and iron up to 42% and 85%, respectively. The process using sulfuric acid could produce high purity silica with 96.44% purity.

In environmentally friendly chemical extraction methods, there are acid alkaline treatment and washing steps aimed at controlling the size and pores in the silica particles [25,26]. Febriana, et al. [30] has succeeded in synthesizing silica by a direct leaching method at atmospheric pressure. Precipitated silica content of 13.6% was obtained under optimum conditions at a temperature of 90 °C and a stirring speed of 800 rpm, with a sodium hydroxide solution of 7.5 M, which produced amorphous precipitate silica with a purity of ±96%. Azat, et al. [25] conducted an experiment to extract silica from rice husk. The leaching process used a hydrochloric acid solution and was calcined at 600 °C. The amorphous silica with a purity of 98.2–99.7% and a certain surface area of 120–980 m$^2$/G was produced from this method.

Another study by Park et al. [28] demonstrating the process of extracting silica from rice husks uses a two-stage continuous process consisting of a friction ball attrition and an alkaline washing method. With the use of NaOH, the silica yield becomes saturated, starting from 0.2 M and a yield of 79%. With the use of KOH, silica yields a saturation starting from 0.5 M with a yield of about 77%. The optimum reaction conditions were a concentration of 0.2 M at 80 °C for 3 h, and a solid content of 6% ($w/v$). This extraction method produces amorphous silica with 98.5% purity. Synthesis of amorphous and crystalline silica from rice husk was also carried out by Zainal et al. [15] using chemical treatment with hydrochloric acid at a temperature of 60 °C. The combustion was carried out at a temperature of 700 °C and 1000 °C for 2 h. From the experiment, it was found that amorphous silica is formed at 700 °C and crystalline silica is formed at 1000 °C. Chemical treatment before combustion increased the silica content from 95.7% to 98.7%.

2.1.4. Pyrolysis

Pyrolysis is one of the methods used to synthesize silica. Catalyst-assisted pyrolysis of polymeric precursors for nanostructures is a simple and easy controlling method. Moreover, its products are of a high purity. It can be applied in fabricating Si-based nanostructures by adjusting the composition of polymeric precursors, catalysts, and atmospheres [31]. In the study of Cho et al. [32], silica particles were synthesized by the flame spray pyrolysis (FSP) method from two precursors of tetraethylorthosilicate (TEOS) and silicate acid. When the concentration of TEOS is increased from 0.1 to 0.5 M, the specific surface area of silica powder is decreased from 285 to 81.4 $m^2/g$ and the average particle size is increased from 9.6 to 33.5 nm.

Research conducted by Gao, et al. [31] nano-/submicron silica spheres were successfully synthesized by the pyrolysis method from amorphous polysilazane preceramic powder with $FeCl_2$ catalyst. The perhydropolysilazane precursor was solidified by heating at 260 °C for 0.5 h in $N_2$. The mixture of amorphous SiCN with $FeCl_2$ was heated to 1250 °C and pyrolyzed there for 2 h. This experiment resulted in an amorphous silica with a diameter of 600–800 nm and a smooth clean surface without any flaws.

A comparison of the result characteristic from several different synthesis methods can be seen in Table 1.

Table 1. Comparison of the result and method of silica synthesis.

| Paper | Source of Silica | Method | Purity (%) | Particle Size (nm) | Product | Surface Area ($m^2/g$) |
|---|---|---|---|---|---|---|
| Abbas, et al. (2019) [20] | Rice husk | Sol-gel | - | - | Mesoporous | 760 |
| Ismail, et al. (2021) [22] | Silica sand | Sol-gel | - | 170.3 ± 14.3 | Nanoparticle | - |
| Sdiri, et al. (2014) [21] | Siliceous sand | Sol-gel | 88.8–97.5 | - | Micro and mesoporous | 340 |
| Zhu Y, et al. (2019) [26] | Silicate glass | Hydrothermal | - | 20–100 | Amorphous nanowires | - |
| Ortiz, et al. (2013) [27] | Sodium silicate | Hydrothermal | - | 200–500 | Mesoporous | 860–1028 |
| Azat, et al. (2019) [25] | Rice husk | Leaching | 98.2–99.5 | - | Amorphous | 120–980 |
| Park, et al. (2021) [28] | Rice husk | Leaching | 98,5 | - | Amorphous silica | 1973 |
| Zainal, et al. (2019) [15] | Rice husk | Leaching | 98 | - | Amorphous and crystalline | - |
| Gao, et al. (2013) [31] | Polylazane preceramic powder | Pyrolysis | - | 600–800 | Nano-/submicron spheres | - |
| Cho, et al. (2009) [32] | TEOS and silicate acid | Pyrolysis | - | 9.6–33.5 | Particle | 81,4 |

*2.2. Synthesis Method of Graphene*

About two past decades, since its first mechanically exfoliation, graphene has been widely synthesized with many various methods and materials. Graphene synthesis can be divided into two different methods, there are: 'Bottom-up' and 'Top-down' [11,33]. The bottom-up method is by using another substrate as a field for planting the graphene by putting the carbon precursor vapor into them, or in other words, using a different source that contains carbon rather than graphite. Meanwhile, for top-down methods using mechanical or chemical steps to obtain one single layer graphene from the structure of the graphite and derivatives (graphite oxide and graphite fluoride), or crushing apart multiple layer graphite into a single graphene sheet [11,34]. The detailed explanation of synthesis graphene is described in the following sections.

2.2.1. Top-Down Approaches

The Top-Down method processing raw material into final product that can be used to widely apply in industry. Starting from its first exfoliation in 2004, a thousand trials have been elevated to produce graphene from graphite. This method has several benefits compared with bottom-up methods, such as its potential to scale up, optimization of cost revenue, no need for substrate transfer, and its high productivity [35]. Therefore, the top-down method is mainly used on large-scale implementations. Some methods are discussed in the following segments.

Exfoliation Method

Basically, this method is the easiest and most widely used to synthesize graphite into graphene. An exfoliation method is divided into several types: mechanical exfoliation (scotch tape), chemical/electrochemical exfoliation, thermal exfoliation, and electrical exfoliation [35,36]. Mechanical exfoliation, also known as scotch tape, was developed by Geim and Novoselov in 2004 by using highly oriented pyrolytic graphite (HOPG) as a precursor and subjected it into oxygen plasma etching. As a result, 5 µm deep mesas existed and then it pressed into a layer of photoresist to the baking process. To peel out flakes of graphite from the mesas, scotch tape was used. After that, the thin flakes obtained are deposited to acetone and framed on the surface of $Si/SiO_2$ wafer. SEM and optical microscopy were used to analyze the few-layer graphene (FLG) formed. From this, Geim and Novoselov gained few- and single-layer graphene flakes up to 10 µm in size [5,37].

GICs, graphite intercalation compounds, were structured by graphite with atoms or molecules in compounds which were intercalated between the carbon layers. Mechanical and thermal exfoliation was served by this intercalation because of its characteristics. The increase in space between the layers weakened the interlayer interactions. The arrangement of the layer involved stacking from first stage GICs to the higher stage to complete formed monolayer platelets. Then, the expanded graphite (EG) could be produced from the higher stage GIC by exfoliating in rapid heating [36]. Graphene oxide (GO) could be formed from graphite oxide by using mechanical exfoliation (i.e., ultrasonication and stirring), since graphite oxide has natural hydrophilic and larger space interlayers along concentration 3 mg/mL [36].

Eswaraiah et al. (2011) implemented this method using focused solar radiation [38]. From a convex lens of 90 mm diameter, solar radiation is intensified and is directed to graphite oxide. It will increase the temperature of solar radiation by 150–200 °C from ambient temperature and a change in power from 60 mW to 2 W. The color of graphite oxide is also changed from light brown to dark black due to the high intensity of radiation. Exfoliation happens at a low temperature (150–200 °C) because of the rapid heating rate (>100 °C s$^{-1}$), then decomposition of the functional group occurs with the evolution of $CO_2$. The synthesis result depicts most of the GO as being efficiently exfoliated to form separated, ultrathin, and transparent sheets. Moreover, TEM images show the wrinkled nature of the graphene sheets. The thickness of graphene sheets calculated by HRTEM lattice imaging came out to be ~1 nm. The average step heights measured between the surface of the sheets and the substrate were found to be ~0.9 to 1.4 nm, proving them to be two atoms thick.

Chemical Reduction of Graphene Oxide/Organic Treatment

Chemical reduction triggered the exfoliated graphene oxide sheets to produce a balanced colloidal dispersion of graphene oxide (GO) [39]. Graphite oxide has hydrophilic characteristics in several solvents, including water, alcohol, etc. The same as graphite oxide, graphene oxide is suspended by sonication in solvents and acts as a precursor for graphene synthesis [37,40]. Purposing graphene formed by oxidizing graphite into graphene is one step ahead of producing a bulk graphene process [41]. Single- or multi-layer GO is created by the modification of Hummer's original method, and is adjusted in a thermal or chemical process [11,37].

Handayani et al. used sulphuric acid, sodium nitrite, and hydrogen peroxide potassium permanganate as chemical reagents in this method [42]. An exfoliation and intercalation process are performed by oxidizing graphite powder into graphene oxide and reducing it into graphene in polar aprotic solvents. Brownish graphene oxide from chemical treatment is the product and then sodium borohydride is added as a reducing agent to dispersing graphene oxide, turning the color black and homogenizing it. Abbas, et al. synthesized graphene oxide from Spent coffee beans via modified Hummer's method and reduced graphene oxide to form graphene using hydrazine as a reducing agent [43]. The electrochemical performance of the rGO-based modified electrode showed that the energy, voltage, and coulombic efficiency of the rGO-based electrode were more than 90% with a stability of up to 65 cycles. The efficiency is comparable to that of pure graphite electrodes used commercially. This is so that the material from biomass-derived rGO has great potential as a substitute for commercial graphite as an electrode material for VRB applications. He et al. in 2011 investigated the properties of graphene paper using chemical reduction [44]. Graphene was prepared in a paper form after synthesis using exfoliation, sonication, homogenization, and the addition of a hydrazine process. Furthermore, it annealed at 80 °C in a vacuum for 24 h before being cooled to room temperature. Characterization methods such as Ultraviolet-Visible Spectroscopy (UV-Vis) absorption spectra, Fourier Transform Infrared (FT-IR) spectra, X-Ray Diffraction (XRD), Thermogravimetric Analysis (TGA), Cyclic Voltammetry (CV) and others were implemented to analyze the properties of chemically reduced graphene (CRG). As a result of this research there is the fabrication of stable CRG aqueous dispersions and papers with different reduction levels. Properties of CRG are associated with levels of chemical reduction that can be adjusted and monitored to obtain the desired thermal stability, electrical property or tensile strength, and advanced controlled properties. Chemical reduction of graphene oxide using sodium acetate trihydrate presents its advantages in low-cost, use of non-toxic agent, no hazardous waste and simple product separation processes [45]. Sodium acetate trihydrate is an effective reducing agent in removing most of the oxygen-containing groups from GO for restoring the conjugated electronic structure of graphene.

Electrochemical Method

Electrochemical reactions mainly consist of cathode, anode, electrolyte, and metallic contact (standard electrochemical cell). The synthesis method of electrochemical reduction has the purpose of returning the actual properties of pristine graphene and of utilizing the reduced graphene (rGO) capability. This method has a resulting product, namely electrochemically reduced graphene oxide (ErGO), which contains graphene structure and properties that are different from pristine graphene. The reduction process is controlled by its potential value, and group of oxygen removal in graphene oxide (GO) by working of the electrode surface. Part of the electrochemical system affected the properties of the ErGO [46].

The pencil core is used as an electrode, a cathode and an anode in research by Liu et al. [47] Electrochemical exfoliation is configured in 1 M aqueous electrolytes ($H_2SO_4$ or $H_3PO_4$) and is potentially ramped for between +7 V and −7 V for 5–8 min. The result of this study is that the exfoliated graphene oxide flakes are quite large and have great electrocatalytic activity and toxicity tolerance for an oxygen reduction reaction in alkaline solution. Different electrode materials were conducted by Parvez et al. in 2013 [48] using Platinum (Pt) as a cathode and graphite as an anode. Electrochemical exfoliation is configured in 0.1 M $H_2SO_4$ solutions and is potentially ramped at +10 V for 10 min. The result of this study is that the exfoliated graphene has yielded about 60% and with multiple layers. Moreover, the product has a large sheet size, low oxygen content, and/or high C/O ratio as well as excellent electronic properties. Referring to the two studies above, the perfect electrolyte for an electromechanical system is a high-grade acid such as $H_2SO_4$. Flaking process on graphite is supported by acid electrolyte.

Ball Milling

Recently, the current development on the graphene synthesis method has a higher potential, which is improved from another step-like ball milling process. The product actually has a chance to enhance and also be an efficient stage for the synthesis method. Although this method is still being developed, high quality graphene from grinding bulk is hopefully produced on a large scale. Graphene, a ball-milling product, is influenced by the wet and dry conditions. Two forces play an important role in this process, such as shear forces exfoliating large graphene sheets, while normal forces break down graphite flakes [35].

Wu et al. show the accomplished mechanochemical reactions from the suspension of graphite and polystyrene (PS) solution by a ball-milling process [49]. High electrical conductive PS/graphene nanocomposites are obtained with homogenous mono- or few-layer graphene sheets. The process is operated by dispersed graphite nanoplatelets in N,N-Dimethylformamide (DMF) solvent to sonicate and generate it into a ball mill with 300 rpm. Mondall et al. used a modified Hummers method and a ball mill with zirconium ball (5 mm diameter) with 800 rpm angular speed to prove the amorphisms of reduced graphene oxides and the removal of oxygen function groups [50].

Lin et al. in 2017 found a new method on graphene synthesis, especially few-layer graphene (FLG) using plasma-assisted ball milling (P-milling) with various media [51]. The ball-milling media used in this experiment are boron nitride (BN), tungsten carbide (WC), zinc oxide (ZnO), iron oxide ($Fe_2O_3$), and germanium oxide ($GeO_2$). Preparation is done by calcining the expandable graphite at 1000 °C for 15 min with a heating rate at 5 °C/min under Ar atmosphere. Then, the sample is mixed with tungsten carbide (WC) with a ratio of 1:4 to place it in the ball-milling process. The P-milling process is set up with vibration on a 7 mm amplitude, 16 Hz frequency, 15 kV voltage, 1.5 A current, and 60 kHz discharge frequency. Treatment time is configured at 2 h, 5 h, 8 h and 10 h. The result of this process is the 6-layered FLG nanosheets with a high quality after the 8 h process with WC medium. The layers were formed due to the influence of inductive capacity of the ball-milling media, where it will affect the quality of FLG. The higher the inductive capacity, the lower the layer formed, and the higher the FLG quality. However, the ideal range for the inductive capacity is around 7–8.

2.2.2. Bottom-Up Approaches

A bottom-up graphene synthesis method that adjusts to the capability of the material from the building part is by part of the structure. The electron character is influenced by the configurations of atoms that are based on carbon precursors. The benefit of this method is that the size of its setting can be adjusted, so the dimension of the product is predictable. However, this method has several limitations compared to the top-down method, including the low properties obtained (i.e., yield), high cost, and the difficulties in scaling up. The following sections explain the examples of the bottom-up method [35].

Chemical Vapor Deposition (CVD)

Since 2009, the CVD method has become popular among researchers and has been often used until now. The carrier of this system is an inert gas with a high temperature and vacuum environment to process highly volatile carbon sources. There are many sources of carbon, such as precursors in gases (methane, acetylene, ethylene), liquids (ethanol, methanol), and solids (bio-carbon, polymer, waste plastic). The opportunity of this method lies in the high electrical conductivity that can be widely applied for electronic devices [11].

Li et al. in 2012 implemented a CVD method by keeping electropolished Cu foils under low pressure and toluene at 500–600 °C to grow continuous single layer graphene films [52]. As a result, the graphene has been produced with a high sheet resistance, a good transmittance of around 97.33% at 550 nm and a good electron mobility of 190 $cm^2$/(V.s). Li et al. also reported this, but from a different precursor, methane [53]. Copper foil was analyzed at a high temperature (1035 °C) under low pressure. The product obtained had a single crystallographic orientation, with a high electron mobility of 4000 $cm^2/(V.s)$.

Principally, the precursor carbon source and the rate of growth graphene determined the form of the graphene product. Big or small, the graphene size could be adjusted by controlling those two factors. Moreover, the large-area, high quality graphene and the large size of the graphene single crystals with different shapes and layers could be fabricated in case of the managed synthesis parameter due to Liu et al. in 2017 [54].

Arc Discharge

Arc discharge is a low cost and environmentally friendly synthesized graphene method. The obtained graphene with this method could be done under several circumstances, such as in a hydrogen, helium, or nitrogen state. Few-layered graphene (FLG) is produced under a helium and carbon dioxide mixture as it was in an experiment by Wu et al. in 2010 [55]. Synthesis method is controlled by directing current around 100–200 A, discharging voltage 30 V, and setting up diameter of anode 13 mm and cathode 40 mm. The result exhibits FLG sheets in organic solvents that can be easily dispersed and have fewer defects than other chemical steps. The capability of the solution used is important for fabricating electrical devices and composite materials.

In 2016, Kim et al. showed the arc discharge process to produce bi- and trilayer graphene using water medium as the dielectric [56]. A DC power supply is used to flow the direct current with amounts 1 A to 4 A between cathode and anode, keeping the voltage at about 25 V to initiate the arc discharge process. To remove the residual solvent, a drying process is needed with heating at around 85 °C for 2 h. The resulting product has the probability to become a good electrode because it has a high transmittance (84.5% at 550 $cm^{-1}$) and an electrical resistivity of 27.7 $k\Omega\ cm^{-2}$.

CNT Unzipping

Multiwalled carbon nanotubes were cut longitudinally by first suspending them in sulphuric acid and then treating them with $KMnO_4$. This produced oxidized graphene nanoribbons, which were subsequently reduced chemically. The resulting graphene nanoribbons were found to be conducting, but were electronically inferior to large-scale graphene sheets due to the presence of oxygen defect sites [37].

Kosynkin et al. in 2009 [57] explained how to get graphene nanoribbons by unzipping CNT. The product obtained was approximately near a 100% yield of the nanoribbon structure. Mechanism opening is dependent on oxidation alkenes by permanganate acid. Graphene structures were found subsequently in acid conditions ($H_2SO_4$) in exfoliating nanotubes.

*2.3. Synthesis Method of Silica–Graphene Hybrid Composites*

A $SiO_2$/graphene composite is a material that has a high specific surface area, good mechanical properties, and good electrical conductivity. Appropriate fabrication methods must be chosen to produce composites with improved physical and chemical properties in order to obtain a high performance [24].

2.3.1. Hydrothermal

The hydrothermal method is one of the simplest methods of graphene/silica ($G/SiO_2$) composite synthesis [23]. Graphene/mesoporous silica composites have been successfully synthesized by Qian et al. with one-step hydrothermal method from Tetraethyl Orthosilicate (TEOS) and organic solvents as a carbon source. Hexadecyltrimethylammonium bromide (CTAB) was dissolved with urea and was then added to an organic solvent, isopropanol, and TEOS. It was reacted for 4 h at 180 °C. The formation mechanism of $G/SiO_2$ is shown in Figure 1. The result of this experiment shows that graphene and silica layer are simultaneously produced and the distribution of graphene is uniform in the composite. In addition, organic solvents such as hexane, heptane, toluene, benzene, and cyclohexane can be used as a carbon source, therefore graphene does not need to use graphite as a carbon source.

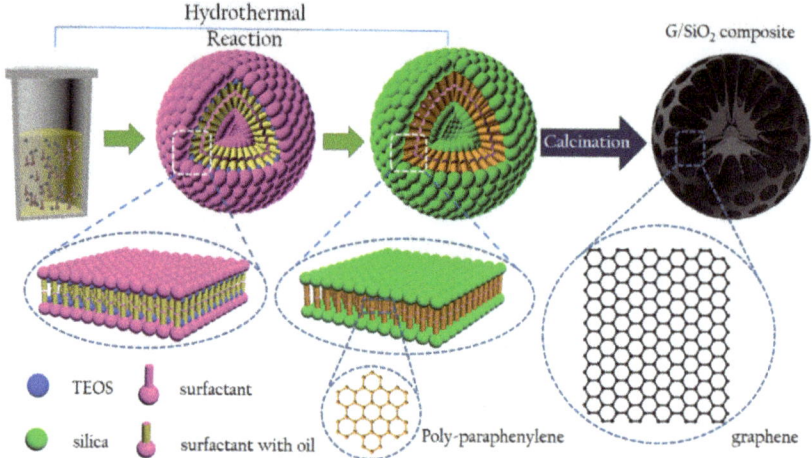

**Figure 1.** Schematic of G/SiO$_2$ composite growth mechanism. This figure is reproduced from ref. [23] with the required copyright permission.

TEM image of G/SiO$_2$ composites prepared with different organic solvents as a carbon source are shown in Figure 2. It is shown that all mesoporous spherical particles with a radial channel structure in the diameter of 0.2–1 µm and uniform distribution of graphene in composite were produced from this experiment. The pore size of the prepared SiO$_2$@G/SiO$_2$ particles reaches 17.7 nm.

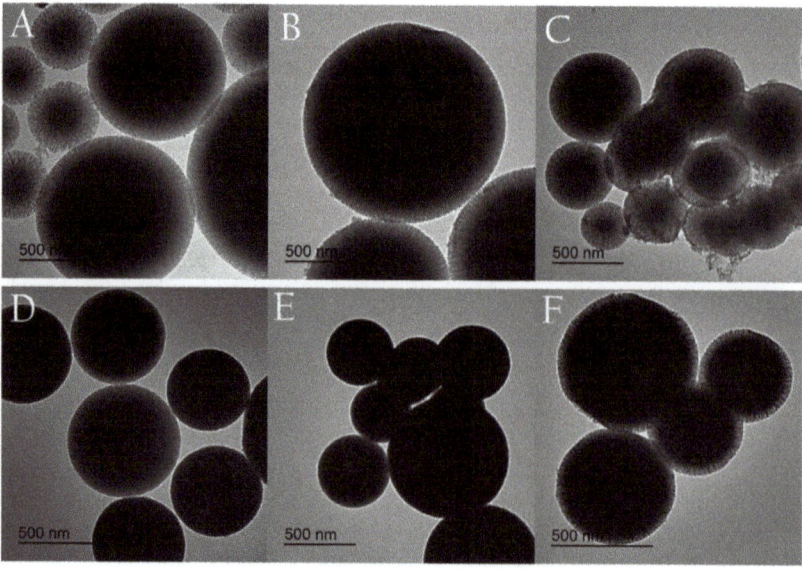

**Figure 2.** TEM images of G/SiO$_2$ composites prepared with different organic solvents as carbon sources. (**A**), toluene; (**B**), xylene; (**C**), mesitylene; (**D**), n-hexane; (**E**), n-heptane; (**F**), cyclohexane. This figure is reproduced from ref. [23] with required copyright permission.

The synthesis of SiO$_2$ nanocomposite/reduced graphene oxide (RGO) has also been successfully synthesized by Yi et al. [58] with a one-step hydrothermal method under acidic conditions using tetraethoxysilane (TEOS) and graphene oxide (GO). Mixtures of GO and SiO$_2$ will undergo a hydrothermal reaction in a cylindrical stainless steel reactor at 120 °C for 12 h. Along with the TEOS hydrolysis process, silica is loaded on a GO sheet surface with covalent bonds. Experimental results show that SiO$_2$ nanoparticles can be dispersed uniformly on the surface of the RGO. The composite containing 75 wt.% SiO$_2$ has a micro-mesopore structure with a surface area of 676 m$^2$/g. The synthesized SiO$_2$/RGO samples had an adsorption performance with the efficiency of Cr(VI) ion removal in wastewater reached equilibrium in 30 min and the adsorption efficiency of Cr(VI) reached 98.8% at pH = 2 and temperature 35 °C.

### 2.3.2. Sol-Gel

Sol-gel is one of the methods used to synthesize graphene/silica composites with the advantage of easy control of chemical composition and the ability to form composites with various filling materials [59]. The synthesis of silica/graphene oxide hydrogel has been successfully carried out by Oh Byeolnim et al. using the sol-gel method based on silica hydrogel with a combination of an acid/base catalyst system. From this experiment, it was found that GO was dispersed and did not suffer structural damage. The higher the GO volume, the faster the gelatinization process so as to minimize the occurrence of GO particle agglomeration. The presence of a catalyst (NaOH) makes the composition in the composite evenly distributed due to the acceleration of the gelatinization process. On the other hand, the hydrogel weakens with increasing GO, but the flexibility increases. It has a modulus of elasticity of 10 kPa–4.5 MPa, the highest compressive strength, compressive strain, and young modulus is about 0.3 Mpa, 0.35, 4.7 Mpa.

Haeri et al. in 2017 exhibited two route sol-gel methods for synthesized silica-functionalized graphene oxide nanosheets (GONs) [60]. The first route of the sol-gel method is running. 5 wt.% silane mixture (TEOS-60 wt.% and APTES-40 wt.%), 80 wt.% alcohol and 15 wt.% deionized water, then added GO nanosheets with the control pH of 4. For the second route, GO nanosheets are put into a mixture after oligomers formed from interaction between the hydrolyzed TEOS and APTES silane precursors at specific circumstances with pH 4 for 72 h, followed by the sonication process. The resulting surface GONs has coarse and unequal characteristics, and also SiO$_2$ nanoparticles appear a lot for the second method compared with the first method. This is because in second method, silane deposition is converted into SiO$_2$ nanoparticles, whereas in method 1, it was grafted in a GO sheet.

### 2.3.3. Hydrolysis

Wang et al. [61] have successfully synthesized silica/graphene oxide sheets for epoxy composites using one step process with assistance of diethylenetriamine (DETA) or ammonium hydroxide (NH$_4$OH) as catalyst. GO was prepared from graphite with a modified Hummer's method and SiO$_2$ was prepared from TEOS with hydrolysis and condensation. The SiO$_2$ could be easily formed onto the GO surface by the hydrolysis of TEOS which improved the dispersibility of GO and enhanced interaction with epoxy matrix. It was revealed that the GO-SiO$_2$/epoxy composites exhibited higher tensile, flexural and impact strength and modulus than that of GO.

Another work of Silica-Graphene Oxide Hybrid Composite Particles synthesizedwith hydrolysis method was performed by Zhang et al. [62]. Preparation of Si-GO Hybrid Composites were conducted by dispersion GO and TEOS in ethanol and hydrated ammonia was added to the mixture as a catalyst. Successful decoration of silica nanoparticles on layered GO surface by hydrolysis of TEOS in Si-GO composite were ensured by SEM and TEM images depicted in Figure 3c,f, respectively. The results show the success of the synthesis of silica particles on the GO surface. Si-GO hybrid composite particles showed a better thermal stability than GO. The Si-GO hybrid composite-based ER (electrorheology) fluid exhibits typical ER characteristics and behaves as a Bingham fluid in the presence of an electric field. ER fluids present a very short relaxation time in dielectric analysis.

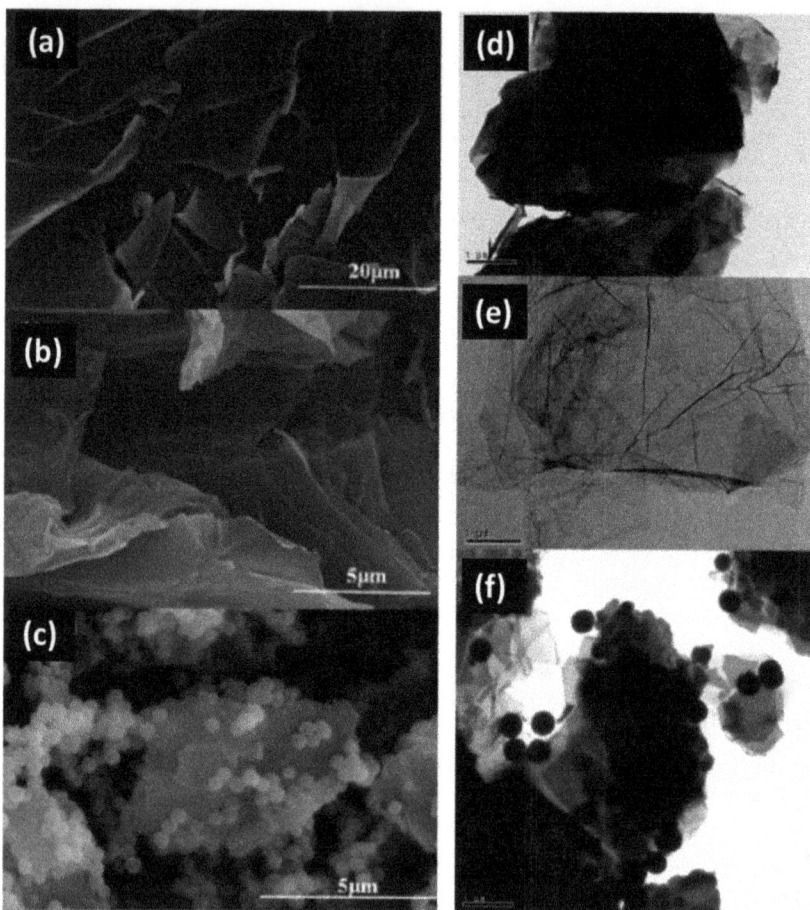

**Figure 3.** (**a**) SEM image of pure graphite, (**b**) SEM image of GO, (**c**) SEM image of Si-GO hybrid composite (**d**) TEM image of pure graphite, (**e**) TEM image of GO, (**f**) TEM image of Si-GO hybrid composite. This figure is reproduced from ref. [62] with the required copyright permission.

2.3.4. Encapsulation

Encapsulation is one of the graphene/silica composite synthesis methods that can produce advantages in increasing silica−epoxy interface adhesion. The GO (graphene) encapsulation core shell hybrid made with an ultra-thin layer GO (graphene) on organic/inorganic objects can provide special properties and applications [63].

The experiment about epoxy/silica composites by introducing graphene oxide to the interface was done by Chen et al. [63] with the core shell method. GO was prepared from natural graphite by the Hummers method and surface modification of $SiO_2$ with an APS coupling agent. The $SiO_2$−GO hybrid was fabricated by mixing the suspension of $SiO_2$−$NH_2$ and the GO solution. Preparation of epoxy/$SiO_2$−GO composites by dispersion of $SiO_2$−GO in tetrahydrofuran solvent. The $SiO_2$ and $SiO_2$−GO composite morphology investigated by SEM and HR-TEM reveal that $SiO_2$ particles exhibits smooth surface (Figure 4a), whereas $SiO_2$−GO shows that the silica surfaces are intimately covered by ultrathin GO with the shell thicknesses less than 3 nm as depicted in Figure 4 b-d The presence of flexible and very thin GO sheets can be attributed to the creases and rough texture of the composite. Mechanical properties such as Young's modulus, tensile strength, fracture toughness, and elongation are 1.36 GPa, 51 MPa, 1.81 MPa m1/2, and 7.36% respectively. It also has a Tg of 181.5–209.1 °C. The relatively reduced Tg values are due to the reduced network density, particle confinement of the filler−matrix interface, and the reduction in organic network density, which dominate the relaxation behavior of epoxy segments.

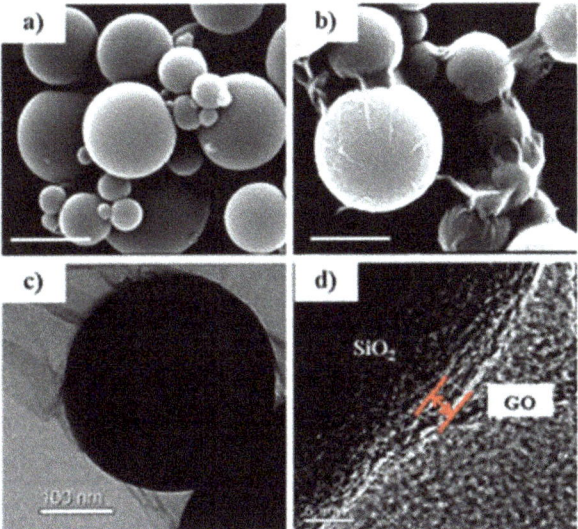

**Figure 4.** SEM image of (**a**) Raw $SiO_2$, (**b**) Created $SiO_2$-GO hybrid, TEM of (**c**) Core shell structured $SiO_2$-GO hybrid, (**d**) Graphene encapsulating silica sphere. This figure is reproduced from ref. [63] with required copyright permission.

The $SiO_2$@poly(methylmethacrylate)–reduced graphene oxide ($SiO_2$@PMMA–rGO) composite was successfully synthesized by Ye et al. [64] with the encapsulation method. The synthesis was carried out by dispersion method polymerization and mixing of colloids based on electrostatic assembly. Monodispersed $SiO_2$ nanoparticles with an average diameter of 300 nm were synthesized by hydrolysis and condensation of TEOS. An aqueous graphene oxide suspension was added to a positively charged $SiO_2$@PMMA nanoparticle dispersion with stirring. The resulting product exhibits a high thermal stability with a decomposition temperature increased by 80 °C. Besides that, it also showed strong mechanical properties with a 108% increase in modulus up to a 125% increase in hardness. This composite has advantages in thermal stability, a strong mechanical performance, and an excellent conductivity. This method can efficiently avoid the agglomeration of the fused nanofillers as well as improving the interfacial adhesion between the PMMA filler and matrix, so that the resulting composite has a high performance. The strength and weaknesses of several processing methods for graphene/silica composites are compared, as seen in Table 2.

Table 2. Several processing methods for synthesis of graphene–silica hybrid composites.

| No | Composites | Processing Method | Strength | Weakness | Ref. |
|---|---|---|---|---|---|
| 1 | Graphene/mesoporous silica (G/SiO$_2$) | Hydrothermal | Graphene does not need to be prepared in advance, graphene and silica layer overlapped to form intercalation, uniformly distribution, organic solvents can be used as carbon sources, no toxic gas is generated during the reaction. It does not require the use of a catalyst | Graphene can only be made by adding TEOS as a precursor, if it is reacted in big open space then the rate of chemical reactions is too slow to produce graphene. | [23] |
| 2 | SiO$_2$/RGO | Hydrothermal | Efficient method, easy-to-synthesize process, low cost, composites stability | Control sheet restacking and aggregation of SiO$_2$ nanoparticles is required | [58] |
| 3 | Silica/graphene oxide hydrogel | Sol-gel | Mechanical properties of the composite hydrogel such as stiffness can be adjusted by adjusting the GO contents | Increasing the addition of GO can weaken and decrease the mechanical properties of the hydrogel | [59] |
| 4 | Silica-functionalized graphene oxide (GO) nanosheets (GONs) | Sol-gel | Using two different route methods which produce various results | - | [60] |
| 5 | Silica/graphene oxide sheets epoxy composites | Hydrolysis | Catalysts (DETA and NH4OH) improving mechanical properties of composites by functionalization GO and forming SiO$_2$ from a promotion of the hydrolysis of TEOS on the GO surface | The mechanical properties and distribution of the resulting particles are highly dependent on the use of a catalyst | [61] |
| 6 | Silica-Graphene Oxide | Hydrolysis | Relatively simple, inexpensive, and fast method | - | [62] |
| 7 | Epoxy/silica composites by introducing graphene oxide | Encapsulation | Interfacial structures and properties can control by using GO as a novel coupling agent | - | [63] |
| 8 | SiO$_2$@poly(methylmethacrylate)–reduced graphene oxide (SiO$_2$@PMMA–rGO) | Encapsulation | Covalent molecular binding and strongly electrical interaction produce outstanding thermal stability, hardness, and electrical conductivity | The morphology of the composites are strongly influenced by the synthesis conditions | [64] |

## 3. Applications of Silica-Graphene-Based Hybrid Composites

### 3.1. Adsorbent

Graphene and its derivatives have the advantage of having a high specific surface area and degrade persistent organic matter chemicals in water, which gives it great potential as an adsorption material for environmental pollutants. Meanwhile, SiO$_2$ has the advantages of being a cheap, non-toxic, and chemically stable material that can overcome the problem of GO aggregation and can increase the specific surface area and adsorption properties. GO has many functional groups containing oxygen, which can absorb heavy metals from wastewater by electrostatic attraction, ion exchange, or surface complexation. Mesoporous silica has good potential in water treatment because of its stable mesoporous structure. Graphene/silica composites as adsorbents for water treatment have advantages over the use of graphene and silica individually because they have a higher adsorption capacity and stability [24].

In the research of Liu et al. [65], synthesis of graphene-silica (GS) composites by anchored nano-zero-valent iron (NZVI) to the surface were used to remove As(III) and As(V) from aqueous solution that the maximum capacity reached 45.57 mg/g and 45.12 mg/g respectively, by electrostatic attraction and complexation. Mesoporous silica ordered graphene oxide with a two-dimensional mesoporous structure and a large surface area has been synthesized by Wang, et al. (2015) [66] through the sol-gel method. This material is applied as a heavy metal adsorbent in water by adsorption separation-inductively coupled plasma mass spectrometry, that have removal efficiency, for metals As, Cd, Cr, Hg, and Pb reaching 97.7, 96.9, 96.0, 98.5, and 78.7%, respectively.

A graphene/silica composite material can be used as an adsorbent for organic pollutants due to their high specific surface area, active site, and good stability [24]. Phenyl-modified magnetic graphene/mesoporous silica (MG-MS-Ph) with a hierarchical pore-bridge structure was synthesized by Wang, et al. (2016) [67], which was applied as a pesticide adsorbent from wastewater. The resulted composite has a surface area of 446.5 m$^2$/g with very regular mesopores, a uniform pore size of 2.8 nm, a pore volume 0.32 cm$^3$/g, and a high saturation magnetization of 25 emu/g. The synthesis of the porous magnetic silica-graphene oxide hybrid composite ($Fe_3O_4$@$mSiO_2$/GO) was successfully carried out by Liu, et al. (2014) [68] through the core-shell method to be applied as a p-nitrophenol adsorbent from an aqueous solution. The maximum adsorption capacity produced reached 1548.78 mg/g at a solution pH of 8 and a temperature of 25 °C.

The adsorbent commonly used for $CO_2$ is an activated porous solid adsorbent such as carbon, zeolites, and metal-organic frameworks. Graphene coupled with silica also has a good performance as a $CO_2$ adsorbent by increasing its pores [24]. Wang, et al. (2019) [69] used a 2D/3D structure of reduced graphene-silica oxide (G-Si) aerogel combination using mesoporous silica SBA-15. These special 2D/3D morphological features result in the high $CO_2$ absorption capacity of 6.02 mmol/g as seen in Figure 5a. In addition, equilibrium is achieved very quickly within the first 12 min for 30-50 wt% of TEPA loading indicating that $CO_2$ gas molecules diffused freely throughout G-Si aerogel and SBA-15 as shown in Figure 5b. Another $CO_2$ adsorbent that has also been successfully synthesized is the reduced amine/graphene oxide (AMS/RGO) modified silica hybrid composite by Vinodh, et al. with a limited growth methodology, which the RGO introduced into the amino alkyl siloxane matrix. This composite shows a good absorption capacity (15% by weight) and a Maximum $CO_2$ adsorption up to 14.7%.

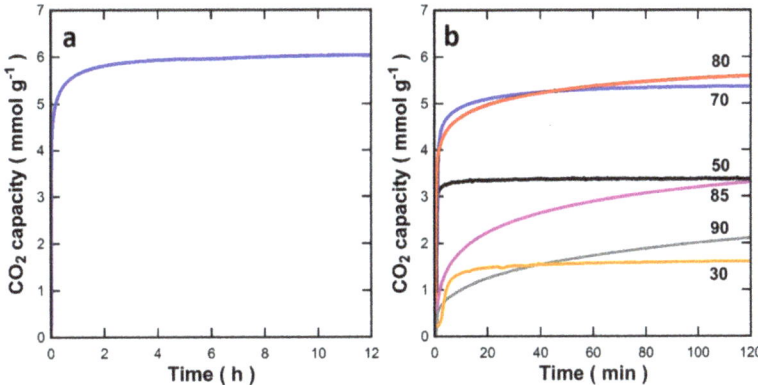

**Figure 5.** $CO_2$ sorption kinetics of (**a**) 2D/3D G-Si aerogel/SBA-15 (4.7 wt.%) loaded with TEPA 80 wt.% and (**b**) short term isotherms by varying the TEPA loading. This figure is reproduced from ref. [69] with the required copyright permission.

### 3.2. Energy Storage

Many sources have stated that graphene has high expectations on building energy storage equipment. Great properties in the electrical field required an important role in constructing it. A combination between graphene structures and other materials such as silica provides the strengthened characteristics of composite [11]. Graphene can be formed into the three-dimensional graphene aerogel type, consisting of interconnected graphene sheets and loaded materials. Furthermore, important properties, such as mechanical and electrical ones, are preserved approaching the structures [70].

In the last decade, power saving technology to store energy has been developed in several ways, such as through batteries, capacitors, and others. Electrochemical capacitors (Ecs), commonly called supercapacitors, are devices with a high power capacity and a long cycle life (>100,000 cycles) which required minimal maintenance, and fast charging [71]. Research in this field has attractive value, due to its potential resource as an energy storage component. Ghosh et al. (2018) [72], by using the sol-gel method, synthesized a layer-by-layer composite that contained reduced graphene oxide (rGO) and iron silicate in glass. From that study it was found that the specific capacitance for the composite is 370 F/g, ranked in high grade for supercapacitor applications. This is because of a multilayer structure from rGO and iron silicate glass supporting the properties, such as a large surface area, swinging bonds, high electrical conductivity, and a racking porosity. A multilayer concept, such as an electrode capacitor, is also applied and was analyzed by Kim et al. (2015) [73] A vacuum channel was performed to understand the electrical and photodetection properties of the resulting product. A high responsivity for the structure of about 1.0 A/W at 633 nm was discovered and there is the hope for a capacitance of around 10 nF/cm$^2$ for 1 μm depletion.

In the research of Abbas, et al., reduced graphene oxide (rGO) from waste coffee bean biomass is used as an alternative electrode material for the VRB (Vanadium redox flow battery) system. The resulting rGO exhibited more than 90% energy, voltage and coulombic efficiency, which was comparable to that of commercially used pure graphite electrodes. This electrode also has a stable cyclic performance for 65 cycles due to its high electrocatalytic activity and its enhanced charge transfer [43].

Another type of graphene form—aerogel—shapes the three-dimensional graphene that maintains interesting properties such as a large specific area, flexibility, and conduction. Furthermore, it could be implemented in catalysis, sensing, energy storage and others [70]. Du et al. in 2018 [74] investigated the effect of the addition of amino-functionalized silica as a template and doping agent for N-doped graphene aerogel. The forming structure is an N-doped rGO aerogel with a filling of $SiO_2$-$NH_2$ particles between the spaces and it has a macroscope diameter of around 50 μm. The number of $SiO_2$-$NH_2$ particles has an opposite value with a pore size of rGO aerogel; when it is higher the pore size it will be smaller, but the number of pores will increase. Electrochemical properties from the structure can be identified as the high specific capacitances of 350 F/g at the current density of 1 A/g with a great reversibility of 88%, a cycling efficiency after 10,000 cycle, and is supported by other properties like a specific large surface area pf 481.8 m$^2$/g, a low series resistance and a high nitrogen doping content of about 4.4 atom% so that the composites are reliable as the components of a capacitor with a high oil-absorbability and recyclability.

Along with the rapid development of technology, the need for energy is also getting higher. An energy-saving concept with promising future prospects is lithium-ion batteries. The allocation of excellent properties other than energy storage, such as a high energy and power density, good durability, and environmental buddies, are impressively used in electric vehicle applications. However, this requires special attention for operating it under the limitation of voltage and temperature [75]. Graphene and silica combined were able to act as the electrode of lithium batteries, both anode and cathode. In 2013, Li et al. [76] mapped out the anode from a combination of graphene oxide and $SiO_2$ nanoparticles by an annealing process. Three-dimensional graphene networks were formed owing to great properties such as reversible capacity (610.9 mAh/g at 50 mA/g after 50 cycles) and superb rate capability (291.5 mAh/g at 5000 mA/g). Considered as the electrode because of high power density, there was a large porosity and a high electrical conductivity based on the attribute of network structure. In addition, the anode can be designed by a self-assembly process as done by Yin et al. (2017) [77], Wang et al. (2018) [78] and Kim et al. (2018) [79].

Self-assembly procedures to construct the anode lithium-ion battery were done by Yin et al. [77] using ultrasonic-assisted hydrothermal and heat treatments. The unification process considered the mass ratio of all the materials used (colloidal silica, sucrose, and graphene oxide), playing an important role in this manner. Setting up the ratio of silica to sucrose of 0.15 exhibited a great electrical performance, such as discharge capacity (906 mA h/g) and reversible capacity (542 mAh/g at 100 mA/g after 216 cycle). A uniformly dispersed order of small $SiO_2$ nanoparticles in the composites display the high properties (good conductivity) with a simple method implied [77]. An applied convenient method by a facile electrostatic self-assembly approach for obtaining nano-Si/reduced graphene oxide porous composite as anode of Li-ion battery application, Wang et al. [78] forfeiting $SiO_2$ as a sacrificial template. The composites consist of nano-$SiO_2$ particles evenly spread across rGO sheets that can intensify electronic conductivity. According to the result as depicted in Figure 6, electrochemical performances increase with the great reversible capacity (1849 mAh/g at 0.2 A/g) along with a decent capacity retention, and a high rate capability (535 mAh/g at 2 A/g) [78]. Kim et al. [79] reported that the novel structure consists of graphene nanocomposite with ordered mesoporous carbon-silica-titania using one-pot evaporation-induced. Mobility of ions is supported by the regular form of mesopores, which can assist in the transport of electrons in the graphene sheets and the penetration process in electrolytes. Furthermore, the structure clarifies the reason why it can be appropriate for electrodes. Within a high reversible capacity (547 mAh/g) and an efficient reversibility (65%), there is an increasing cycle work besides the large structural area [79].

On the other hand, this was also done for cathode of lithium-sulfur batteries by Kim et al. in 2014 [80] by using mesoporous-graphene silica that was infiltrated with sulfur on the organic liquid (polysulfide) electrolyte, resulting in the composite structure with a good electrochemical performance. Respectively, the electrical properties reported for cycling stability and reversible capacity were about 500 mAh/g and 380 mAh/g after 400 cycles. To be more convincing, further research was needed to discover the properties of this structure [80].

For solar energy conversion, alternative power saving, the best strategy for hydrogen production is photoelectrochemical (PEC) water splitting based on semiconductor electrodes. In 2018, Zhao et al. [81] analyzed the function and synergy effect of silane molecules and GO in $WO_3$ nanosheets array ($WO_3$ NS), where the $WO_3$ NS was already grafted by silane molecules to make an external electric field (EEF). The result product possessed a high photocurrent of 1.25 mA/cm$^2$ at 1.23 V vs RHE, rising 1.8 times as before. The important role of silane for this structure was a hole-storage spot together with GO being a hole-transfer pathway, carrier for channel, and increasing reactive site to promote water oxidation kinetics.

*3.3. Biomedical Fields*

GO has good biocompatibility, low toxicity, water solubility, and easy surface modification, which makes it attractive in the biomedical field. Meanwhile, silica nanoparticles have excellent biocompatibility, encapsulation ability in hydrophilic and hydrophobic molecules, high surface area, tunable morphology, and scalable synthetics availability. This makes the combination of graphene and silica in composites have a high potential for application in medical fields such as drug carriers, imaging, diagnosis, and therapy [24].

Research conducted by He, et al. [82] with the assembly of reduced graphene oxide (RGO) and mesoporous silica grafted with an alkyl chain (MSN-C18) for application as a drug carrier on exposure to near-infrared light (NIR). This material has a structure formed by the noncovalent interaction of the RGO cap and the alkyl chain at the MSN-C18 surface. There is an unlocking mechanism on this material that allows the loaded drug molecules to be released by irradiating NIR light. These drug carrier agents can be a promising drug delivery system for cancer therapy.

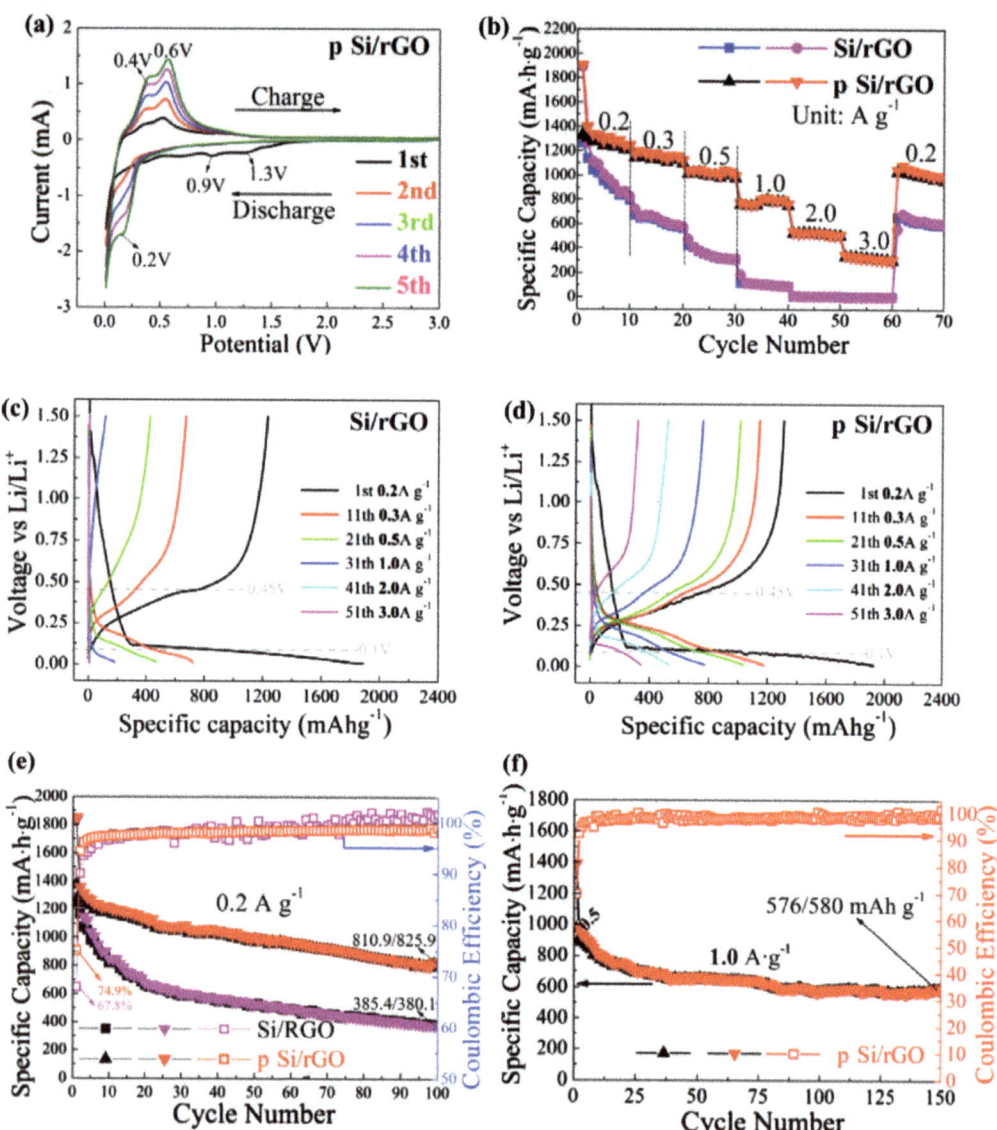

**Figure 6.** (a) CV curves of p Si/rGO at a scanning rate of 0.1 mV s$^{-1}$ in the potential range of 0.01-3.0 V (vs. Li/Li$^+$). (b) Rate capability of p Si/rGO and Si/rGO. Galvanostatic discharge/charge curves of (c) Si/rGO and (d) p Si/rGO; (e) Cycling performance of p Si/rGO and Si/rGO at 0.2 A g$^{-1}$; (f) Cycling performance of p Si/rGO 1.0 A g$^{-1}$ in a potential range of 0.01–1.5 V vs. Li/Li$^+$. The loading mass of active material is 1.1 mg/cm$^2$. This figure is reproduced from ref. [78] with required copyright permission.

Mesoporous silica nanoparticles (MSNPs) coated in blue fluorescent N-graphene quantum dots, loaded with DOX drug, and finally coated with hyaluronic acid (HA), was synthesized by Gui, et al. [83] for intracellular delivery of the cancer drug doxorubicin (DOX) to specific targeting of tumor cells. Imaging of human cervical carcinoma (HeLa) cells may arise as a result of cellular uptake of NPs with HA-DOX-GQD@MSNPs type architecture via fluorescence microscopy. Song, et al. [84] have designed a multifunctional probe incorporating active-targeted fluorescent imaging (FL)/photoacoustic imaging (PA) and chemo-photothermal therapy for tumors. Modified graphene oxide (GO) folate (FA) molecules were used to coat core-shell silver sulfide@mesopore silica (QD@Si) for loading the antitumoral doxorubicin (DOX) on mesoporous channels by the presence of electrostatic adhesion, and a delivery system (QD@Si-D/GO-FA) for active targeted dual-mode imaging and synergistic chemo-photothermal for tumors. The obtained cell survival rate was 76.3 ± 4.6%, which indicates that the probe has good biocompatibility. Tumors can be effectively killed at an increase in temperature to 63.5 °C under laser irradiation with combination chemotherapy due to the presence of GO exfoliation from QD@Si-D/GO-FA after irradiation.

In the study of Shao, et al. [85], mesoporous silica-coated polydopamine (MS)-fused reduced graphene oxide (pRGO) with modified hyaluronic acid (HA) (pRGO@MS-HA) has been used for cancer chemo-photothermal therapy as seen in Figure 7. This material is used to enhance doxorubicin (DOX) loading, with good dispersibility, excellent photothermal and tumor cell killing efficiency, and a specificity for targeting tumor cells, which is better than any monotherapy.

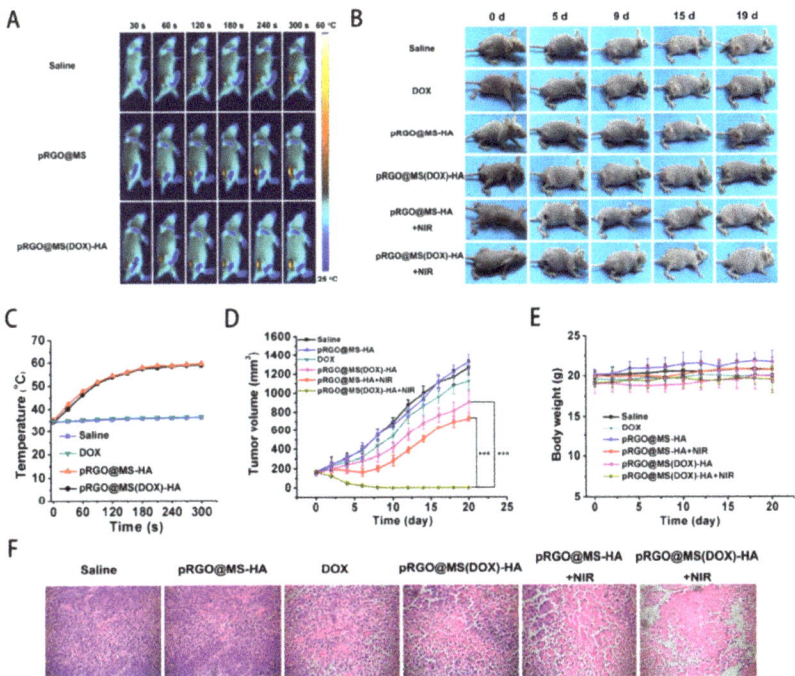

**Figure 7.** In vivo anti-tumor activity of pRGO@MS(DOX)-HA nanocomposite. (**A**) IR thermal images of HeLa tumor-bearing mice upon 808 nm-laser irradiation for different periods of time. (**B**) Representative images of mice bearing HeLa tumors after different treatments for varied time periods. (**C**) Temperature variation curves of tumor regions recorded by the IR thermal camera during NIR laser irradiation. (**D**) Tumor growth curves of mice after various treatments (five mice for each group). (**E**) The average body weights of mice after various treatments. (**F**) Representative H&E sections of tumors after treatment with saline, DOX, pRGO@MS-HA, pRGO@MS-HA+NIR, pRGO@MS(DOX)-HA, pRGO@MS(DOX)-HA+NIR. This figure is reproduced from ref. [85] with the required copyright permission.

### 3.4. Catalysis

By using a two-step reduction method, purposely to make sandwich nanostructure electrocatalyst that contain silica nanosphere filled palladium encapsulated with graphene (denoted Pd/SiO$_2$@RGO), this study has been research by Yang et al. in 2018. [86] The product has several advantages, mainly in methanol electrooxidation. There are SiO$_2$ particles that help the spreading of Pd NPs and prevent the aggregation of rGO, and also moderate the rGO strategy. Pd/SiO$_2$@RGO shows an excellent durability because it has the highest retained current density of 308.5 mA/mg and the lowest current decay speed of 19.3% during 1500 s.

In 2018, Nguyen et al. [87] find out for an photocatalytic application under visible irradiation by using convenient simple self-assembly to mixing lanthanum copper sulfur (LaCuS$_2$) with mesoporous silica and graphene oxide, in order to shape a new ternary catalyst. The result has a pore size of around 5.83 nm with excellent photocatalytic characteristics. Under pH 11, the rate degrades by almost 100% until pH 9, which means an enhanced dye removal percentage. Furthermore, the optimum amount is 0.05 g for the gallic acid photocatalytic, which refers to good photocatalytic performances [87]. On the other hand, Oh et al. [88], uses TEOS and cetyltrimethylammonium bromide (CTAB) on a self-assembly method to create mesoporous SiO$_2$/CdO-graphene composites (SCdOG). The result of the photocatalytic degradation achieves almost 100% Methylene Blue (MB) organic dye removal after the adsorption equilibrium for 2 h, as shown in Figure 8. Furthermore, the adsorption capability was the highest in the case of MB dye, compared with the other dyes. Besides that, this work opens a way to elevate the photocatalytic activity of gallic acid at ambient conditions without any further different oxidation processes, as well as for developing an efficient hetero-system for hydrogen production.

The composite that contains mesoporous silica layers within an encapsulating graphene nanosheet supported by an ultrafine metal was studied by Shang et al. in 2014 [89], and was developed by Sarkar et al. in 2019 [90] by changing metal uses with Cu. Besides producing the structure with the desire properties, their research is also trying to provide a catalyst which can be useful in chemical reactions. Shang et al. found catalysts with high activity and stability, and great recycling and reusability. Catalytic performance is enhanced by SiO$_2$ layers, and can be deactivated by feedstock poisoning.

Here, the applications and challenges for graphene/silica composites are compared as seen in Table 3.

From the application of graphene-silica hybrid composite, we summarized the properties of graphene, silica and graphene/silica composites and its potential applications which is depicted in the Table 4:

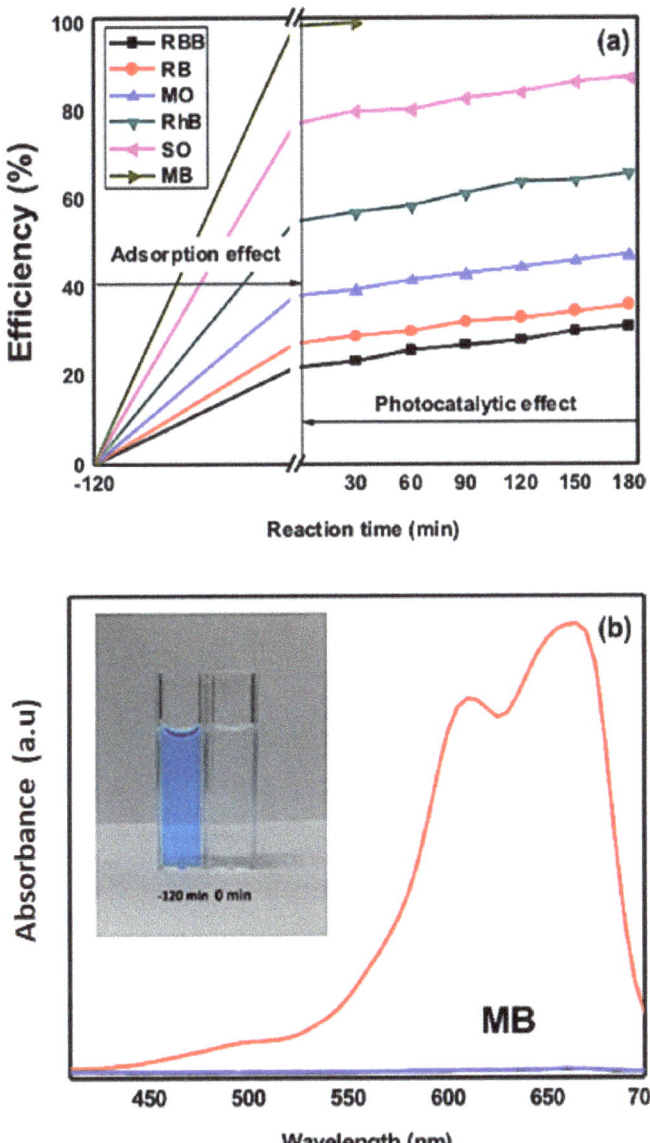

**Figure 8.** (**a**) SCdOG nanocomposite for the different cationic dyes degradation under visible light irradiation. (**b**) Adsorption capability of SCdOG nanocomposites for MB removal. The experiments were carried out with a neutral pH. This figure is reproduced from ref. [88] with the required copyright permission.

Table 3. Applications of graphene-silica hybrid composites, advantages and challenges.

| No | Applications of Graphene/Silica Composites | Method | Advantages | Challenge | Ref. |
|---|---|---|---|---|---|
| 1 | Adsorbent for As(III) and As(V) from aqueous solution | electrostatic attraction and complexation | Can be composited with other materials to increase absorption efficiency | Dependent on pH of the solution, unable to reach WHO drinking water standard | [65] |
| 2 | Adsorbent for heavy metal As, Cd, Cr, Hg, and Pb | Sol-gel | Low-cost, environmental friendly synthesis method, highly efficient adsorption | Complicated manufacturing process | [66] |
| 3 | Adsorbent for pesticides | One-step solvothermal and one-step method | Low cost and efficient adsorbents | low concentrations pesticides in complex wastewater. | [67] |
| 4 | Adsorbent for p-nitrophenol | Grafting and core-shell | High adsorption capacity, composites could be easily separated from solutions through an external magnetic force | The introduction of $SiO_2$ and GO will reduce the magnetization so that an external magnetic field is needed, the rate of diffusion slows down in the first stage | [68] |
| 5 | Adsorbent for $CO_2$ capture | Freeze-drying method | High $CO_2$ sorption capacity, very stable under sorption | Morphological feature of the 2D/3D sorbent assembly is attributed to decreasing surface area and pore volume, very slow sorption kinetics | [69] |
| 6 | Energy storage: electrode material in supercapacitors | Sol-gel | specific capacitance of the composite is considerably higher than that of graphene and has good cyclic stability as electrode material for supercapacitor | The measurement of temperature dependent resistance for the composite in the temperature range from 5 K–300 K was performed under cycle cryostat and high vacuum condition | [72] |
| 7 | Energy storage: supercapacitor electrode | Hydrothermal method | Ultrahigh specific surface area, high capacitance and long lifetime | Need high temperature annealing process | [74] |
| 8 | Energy storage: as anode materials of lithium lithium-ion batteries | Hydrothermal method and heat treatments | enhance the electrical conductivity, and improve the electrochemical performance. |  | [77] |
| 9 | Energy storage: Lithium battery electrode | electrostatic self-assembly method | Enhance the electronic conductivity, provide more transfer channels for Li+, excellent electrochemical performance | The pH value of process needs to be adjusted to help electrostatic self-assembly method | [78] |
| 10 | Biomedical field: drug carrier for near infraredlight-responsive controlled drug release | Capped noncovalent binding | Biocompatible, biofriendly, efficient killing efficacy towards cancer cells | NIR light is needed to control the drug release from mesopores to nucleus | [82] |

Table 3. Cont.

| No | Applications of Graphene/Silica Composites | Method | Advantages | Challenge | Ref. |
|---|---|---|---|---|---|
| 11 | Biomedical field: fluorescent imaging of tumor cells and drug delivery | Coating | enables simultaneous drug release, fluorescent monitoring | Metal ion can quench the intensity of the N-GQDs (N-Doped graphene quantum dot) | [83] |
| 12 | Biomedical field: imaging and Chemo- Photothermal Synergistic Therapy Against Tumor | Coating Core-Shell | Biocompatibility, provide a basis for the early diagnosis and treatment of tumor | Laser radiation are needed to produce a more effective tumor killed | [84] |
| 13 | Biomedical field: Chemo- Photothermal Therapy | Coating | Good biocompatibility, dispersibility, excellent photothermal property, remarkable tumor cell killing efficiency, specificity to target tumor cells | Fluoroscopy results differ in certain body parts due to organ efficiency | [85] |
| 14 | Catalyst: electrocatalysts for methanol oxidation reaction | Hydrothermal method | Improve the electrocatalytic performance, long-time endurance and superior durability. | | [86] |
| 15 | Catalyst: photocatalytic of organic dyes, gallic acid | Hydrothermal method | Enhanced photocatalytic activity for organic dyes and gallic acid, improved the hydrogen evolution process | | [88] |
| 16 | Catalyst: for Oxidation and Reduction Reactions | Deposition–precipitation method. | High catalytic activity and excellent high-temperature stability | Nanosize catalyst can agglomerate and sinter very easily during high temperature calcination | [89] |

Table 4. Properties of graphene, silica and graphene/silica composites and its potential applica-ions.

| No | Properties of Graphene | Properties of Silica | Properties of Graphene/Silica Composites | Potential Applications | Ref. |
|---|---|---|---|---|---|
| 1 | Graphene and its derivatives exhibit high specific surface area, however, graphene oxide will be easily agglomerated in the aqueous solution and re-stack between layers | $SiO_2$ is a non-toxic and chemically stable material which not only easily overcomes the aggregation problem of GO but also improves the specific surface area and adsorption properties | The combination of graphene and silica nanoparticles enhance the specific surface area, prevent restacking of graphene sheet and produce an excellent adsorption capacity | Environment and adsorption material | [65–68] |
| 2 | Graphene-based materials have excellent chemical and physical stability and high electrical conductivity, however, graphene sheets are easy to restack | $SiO_2$ particles could be inserted into the space between graphene sheets to produce a rigid support for flexible graphene sheets to prevent the π-π stacking of graphene sheets | Ultrahigh specific surface area, hierarchical porous structure, high capacitance and long lifetime | Energy storage | [74,77] |

Table 4. *Cont.*

| No | Properties of Graphene | Properties of Silica | Properties of Graphene/Silica Composites | Potential Applications | Ref. |
|---|---|---|---|---|---|
| 3 | Graphene, especially graphene oxide (GO), has good water solubility, low toxicity, good biocompatibility, and easy surface modification | Silica has a high surface area, good biocompatibility, encapsulation capability in hydrophilic and hydrophobic molecules, tunable morphology, and scalable synthetic availability | The combination of graphene and silica nanoparticles exhibit excellent synergistic properties include high surface area, excellent biocompatibility, tunable morphology and low toxicity as biomedical composite materials | Biomedical application includes drug delivery system, imaging and therapy | [84,85] |
| 4 | Graphene shows strong catalytic activity in photocatalysis and electrocatalysis, owing to its large surface area, has excellent conductivity for electron capture and transport | Silica has large surface area, regular pore size, thermal and chemical stability, and variable chemical functional groups. Silica can prevent the agglomeration of graphene, and enhance the electrocatalytic performance of graphene | The combination of graphene and silica nanoparticles integrates the advantages of the two components and shows remarkable application prospects in improving the catalytic performance | Catalysis | [86,88] |

## 4. Conclusions and Future Prospects

Various advances in synthesis methods and applications related to silica, graphene, and graphene/silica hybrid composites have been reviewed in this paper. Graphene and silica have become interesting materials for use in various fields because of their special properties. Graphene and its derivatives have properties such as a large surface area, high electrical conductivity, high thermal conductivity, high mechanical strength, great optical transmittance, high modulus of elasticity, and high electron intrinsic mobility. Silica has a wide application based on its various properties such as low density, tendency to produce the strength and hardness of ceramic materials, high porous, ultra-lightweight, biocompatibility, nanostructured properties, high specific surface area, and gas adsorption capacity due to the presence of its micro and mesopores. The several major findings for this review are (1) the current progress of strategy for synthesis graphene/silica was described with their advantages and potential applications; (2) The combination of graphene with silica in graphene/silica hybrid composite can avoid performance degradation of materials caused by agglomeration of graphene and greatly increasing specific surface area and biocompatibility. On the other hand, graphene can provide excellent mechanical properties and a high electrical and thermal conductivity. (3) The combination of graphene and silica in graphene/silica hybrid composites can form a synergistic effect to produce excellent properties. These excellent properties of graphene/silica contribute to spacious application prospects in many fields such as energy storage, catalysts, adsorbent, and biomedicine.

One of the challenges of silica/graphene-based composites is the difficulty of maximizing electrical conductivity, thermal conductivity, and electromagnetic shielding together because of their high surface areas and chemical stability that tend to resist losses. The development of new fabrication methods need to be carried out by taking into account the optimization of components and the interactions between the components graphene and silica, as well as the structure and property relationships, in order to efficiently produce properties suitable for the desired application. In this way, $SiO_2$/graphene composites can produce extraordinary properties that can open new technological opportunities in various fields.

**Author Contributions:** Conceptualization, M.H.; wrote the paper, M.H., N.N., A.N., A.B.P., E.F., E.S. and F.F.; writing—review and editing, M.H.; supervision, M.H. and A.R.; project administration, M.H.; funding acquisition, M.H. All authors have read and agreed to the published version of the manuscript.

**Funding:** This research was funded by National Research and Innovation Agency (BRIN), grant number 26/A/DT/2021.

**Conflicts of Interest:** The authors declare no conflict of interest.

## References

1. Komarneni, S. Nanocornposites. *J. Mater. Chem.* **1992**, *2*, 1219–1230. [CrossRef]
2. Thostenson, E.; Li, C.; Chou, T. Nanocomposites in context. *Compos. Sci. Technol.* **2005**, *65*, 491–516. [CrossRef]
3. Bogue, R. Nanocomposites: A review of technology and applications. *Assem. Autom.* **2011**, *31*, 106–112. [CrossRef]
4. Sonawane, G.H.; Patil, S.P.; Sonawane, S.H. Nanocomposites and Its Applications. In *Applications of Nanomaterials*; Elsevier Ltd.: Amsterdam, The Netherlands, 2018; pp. 1–22. [CrossRef]
5. Geim, A.K.; Novoselov, K.S. The rise of graphene. *Nanosci. Technol A Collect. Rev. Nat. J.* **2009**, 11–19.
6. Handayani, M.; Sulistiyono, E.; Rokhmanto, F.; Darsono, N.; Fransisca, P.L.; Erryani, A.; Wardono, J.T. Fabrication of Graphene Oxide/Calcium Carbonate/Chitosan Nanocomposite Film with Enhanced Mechanical Properties. *IOP Conf. Series: Mater. Sci. Eng.* **2019**, *578*, 12073. [CrossRef]
7. Alvial-Palavicino, C.; Konrad, K. The rise of graphene expectations: Anticipatory practices in emergent nanotechnologies. *Futures* **2019**, *109*, 192–202. [CrossRef]
8. Latiff, N.M.; Fu, X.; Mohamed, D.K.; Veksha, A.; Handayani, M.; Lisak, G. Carbon based copper(II) phthalocyanine catalysts for electrochemical $CO_2$ reduction: Effect of carbon support on electrocatalytic activity. *Carbon* **2020**, *168*, 245–253. [CrossRef]
9. Krisnandi, Y.K.; Abdullah, I.; Prabawanta, I.B.G.; Handayani, M. In-situ hydrothermal synthesis of nickel nanoparticle/reduced graphene oxides as catalyst on $CO_2$ methanation. In *AIP Conference Proceedings*; AIP Publishing LLC: Melville, NY, USA, 2020; Volume 2242, p. 40046. [CrossRef]
10. Abergel, D.; Apalkov, V.; Berashevich, J.; Ziegler, K.; Chakraborty, T. Properties of graphene: A theoretical perspective. *Adv. Phys.* **2010**, *59*, 261–482. [CrossRef]
11. Tsang, A.C.H.; Huang, H.; Xuan, J.; Wang, H.; Leung, D. Graphene materials in green energy applications: Recent development and future perspective. *Renew. Sustain. Energy Rev.* **2020**, *120*, 109656. [CrossRef]
12. Zhu, Y.; Murali, S.; Cai, W.; Li, X.; Suk, J.W.; Potts, J.R.; Ruoff, R.S. Graphene-based Materials: Graphene and Graphene Oxide: Synthesis, Properties, and Applications (Adv. Mater. 35/2010). *Adv. Mater.* **2010**, *22*, 3906–3924. [CrossRef] [PubMed]
13. Papageorgiou, D.G.; Kinloch, I.A.; Young, R.J. Mechanical properties of graphene and graphene-based nanocomposites. *Prog. Mater. Sci.* **2017**, *90*, 75–127. [CrossRef]
14. Soldano, C.; Mahmood, A.; Dujardin, E. Production, properties and potential of graphene. *Carbon* **2010**, *48*, 2127–2150. [CrossRef]
15. Zainal, N.S.; Mohamad, Z.; Mustapa, M.S.; Badarulzaman, N.A.; Zulkifli, A.Z.; Bhd, S.A.N.S.S. The Ability of Crystalline and Amorphous Silica from Rice Husk Ash to Perform Quality Hardness for Ceramic Water Filtration Membrane. *Int. J. Integr. Eng.* **2019**, *11*, 229–235. [CrossRef]
16. Mulyati, S.; Muchtar, S.; Yusuf, M.; Arahman, N.; Sofyana, S.; Rosnelly, C.M.; Fathanah, U.; Takagi, R.; Matsuyama, H.; Shamsuddin, N.; et al. Production of High Flux Poly(Ether Sulfone) Membrane Using Silica Additive Extracted from Natural Resource. *Membranes* **2020**, *10*, 17. [CrossRef]
17. Sharma, J.; Polizos, G. Hollow Silica Particles: Recent Progress and Future Perspectives. *Nanomaterials* **2020**, *10*, 1599. [CrossRef]
18. Zulfiqar, U.; Subhani, T.; Husain, S.W. Synthesis and characterization of silica nanoparticles from clay. *J. Asian Ceram. Soc.* **2016**, *4*, 91–96. [CrossRef]
19. Chen, J.-J.; Li, H.-J.; Zhou, X.-H.; Li, E.-Z.; Wang, Y.; Guo, Y.-L.; Feng, Z.-S. Efficient synthesis of hollow silica microspheres useful for porous silica ceramics. *Ceram. Int.* **2017**, *43*, 13907–13912. [CrossRef]
20. Abbas, N.; Khalid, H.R.; Ban, G.; Kim, H.T.; Lee, H. Silica aerogel derived from rice husk: An aggregate replacer for lightweight and thermally insulating cement-based composites. *Constr. Build. Mater.* **2019**, *195*, 312–322. [CrossRef]
21. Sdiri, A.; Higashi, T.; Bouaziz, S.; Benzina, M. Synthesis and characterization of silica gel from siliceous sands of southern Tunisia. *Arab. J. Chem.* **2014**, *7*, 486–493. [CrossRef]
22. Ismail, A.; Saputri, L.N.M.Z.; Dwiatmoko, A.A.; Susanto, B.H.; Nasikin, M. A facile approach to synthesis of silica nanoparticles from silica sand and their application as superhydrophobic material. *J. Asian Ceram. Soc.* **2021**, *9*, 665–672. [CrossRef]
23. Qian, H.; Li, W.; Wang, X.; Xie, F.; Li, W.; Qu, Q. Simultaneous growth of graphene/mesoporous silica composites using liquid precursor for HPLC separations. *Appl. Surf. Sci.* **2021**, *537*, 148101. [CrossRef]
24. Ma, H.; Li, H.; Xiong, Y.; Dong, F. Rational design, synthesis, and application of silica/graphene-based nanocomposite: A review. *Mater. Des.* **2021**, *198*, 109367. [CrossRef]
25. Azat, S.; Sartova, Z.; Bekseitova, K.; Askaruly, K. Extraction of high-purity silica from rice husk via hydrochloric acid leaching treatment. *Turk. J. Chem.* **2019**, *43*, 1258–1269. [CrossRef]
26. Zhu, Y.; He, Y.; Li, Z.; Zhang, J.; Shen, T.; Chen, Y.; Yang, D.-Q.; Sacher, E.; Jifan, Z. Synthesis of amorphous $SiO_2$ nanowires by one-step low temperature hydrothermal process. *Mater. Res. Express* **2019**, *6*, 115202. [CrossRef]
27. Meléndez-Ortiz, H.I.; Mercado-Silva, A.; García-Cerda, L.A.; Castruita, G.; Perera-Mercado, Y.A. Hydrothermal Synthesis of Mesoporous Silica MCM-41 Using Commercial Sodium Silicate. *J. Mex. Chem. Soc.* **2017**, *57*, 73–79. [CrossRef]
28. Park, J.; Gu, Y.; Park, S.; Hwang, E.; Sang, B.-I.; Chun, J.; Lee, J. Two-Stage Continuous Process for the Extraction of Silica from Rice Husk Using Attrition Ball Milling and Alkaline Leaching Methods. *Sustainability* **2021**, *13*, 7350. [CrossRef]

29. Sulistiyono, E.; Handayani, M.; Prasetyo, A.B.; Irawan, J.; Febriana, E.; Firdiyono, F.; Yustanti, E.; Sembiring, S.N.; Nugroho, F.; Muslih, E.Y. Implementation of sulfuric acid leaching for aluminum and iron removal for improvement of low-grade silica from quartz sand of Sukabumi, Indonesia. *East. Eur. J. Enterp. Technol.* **2021**, *3*, 32–40. [CrossRef]
30. Febriana, E.; Manurung, U.A.B.; Prasetyo, A.B.; Handayani, M.; Muslih, E.Y.; Nugroho, F.; Sulistiyono, E.; Firdiyono, F. Dissolution of quartz sand in sodium hydroxide solution for producing amorphous precipitated silica. *IOP Conf. Series: Mater. Sci. Eng.* **2020**, *858*, 012047. [CrossRef]
31. Gao, F.; Peng, Z.; Fu, X. One-Step Synthesis and Characterization of Silica Nano-/Submicron Spheres by Catalyst-Assisted Pyrolysis of a Preceramic Polymer. *J. Nanomater.* **2013**, *2013*, 5. [CrossRef]
32. Cho, K.; Chang, H.; Kil, D.S.; Park, J.; Jang, H.D.; Sohn, H.Y. Mechanisms of the Formation of Silica Particles from Precursors with Different Volatilities by Flame Spray Pyrolysis. *Aerosol Sci. Technol.* **2009**, *43*, 911–920. [CrossRef]
33. Avouris, P.; Dimitrakopoulos, C. Graphene: Synthesis and applications. *Mater. Today* **2012**, *15*, 86–97. [CrossRef]
34. Edwards, R.S.; Coleman, K.S. Graphene synthesis: Relationship to applications. *Nanoscale* **2013**, *5*, 38–51. [CrossRef]
35. Zhang, Z.; Fraser, A.; Ye, S.; Merle, G.; Barralet, J.E. Top-down bottom-up graphene synthesis. *Nano Futur.* **2019**, *3*, 042003. [CrossRef]
36. Potts, J.R.; Dreyer, D.R.; Bielawski, C.W.; Ruoff, R.S. Graphene-based polymer nanocomposites. *Polymers* **2011**, *52*, 5–25. [CrossRef]
37. Cooper, D.R.; D'Anjou, B.; Ghattamaneni, N.; Harack, B.; Hilke, M.; Horth, A.; Majlis, N.; Massicotte, M.; Vandsburger, L.; Whiteway, E.; et al. Experimental Review of Graphene. *ISRN Condens. Matter Phys.* **2012**, *2012*, 501686. [CrossRef]
38. Eswaraiah, V.; Aravind, S.S.J.; Ramaprabhu, S. Top down method for synthesis of highly conducting graphene by exfoliation of graphite oxide using focused solar radiation. *J. Mater. Chem.* **2011**, *21*, 6800–6803. [CrossRef]
39. Handayani, M.; Kepakisan, K.A.A.; Anshori, I.; Darsono, N.; Thaha, Y.N. Graphene oxide based nanocomposite modified screen printed carbon electrode for qualitative cefixime detection. *Proc. Int. Semin. Metall. Mater. Accel. Res. Innov. Metall. Mater. Incl. Sustain. Ind.* **2021**, *2382*, 40005. [CrossRef]
40. Kim, H.; Abdala, A.A.; MacOsko, C.W. Graphene/polymer nanocomposites. *Macromolecules* **2010**, *43*, 6515–6530. [CrossRef]
41. Febriana, E.; Handayani, M.; Susilo, D.N.A.; Yahya, M.S.; Ganta, M.; Sunnardianto, G.K. A simple approach of synthesis of graphene oxide from pure graphite: Time stirring duration variation. *Proc. Int. Semin. Metall. Mater. Accel. Res. Innov. Metall. Mater. Incl. Sustain. Ind.* **2021**, *2382*, 40006. [CrossRef]
42. Handayani, M.; Ganta, M.; Susilo, D.N.A.; Yahya, M.S.; Sunnardianto, G.K.; Darsono, N.; Sulistiyono, E.; Setiawan, I.; Lestari, F.P.; Erryani, A. Synthesis of graphene oxide from used electrode graphite with controlled oxidation process. *IOP Conf. Ser. Mater. Sci. Eng.* **2019**, *541*, 12032. [CrossRef]
43. Abbas, A.; Eng, X.E.; Ee, N.; Saleem, F.; Wu, D.; Chen, W.; Handayani, M.; Tabish, T.A.; Wai, N.; Lim, T.M. Development of reduced graphene oxide from biowaste as an electrode material for vanadium redox flow battery. *J. Energy Storage* **2021**, *41*, 102848. [CrossRef]
44. Chen, H.; He, G.; Zhu, J.; Bei, F.; Sun, X.; Wang, X. Synthesis and characterization of graphene paper with controllable properties via chemical reduction. *J. Mater. Chem.* **2011**, *21*, 14631–14638. [CrossRef]
45. Zhang, X.; Li, K.; Li, H.; Lu, J.; Fu, Q.; Chu, Y. Graphene nanosheets synthesis via chemical reduction of graphene oxide using sodium acetate trihydrate solution. *Synth. Met.* **2014**, *193*, 132–138. [CrossRef]
46. Toh, S.Y.; Loh, K.S.; Kamarudin, S.K.; Daud, W.R.W. Graphene production via electrochemical reduction of graphene oxide: Synthesis and characterisation. *Chem. Eng. J.* **2014**, *251*, 422–434. [CrossRef]
47. Liu, J.; Yang, H.; Zhen, S.G.; Poh, C.K.; Chaurasia, A.; Luo, J.; Wu, X.; Yeow, E.K.L.; Sahoo, N.G.; Lin, J.; et al. A Green Approach to the Synthesis of High-Quality Graphene Oxide Flakes via Electrochemical Exfoliation of Pencil Core. *RSC Adv.* **2013**, *207890*, 8669–8679. [CrossRef]
48. Parvez, K.; Li, R.; Puniredd, S.R.; Hernandez, Y.; Hinkel, F.; Wang, S.; Feng, X.; Mullen, K. Electrochemically exfoliated graphene as solution-processable, highly conductive electrodes for organic electronics. *ACS Nano* **2013**, *7*, 3598–3606. [CrossRef]
49. Wu, H.; Zhao, W.; Hu, H.; Chen, G. One-step in situ ball milling synthesis of polymer-functionalized graphene nanocomposites. *J. Mater. Chem.* **2011**, *21*, 8626–8632. [CrossRef]
50. Mondal, O.; Mitra, S.; Pal, M.; Datta, A.; Dhara, S.; Chakravorty, D. Reduced graphene oxide synthesis by high energy ball milling. *Mater. Chem. Phys.* **2015**, *161*, 123–129. [CrossRef]
51. Lin, C.; Yang, L.; Ouyang, L.; Liu, J.; Wang, H.; Zhu, M. A new method for few-layer graphene preparation via plasma-assisted ball milling. *J. Alloys Compd.* **2017**, *728*, 578–584. [CrossRef]
52. Li, X.; Magnuson, C.W.; Venugopal, A.; Tromp, R.M.; Hannon, J.B.; Vogel, E.M.; Colombo, L.; Ruoff, R.S. Large-Area Graphene Single Crystals Grown by Low-Pressure Chemical Vapor Deposition of Methane on Copper. *J. Am. Chem. Soc.* **2011**, *133*, 2816–2819. [CrossRef]
53. Li, X.; Magnuson, C.W.; Venugopal, A.; An, J.; Suk, J.W.; Han, B.; Borysiak, M.; Cai, W.; Velamakanni, A.; Zhu, Y.; et al. Graphene Films with Large Domain Size by a Two-Step Chemical Vapor Deposition Process. *Nano Lett.* **2010**, *10*, 4328–4334. [CrossRef]
54. Liu, H.; Liu, Y. Controlled Chemical Synthesis in CVD Graphene. *Phys. Sci. Rev.* **2017**, *2*, 1–28. [CrossRef]
55. Wu, Y.; Wang, B.; Ma, Y.; Huang, Y.; Li, N.; Zhang, F.; Chen, Y. Efficient and large-scale synthesis of few-layered graphene using an arc-discharge method and conductivity studies of the resulting films. *Nano Res.* **2010**, *3*, 661–669. [CrossRef]
56. Kim, S.; Song, Y.; Wright, J.; Heller, M.J. Graphene bi- and trilayers produced by a novel aqueous arc discharge process. *Carbon* **2016**, *102*, 339–345. [CrossRef]

57. Wu, Z.-S.; Ren, W.; Gao, L.; Zhao, J.; Chen, Z.; Liu, B.; Tang, D.; Yu, B.; Jiang, C.; Cheng, H.-M. Synthesis of Graphene Sheets with High Electrical Conductivity and Good Thermal Stability by Hydrogen Arc Discharge Exfoliation. *ACS Nano* **2009**, *3*, 411–417. [CrossRef] [PubMed]
58. Yi, G.; Xing, B.; Zeng, H.; Wang, X.; Zhang, C.; Cao, J.; Chen, L. One-Step Synthesis of Hierarchical Micro-Mesoporous $SiO_2$/Reduced Graphene Oxide Nanocomposites for Adsorption of Aqueous Cr(VI). *J. Nanomater.* **2017**, *2017*, 6286549. [CrossRef]
59. Oh, B.; Oh, J.-S.; Lee, E.-J.; Han, C.-M. Synthesis of uniformly dispersed silica/graphene oxide composite hydrogel using acid/base combinatorial catalysts system. *Mater. Today Commun.* **2021**, *26*, 101841. [CrossRef]
60. Haeri, S.; Asghari, M.; Ramezanzadeh, B. Enhancement of the mechanical properties of an epoxy composite through inclusion of graphene oxide nanosheets functionalized with silica nanoparticles through one and two steps sol-gel routes. *Prog. Org. Coatings* **2017**, *111*, 1–12. [CrossRef]
61. Wang, M.; Ma, L.; Li, B.; Zhang, W.; Zheng, H.; Wu, G.; Huang, Y.; Song, G. One-step generation of silica particles onto graphene oxide sheets for superior mechanical properties of epoxy composite and scale application. *Compos. Commun.* **2020**, *22*, 100514. [CrossRef]
62. Zhang, W.L.; Choi, H.J. Silica-Graphene Oxide Hybrid Composite Particles and Their Electroresponsive Characteristics. *Langmuir* **2012**, *28*, 7055–7062. [CrossRef] [PubMed]
63. Chen, L.; Chai, S.; Liu, K.; Ning, N.; Gao, J.; Liu, Q.; Chen, F.; Fu, Q. Enhanced Epoxy/Silica Composites Mechanical Properties by Introducing Graphene Oxide to the Interface. *ACS Appl. Mater. Interfaces* **2012**, *4*, 4398–4404. [CrossRef]
64. Ye, W.; Zhang, L.; Li, C. Facile fabrication of silica–polymer–graphene collaborative nanostructure-based hybrid materials with high conductivity and robust mechanical performance. *RSC Adv.* **2015**, *5*, 25450–25456. [CrossRef]
65. Liu, P.; Liang, Q.; Luo, H.; Fang, W.; Geng, J. Synthesis of nano-scale zero-valent iron-reduced graphene oxide-silica nano-composites for the efficient removal of arsenic from aqueous solutions. *Environ. Sci. Pollut. Res.* **2019**, *26*, 33507–33516. [CrossRef] [PubMed]
66. Wang, X.; Pei, Y.; Lu, M.; Lu, X.; Du, X. Highly efficient adsorption of heavy metals from wastewaters by graphene oxide-ordered mesoporous silica materials. *J. Mater. Sci.* **2015**, *50*, 2113–2121. [CrossRef]
67. Wang, X.; Wang, H.; Lu, M.; Teng, R.; Du, X. Facile synthesis of phenyl-modified magnetic graphene/mesoporous silica with hierarchical bridge-pore structure for efficient adsorption of pesticides. *Mater. Chem. Phys.* **2017**, *198*, 393–400. [CrossRef]
68. Liu, F.; Wu, Z.; Wang, D.; Yu, J.; Jiang, X.; Chen, X. Magnetic porous silica–graphene oxide hybrid composite as a potential adsorbent for aqueous removal of p-nitrophenol. *Colloids Surfaces A: Physicochem. Eng. Asp.* **2016**, *490*, 207–214. [CrossRef]
69. Wang, W.; Motuzas, J.; Zhao, X.S.; da Costa, J.C.D. 2D/3D amine functionalised sorbents containing graphene silica aerogel and mesoporous silica with improved $CO_2$ sorption. *Sep. Purif. Technol.* **2019**, *222*, 381–389. [CrossRef]
70. Zhao, B.; Sun, T.; Zhou, X.; Liu, X.; Li, X.; Zhou, K.; Dong, L.; Wei, D. Three-dimensional graphene composite containing graphene-$SiO_2$ nanoballs and its potential application in stress sensors. *Nanomaterials* **2019**, *9*, 438. [CrossRef]
71. Zhang, L.L.; Zhou, R.; Zhao, X.S. Graphene-based materials as supercapacitor electrodes. *J. Mater. Chem.* **2010**, *20*, 5983–5992. [CrossRef]
72. Ghosh, A.; Miah, M.; Majumder, C.; Bag, S.; Chakravorty, D.; Saha, S.K. Synthesis of multilayered structure of nano-dimensional silica glass/reduced graphene oxide for advanced electrochemical applications. *Nanoscale* **2018**, *10*, 5539–5549. [CrossRef]
73. Kim, M.; Kim, H.K. Ultraviolet-enhanced photodetection in a graphene/$SiO_2$/Si capacitor structure with a vacuum channel. *J. Appl. Phys.* **2015**, *118*, 104504. [CrossRef]
74. Du, Y.; Liu, L.; Xiang, Y.; Zhang, Q. Enhanced electrochemical capacitance and oil-absorbability of N-doped graphene aerogel by using amino-functionalized silica as template and doping agent. *J. Power Sources* **2018**, *379*, 240–248. [CrossRef]
75. Lu, L.; Han, X.; Li, J.; Hua, J.; Ouyang, M. A review on the key issues for lithium-ion battery management in electric vehicles. *J. Power Sources* **2013**, *226*, 272–288. [CrossRef]
76. Li, H.; Lu, C. Preparation of three-dimensional graphene networks for use as anode of lithium ion batteries. *Funct. Mater. Lett.* **2013**, *6*, 2–5. [CrossRef]
77. Yin, L.-H.; Wu, M.; Li, Y.-P.; Wu, G.-L.; Wang, Y.-K.; Wang, Y. Synthesis of $SiO_2$@carbon-graphene hybrids as anode materials of lithium-ion batteries. *New Carbon Mater.* **2017**, *32*, 311–318. [CrossRef]
78. Wang, M.-S.; Wang, Z.-Q.; Jia, R.; Yang, Y.; Zhu, F.-Y.; Yang, Z.-L.; Huang, Y.; Li, X.; Xu, W. Facile electrostatic self-assembly of silicon/reduced graphene oxide porous composite by silica assist as high performance anode for Li-ion battery. *Appl. Surf. Sci.* **2018**, *456*, 379–389. [CrossRef]
79. Kim, J.; Kim, D.; Ryu, J.H.; Yoon, S. One pot synthesis of ordered mesoporous carbon–silica–titania with parallel alignment against graphene as advanced anode material in lithium ion batteries. *J. Ind. Eng. Chem.* **2019**, *71*, 93–98. [CrossRef]
80. Kim, K.H.; Jun, Y.-S.; Gerbec, J.A.; See, K.A.; Stucky, G.D.; Jung, H.-T. Sulfur infiltrated mesoporous graphene–silica composite as a polysulfide retaining cathode material for lithium–sulfur batteries. *Carbon* **2014**, *69*, 543–551. [CrossRef]
81. Zhao, Z.; Zheng, L.; Hu, W.; Zheng, H. Synergistic effect of silane and graphene oxide for enhancing the photoelectrochemical water oxidation performance of WO3NS arrays. *Electrochim. Acta* **2018**, *292*, 322–330. [CrossRef]
82. He, D.; Li, X.; He, X.; Wang, K.; Tang, J.; Yang, X.; He, X.; Yang, X.; Zou, Z. Noncovalent assembly of reduced graphene oxide and alkyl-grafted mesoporous silica: An effective drug carrier for near-infrared light-responsive controlled drug release. *J. Mater. Chem. B* **2015**, *3*, 5588–5594. [CrossRef]

83. Gui, W.; Zhang, J.; Chen, X.; Yu, D.; Ma, Q. N-Doped graphene quantum dot@mesoporous silica nanoparticles modified with hyaluronic acid for fluorescent imaging of tumor cells and drug delivery. *Microchim. Acta* **2018**, *185*, 66. [CrossRef] [PubMed]
84. Song, Y.-Y.; Li, C.; Yang, X.-Q.; An, J.; Cheng, K.; Xuan, Y.; Shi, X.-M.; Gao, M.-J.; Song, X.-L.; Zhao, Y.-D.; et al. Graphene oxide coating core–shell silver sulfide@mesoporous silica for active targeted dual-mode imaging and chemo-photothermal synergistic therapy against tumors. *J. Mater. Chem. B* **2018**, *6*, 4808–4820. [CrossRef] [PubMed]
85. Shao, L.; Zhang, R.; Lu, J.; Zhao, C.; Deng, X.; Wu, Y. Mesoporous Silica Coated Polydopamine Functionalized Reduced Graphene Oxide for Synergistic Targeted Chemo-Photothermal Therapy. *ACS Appl. Mater. Interfaces* **2017**, *9*, 1226–1236. [CrossRef] [PubMed]
86. Yang, F.; Zhang, B.; Dong, S.; Tang, Y.; Hou, L.; Chen, Z.; Li, Z.; Yang, W.; Xu, C.; Wang, M.; et al. Silica nanosphere supported palladium nanoparticles encapsulated with graphene: High-performance electrocatalysts for methanol oxidation reaction. *Appl. Surf. Sci.* **2018**, *452*, 11–18. [CrossRef]
87. Nguyen, D.C.T.; Woo, J.-H.; Cho, K.Y.; Jung, C.-H.; Oh, W.-C. Highly efficient visible light driven photocatalytic activities of the LaCuS2-graphene composite-decorated ordered mesoporous silica. *Sep. Purif. Technol.* **2018**, *205*, 11–21. [CrossRef]
88. Oh, W.-C.; Nguyen, D.C.T.; Areerob, Y. Novel cadmium oxide-graphene nanocomposite grown on mesoporous silica for simultaneous photocatalytic H2-evolution. *Chemosphere* **2020**, *239*, 124825. [CrossRef] [PubMed]
89. Shang, L.; Bian, T.; Zhang, B.; Zhang, D.; Wu, L.-Z.; Tung, C.-H.; Yin, Y.; Zhang, T. Graphene-Supported Ultrafine Metal Nanoparticles Encapsulated by Mesoporous Silica: Robust Catalysts for Oxidation and Reduction Reactions. *Angew Chem.* **2014**, *126*, 254–258. [CrossRef]
90. Sarkar, C.; Pendem, S.; Shrotri, A.; Dao, D.Q.; Mai, P.P.T.; Ngoc, T.N.; Chandaka, D.R.; Rao, T.V.; Trinh, Q.T.; Sherburne, M.P.; et al. Interface Engineering of Graphene-Supported Cu Nanoparticles Encapsulated by Mesoporous Silica for Size-Dependent Catalytic Oxidative Coupling of Aromatic Amines. *ACS Appl. Mater. Interfaces* **2019**, *11*, 11722–11735. [CrossRef]

Article

# Preparation and Evaluation of Cu-Zn-GNSs Nanocomposite Manufactured by Powder Metallurgy

A. T. Hamed [1], E. S. Mosa [2], Amir Mahdy [2], Ismail G. El-Batanony [3] and Omayma A. Elkady [4,*]

[1] 15th of May Higher Institute of Engineering, Cairo 14531, Egypt; ashraftalaat@hccae.edu.eg
[2] Mining, Metallurgy and Petroleum Engineering Department, Faculty of Engineering, AL-Azhar University, Cairo 11651, Egypt; eisasalem.12@azhar.edu.eg (E.S.M.); amirmahdy.12@azhar.edu.eg (A.M.)
[3] Mechanical Engineering Department, Faculty of Engineering, Al-Azhar University, Cairo11651, Egypt; ismailghazy2017@azhar.edu.eg
[4] Powder Technology Department, Central Metallurgical Research and Development Institute (CMRDI), Cairo 11912, Egypt
* Correspondence: o.alkady68@gmail.com

**Citation:** Hamed, A.T.; Mosa, E.S.; Mahdy, A.; El-Batanony, I.G.; A. Elkady, O. Preparation and Evaluation of Cu-Zn-GNSs Nanocomposite Manufactured by Powder Metallurgy. *Crystals* **2021**, *11*, 1449. https://doi.org/10.3390/cryst11121449

Academic Editors: Cyril Cayron, Walid M. Daoush, Fawad Inam, Mostafa Ghasemi Baboli and Maha M. Khayyat

Received: 27 October 2021
Accepted: 18 November 2021
Published: 24 November 2021

**Publisher's Note:** MDPI stays neutral with regard to jurisdictional claims in published maps and institutional affiliations.

**Copyright:** © 2021 by the authors. Licensee MDPI, Basel, Switzerland. This article is an open access article distributed under the terms and conditions of the Creative Commons Attribution (CC BY) license (https://creativecommons.org/licenses/by/4.0/).

**Abstract:** Room-temperature ball milling technique has been successfully employed to fabricate copper-zinc graphene nanocomposite by high-energy ball milling of elemental Cu, Zn, and graphene. Copper powder reinforced with 1-wt.% nanographene sheets were mechanically milled with 5, 10, 15, and 20 wt.% Zn powder. The ball-to-powder weight ratio was selected to be 10:1 with a 400-rpm milling speed. Hexane and methanol were used as a process control agent (PCA) during composite fabrication. The effect of PCA on the composite microstructure was studied. The obtained composites were compacted by a uniaxial press under 700 MPa. The compacted samples were sintered under a controlled atmosphere at 1023 K for 90 min. The microstructure, mechanical, and tribological properties of the prepared Cu-Zn GrNSs nanocomposites were studied. All results indicated that composites using hexane as PCA had a uniform microstructure with higher densities. The densities of sintered samples were decreased gradually by increasing the Zn percent. The obtained composite contained 10 wt.% Zn had a more homogeneous microstructure, low porosity, higher Vickers hardness, and compression strength, while the composite contained 15 wt.% Zn recorded the lowest wear rate. Both the electrical and thermal conductivities were decreased gradually by increasing the Zn content.

**Keywords:** copper-zinc alloy; graphene nanosheets; microstructure; mechanical properties; electrical conductivity; thermal conductivity; wear rate

## 1. Introduction

The fabrication of Cu-matrix composite has attracted an increased interest, especially the fabrication of Cu-Zn alloy (brass), which has been widely used as an industrial material due to its excellent characteristics, such as: high corrosion resistance, non-magnetism, and good plasticity [1]. Cu-Zn alloy is significantly less expensive than copper, but unfortunately, has low strength properties, which can negate the economic advantage of brass [2,3]. Zinc plays a crucial role in the mechanical properties of copper-zinc alloys. For many working conditions, the copper-zinc alloys are subjected to the static tensile load or dynamic fatigue load. During the past decades, much of the research on the microstructure and mechanical properties of the copper-zinc alloys has been carried out, including the microstructure evolution and tensile plastic deformation, by using equal-channel angular pressing process, which demonstrated the increase in both plasticity and strength of refining grain of copper-zinc alloy with grain sizes less than 100 nm [4–6]. In recent years, engineering applications of copper-zinc alloys have increased steadily due to their attractive properties, such as high specific strength and good machinability. In order to improve the strength of Cu-Zn alloy, previous researchers have completed the addition of one or more alloying elements, such as Sn, Mn, Ni, Al, or Co.

An excessive effort has been made to develop a high-strength duplex brass alloy with small amounts of a chromium (Cr) additive by powder metallurgy technique [7]. After iron, aluminum, and copper, zinc is the fourth most widely used metal globally. In 2018, the global zinc supply increased to 13.4 Mt, with a global demand of 13.77 Mt. [8]. Significant amounts of zinc are recycled, and secondary zinc production is estimated to be 20–40% of global consumption [9]. However, due to of the strict limitation on impurities in die-casting composition standards, almost all zinc die-casting alloys are prepared from primary zinc production. In general, about one half of the consumed zinc finds its application in galvanizing steel, to prevent corrosion [10]. Other essential applications involve using zinc for other coatings, or as an alloying element in brasses, bronzes, aluminum, and magnesium alloys. Zinc is exploited as an oxide in chemical, pharmaceutical, cosmetics, paint, rubber, and agricultural industries. Zinc-based alloys offer a series of properties that make them particularly attractive for die-casting manufacturing and, in general, for foundry technologies [11]. The large precipitates reduce the ductility of this alloy. On the other hand, less than one mass% Cr addition in the brass alloy can prevent many precipitations, which causes a remarkable decrease in ductility and machinability. It is possible to produce a high-strength brass alloy with a supersaturated solid solution using the rapid solidification method. Brass alloys were prepared using rapid solidification of the ternary Cu-40 mass% Zn-0.5 mass% Cr alloy powder. The effect of solid solute Cr behavior in the consolidated materials on microstructures and mechanical properties was investigated [12]. Mechanical milling has received an increased interest as a simple and environmentally friendly alternative to high temperature [13]. It is well known that ball milling elemental powders induces a solid-state reaction through the atomic mixing of the components, which has been used to synthesize various equilibrium and non-equilibrium alloy powders with extremely fine microstructures. To avoid the formation of an oxide in the preparation process, mechanochemical reactions of Cu and its oxides under several atmospheres have been investigated separately. The advantage of this technique is the comparison with other fabrication techniques in the ability to synthesis Cu-Zn alloy without oxide phases, and the uniform distribution of the reinforcement particles in one-step [14]. Cu-Zn alloy was prepared by high-energy ball milling of elemental copper and zinc powders by the attrition mill. The different parameters, such as milling time, ball-to-powder ratio, and milling speeds, were optimized. The results show that different milling parameters can produce the different Cu-Zn alloy phases. It has been found that milling time is highly significant to the refining process, and the ratios of the ball to powder also benefit the new phases formed. Copper-based composites are widely used in various applications, and the Cu-Zn system is rich in equilibrium and non-equilibrium phases such as α, β, γ, η, and β′ [15]. In addition, different compositions of many copper-zinc alloys, such as 70 Cu-30 Zn, 75 Cu-25 Zn, and 80 Cu-20Zn alloys, have already been studied as a function of strain amplitude during the load interruptions, and the characteristic shapes of these loops were considered to arise from zinc segregation to dislocations as a function of cycle strain. Although the damage mechanism of copper-zinc alloys has already been obtained by many researchers, the quantitative effects of the proportion of zinc on the deformation behavior, plastic work consumption, and strain-life of copper-zinc alloys during static tensile tests and dynamic fatigue tests were seldom mentioned. C11000 copper and H63 copper-zinc alloy were adopted to investigate the quantitative effects of zinc on the static and dynamic mechanical properties of copper-zinc alloys. The deformation behavior, plastic work consumption, and strain-life of the C11000 copper and H63 copper-zinc alloy during static tensile and dynamic fatigue tests were investigated. During the tensile testing, C11000 copper shows apparent plastic deformation behavior with a tensile strength of 270.1 MPa and elongation of 19.6%. H63 copper-zinc alloy shows the obvious brittle deformation behavior with a tensile strength of 396.8 MPa and elongation of 3.4% [16].

The present work aims to offer simple and suitable techniques for fabrication Cu-Zn GNSs nanocomposites by powder metallurgy techniques, including mechanical milling, cold compaction, sintering process, and studying the effect of hybrid reinforcement of copper with constant percent of GNSs and a variable percentage of Zn for different mechanical applications.

## 2. Experimental Work

Elemental powders of high purity (99.94%) Cu (75 μm), Zinc 99% purity (75 μm), and 1 wt.% of high purity (99.95%) Graphene nanosheets with 50 nm particle size were used as starting reactant materials. Figure 1 showed the microstructure of the used powders, in which graphene had a nanosheet structure (Figure 1a), while copper had a dendritic structure (Figure 1b), and zinc had irregular spheres-like morphology (Figure 1c).

(a)

(b)

(c)

**Figure 1.** SEM micrographs for Graphene nanosheets (**a**), Copper (**b**), and Zinc powder (**c**).

The powders were mixed to give the desired composition and then sealed in a stainless steel (SUS 316) vial (80 mL in volume) together with 10 stainless steel (SUS 316) balls (12 mm in diameter) in a glove box under an Ar gas atmosphere ($O_2$ and $H_2O$ are less than 10 ppm). The ball-to-powder weight ratio was selected to be 10:1. The milling procedure was carried out at room temperature, using a high-energy ball mill (FRITACH P6) for 12 hr. milling time with 400-rpm milling speed.

Hexane or methanol was used as a PCA (to choose the preferable one, Cu with 1wt.% GNSs was mixed mechanically with 10 wt.% of hexane or methanol). Copper GNSs was reinforced with 5, 10, 15, and 20-wt % Zinc powder (Table 1). The mixed composite powders were compacted in a uniaxial single hydraulic press using stainless steel die under 700 MPa pressure. Next, the compacted samples were sintered in a vacuum furnace at 1023 K for 90 min., in which there were three holding steps. During the first step (degassing step), the applied heating rate was 276 K/min. and holed in 523 K. The second step (melting of Zn) was headed with the same heating rate and holed in 673 K. Finally, the third step (sintering process) was headed with a heating rate of 278 K/min. and holed in 1023 K for the complete sintering process.

**Table 1.** The composition of fabricated composites.

| Alloys | Cu wt.% | Zn wt.% | GNSs wt.% |
| --- | --- | --- | --- |
| Cu94Zn5GNSs1 | 94 | 5 | 1 |
| Cu89Zn10GNSs1 | 89 | 10 | 1 |
| Cu84Zn15GNSs1 | 84 | 15 | 1 |
| Cu79Zn20GNSs1 | 79 | 20 | 1 |

The phase structure and composition of the sintered Cu-Zn GNSs were characterized by scanning electron microscope (SEM) and X-ray diffraction (XRD). For microstructure investigation, the samples were ground with 220, 400, 600, 800, 1000, 1200, 2000, and 3000 grit SiC paper and polished with 6-micron alumina paste. The microstructure was studied using an Optical microscope (model Leco LX 31, camera PAX-Cam). Additionally, a field emission scanning electron microscope (models FEI Inspect S 50) was used. The densities of the sintered samples were estimated according to Archimedes' principle, using water as a floating liquid. The sintered samples were weighed in air and in distilled water, and the actual density ($\rho_{act.}$) was calculated according to the Equation (1):

$$\rho_{act.} = \frac{W_a}{W_a - W_w} \quad (1)$$

where $W_a$ and $W_w$ were the weight of the sample in air and water, respectively. The theoretical density ($\rho_{th.}$) for the investigated composite was determined according to the following Equation (2):

$$\rho_{th.} = (V_M * \rho_M) + (V_R * \rho_R) \quad (2)$$

where $V_M$ and $R_M$ were the volume fraction and density of the matrix, respectively, while $V_R$ and $\rho_R$ were those for the reinforcement sample, respectively [17].

$$\text{Relative Density} = \rho_{act.}/\rho_{th.} \quad (3)$$

XRD device of the (model D8 XPORT) was used to emphasize the chemical composition and any new chemical compound or intermetallic formed between the constituent of the composition, and to study the crystal structure of the sintered composition. Vickers hardness was measured using Vickers microhardness tester (D-6700 Wolpert, Meisenweg, Germany) at a load of 10 kg/f, and an indentation time of 10 s for all specimens. The reported Vickers hardness values of the specimens were represented by the average of five readings of each sample. Compression strength test of the investigated samples was performed using a micro-computer controlled uniaxial universal testing machine (WDW300). The samples used for compression tests had a 10 mm cross-section and a height of 15 mm. The applied crosshead speed of the universal test machine used was 2 mm/min, and the tests were conducted at room temperature.

The electrical conductivity, resistivity, and (International Annealed Copper Standard) IACS% were estimated for the sintered sample. The test was established using Material Tester for Metal, PCE-COM20. Next, the thermal conductivity was calculated using the Wiedemann and Franz equation, which is a relationship between electrical and thermal conductivity [14]. The Wiedmann-Franz relation is shown in the following equation:

$$K/\sigma = LT$$

where K is the thermal conductivity in w/m·k, $\sigma$ is the electrical conductivity s/m, L is Lorenz constant which equals $2.44 \times 10^{-8}$ w x $\Omega$ x $k^{-2}$ value, and T is the absolute temperature in Kelvin.

The adhesive wear was carried out using the Tribometer pin on a ring testing machine under normal loads of 10, 20, and 30 N, at 150, 300, and 450 rpm, respectively, during the sliding process. The adhesive wear of the pin was determined as the weight loss divided the time to determination wear rate per unit second. A sensitive electronic balance was used to measure weight loss.

## 3. Results and Discussions

The density results of Cu-1% GNSs with hexane and methanol indicated that the hexane sample had a higher density than the methanol sample, as shown in Table 2.

**Table 2.** Relative densities and process control agent (PCA) used.

| Sample | Process Control Agent (PCA) | Relative Density |
|---|---|---|
| Cu-1 wt.% GNSs | Hexane | 86.35 |
| Cu-1 wt.% GNSs | Methanol | 76.32 |

Furthermore, microstructure indicated the good homogeneity of the hexane sample rather than the methanol sample, as shown in Figure 2.

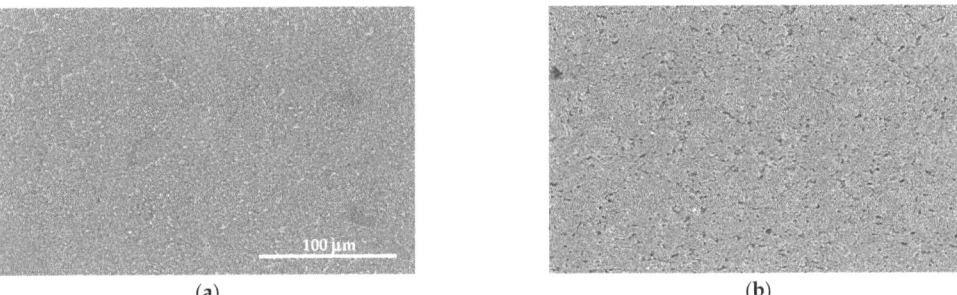

Figure 2. The microstructure of the Cu-1wt.% GNSs for hexane (a) and methanol (b), respectively.

From the above results (Table 2), we found the use of hexane was better than using methanol as a PCA. In the present study, all CU-Zn GNSs composites were manufactured using hexane as a PCA. This result was mentioned previous in prior work [18].

*3.1. Optical Micrographs*

Figure 3 shows the microstructure of Cu-Zn GNSs composites estimated by optical microscope, in which a, b, c, and d represented Cu-Zn-1% GNSs with 5, 10, 15, and 20 wt.% Zn percentage, respectively. GNSs were well distributed throughout the copper matrix. Samples containing 10 wt.% Zn had the most homogenous and uniform microstructure, which was confirmed by the SEM images and the density values.

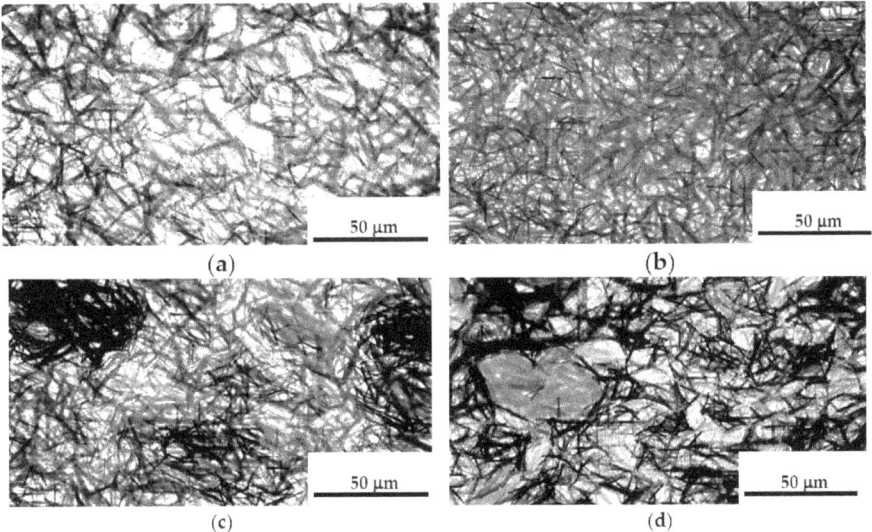

Figure 3. Optical micrographs of sintered Cu-Zn GNSs nanocomposites: (a) Cu-Zn 5 wt.% + Graphene 1 wt.%; (b) Cu-Zn 10 wt.% + Graphene 1 wt.%; (c) Cu-Zn 15 wt.% + Graphene 1 wt.%; and (d) Cu-Zn 20 wt.% + Graphene 1 wt.%.

*3.2. Microstructure Examination*

Figure 4 showed the microstructure of Cu-Zn (5, 10, 15, and 20 wt.% zinc)/1 wt.% GNSs nanocomposites (a, b, c, and d), respectively. For all samples there were three areas: white, gray, and black. The white area represented the Zn metal; the gray area represented the Cu matrix; and the black area represented the GNSs and the pores. It could be noted that, samples containing 10 wt.% Zn had a good microstructure, good homogeneity between GNSs, and a Cu-Zn matrix with the lowest porosity. By increasing the Zn contents, some aggregations transpired that causes the formation of pores, as recorded in the density results.

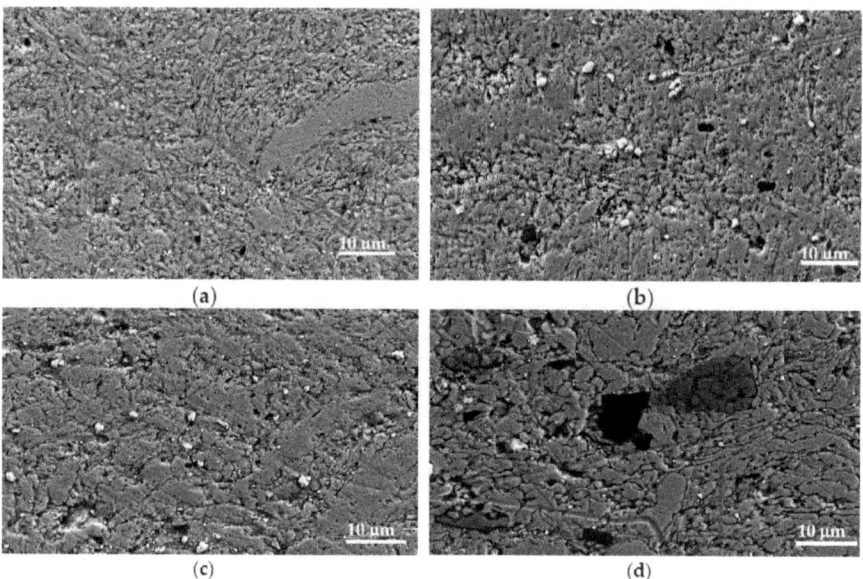

**Figure 4.** SEM of sintered Cu-Zn GNSs composites: (**a**) Zn 5 wt.%, (**b**) Zn 10 wt.%, (**c**) Zn 15 wt.%, and (**d**) Zn 20 wt.%.

*3.3. EDX Analysis*

Figure 5a–d showed the EDX analysis of Cu-Zn (5, 10, 15, and 20)/1 wt.% GNSs samples, respectively. It was clear that the specimens had perfect homogenous dispersion with a smaller number of GNSs agglomerations, due to the good mixing process between Cu-Zn and GNSs. Furthermore, the EDX patterns of all samples referred to the presence of Cu, Zn and C (that belong to GNSs) in a good homogeneity (Figure 5). The percent of each constitute were near to the added ones, which indicated the good processing parameters. This could be due to the suitable PCA in the mechanical milling process with good sintering conditions suitable for Cu-Zn solid solution formation.

*3.4. Relative Density*

Table 3 Showed the effect of Zn metal additions on the relative density of Cu matrix reinforced with 1 wt.% GNSs. The density increased by increasing Zn value from 5 up to 10 wt.%, then decreased gradually by increasing Zn up to 20 wt.%. Generally, decreasing the density could be attributed to the lower density values of Zn (7.14 g/cc) and GNSs (2.2 g/cc) than that of Cu (8.96 g/cc). As such, the addition of light material to a heavier one decreased the overall density [19]. Furthermore, the mismatch between GNSs and the Cu-Zn brass alloy was due to the high surface energy and the non-wettability problem between GNSs and its ceramic nature with the metallic brass Cu-Zn alloy [20]. This was causing a gap between the internal particles, which formed voids that decreased the density. However, the 10 wt.% Zn sample had the highest density value, as it had the best homogenous microstructure; both Zn and GNSs were well distributed on the Cu matrix, and 10 wt.% Zn percentage was the most suitable for Cu-Zn alloy formation.

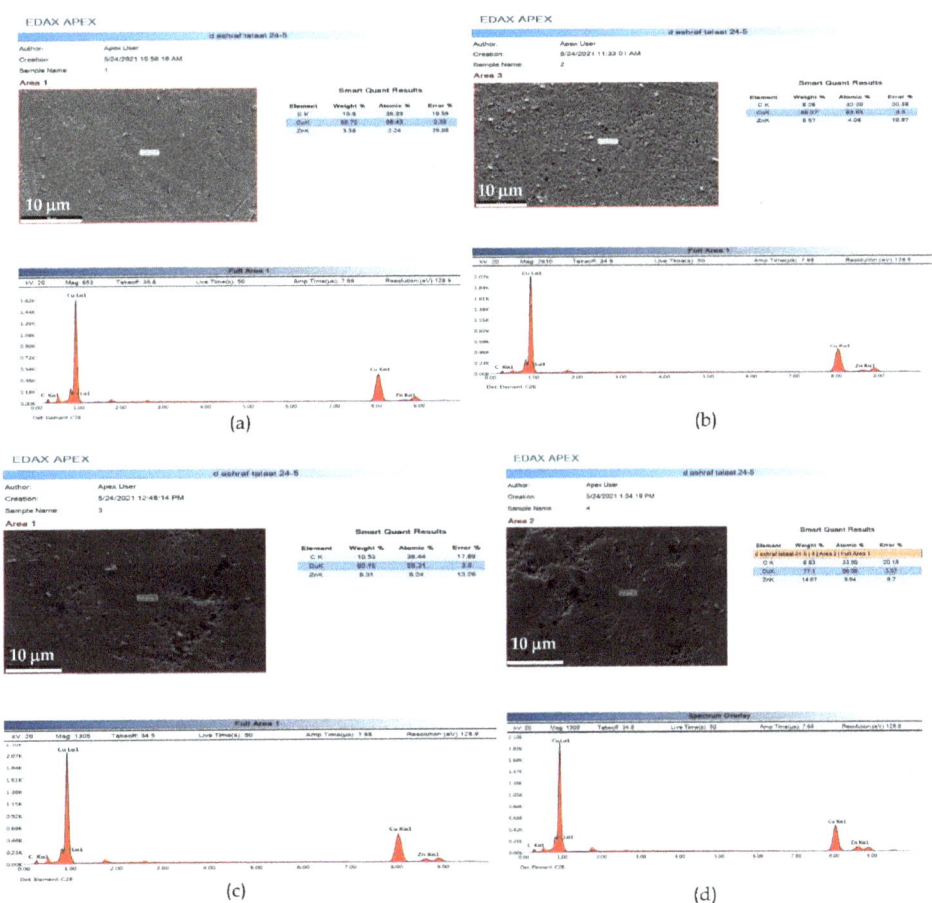

**Figure 5.** EDX of sintered composite samples: (**a**) Cu94Zn5GNSs1, (**b**) Cu89Zn10GNSs1, (**c**) Cu84Zn15GNSs1, and (**d**) Cu79Zn20GNSs1.

**Table 3.** Relative density measured value for obtained composites.

| Alloys | Relative Density % |
| --- | --- |
| Cu94Zn5GNSs1 | 89.90 |
| Cu89Zn10GNSs1 | 90.64 |
| Cu84Zn15GNSs1 | 87.41 |
| Cu79Zn20GNSs1 | 85.82 |

### 3.5. XRD Analysis

XRD pattern for obtained nanocomposites (Figure 6) showed the phase composition and structure of the manufactured Cu-Zn GNSs. For 5 and 10 wt.% Zn samples, only the phases corresponding to Cu and Zn were observed, while for the higher percentages of Zn (15 and 20 wt.%), new peaks were recorded corresponding to the β (Cu-Zn) solid solution. This could be explained according to the phase rule and phase diagram between Cu and Zn. Cu-Zn was an important binary alloy system. In the interested temperature ranges from 300 to 1500 K, there were eight phases: liquid, Cu, β, β′, γ, δ, ϵ, and Zn phases, as shown in Figure 7. A new description of the liquid phase and a simplified description of the body-centered cubic (bcc) phase was proposed, in which a solution of a solid in a solid occurred for a certain concentration of both alloys at a certain temperature. Two metals were combined in a solid solution, such as zinc in copper, where the Zn atoms replaced the Cu atoms in the unit cell, leading to the formation of Cu-Zn solid solution. For the small Zn percent, the Cu-Zn beta phase was formed, but its percent was lower than the XRD device limits.

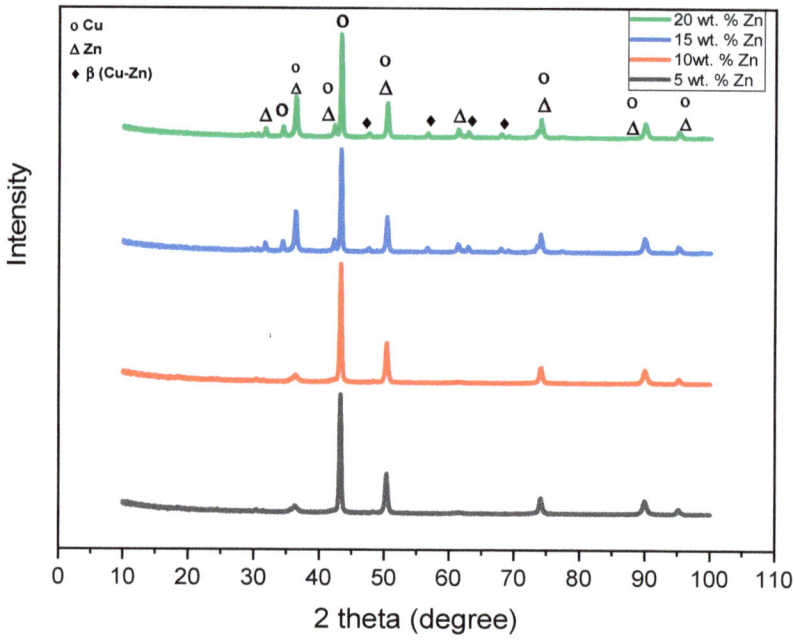

**Figure 6.** XRD patterns of the sintered Cu-Zn-GNSs composite.

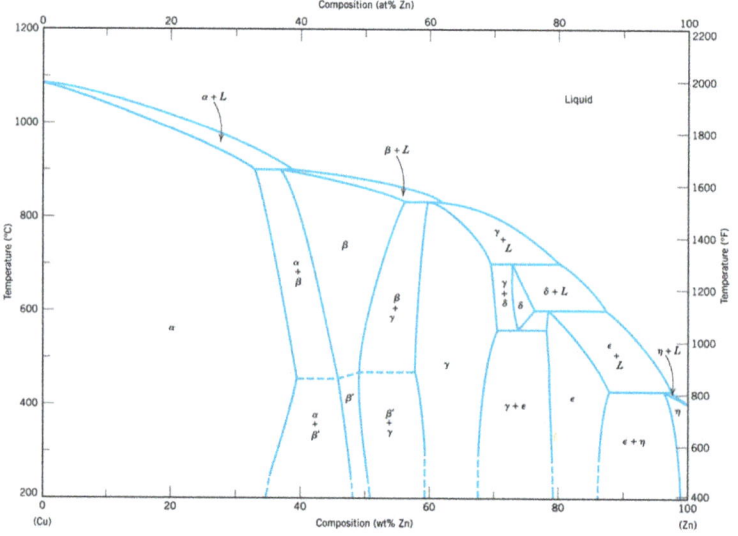

**Figure 7.** Phase diagram of Cu-Zn alloy.

## 3.6. Vickers Hardness

Figure 8 presented the effect of Zn additions on Vickers hardness values of Cu matrix reinforced with 1 wt.% GNSs. The Figure demonstrated that the hardness of all prepared samples was higher than that of pure annealed Cu (40 Hv). As such, reinforcing Cu with 1 wt.% GNSs and Zn metal improved the hardness of Cu up to 78.1 Hv for 10 wt.% Zn sample. This may be attributed to the high strength of GNSs, which dispersed homogeneously on the Cu matrix. GNSs had superior properties, such as being super-flexible, super-strong, super-light, and super-thin. Owing to all these extraordinary properties, reinforcing any ductile metal with GNSs improved its mechanical properties, especially hardness. GNSs were the hardest material known, until now. It had a tensile strength of 130 GPa, and as such, the addition of GNSs to Cu-Zn alloy enhanced the microhardness. A Zn sample has the highest hardness with a 10 wt.%, possibly attributed to its highest density and more uniform microstructure. Although the hardness of 15 and 20 wt.% Zn was decreased, it was still higher than that of pure Cu. This could be attributed to reinforcing Cu with a hard ceramic GNSs and Zn metal that formed a solid solution that gave strength to the samples [21].

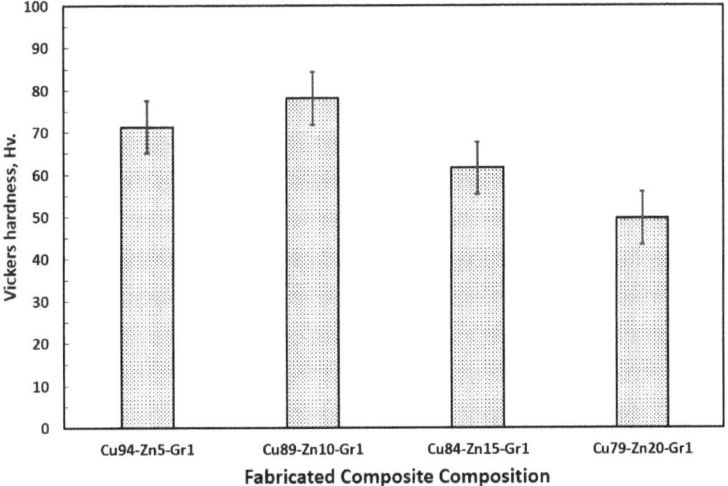

**Figure 8.** Dependence of the Vickers hardness of bulk composite samples.

## 3.7. Compression Strengths Estimation

Figure 9 showed the effect of 1 wt.% GNSs and Zn additions on the compression strength of Cu-Zn-GNSs nanocomposites. The Figure showed that compression strength increased by Zn addition up to 10 wt.% sample, which subsequently decreased by increasing the Zn percentage up to 20 wt.% Zn. This may be explained according to the higher density and hardness of the 10 wt.% Zn sample. Furthermore, the good homogeneous structure of GNSs and Zn was throughout the Cu matrix. Meanwhile, by increasing the Zn percentage, the density decreased, and consequently, porosity increased. As such, the availability of material cracking was increased as the internal pores were considered as a center for crack initiation and spreading [22]. It must be mentioned that dispersing GNSs in the Cu-Zn alloy improved the overall strength of the manufactured samples due to the good mechanical milling parameters that helped in its dispersing and distribution in the Cu matrix. The strength of the Cu-Zn alloy matrix was improved, which made the sample able to withstand the high mechanical loads.

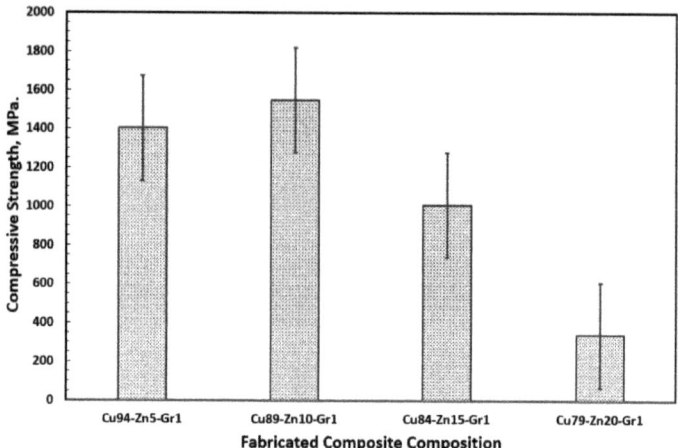

**Figure 9.** Effect of compositions on compressive strengths of obtained composite samples.

*3.8. Electrical Conductivity Measurements*

Figure 10 showed the effect of Zn additions and graphene nanosheets on the electrical conductivity of Cu-Zn-GNSs composites. A gradual decrease in the electrical conductivity occurred by increasing Zn percentages. This could be attributed to the lower electrical conductivity value of Zn than that of Cu ($1.63 \times 10^7$ and $5.96 \times 10^7$ s/m for Zn and Cu, respectively). The incorporation of low conductive particles with low free electrons to a conductive matrix restricted the electron motion and decreased the conductivity value. However, although the conductivity of Cu-Zn GNSs composites decreased, the 5 and 10 wt.% Zn samples had a high conductivity value, which was already in the application ranges of Cu. Furthermore, the presence of GNSs helped in enhancing the electrical conductivity of the prepared samples.

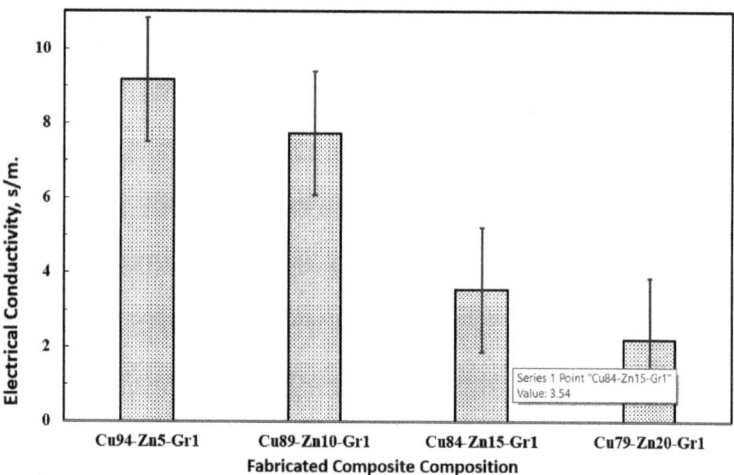

**Figure 10.** Effect Zn additions have on the electrical conductivity of Cu-Zn-GNSs nanocomposites.

*3.9. Thermal Conductivity Estimation*

Figure 11 showed the effect of Zn additions on the thermal conductivity of Cu-Zn GNSs composites. The Figure showed gradual decreases in the thermal conductivity by Zn increment. This could be attributed to more than one reason; the first is a decreased thermal conductivity value of Zn than that of Cu, which was 401 (w/m·k) for Cu and 116 (w/m·k) for Zn. As such, the presence of Zn particles with low thermal conductivity decreased the conductivity of the prepared samples [23]. The second reason was the formation of pores by the addition of Zn and GNSs, which also had a negative effect on the thermal conductivity as the conductivity of any pore is zero, which restricted the heat carrier mobility.

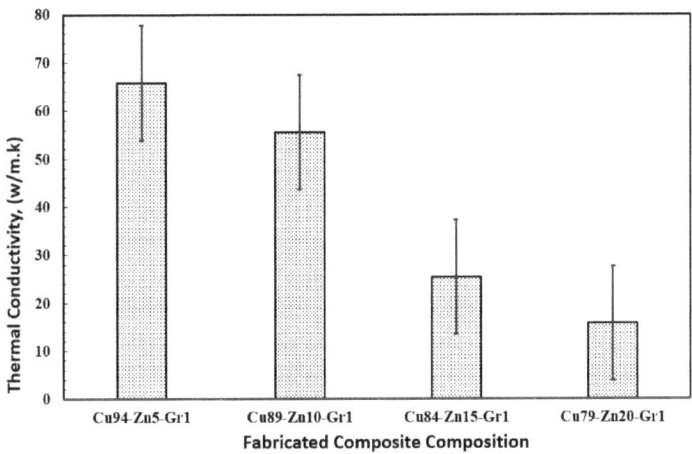

**Figure 11.** Effect of Zn additions on the thermal conductivity of Cu-Zn-GNSs composites.

*3.10. Wear Behavior*

Figures 12–14 showed the effects of Zn additions, applied load, and milling speed on the wear rate of the manufactured Cu-Zn GNSs composites. From Figures 12–14, it was clear that as the applied load increased, the wear rate increased, and by increasing the milling speed (rpm), the wear rate also increased. This could be explained by the fact that the contact area between the pin and the sample's surface increased, giving the pin more chance to wear more area from the contact surface, and consequently, the wear rate increased [24]. By increasing the rpm, the increased wear rate could be explained by the fact which stated that increasing (rpm) led to an increase in the frictional track on the sample's surface that increases the grooves and wear rate [25]. Another phenomenon from the wear rate figures could be observed: the decreased wear rate from increasing the zinc ratio up to the 15 wt.% that recorded the lowest wear rate value, then increased for the 20 wt.% sample. This was for both the 150 and the 450 rpm groups for the three applied loads (10, 20, and 30 N). This could be explained by the addition of a ceramic or lubricant material, such as the kind GNSs gives to increase strength to the ductile Cu with its good distribution [19,26]. Furthermore, low density (2.2 g/cc) caused the collection of it on the sample's surface, increasing the sliding of the pin on the surface, consequently decreasing the wear rate. The presence of GNSs caused the formation of tribo-layer on the sample's surface that decreased the wear rate. For the 20 wt.% Zn sample, some agglomerations of GNSs occurred that increased the porosity, which consequently caused the wear rate to increase. Zinc formed a solid solution with Cu and provided more strength by forming an alloy that resisted wear and corrosion. However, for 300 rpm, the wear rate decreased by increasing the Zn percentage gradually to 20 wt.% Zn sample, which recorded the lowest wear rate [27].

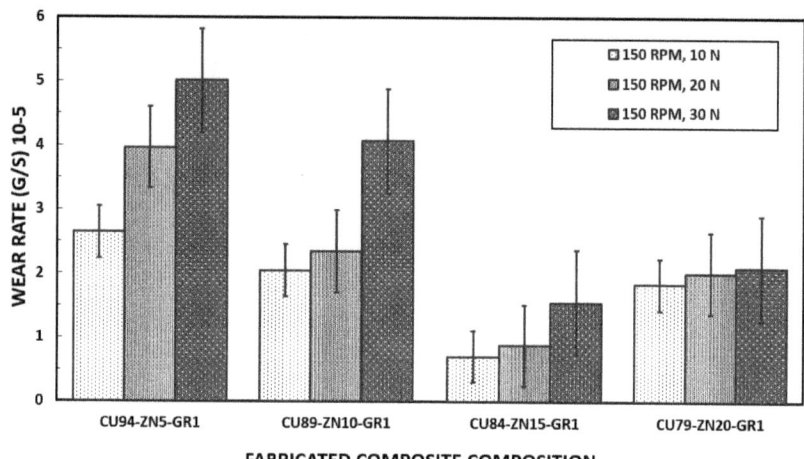

Figure 12. Effect of different loads and rpm on the wear rate of obtained composite sample.

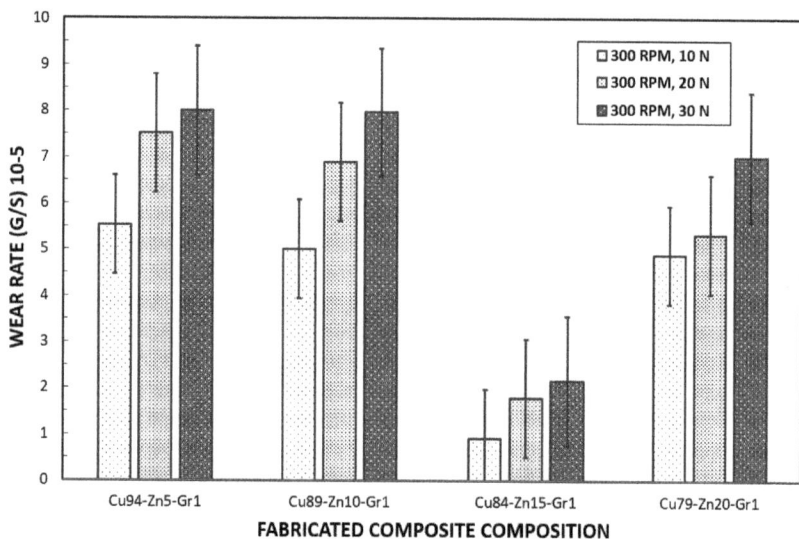

Figure 13. Effect of different loads and rpm on the wear rate of obtained composite samples.

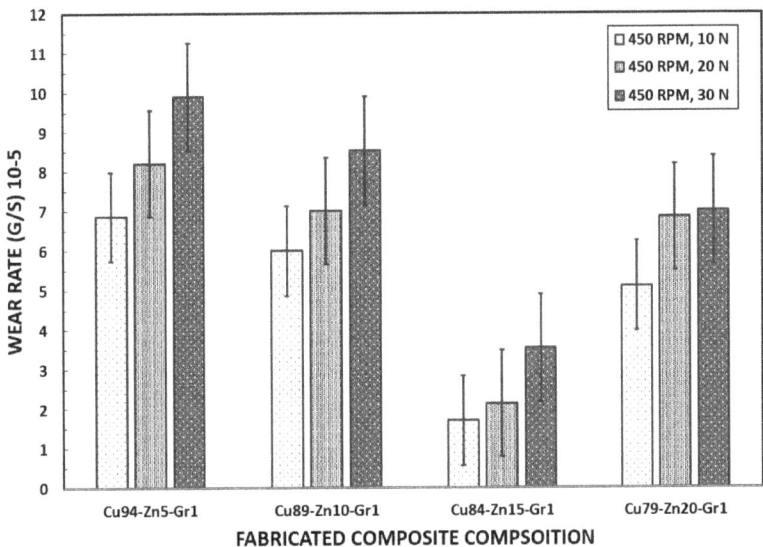

**Figure 14.** Effect of different loads and rpm on the wear rate of obtained composite samples.

## 4. Conclusions

In this research, Cu-Zn GNSs nanocomposites were successfully produced by powder metallurgy methods, including ball milling technique, followed by compacting steps using uniaxial press. The results showed that bulk nanocomposites prepared using hexane as PCA had a uniform microstructure with no evidence for the presence of voids. Furthermore, the addition of 10 wt.% zinc increased the densities of obtained nanocomposite materials, which ultimately decreased with further additions of zinc.

XRD revealed the formation of beta phase Cu-Zn solid solution for higher percentages of Zn (15 and 20 wt.%). Obtained nanocomposites containing 10 wt.% zinc recorded a higher value of Vickers hardness (78.1 Hv), while those with 20 wt.% zinc recorded (49.6 Hv).

Good dispersing GNSs in the Cu-Zn alloy improved the overall strength of the manufactured nanocomposites, with higher values of compressive strength for composite containing 10% zinc.

Both the electrical and thermal conductivities for obtained nanocomposites decreased gradually by increasing zinc content. In addition, the nanocomposite having 15 wt.% zinc recorded the lowest wear rate.

Cu-Zn GNSs-CNTs hybrid nanocomposites were prepared using mechanical ball milling technique. We will discuss the effect of the addition of CNTs with GNSs in the near future.

**Author Contributions:** Writing—original draft preparation, A.T.H., E.S.M. and A.A.M.; writing—review and editing, I.G.E.-B. and O.A.E. All authors have read and agreed to the published version of the manuscript.

**Funding:** This research received no external funding.

**Data Availability Statement:** Not applicable.

**Conflicts of Interest:** The authors declare no conflict of interest.

## References

1. Yusoff, M.; Mohamad, M.; Abu Bakar, M.B.; Masri, M.N.; Noriman, N.Z.; Dahham, O.S.; Umar, M.U. Copper Alloy Reinforced by Graphene by Powder Metallurgy Technique. *AIP Conf. Proc.* **2020**, *2213*, 020264.
2. Pola, A.; Rollez, D.; Prenger, F. Zinc Alloy Family for Foundry Purposes. *World Metall. Erzmetall* **2015**, *68*, 354–358.
3. Yang, G.S.; Lee, J.K.; Jang, W.Y. Mechanical Properties and Microstructure Observation with Grain Refinement in CuZnAl Alloy. *Mater. Sci. Forum Jan.* **2008**, *569*, 173–176. [CrossRef]
4. Rajeshkumara, L.; Suriyanarayanan, R.; Hari, K.S.; Babu, S.V.; Bhuvaneswari, V.; Karunan, M.P.J. Influence of boron carbide addition on particle size of copper zinc alloys synthesized by powders metallurgy. *Mater. Sci. Eng.* **2020**, *954*, 1–9. [CrossRef]
5. Traleski, A.V.; Vurobi, S., Jr.; Cintho, O.M. Osvaldo Mitsuyuki Cintho, Processing of Cu-Al-Ni and Cu-Zn-Al Alloys by Mechanical Alloying. *Mat. Sci. Forum* **2012**, *727–728*, 200–205.
6. Da Silva, F.C.; Kazmierczak, K.; da Costa, C.-E.; Milan, J.C.G.; Torralba, J.M. Zamak 2 Alloy Produced by Mechanical Alloying and Consolidated by Sintering and Hot Pressing. *J. Manuf. Sci. Eng.* **2017**, *139*, 1–7. [CrossRef]
7. Imai, H.; Li, S.; Kondoh, K.; Kosaka, Y.; Okada, K.; Yamamoto, K.; Takahashi, M.; Umeda, J. Microstructure and Mechanical Properties of Cu-40%Zn-0.5%Cr Alloy by Powder Metallurgy. *Mater. Trans.* **2014**, *55*, 528–533. [CrossRef]
8. Global Zinc Market to Grow at 3.8% in 2022. Available online: https://www.mining-technology.com/comment/zinc-outlook-2019/ (accessed on 5 January 2020).
9. Lynch, R.F. Zinc: Alloying, Thermomechanical Processing, Properties, and Applications. In *Encyclopedia of Materials: Science and Technology*; Elsevier: Amsterdam, The Netherlands, 2001; pp. 9869–9883.
10. Marder, A.R. Metallurgy of zinc-coated steel. *Prog. Mater. Sci.* **2000**, *45*, 191–271. [CrossRef]
11. Levy, G.K.; Goldman, J.; Aghion, E. The prospects of zinc as a structural material for biodegradable implants: A review paper. *Metals* **2017**, *7*, 402. [CrossRef]
12. Pola, A.; Tocci, M.; Goodwin, F.E. Review of Microstructures and Properties of Zinc Alloys. *Metals* **2020**, *10*, 253. [CrossRef]
13. Panigrahi, M.; Avar, B. Influence of mechanical alloying on structural, thermal, and magnetic properties of Fe50Ni10Co10Ti10B20 high entropy soft magnetic alloy. *J. Mater. Sci. Mater. Electron.* **2021**, *32*, 21124–21134. [CrossRef]
14. Wang, K.J.; Cai, X.L.; Wang, H.; Hu, J.; Zhang, Y.F. Preparation of Cu-Zn Alloy by Different High Energy Ball Milling. *Adv. Mater. Res.* **2012**, *12*, 259–262.
15. Kaijun, W. Preparation of Cu-Zn Alloy by High Energy Ball Milling of Elemental Copper and Zinc. *Adv. Mater. Res.* **2011**, *148–149*, 1413–1416.
16. El-Khatib, S.; Shash, A.Y.; Elsayed, A.H.; El-Habak, A. Effect of carbon nano-tubes and dispersions of SiC and $Al_2O_3$ on the mechanical and physical properties of copper-nickel alloy. *Heliyon* **2018**, *4*, e00876. [CrossRef]
17. Zhi-Chao, M.A. Effects of zinc on static and dynamic mechanical properties of copper-zinc alloy. *J. Cent. South Univ.* **2015**, *22*, 2440–2445.
18. Hamed, A.; Mosa, E.; Mahdy, A.; El-Batanony, I.; Alkady, O. Effect of process controlling agent on the microstructure, and mechanical properties of copper/graphene composite. In Proceedings of the Al Azhar Engineering Fifteenth International Conference, Cairo, Egypt, 13–15 March 2021.
19. Fathy, A.; Elkady, O.; Abu-Oqail, A. Microstructure, mechanical and wear properties of Cu-$ZrO_2$ nanocomposites. *Mater. Sci. Technol.* **2017**, *33–17*, 2138–2146. [CrossRef]
20. Yehia, H.M. Effect of graphene nano-sheets content and sintering time on the microstructure, coefficient of thermal expansion, and mechanical properties of (Cu/WC-TiC-Co) nano-composites. *J. Alloys Compd.* **2018**, *764*, 36–43. [CrossRef]
21. Fathy, A.; Elkady, O.; Abu-Oqail, A. Synthesis and characterization of Cu-$ZrO_2$ nanocomposite produced by thermochemical process. *J. Alloys Compd.* **2017**, *719*, 411–419. [CrossRef]
22. Fathy, A.; Elkady, O.; Abu-Oqail, A. Effect of high energy ball milling on strengthening of Cu-$ZrO_2$ nanocomposites. *Ceram. Int.* **2019**, *45*, 5866–5875. [CrossRef]
23. El-Kady, O.; Yehia, H.M.; Nouh, F. Preparation and characterization of Cu/(WC-TiC-Co)/graphene nanocomposites as a suitable material for heat sink by powder metallurgy method. *Int. J. Refract. Met. Hard Mater.* **2019**, *79*, 108–114. [CrossRef]
24. Elkady, O.A.M.; Abu-Oqail, A.; Ewais, E.M.M.; El-Sheikh, M. Physico-mechanical and tribological properties of Cu/h-BN nanocomposites synthesized by PM route. *J. Alloys Compd.* **2015**, *625*, 309–317. [CrossRef]
25. Abu-Oqail, A.; Wagih, A.; Fathy, A.; Elkady, O.; Kabeel, A.M. Production and properties of Cu-ZrO2 nanocomposites. *J. Compos. Mater.* **2017**, *52–11*, 1519–1529.
26. Yehia, H.M.; Abu-Oqail, A.; Elmaghraby, M.A.; Elkady, O.A. Microstructure, hardness, and tribology properties of the (Cu/MoS2)/graphene nanocomposite via the electroless deposition and powder metallurgy technique. *J. Compos. Mater.* **2020**, *54–23*, 3435–3446. [CrossRef]
27. Sadoun, A.M.; Mohammed, M.M.; Elsayed, E.M.; Meselhy, A.F.; El-Kady, O. A Effect of nano $Al_2O_3$ coated Ag addition on the corrosion resistance and electrochemical behavior of Cu-$Al_2O_3$ nanocomposites. *J. Mater. Res. Technol.* **2020**, *9–3*, 4485–4493. [CrossRef]

Article

# CO$_2$ and CH$_2$ Adsorption on Copper-Decorated Graphene: Predictions from First Principle Calculations

Oleg Lisovski, Sergei Piskunov, Dmitry Bocharov *, Yuri F. Zhukovskii †, Janis Kleperis, Ainars Knoks and Peteris Lesnicenoks

Institute of Solid State Physics, University of Latvia, Kengaraga Street 8, LV-1063 Riga, Latvia; Olegs.Lisovskis@cfi.lu.lv (O.L.); piskunov@cfi.lu.lv (S.P.); janis.kleperis@cfi.lu.lv (J.K.); ainars.knoks@cfi.lu.lv (A.K.); Peteris.Lesnicenoks@cfi.lu.lv (P.L.)
* Correspondence: bocharov@cfi.lu.lv
† Deceased.

**Abstract:** Single-layer graphene decorated with monodisperse copper nanoparticles can support the size and mass-dependent catalysis of the selective electrochemical reduction of CO$_2$ to ethylene (C$_2$H$_4$). In this study, various active adsorption sites of nanostructured Cu-decorated graphene have been calculated by using density functional theory to provide insight into its catalytic activity toward carbon dioxide electroreduction. Based on the results of our calculations, an enhanced adsorption of the CO$_2$ molecule and CH$_2$ counterpart placed atop of Cu-decorated graphene compared to adsorption at pristine Cu metal surfaces was predicted. This approach explains experimental observations for carbon-based catalysts that were found to be promising for the two-electron reduction reaction of CO$_2$ to CO and, further, to ethylene. Active adsorption sites that lead to a better catalytic activity of Cu-decorated graphene, with respect to general copper catalysts, were identified. The atomic configuration of the most selective CO$_2$ toward the reduction reaction nanostructured catalyst is suggested.

**Keywords:** graphene; nanodecoration; first-principles calculations; adsorption; CO$_2$ electroreduction

**Citation:** Lisovski, O.; Bocharov, D.; Piskunov, S.; Zhukovskii, Y.F.; Kleperis, J.; Knoks, A.; Lesnicenoks, P. CO$_2$ and CH$_2$ Adsorption on Copper-Decorated Graphene: Predictions from First Principle Calculations. *Crystals* **2022**, *12*, 194. https://doi.org/10.3390/cryst12020194

Academic Editors: Walid M. Daoush, Fawad Inam, Mostafa Ghasemi Baboli and Maha M. Khayyat

Received: 12 November 2021
Accepted: 25 January 2022
Published: 28 January 2022

**Publisher's Note:** MDPI stays neutral with regard to jurisdictional claims in published maps and institutional affiliations.

**Copyright:** © 2022 by the authors. Licensee MDPI, Basel, Switzerland. This article is an open access article distributed under the terms and conditions of the Creative Commons Attribution (CC BY) license (https://creativecommons.org/licenses/by/4.0/).

## 1. Introduction

The electroreduction of CO$_2$ from exhausts to hydrocarbons can provide a sustainable supply of valuable raw materials for the chemical industry and fuels for transport and energetics [1]. The reduction of captured excessive carbon dioxide from the atmosphere could lead to a decrease in the greenhouse effect. CO$_2$ can be reduced to hydrocarbons—in particular, ethylene and methane (CH$_4$)—by electrochemical reactions 2CO$_2$ + 12$e^-$ + 8H$_2$O → C$_2$H$_4$ + 12OH$^-$ and CO$_2$ + 8$e^-$ + 6H$_2$O → CH$_4$ + 8OH$^-$, respectively. Ethylene has a wide range of applications in industry, polymer production, and agriculture. One of the most promising catalysts that can electroreduce CO$_2$ to C$_2$H$_4$ is copper metal [2,3]. However, along with ethylene (C$_2$H$_4$), many other carbon side-products are formed, including methane (CH$_4$), carbon monoxide (CO), and formate anion (HCOO$^-$) [4–7]. Besides, copper catalysts are very susceptible to poisoning and deactivation, usually, within half an hour after the start of the reduction process of carbon dioxide [8,9]. For the aforementioned reasons, significant efforts have recently been made to develop catalysts that can selectively reduce CO$_2$ to ethylene over long-lasting time periods. [7,10,11]. Polycrystalline Cu surfaces do not show a significant preference towards ethylene formation, with a C$_2$H$_4$/CH$_4$ product ratio of around 1:2 [3–5,12,13]. Their insufficient selectivity is considered to be due to the large heterogeneity of the centers of different catalytic activities on the polycrystalline surface. This is confirmed by the study of the influence of different copper planes on the selectivity of the electroreduction of carbon dioxide [7,14]. It was found that the (100) surface of single crystals of Cu favors the formation of ethylene more than Cu (111), as indicated by their ratios C$_2$H$_4$/CH$_4$ 1.3 and 0.2, respectively [7]. Interestingly, when the high index Cu (711), Cu (911), and Cu (810) planes formed by cleaving Cu (100) were examined, an even higher selectivity was displayed toward $C_2H_4$, with the $C_2H_4/CH_4$ ratio increasing to 10 for Cu (711). The reason for the catalyst's selectivity toward certain hydrocarbons is the increased number of surface steps in the high index facets or periodic formation of Cu terraces. This means that the system that is modified in such a way may exhibit an increased efficiency.

It was shown that copper nanoparticles with a large surface area have good selectivity for the formation of hydrocarbons, especially of ethylene [6]. It has been suggested that the edges and numerous steps formed on the surface of copper nanoparticles may be of decisive importance for the selective formation of ethylene. In favor of this, quantum chemical modeling has shown that intermediate reaction products, such as *CHO, are more stable at the steps of the Cu (211) surface than at the Cu (100) terraces. (Hereafter, an asterisk indicates that a species is adsorbed on a surface). This can lead to an increase in their concentration, and, ultimately dimerization to $C_2H_4$ [15].

Composites and hybrid structures, as well as nanoobjects based on graphene, has attracted huge attention in experimental and theoretical studies during the last decade [16–22] after this material was discovered in 2004 by Novoselov and Geim [23]. The decoration of graphene with metals (e.g., Fe [24], Pt [25,26], Pd [27]), as well as organic (e.g., tetracyanoethylene [28]) and inorganic compounds (e.g., $Bi_2O_3$ [29]), could improve electrocatalytic adsorption and gas sensing properties toward different gases

Different graphene-based catalysts for direct electrochemical $CO_2$ reduction were reported in the literature, e.g., atomic Fe dispersed on nitrogen-doped graphene [30], B-doped graphene [31], N-doped graphene [32], defective graphene produced by a nitrogen removal procedure from N-doped graphene [33], Ni-decorated graphene [34], $Co_3O_4$ spinel nanocubes on N-doped graphene [35], etc. The review of graphene-based materials for electrochemical $CO_2$ reduction was published by Ma et al. recently [36]. Several recent studies show that the modification of the graphene surface by copper nanoparticles or by creating Cu-contained heterostructures is an interesting approach in the development of efficient electrocatalysts [37–51].

These preceding studies have led us to elaborate on a theoretical model for a stable $C_2H_4$-selective electrocatalyst based on copper-nanocluster-decorated graphene and to understand how this selectivity can be increased. Carbon-based materials are potentially interesting catalysts for the $CO_2$ reduction reaction due to their low cost and especially due to their ability to form a wide range of hybrid nanostructures [52–57]. Carbon-based catalysts are chemically inactive at negative bias potentials and provide high overpotentials for the hydrogen evolution reaction compared to metal surfaces [58]. Pristine graphene does not exhibit any catalytic activity. However, by introducing dopants [59–61] and defects [62] during the synthesis, the electronic structure and catalytic properties of nanostructured carbon materials [62] are tailored. In particular, N-doping has been shown to significantly enhance the $CO_2$ reduction activity [39,41,61,63–66].

Experimental results obtained recently [67] suggest that the reaction pathways of the $CH_4$ and $C_2H_4$ formation are separated at an early stage of CO reduction. Results from a recent experimental study of CO electroreduction on single-crystal copper electrodes [68] imply that there are two separate pathways for $C_2H_4$ formation: one (i) that shares an intermediate with the pathway to $CH_4$, and a second one (ii) that occurs mainly on Cu (100) and probably involves the formation of a CO dimer as the key intermediate [69]. Considering the pathway (i), it is obvious that the *$CH_2$ dimerization is a crucial step for the final $C_2H_4$ production (the so-called "carbene" mechanism). *$CH_2$ can be produced by the protonation and deoxygenation of *CO [5]. *$CH_2$ can be also obtained from subsequent reductions of *HCO, *C, and *CH. The further reduction of a single *$CH_2$ gives rise to *$CH_3$ and finally to $CH_4$.

In this study, the adsorption of $CO_2$ is considered to typically be the rate-determining step in the $CO_2$ reduction reaction, and thus it is desirable to find/design catalyst sites that bond $CO_2$ strongly—preferably stronger than H adsorption [70]. To shed light on the trends in the catalytic activity of the Cu-decorated pristine and N-doped graphene system, in this work, systematic density functional theory (DFT) calculations of the adsorption of $CO_2$ and intermediates on Cu-decorated graphene are performed. The first principle calculations is performed for the monodisperse $Cu_7$ nanocluster deposited at the 5 × 5 supercell of graphene to predict the electronic properties of the $Cu_7$ facet in light of its different affinities for *$CO_2$ and *$CH_2$, and thus to provide deeper insights into its intrinsic activities for $CO_2$ electroreduction. The reaction energies for the formation of intermediates on $Cu_7$/graphene-nanostructured surfaces have been calculated using the hybrid DFT approach. In general, this work may not only give a deep insight into the reaction mechanisms toward $C_2H_4$ formation on Cu-decorated carbon nanomaterials, but may also provide guidelines for designing Cu-based catalysts to effectively produce multicarbon compounds.

## 2. Computational Details

Modeling was carried out at the DFT level of theory. This approach is based on the linear combination of atomic orbitals (LCAO) method with atom-centered localized Gaussian-type functions (GTFs) forming the basis sets (BS). Fully relaxed $Cu_7$/graphene nanostructures were calculated using hybrid exchange-correlation functional HSE06 according to the prescription given in Refs. [71,72]. Its particular feature is the use of an error-function-screened Coulomb potential for calculating the exchange energy. This functional was chosen to reproduce the basic atomic and electronic properties of both graphene and the most stable Cu (111) qualitatively close to those experimentally observed. The calculations were executed with CRYSTAL17 computational code [73], which was developed for the atomistic modeling of solid state chemistry. Using such a computation strategy, the geometries have been optimized with various species adsorbed on the graphene and metal Cu catalyst, and the adsorption energies of various species that are considered in this study have been calculated for nitrogen, oxygen, the $CO_2$ molecule, and the $CH_2$ radical. Besides the graphene and copper catalyst, basis sets are required for atoms of adsorbed species. For all atoms in the studied materials, full electron valence BSs [73] were used. For Cu, C, O, and H atoms, the triple-zeta BSs were obtained from Ref. [74]; on the other side, for the N atom, the basis set in the form of 6s-31sp-1d was obtained from Ref. [75].

To evaluate the Coulomb and exchange series appearing in the SCF equations for periodic systems, five tolerances were controlled: $10^{-8}$, $10^{-8}$, $10^{-8}$, $10^{-8}$, $10^{-16}$ (related to estimates of overlap or penetration for integrals of Gaussian functions on different centers, which define cut-off limits for series summation). To provide the correct summation in both direct and reciprocal lattices, the reciprocal space was integrated by sampling the interface Brillouin zone (BZ) with the 8 × 8 × 1 Monkhorst–Pack meshes [76] for slab calculations, which gives, in total, 34 k-points evenly distributed in the BZ. The calculations are considered to be convergent if the total energy differs by $10^{-7}$ a.u. or less in two successive cycles of the self-consistent-field (SCF) procedure [73].

The adsorption energy $E_{ads}$ was calculated with the following equation:

$$E_{ads} = E_{ads/sub} - E_{molecule} - E_{sub} \qquad (1)$$

where $E_{ads/sub}$ and $E_{sub}$ are the total energy of the $Cu_7$/graphene nanostructure with the adsorbed $CO_2$, molecule or *$CH_2$ intermediate, and $Cu_7$/graphene nanostructure slab, respectively, and $E_{molecule}$ is the total energy of the isolated $CO_2$ molecule or *$CH_2$ intermediate, analogously to Ref. [77]. The energetically favorable adsorption (chemisorption) takes place if the adsorption energy $E_{ads}$ is negative [78].

## 3. Results and Discussion

### 3.1. Cu/Graphene Cluster

Within the framework of this study, an efficient and reliable model of the monodisperse $Cu_7$ cluster deposited on single-layered graphene is constructed. The model consists of a 5 × 5 graphene supercell periodically repeated in the $xy$ plane, with seven Cu atoms forming a nanodot deposited in every supercell. Such a model is a balanced solution for the efficient use of computer resources and reliable prediction of the electronic structure and energetics of the nanostructures under study. Figure 1 shows schematic views (aside and atop) for the fully optimized two-dimensional Cu$_7$/graphene nanostructure containing the faceted Cu nanodot. For this cluster, a complete relaxation of the atomic coordinates was carried out and the binding energy of Cu atoms was estimated for this model.

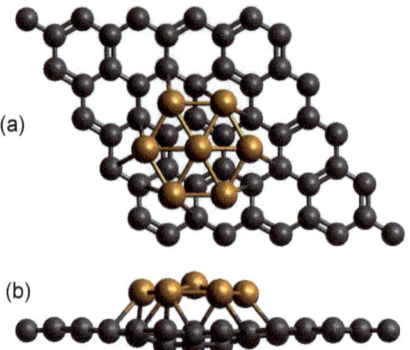

**Figure 1.** Top (**a**) and aside (**b**) views of equilibrium structure of six-faceted Cu nanopyramid deposited on graphene monolayer. Grey balls stand for carbon atoms and orange for copper.

The Cu$_7$ cluster is quite strongly physisorbed to the graphene layer with the binding energy of $-1.54$ eV/Cu atom. The negative binding energy means that energy is released after the substrate–adsorbate coupling. Single Cu atoms tend to adsorb at the hollow sites of graphene with the binding energy of $-2.65$ eV/Cu atom. Thus, Cu atoms deposited at graphene could reproduce the facets of the most stable Cu (111) surface. Nevertheless, a single Cu atom deposited at graphene forms quite weak Cu-C graphene bonds, with a bond population of 80 milli electrons.

The strongest bonding between Cu and graphene takes place at the defective graphene layer containing a carbon vacancy (Figure 2). The binding energy of the Cu-C$_{vacancy}$ complex is approximately $-6.63$ eV/Cu atom. However, the energy of vacancy formation is quite high (17.5 eV) and such a mechanism of Cu cluster adsorption at graphene is energetically unfavorable.

**Figure 2.** Equilibrium structure of C$_{graphene}$ atom substituted for Cu with the binding energy of $-3.59$ eV/Cu atom. Grey balls stand for carbon atoms and orange for copper.

Since the N-doping has been shown to significantly enhance the CO$_2$ reduction activity of graphene [39,41,61,63–66], the six-faceted Cu nanopyramid deposited atop the N-saturated graphene monolayer (Figure 3) is considered as well. The presence of the nitrogen atoms at the graphene support allows for the strong chemisorption of the Cu atom with the bond population of Cu–N = 303 milli electrons and N–C$_{graphene}$ = 344 milli electrons. The presence of a nitrogen monolayer may lead to a stronger adsorption of the Cu nanocluster at graphene; however, the presence of N practically does not influence CO$_2$ adsorption at Cu$_7$/graphene. Therefore, an N layer atop graphene is not considered in the further modeling of CO$_2$ reduction.

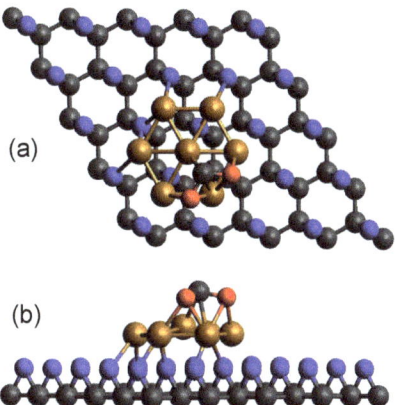

**Figure 3.** Schematic representation of top (**a**) and aside (**b**) views of six-faceted Cu nanopyramid deposited on N-saturated graphene monolayer. Grey balls stand for carbon atoms, orange for copper, red balls are oxygen atoms, and blue ones are nitrogen atoms.

Due to the relatively large distance between carbon layers in double-layered graphene (∼6.94 Å), the layer-to-layer interaction is negligible and does not influence the Cu nanocluster adhesion to the graphene layer. Therefore, in further modeling, it is assumed that the Cu cluster deposited at single-layered graphene can mimic the Cu cluster deposited at double-layered graphene.

Therefore, the constructed model of the six-faceted Cu nanopyramid deposited on the graphene monolayer (Figure 1) is considered as the most appropriate for large-scale ab initio total energy calculations of $CO_2$ and $CH_2$ molecules atop periodic $Cu_7$/graphene nanostructures (an electrically neutral system) using state-of-the-art total energy codes to estimate the energetics of a chain of elemental reactions under the influence of the copper nanocatalyst and graphene support.

*3.2. $CO_2$ Adsorption*

Taking into account that the adsorption of $CO_2$ is typically assumed to be the rate-determining step in $CO_2$ reduction, we pay major attention to the free energies of $CO_2$ adsorption at the $Cu_7$/graphene nanostructure, and, for comparative reasons, we have modeled $CO_2$ adsorption on its constituents, the most stable Cu (111) surface, represented by a three-layer and six-layer slab and pristine graphene (Figure 4). For the slabs, we considered the non-symmetrical one-sided and symmetrical two-sided deposition of absorbed $CO_2$ molecules. The adsorption energies for the $CO_2$ molecule and *$CH_2$ intermediate for the $Cu_7$/graphene nanostructure, the Cu (111) surface, represented by the three-layer and six-layer slab and pristine graphene, are given in Table 1.

**Table 1.** Calculated adsorption energies (eV) of $CO_2$ molecule and *$CH_2$ intermediate on $Cu_7$/graphene nanostructure, Cu (111) surface, represented as three-layer and six-layer slab, and pristine graphene layer with 5 × 5 supercell.

|  |  | $CO_2$ | *$CH_2$ |
|---|---|---|---|
| $Cu_7$/graphene |  | −6.04 | −7.31 |
| three-layer slab | one-sided | −6.69 | −6.53 |
|  | two-sided | −6.74 | −6.43 |
| six-layer slab | one-sided | −6.99 | −6.38 |
|  | two-sided | −6.89 | −6.46 |
| pristine graphene |  | −0.43 | −3.12 |

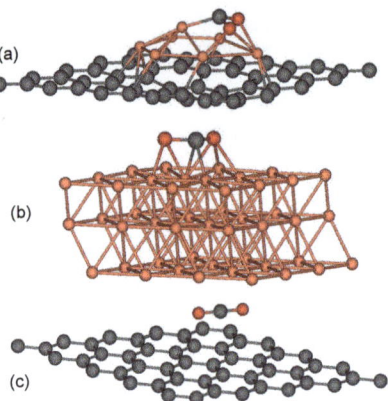

**Figure 4.** Schematic representation of the most energetically favorable adsorption positions of $CO_2$ molecule on (**a**) $Cu_7$/graphene nanostructure, (**b**) Cu (111) surface, and (**c**) pristine graphene monolayer. Grey balls stand for carbon atoms, orange for copper, and red balls are oxygen.

For all materials under consideration, the most energetically favourable adsorption site for the $CO_2$ molecule is the bridge position. Only the weak physisorption of $CO_2$ on pristine graphene is predicted from our calculation, with a relatively small free adsorption energy of −0.43 eV, which is in agreement with trends reported in Ref. [28]. The free adsorption energy calculated for the Cu (111) surface is in the range between −6.69 and −6.99 eV depending on the slab thickness and one-sided or two-sided adsorption of $CO_2$ molecules, whereas the adsorption energy of −6.04 eV per $CO_2$ molecule is predicted for the $Cu_7$/graphene nanostructure. A lower number points to a stronger chemical binding. The stronger binding of $CO_2$ to the $Cu_7$ nanocluster at graphene (similar to the adsorption at the pristine Cu (111) surface) can be explained by the presence of <111> grain boundaries of the $Cu_7$ nanocluster, which are known to be chemically more reactive. Figure 5a shows the projected density of states (PDOS) calculated for the $CO_2$ molecule adsorbed at the Cu (111) surface. The strong adsorption of the $CO_2$ molecule can be explained by Cu $3d$–O $2p$ orbitals hybridization seen in Figure 5a by the peaks at approximately −4 eV. The Cu–O bond population calculated by Mulliken population analysis is equal to 434 milli electrons. Both O atoms of the $CO_2$ molecule are strongly bonded to the Cu atoms of the Cu (111) surface. The only weak physisorption of $CO_2$ is predicted at graphene (PDOS in Figure 5b), with a $C_{graphene}$–$C_{CO_2}$ bond population of 0.012 milli electrons Our prediction is in agreement with the recent experimental observations. According to the data available in the literature, to improve the selectivity of $CO_2$ electrochemical reduction in producing $C_2$ products, Kanan et al. synthesized Cu nanoparticles containing grain boundaries and observed a substantial enhancement in the Faradaic efficiency of generating multi-carbon hydrocarbons [79]. This enhancement is correlated with the density of the grain boundary areas [80]. Cheng et al. conducted the atomistic modeling for the chemical vapor deposition process of Cu nanoparticles and found that strong CO binding with under-coordinated surface square sites could promote C–C coupling ("carbene" mechanism) [81]. According to our predictions, the boundary between the $Cu_7$ cluster and graphene can demonstrate the best catalytic ability for the $C_2H_4$ formation. This is due to the adsorption properties of neighboring Cu sites that are significantly perturbed by the presence of the nearest C, and the stronger Cu–O bonding is formed on the catalyst surface, which can also enhance $H_2C=CH_2$ evolution [14].

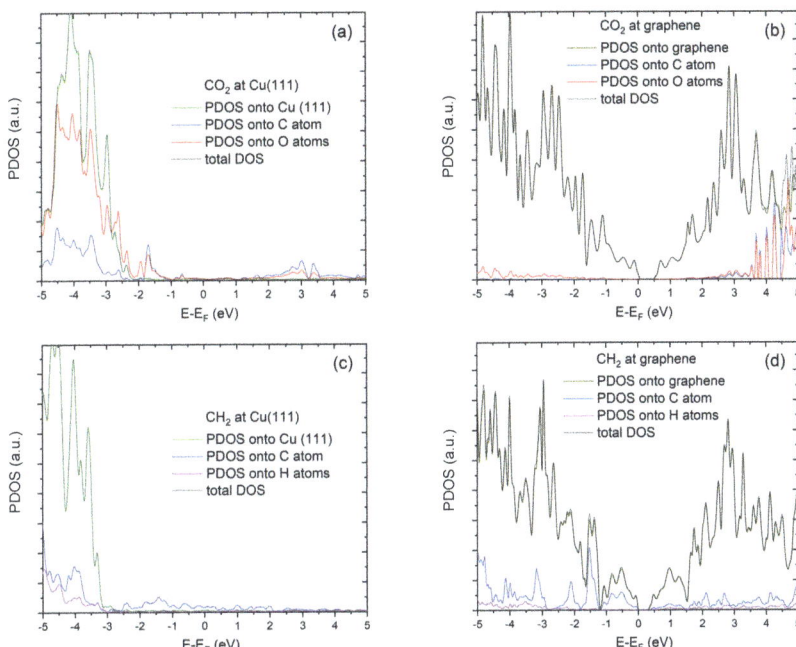

**Figure 5.** Projected density of states (PDOS) calculated for (**a**) $CO_2$ adsorbed at Cu (111) surface, (**b**) $CO_2$ adsorbed at graphene, (**c**) $CH_2$ adsorbed at Cu (111) surface, and (**d**) $CH_2$ adsorbed at graphene. PDOS onto all orbitals of H, C, and O atoms are magnified 10 times.

### 3.3. $CH_2$ Adsorption

Results from a recent experimental study of CO electroreduction on single-crystal copper electrodes [5] further implied that one of the most probable pathways for $C_2H_4$ formation is one that shares an intermediate with the pathway to $CH_4$. Considering that pathway, it is reasonable that the *$CH_2$ dimerization is a crucial step for the final $C_2H_4$ production ("carbene" mechanism). The further reduction of single *$CH_2$ gives rise to *$CH_3$ and finally to $CH_4$. Therefore, in this study, the adsorption energy of *$CH_2$ on $Cu_7$/graphene, pristine Cu (111), and pristine graphene (Figure 6) is calculated. For both pristine Cu (111) and $Cu_7$/graphene nanostructures, the most energetically favorable adsorption site of *$CH_2$ adsorbate is the hollow position between neighboring copper atoms, whereas the bridge position between neighboring C–C atoms is the most energetically preferable for the *$CH_2$ adsorption on pristine graphene. Figures 5c,d show the PDOS calculated for the $CH_2$ component adsorbed on the Cu (111) surface and graphene, respectively. $CH_2$ relatively strongly adsorbed at Cu (111) with hybrydized Cu $3d$–C $2p$ orbitals, forming square planar $sp^2d$ hybridisation (Figure 5c, peaks at approximately $-4$ eV) and a Cu–C bond population of 390 milli electrons, whereas the $C_{graphene}$–$C_{CH_2}$ bond population of 0.240 milli electrons is calculated for $CH_2$ at graphene.

Only a weak physisorption of *$CH_2$ on pristine graphene is predicted from our calculation, with a relatively small adsorption energy of $-3.12$ eV (PDOS in Figure 5d), which is in qualitative agreement with the observation reported in Ref. [28]. This may consequently lead to the relatively small barrier of $CH_2$–$CH_2$ dimerization. The adsorption energy calculated for the Cu (111) surface is in the range between $-6.38$ and $-6.53$ eV for the different thicknesses of slabs and the symmetrical/non-symmetrical deposition of *$CH_2$ intermediates, whereas a free adsorption energy of $-7.31$ eV is predicted for the $Cu_7$/graphene nanostructure (Table 1). From the calculated adsorption energies, we predict that the most energetically preferable $CH_2$ dimerization can take place on pristine graphene, whereas only a small difference in $CH_2$ dimerization can be predicted for the Cu (111) and $Cu_7$/graphene nanocluster.

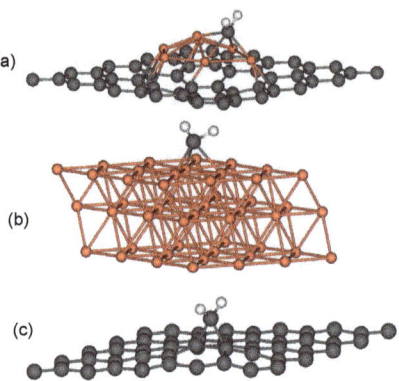

**Figure 6.** Schematic representation of the most energetically favourable adsorption positions of *$CH_2$ intermediate on the top of (**a**) $Cu_7$/graphene nanostructure, (**b**) Cu (111) surface, and (**c**) pristine graphene monolayer. Grey balls stand for carbon atoms, orange for copper, and white balls are hydrogen.

## 4. Conclusions

The main goal of this study is to contribute to the description of experimentally achievable results, allowing for the further optimization of the cathode composition and structure. In order to give theoretical predictions, we have constructed an efficient and reliable model that can be considered as the most appropriate for large-scale ab initio total energy calculations of $CO_2$ and $CH_2$ elements atop periodic $Cu_7$/graphene nanostructures. In these calculations, in order to examine a chain of elemental reactions under the influence of the copper nanocatalyst and graphene support, the state-of-the-art total energy codes were used. In the modeled nanocluster, Cu atoms reproduce the facets of the most stable Cu (111) surface. Adatoms and/or defects, e.g., vacancies, at the graphene support may make Cu–C graphene bonds stronger, facilitating the growth of the nanocluster. Assuming that the adsorption of $CO_2$ is typically the rate-determining step in the $CO_2$ reduction reaction, the energies of $CO_2$ adsorption at the $Cu_7$/graphene nanostructure have been calculated and compared to the adsorption energies of species placed on the pristine Cu (111) surface and pristine graphene. The strong binding of $CO_2$ to the $Cu_7$ nanocluster at graphene is close to binding to the pristine Cu (111) surface. This is explained by the presence of <111> facets at the $Cu_7$ nanocluster. This prediction is in agreement with the recent experimental observations to improve the selectivity of $CO_2$ electrochemical reduction in producing the $CH_2$–$CH_2$ intermediates. Cu nanoparticles containing grain boundaries were synthesized and a substantial enhancement in the Faradaic efficiency of generating multi-carbon hydrocarbons was observed [79]. This enhancement is correlated with the density of the grain boundary areas [80]. According to predictions obtained in this study, the Cu cluster at graphene demonstrates the best catalytic ability for $C_2H_4$ formation. This is due to the fact that the adsorption properties of neighboring Cu sites are significantly perturbed by the presence of the nearest C, and the stronger Cu–O bonding is formed on the catalyst surface, which also can enhance $H_2C=CH_2$ evolution. Based on this, it is predicted that the larger the length of the grain boundaries of the $Cu_n$ nanocluster deposited at graphene, the more selective the catalyst is to the $C_2H_4$. According to a recent experimental study of $CO_2$ electroreduction on copper electrodes [81], it is expected that one of the most probable pathways for $C_2H_4$ formation is one that shares an intermediate with the pathway to $CH_4$. Considering this pathway, it is obvious that *$CH_2$ dimerization is a crucial step for the final $C_2H_4$ production ("carbene" mechanism). In this respect, the $CH_2$–$CH_2$ dimerization reaction is responsible for the $C_2H_4$ evolution. The lowest dimerization barrier can be predicted for the pristine graphene due to the lowest adsorption energy, meaning that the whole $CO_2$ reduction reaction taking place at the grain boundary of the $Cu_n$/graphene nanocluster may lead to an improved selectivity to ethylene.

**Author Contributions:** O.L.: conceptualization, investigation, methodology, visualization, formal analysis, writing—review and editing; S.P.: conceptualization, investigation, methodology, formal analysis, writing—original draft, writing—review and editing; D.B.: investigation, resources, formal analysis, writing—original draft, writing—review and editing; Y.F.Z.: conceptualization, supervision; J.K.: conceptualization, funding acquisition, project administration, writing—review and editing, supervision; A.K.: conceptualization, investigation, writing—review and editing; P.L.: conceptualization, investigation. All authors have read and agreed to the published version of the manuscript.

**Funding:** The authors would like to express their gratitude for funding from the European Union's Horizon 2020 research and innovation programme under grant agreement No. 768789 (CO2EXIDE project). In the last stage of investigation and during the preparation of the publication, the authors were assisted by the postdoc D.B. with his own funding from project No. 1.1.1.2/VIAA/1/16/147 (1.1.1.2/16/I/001) under the activity "Post-doctoral research aid" realized at the Institute of Solid State Physics, University of Latvia.

**Institutional Review Board Statement:** Not applicable.

**Informed Consent Statement:** Not applicable.

**Data Availability Statement:** The data presented in this study are available in article.

**Acknowledgments:** Calculations were performed using Latvian Super Cluster (LASC), located in Center of Excellence at Institute of Solid State Physics, the University of Latvia, which is supported by European Union Horizon2020 Framework Programme H2020-WIDESPREAD-01-2016-2017-TeamingPhase2 under Grant Agreement No. 739508, project CAMART2.

**Conflicts of Interest:** The authors declare no conflict of interest. The funders had no role in the design of the study; in the collection, analyses, or interpretation of data; in the writing of the manuscript, or in the decision to publish the results.

# References

1. Whipple, D.T.; Kenis, P.J.A. Prospects of $CO_2$ utilization via direct heterogeneous electrochemical reduction. *J. Phys. Chem.* **2010**, *1*, 3451–3458. [CrossRef]
2. Hori, Y.; Wakebe, H.; Tsukamoto, T.; Koga, O. Electrocatalytic process of CO selectivity in electrochemical reduction of $CO_2$ at metal electrodes in aqueous media. *Electrochim. Acta* **1994**, *39*, 1833–1839. [CrossRef]
3. Hori, Y. Electrochemical $CO_2$ reduction on metal electrodes. In *Modern Aspects of Electrochemistry*; Vayenas C., White, R., Gamboa-Aldeco, M., Eds.; Springer: New York, NY, USA, 2008; pp. 89–189.
4. Kuhl, K.P.; Cave, E.R.; Abram, D.N.; Jaramillo, T.F. New insights into the electrochemical reduction of carbon dioxide on metallic copper surfaces. *Energy Environ. Sci.* **2012**, *5*, 7050–7059. [CrossRef]
5. Hori, Y.; Murata, A.; Takahashi, R. Formation of hydrocarbons in the electrochemical reduction of carbon dioxide at a copper electrode in aqueous solution. *J. Chem. Soc. Faraday Trans. 1* **1989**, *85*, 2309–2326. [CrossRef]
6. Tang, W.; Peterson, A.A.; Varela, A.S.; Jovanov, Z.P.; Bech, L.; Durand, W.J.; Dahl, S.; Nørskov, J.K.; Chorkendorff, I. The importance of surface morphology in controlling the selectivity of polycrystalline copper for $CO_2$ electroreduction. *Phys. Chem. Chem. Phys.* **2012**, *14*, 76–81. [CrossRef]
7. Hori, Y.; Takahashi, I.; Koga, O.; Hoshi, N. Electrochemical reduction of carbon dioxide at various series of copper single crystal electrodes. *J. Mol. Catal. Chem.* **2003**, *199*, 39–47. [CrossRef]
8. Hori, Y.; Konishi, H.; Futamura, T.; Murata, A.; Koga, O.; Sakurai, H.; Oguma, K. Deactivation of copper electrode in electrochemical reduction of $CO_2$. *Electrochim. Acta* **2005**, *50*, 5354–5369. [CrossRef]
9. DeWulf, D.W.; Jin, T.; Bard, A.J. Electrochemical and surface studies of carbon dioxide reduction to methane and ethylene at copper electrodes in aqueous solutions. *J. Electrochem. Soc.* **1989**, *136*, 1686–1691. [CrossRef]
10. Schouten, K.J.P.; Pérez Gallent, E.; Koper, M.T. The influence of pH on the reduction of CO and $CO_2$ to hydrocarbons on copper electrodes. *J. Electrochem. Soc.* **2014**, *716*, 53–57. [CrossRef]
11. Schouten, K.J.P.; Pérez Gallent, E.; Koper, M.T.M. Structure sensitivity of the electrochemical reduction of carbon monoxide on copper single crystals. *ACS Catal.* **2013**, *3*, 1292–1295. [CrossRef]
12. Li, C.W.; Kanan, M.W. $CO_2$ reduction at low overpotential on Cu electrodes resulting from the reduction of thick $Cu_2O$ films. *J. Am. Chem. Soc.* **2012**, *134*, 7231–7234. [CrossRef] [PubMed]
13. Gattrell, M.; Gupta, N.; Co, A. A review of the aqueous electrochemical reduction of $CO_2$ to hydrocarbons at copper. *J. Electroanal. Chem.* **2006**, *594*, 1–19. [CrossRef]

14. Hori, Y.; Takahashi, I.; Koga, O.; Hoshi, N. Selective formation of C2 compounds from electrochemical reduction of $CO_2$ at a series of copper single crystal electrodes. *J. Phys. Chem. B* **2002**, *106*, 15–17. [CrossRef]
15. Durand, W.J.; Peterson, A.A.; Studt, F.; Abild-Pedersen, F.; Nørskov, J.K. Structure effects on the energetics of the electrochemical reduction of $CO_2$ by copper surfaces. *Surf. Sci.* **2011**, *605*, 1354–1359. [CrossRef]
16. Fu, P.; Jia, R.; Wang, J.; Eglitis, R.I.; Zhang, H. 3D-graphene/boron nitride-stacking material: A fundamental van der Waals heterostructure. *Chem. Res. Chin. Univ.* **2018**, *34*, 434–439. [CrossRef]
17. Sergeyev, D.; Ashikov, N.; Zhanturina, N. Electric transport properties of a model nanojunction "Graphene–Fullerene $C_{60}$–Graphene". *Int. J. Nanosci.* **2021**, *20*, 2150007, [CrossRef]
18. Krasnenko, V.; Kikas, J.; Brik, M.G. Modification of the structural and electronic properties of graphene by the benzene molecule adsorption. *Phys. B Condens. Matter* **2012**, *407*, 4557–4561. [CrossRef]
19. Krasnenko, V.; Boltrushko, V.; Hizhnyakov, V. Vibronic interactions proceeding from combined analytical and numerical considerations: Covalent functionalization of graphene by benzene, distortions, electronic transitions. *J. Chem. Phys.* **2016**, *144*, 134708, [CrossRef]
20. Krasnenko, V.; Boltrushko, V.; Klopov, M.; Hizhnyakov, V. Conjoined structures of carbon nanotubes and graphene nanoribbons. *Phys. Scr.* **2014**, *89*, 044008, [CrossRef]
21. Bystrov, V.S.; Bdikin, I.K.; Silibin, M.V.; Meng, X.J.; Lin, T.; Wang, J.L.; Karpinsky, D.V.; Bystrova, A.V.; Paramonova, E.V. Pyroelectric properties of ferroelectric composites based on polyvinylidene fluoride (PVDF) with graphene and graphene oxide. *Ferroelectrics* **2019**, *541*, 17–24. [CrossRef]
22. Xi, J.Y.; Jia, R.; Li, W.; Wang, J.; Bai, F.Q.; Eglitis, R.I.; Zhang, H.X. How does graphene enhance the photoelectric conversion efficiency of dye sensitized solar cells? An insight from a theoretical perspective. *J. Mater. Chem. A* **2019**, *7*, 2730–2740. [CrossRef]
23. Novoselov, K.S.; Geim, A.K.; Morozov, S.V.; Jiang, D.; Zhang, Y.; Dubonos, S.V.; Grigorieva, I.V.; Firsov, A.A. Electric field effect in atomically thin carbon films. *Science* **2004**, *306*, 666–669. [CrossRef] [PubMed]
24. Ali, M.; Tit, N.; Yamani, Z.H. First principles study on the functionalization of graphene with Fe catalyst for the detection of $CO_2$: Effect of catalyst clustering. *Appl. Surf. Sci.* **2020**, *502*, 144153. [CrossRef]
25. Salih, E.; Ayesh, A.I. Pt-doped armchair graphene nanoribbon as a promising gas sensor for CO and $CO_2$: DFT study. *Phys. E Low-Dimens. Syst. Nanostruct.* **2021**, *125*, 114418. [CrossRef]
26. Rad, A.S. Adsorption of $C_2H_2$ and $C_2H_4$ on Pt-decorated graphene nanostructure: Ab-initio study. *Synth. Met.* **2016**, *211*, 115–120. [CrossRef]
27. Ma, L.; Zhang, J.M.; Xu, K.W.; Ji, V. A first-principles study on gas sensing properties of graphene and Pd-doped graphene. *Appl. Surf. Sci.* **2015**, *343*, 121–127. [CrossRef]
28. Osouleddini, N.; Rastegar, S.F. DFT study of the $CO_2$ and $CH_4$ assisted adsorption on the surface of graphene. *J. Electron Spectrosc. Relat. Phenom.* **2019**, *232*, 105–110. [CrossRef]
29. Mulik, B.B.; Bankar, B.D.; Munde, A.V.; Biradar, A.V.; Sathe, B.R. Bismuth-oxide-decorated graphene oxide hybrids for catalytic and electrocatalytic reduction of $CO_2$. *Chem. Eur. J.* **2020**, *26*, 8801–8809. [CrossRef]
30. Zhang, C.; Yang, S.; Wu, J.; Liu, M.; Yazdi, S.; Ren, M.; Sha, J.; Zhong, J.; Nie, K.; Jalilov, A.S.; et al. Electrochemical $CO_2$ reduction with atomic iron-dispersed on nitrogen-doped graphene. *Adv. Energy Mater.* **2018**, *8*, 1703487. [CrossRef]
31. Sreekanth, N.; Nazrulla, M.A.; Vineesh, T.V.; Sailaja, K.; Phani, K.L. Metal-free boron-doped graphene for selective electroreduction of carbon dioxide to formic acid/formate. *Chem. Commun.* **2015**, *51*, 16061–16064. [CrossRef]
32. Wang, H.; Chen, Y.; Hou, X.; Ma, C.; Tan, T. Nitrogen-doped graphenes as efficient electrocatalysts for the selective reduction of carbon dioxide to formate in aqueous solution. *Green Chem.* **2016**, *18*, 3250–3256. [CrossRef]
33. Han, P.; Yu, X.; Yuan, D.; Kuang, M.; Wang, Y.; Al-Enizi, A.M.; Zheng, G. Defective graphene for electrocatalytic $CO_2$ reduction. *J. Colloid Interface Sci.* **2019**, *534*, 332–337. [CrossRef] [PubMed]
34. Jiang, K.; Siahrostami, S.; Zheng, T.; Hu, Y.; Hwang, S.; Stavitski, E.; Peng, Y.; Dynes, J.; Gangisetty, M.; Su, D.; et al. Isolated Ni single atoms in graphene nanosheets for high-performance $CO_2$ reduction. *Energy Environ. Sci.* **2018**, *11*, 893–903. [CrossRef]
35. Sekar, P.; Calvillo, L.; Tubaro, C.; Baron, M.; Pokle, A.; Carraro, F.; Martucci, A.; Agnoli, S. Cobalt spinel nanocubes on N-doped graphene: A synergistic hybrid electrocatalyst for the highly selective reduction of carbon dioxide to formic acid. *ACS Catal.* **2017**, *7*, 7695–7703. [CrossRef]
36. Ma, T.; Fan, Q.; Li, X.; Qiu, J.; Wu, T.; Sun, Z. Graphene-based materials for electrochemical $CO_2$ reduction. *J. $CO_2$ Util.* **2019**, *30*, 168–182. [CrossRef]
37. Ma, Z.; Tsounis, C.; Kumar, P.V.; Han, Z.; Wong, R.J.; Toe, C.Y.; Zhou, S.; Bedford, N.M.; Thomsen, L.; Ng, Y.H.; et al. Enhanced electrochemical $CO_2$ reduction of Cu@Cu$_x$O nanoparticles decorated on 3D vertical graphene with intrinsic sp3-type defect. *Adv. Funct. Mater.* **2020**, *30*, 1910118. [CrossRef]
38. Fazel Zarandi, R.; Rezaei, B.; Ghaziaskar, H.; Ensafi, A. Modification of copper electrode with copper nanoparticles@reduced graphene oxide–Nile blue and its application in electrochemical $CO_2$ conversion. *Mater. Today Energy* **2020**, *18*, 100507. [CrossRef]
39. Yuan, J.; Yang, M.P.; Zhi, W.Y.; Wang, H.; Wang, H.; Lu, J.X. Efficient electrochemical reduction of $CO_2$ to ethanol on Cu nanoparticles decorated on N-doped graphene oxide catalysts. *J. $CO_2$ Util.* **2019**, *33*, 452–460. [CrossRef]
40. Ni, W.; Li, C.; Zang, X.; Xu, M.; Huo, S.; Liu, M.; Yang, Z.; Yan, Y.M. Efficient electrocatalytic reduction of $CO_2$ on Cu$_x$O decorated graphene oxides: An insight into the role of multivalent Cu in selectivity and durability. *Appl. Catal. B* **2019**, *259*, 118044. [CrossRef]

41. Dongare, S.; Singh, N.; Bhunia, H. Nitrogen-doped graphene supported copper nanoparticles for electrochemical reduction of $CO_2$. *J. $CO_2$ Util.* **2021**, *44*, 101382. [CrossRef]
42. Zhu, G.; Li, Y.; Zhu, H.; Su, H.; Chan, S.H.; Sun, Q. Enhanced $CO_2$ electroreduction on armchair graphene nanoribbons edge-decorated with copper. *Nano Res.* **2021**, *10*, 1641–1650. [CrossRef]
43. Li, Q.; Zhu, W.; Fu, J.; Zhang, H.; Wu, G.; Sun, S. Controlled assembly of Cu nanoparticles on pyridinic-N rich graphene for electrochemical reduction of $CO_2$ to ethylene. *Nano Energy* **2016**, *24*, 1–9. [CrossRef]
44. Yuan, J.; Yang, M.P.; Hu, Q.L.; Li, S.M.; Wang, H.; Lu, J.X. $Cu/TiO_2$ nanoparticles modified nitrogen-doped graphene as a highly efficient catalyst for the selective electroreduction of $CO_2$ to different alcohols. *J. $CO_2$ Util.* **2018**, *24*, 334–340. [CrossRef]
45. Geioushy, R.; Khaled, M.; Alhooshani, K.; Hakeem, A.; Rinaldi, A. Graphene/$ZnO$/$Cu_2O$ electrocatalyst for selective conversion of $CO_2$ into n-propanol. *Electrochim. Acta* **2017**, *245*, 448–454. [CrossRef]
46. Song, Y.; Peng, R.; Hensley, D.K.; Bonnesen, P.V.; Liang, L.; Wu, Z.; Meyer, H.M., III; Chi, M.; Ma, C.; Sumpter, B.G.; et al. High-Selectivity Electrochemical Conversion of $CO_2$ to Ethanol using a Copper Nanoparticle/N-Doped Graphene Electrode. *ChemistrySelect* **2016**, *1*, 6055–6061. [CrossRef]
47. Cao, C.; Wen, Z. Cu nanoparticles decorating rGO nanohybrids as electrocatalyst toward $CO_2$ reduction. *J. $CO_2$ Util.* **2017**, *22*, 231–237. [CrossRef]
48. Hossain, M.N.; Wen, J.; Chen, A. Unique copper and reduced graphene oxide nanocomposite toward the efficient electrochemical reduction of carbon dioxide. *Sci. Rep.* **2017**, *7*, 3184. [CrossRef]
49. Geioushy, R.; Khaled, M.M.; Hakeem, A.S.; Alhooshani, K.; Basheer, C. High efficiency graphene/$Cu_2O$ electrode for the electrochemical reduction of carbon dioxide to ethanol. *J. Electroanal. Chem.* **2017**, *785*, 138–143. [CrossRef]
50. Liu, S.; Huang, S. Structure engineering of Cu-based nanoparticles for electrochemical reduction of $CO_2$. *J. Catal.* **2019**, *375*, 234–241. [CrossRef]
51. Legrand, U.; Boudreault, R.; Meunier, J. Decoration of N-functionalized graphene nanoflakes with copper-based nanoparticles for high selectivity $CO_2$ electroreduction towards formate. *Electrochim. Acta* **2019**, *318*, 142–150. [CrossRef]
52. Dai, L. Functionalization of graphene for efficient energy conversion and storage. *Acc. Chem. Res.* **2013**, *46*, 31–42. [CrossRef] [PubMed]
53. Wang, H.; Yuan, X.; Zeng, G.; Wu, Y.; Liu, Y.; Jiang, Q.; Gu, S. Three dimensional graphene based materials: Synthesis and applications from energy storage and conversion to electrochemical sensor and environmental remediation. *Adv. Colloid Interface Sci.* **2015**, *221*, 41–59. [CrossRef] [PubMed]
54. Siahrostami, S.; Jiang, K.; Karamad, M.; Chan, K.; Wang, H.; Nørskov, J. Theoretical investigations into defected graphene for electrochemical reduction of $CO_2$. *ACS Sustain. Chem. Eng.* **2017**, *5*, 11080–11085. [CrossRef]
55. Varela, A.S.; Ranjbar Sahraie, N.; Steinberg, J.; Ju, W.; Oh, H.S.; Strasser, P. Metal-doped nitrogenated carbon as an efficient catalyst for direct $CO_2$ electroreduction to CO and hydrocarbons. *Angew. Chem. Int. Ed.* **2015**, *54*, 10758–10762. [CrossRef]
56. Cheng, M.J.; Kwon, Y.; Head-Gordon, M.; Bell, A.T. Tailoring metal-porphyrin-like active sites on graphene to improve the efficiency and selectivity of electrochemical $CO_2$ reduction. *J. Phys. Chem. C* **2015**, *119*, 21345–21352. [CrossRef]
57. Pérez-Sequera, A.C.; Díaz-Pérez, M.A.; Serrano-Ruiz, J.C. Recent advances in the electroreduction of $CO_2$ over heteroatom-doped carbon materials. *Catalysts* **2020**, *10*, 1179. [CrossRef]
58. Yang, N.; Waldvogel, S.R.; Jiang, X. Electrochemistry of carbon dioxide on carbon electrodes. *ACS Appl. Mater. Interfaces* **2016**, *8*, 28357–28371. [CrossRef]
59. Liu, J.; Song, P.; Ning, Z.; Xu, W. Recent advances in heteroatom-doped metal-free electrocatalysts for highly efficient oxygen reduction reaction. *Electrocatalysis* **2015**, *6*, 132–147. [CrossRef]
60. Wong, W.; Daud, W.; Mohamad, A.; Kadhum, A.; Loh, K.; Majlan, E. Recent progress in nitrogen-doped carbon and its composites as electrocatalysts for fuel cell applications. *Int. J. Hydrogen. Energy* **2013**, *38*, 9370–9386. [CrossRef]
61. Li, W.; Seredych, M.; Rodríguez-Castellón, E.; Bandosz, T.J. Metal-free nanoporous carbon as a catalyst for electrochemical reduction of $CO_2$ to CO and $CH_4$. *ChemSusChem* **2016**, *9*, 606–616. [CrossRef]
62. Terrones, H.; Lv, R.; Terrones, M.; Dresselhaus, M.S. The role of defects and doping in 2D graphene sheets and 1D nanoribbons. *Rep. Prog. Phys.* **2012**, *75*, 062501. [CrossRef] [PubMed]
63. Sharma, P.P.; Wu, J.; Yadav, R.M.; Liu, M.; Wright, C.J.; Tiwary, C.S.; Yakobson, B.I.; Lou, J.; Ajayan, P.M.; Zhou, X.D. Nitrogen-doped carbon nanotube arrays for high-efficiency electrochemical reduction of $CO_2$: On the understanding of defects, defect density, and selectivity. *Angew. Chem. Int. Ed.* **2015**, *54*, 13701–13705. [CrossRef] [PubMed]
64. Chai, G.L.; Guo, Z.X. Highly effective sites and selectivity of nitrogen-doped graphene/CNT catalysts for $CO_2$ electrochemical reduction. *Chem. Sci.* **2016**, *7*, 1268–1275. [CrossRef] [PubMed]
65. Liu, Y.; Chen, S.; Quan, X.; Yu, H. Efficient electrochemical reduction of carbon dioxide to acetate on nitrogen-doped nanodiamond. *J. Am. Chem. Soc.* **2015**, *137*, 11631–11636. [CrossRef] [PubMed]
66. Wu, J.; Yadav, R.M.; Liu, M.; Sharma, P.P.; Tiwary, C.S.; Ma, L.; Zou, X.; Zhou, X.D.; Yakobson, B.I.; Lou, J.; et al. Achieving highly efficient, selective, and stable $CO_2$ reduction on nitrogen-doped carbon nanotubes. *ACS Nano* **2015**, *9*, 5364–5371. [CrossRef] [PubMed]
67. Schouten, K.J.P.; Kwon, Y.; van der Ham, C.J.M.; Qin, Z.; Koper, M.T.M. A new mechanism for the selectivity to C1 and C2 species in the electrochemical reduction of carbon dioxide on copper electrodes. *Chem. Sci.* **2011**, *2*, 1902–1909. [CrossRef]
68. Schouten, K.J.P.; Qin, Z.; Pérez Gallent, E.; Koper, M.T.M. Two pathways for the formation of ethylene in CO reduction on single-crystal copper electrodes. *J. Am. Chem. Soc.* **2012**, *134*, 9864–9867. [CrossRef]

69. Hori, Y.; Murata, A.; Takahashi, R.; Suzuki, S. Electroreduction of carbon monoxide to methane and ethylene at a copper electrode in aqueous solutions at ambient temperature and pressure. *J. Am. Chem. Soc.* **1987**, *109*, 5022–5023. [CrossRef]
70. Back, S.; Lim, J.; Kim, N.Y.; Kim, Y.H.; Jung, Y. Single-atom catalysts for $CO_2$ electroreduction with significant activity and selectivity improvements. *Chem. Sci.* **2017**, *8*, 1090–1096. [CrossRef]
71. Heyd, J.; Scuseria, G.E.; Ernzerhof, M. Hybrid functionals based on a screened Coulomb potential. *J. Chem. Phys.* **2003**, *118*, 8207–8215. [CrossRef]
72. Krukau, A.V.; Vydrov, O.A.; Izmaylov, A.F.; Scuseria, G.E. Influence of the exchange screening parameter on the performance of screened hybrid functionals. *J. Chem. Phys.* **2006**, *125*, 224106. [CrossRef]
73. Dovesi, R.; Saunders, V.R.; Roetti, C.; Orlando, R.; Zicovich-Wilson, C.M.; Pascale, F.; Civalleri, B.; Doll, K.; Harrison, N.M.; Bush, I.J.; et al. *CRYSTAL17 User's Manual*; University of Torino: Torino, Italy, 2017.
74. Peintinger, M.F.; Oliveira, D.V.; Bredow, T. Consistent Gaussian basis sets of triple-zeta valence with polarization quality for solid-state calculations. *J. Comput. Chem.* **2013**, *34*, 451–459. [CrossRef]
75. Gatti, C.; Saunders, V.R.; Roetti, C. Crystal field effects on the topological properties of the electron density in molecular crystals: The case of urea. *J. Chem. Phys.* **1994**, *101*, 10686–10696. [CrossRef]
76. Monkhorst, H.J.; Pack, J.D. Special points for Brillouin-zone integrations. *Phys. Rev. B* **1976**, *13*, 5188–5192. [CrossRef]
77. Usseinov, A.B.; Akilbekov, A.T.; Kotomin, E.A.; Popov, A.I.; Seitov, D.D.; Nekrasov, K.A.; Giniyatova, S.G.; Karipbayev, Z.T. The first principles calculations of $CO_2$ adsorption on $(10\bar{1}0)$ ZnO surface. *AIP Conf. Proc.* **2019**, *2174*, 020181. [CrossRef]
78. Norsko, J. Chemisorption on metal surfaces. *Rep. Prog. Phys.* **1990**, *53*, 1253. [CrossRef]
79. Li, C.; Ciston, J.; Kanan, M. Electroreduction of carbon monoxide to liquid fuel on oxide-derived nanocrystalline copper. *Nature* **2014**, *508*, 504–507. [CrossRef]
80. Feng, X.; Jiang, K.; Fan, S.; Kanan, M.W. A direct grain-boundary-activity correlation for CO electroreduction on Cu nanoparticles. *ACS Cent. Sci.* **2016**, *2*, 169–174. [CrossRef]
81. Cheng, T.; Xiao, H.; Goddard, W.A. Nature of the active sites for CO reduction on copper nanoparticles; suggestions for optimizing performance. *J. Am. Chem. Soc.* **2017**, *139*, 11642–11645. [CrossRef]

Article

# Direct Observation of Induced Graphene and SiC Strengthening in Al–Ni Alloy via the Hot Pressing Technique

Omayma A. Elkady [1,*], Hossam M. Yehia [2], Aya A. Ibrahim [3], Abdelhalim M. Elhabak [3], Elsayed. M. Elsayed [4] and Amir A. Mahdy [5]

[1] Powder Technology Department, Central Metallurgical Research and Development Institute (CMRDI), Helwan 11421, Egypt
[2] Mechanical Department, Faculty of Technology and Education, Helwan University, Cairo 11795, Egypt; Hossamelkeber@techedu.helwan.edu.eg
[3] Faculty of Engineering, Cairo University, Cairo 12613, Egypt; eng.aya.ahmed.90@gmail.com (A.A.I.); Elhabak@gmail.com (A.M.E.)
[4] Mineral Processing Technology Department, Central Metallurgical Research and Development Institute (CMRDI), Helwan 11421, Egypt; Elsayed@gmail.com
[5] Mining, Metallurgy and Petroleum Engineering Department, Faculty of Engineering, AL-Azhar University, Cairo 11651, Egypt; Mahdy@gmail.com
* Correspondence: o.alkady68@gmail.com; Tel.: +20-010-6233-1896

**Citation:** Elkady, O.A.; Yehia, H.M.; Ibrahim, A.A.; Elhabak, A.M.; Elsayed, E.M.; Mahdy, A.A. Direct Observation of Induced Graphene and SiC Strengthening in Al–Ni Alloy via the Hot Pressing Technique. *Crystals* **2021**, *11*, 1142. https://doi.org/10.3390/cryst11091142

Academic Editor: John Parthenios

Received: 19 August 2021
Accepted: 9 September 2021
Published: 18 September 2021

**Publisher's Note:** MDPI stays neutral with regard to jurisdictional claims in published maps and institutional affiliations.

**Copyright:** © 2021 by the authors. Licensee MDPI, Basel, Switzerland. This article is an open access article distributed under the terms and conditions of the Creative Commons Attribution (CC BY) license (https://creativecommons.org/licenses/by/4.0/).

**Abstract:** In this study, Al/5 Ni/0.2 GNPs/x SiC (x = 5, 10, 15, and 20 wt%) nanocomposites were constituted using the powder metallurgy–hot pressing technique. The SiC particles and GNPs were coated with 3 wt% Ag using the electroless deposition technique then mixed with an Al matrix and 5% Ni using ball milling. The investigated powders were hot-pressed at 550 °C and 600 °C and 800 Mpa. The produced samples were evaluated by studying their densification, microstructure, phase, chemical composition, hardness, compressive strength, wear resistance, and thermal expansion. A new intermetallic compound formed between Al and Ni, which is aluminum nickel (Al$_3$Ni). Graphene reacted with the Ni and formed the nickel carbide Ni$_3$C. Additionally, it reacted with the SiC and formed the nickel–silicon composite Ni$_{31}$Si$_{12}$ at different percentages. A proper distribution for Ni, GNs, and SiC particles and excellent adhesion were observed. No grain boundaries between the Al matrix particles were discovered. Slight increases in the density values and quite high convergence were revealed. The addition of 0.2 wt% GNs to Al-5Ni increased the hardness value by 47.38% and, by adding SiC-Ag to the Al-5Ni-0.2GNs, the hardness increased gradually. The 20 wt% sample recorded 121.6 HV with a 56.29% increment. The 15 wt% SiC sample recorded the highest compressive strength, and the 20 wt% SiC sample recorded the lowest thermal expansion at the different temperatures. The five Al-Ni-Gr-SiC samples were tested as an electrode for electro-analysis processes. A zinc oxide thin film was successfully prepared by electrodeposition onto samples using a zinc nitrate aqueous solution at 25 °C. The electrodeposition was performed using the linear sweep voltammetric and potentiostatic technique. The effect of the substrate type on the deposition current was fully studied. Additionally, the ohmic resistance polarization values were recorded for the tested samples in a zinc nitrate medium. The results show that the sample containing the Al-5 Ni-0.2 GNs-10% SiC composite is the most acceptable sample for these purposes.

**Keywords:** hot pressing; aluminum matrix composites; electroless silver and nickel precipitation; hardness; wear resistance

## 1. Introduction

Recently, the requirement for high-strength and light-weight components has greatly increased for aerospace and automotive applications. This induced us to study and investigate new materials that satisfy our needs. Nowadays, aluminum matrix composites (AMCs) play a vital role because of their low density (2.7 g cm$^{-3}$), high specific strength (i.e., strength-to-density ratio), high-temperature creep resistance, high thermal conductivity, and low electrical resistivity [1,2]. Al metal's high coefficient of thermal expansion and low strength cause a restriction on its applications in many fields, such as heat sink materials and other applications requiring mechanical and thermal resistance. To improve these disadvantages, Al has been reinforced with ceramic particles that have a low coefficient of thermal expansion, a high melting point, and high strength, such as $Al_2O_3$, SiC, GNs, and $Y_2O_3$ [3,4]. Additionally, aluminum metal suffers from low wear resistance, limiting its use in tribological applications [2]. Researchers have investigated different materials in the Al matrix that have a self-lubricating property [5,6]. Graphene is considered to be one of the most important materials widely used to obtain self-lubricating composites [6,7]. Graphene is a two-dimensional material that consists of a single layer of sp2-hybridized carbon atoms with a 0.34 nm thickness [8–10]. It has excellent physical properties, such as high strength (~130 Gpa), a high Young's modulus (~1.0 TPa), high thermal conductivity (~5000 W $m^{-1}$ $K^{-1}$), and high electronic mobility (~15,000 $cm^2$ $V^{-1}$ $s^{-1}$), at room temperature [11–14]. Graphene's unique 2D structure and its properties provide an ideal AMC reinforcing agent. In other literature, it was shown that the addition of GNs caused an increase in the tribological and mechanical properties of AMCs [15–23]. Generally, higher graphene contents lead to a decline in the mechanical and tribological properties due to GN particles' non-uniform distribution within the matrices, which increases clumping [24] and produces extensive carbide formation. Hot extrusion, hot rolling [25], friction stir processing [14], and wet mixing fabrication methods [26] are advanced techniques used for the fabrication of metal matrix composites reinforced with graphene.

Ceramic materials have been used to enhance the Al matrix composite's mechanical and physical properties [27,28]. The advantages of all forms of SiC include high radiation and chemical tolerance, high thermal conductivity, high hardness and Young's modulus values (typically ~450 Gpa compared with ~130 GPa for Si), and, for some polytypes (notably 4H and 6H), a high critical electric field (above 2 $MVcm^{-1}$). These properties offer many possibilities for using SiC as a material for various devices and sensors, particularly in applications featuring high temperatures or power [29,30]. It has unique properties, such as superior hardness, wear resistance, and corrosion resistance, making it a good reinforcing material in metal matrix composites such as Fe, Al, and Mg-based materials [31]. Additionally, the elasticity of Al/SiC decreases when increasing the volume fraction of SiC [32]. So, modification of the metal on the SiC surface is usually utilized to control the interface reaction and enhance the interface combination [33,34]. Amirkhanlou et al. (2010) used SiC as a reinforcement in Al (A356) and found an increase in the hardness and impact energy of the composite [25]. Nickel has a positive effect on the mechanical and physical properties of Al composites in liquidus and solidus temperatures and corrosion resistance [35]. The intermetallic compounds of Al–Ni have attracted attention as materials for applications that need high temperatures due to properties such as a high melting temperature, high creep strength, a low density, high corrosion resistance, and high oxidation resistance [36–39].

The low wettability between ceramics and Al is a challenge because interfacial bonding is not easy to obtain. It causes the formation of pores inside the prepared composite, which harm the samples' physical, chemical, and mechanical properties. The electroless deposition technique has been used to produce a wettability layer on the metal surface to enhance the interfacial bond between the metal and ceramic phases [40]. To improve their wettability, Ni powder obtained by using the electroless deposition technique has been used and mixed with Al, SiC, and GNPs.

Several techniques, such as casting [41], mechanical alloying [41,42], mechanical milling [43,44], powder metallurgy [45,46], and powder metallurgy–hot pressing [35,47], have been used in the fabrication of Al matrix composites. Hossam et al. [47] showed that full densification and high diffusion in the Al matrix can be achieved by using the powder metallurgy–hot pressing technique. Powder metallurgy–hot pressing is a technique established in four steps that mix the mixture, press it into a suitably shaped die, heat the die to a predetermined temperature, and perform the final hot pressing at the predetermined pressure [47].

Shimaa A. Abolkassem et al. [35] studied the influence of the vacuum, hot isotactic (HIP), and hot compaction powder metallurgy techniques on an Al matrix composite reinforced with a SiC–Ni hybrid. The results revealed that that the powder metallurgy hot compaction technique achieved the highest densification (97–100%) for the Al matrix composite compared with the HIP samples (94–98%) and the vacuum-sintered ones (92–96%).

A novel hot pressing technique was used to fabricate Al-Ni/Graphene/SiC composites. Mixed powder was placed in a W320 steel die and then cold-pressed at the predetermined pressure. After that, the die was heated in a furnace to a suitable temperature for a suitable time and then hot-pressed. The effects of graphene and different amounts of SiC on the chemical composition, microstructure, densification, hardness, compressive strength, wear resistance, and thermal expansion of the Al/5Ni matrix were investigated.

This study also tested five Al-Ni-Gr-SiC electrodes by electrochemical deposition of a ZnO thin film on these electrodes. The ultimate goal of this work is to determine the suitability of the manufactured Al-Ni-GNs/x SiC composites for electrochemical applications by studying the influence of SiC ratios on the electrochemical behavior of Al-Ni-Gr-SiC electrodes in a traditional three-electrode cell and its ohmic resistance to oxidation and abrasion.

## 2. Materials and Methods

### 2.1. Materials

In this study, four different elements (Al, Ni, GNs, and SiC) were used to prepare new aluminum nanocomposites without porosity and with suitable physical and mechanical properties by the hot pressing technique. Fine pure aluminum powder with particle size of 0.5–2 µm (98.9% purity) was supplied by LOBA CHEMIE Pvt. Ltd. Colaba, Mumbai, India. The Al was used as the matrix of the composites. The GNs, Ni, and different amounts of SiC were used as reinforcements. GNs with a purity of 99.99% and a sheet thickness of 2–10 nm were supplied by Advanced Chemical Supply (ACS). To clean the GNs and SiC powders, sodium hydroxide and acetone were used. Nickel chloride ($NiCl_2 \cdot 6H_2O$), potassium sodium tartrate, ammonium chloride, and 33% ammonia were used to adjust the PH, and sodium hypophosphite was used to precipitate the nano-nickel powder by the electroless process. All chemicals were purchased from El-Naser Company, Helwan, Egypt.

### 2.2. Fabrication Procedures

The SiC powder was milled for 20 h to prepare it on the nanoscale. The milling process was performed with a 10:1 ball-to-powder ratio at 450 rpm. Ceramic alumina balls with a 12 mm diameter were used. Before the GNs and SiC powders were put in the aluminum matrix, they were cleaned of manufacturing contamination by stirring in a 10 wt% sodium hydroxide solution and acetone for 1 h, respectively. To improve the adhesion between the Al and the ceramic particle reinforcements, a 3 wt% nano-silver metal was precipitated by the electroless plating process onto the GN and SiC particles' surface individually [48–52]. To achieve this, a bath containing 3 g/L of silver nitrate (as a silver metal source) and 5 mL of formaldehyde/10 mL of water (as a reducing agent) was prepared. The pH of the solution was adjusted to 11.

Table 1 illustrates the chemical composition of the bath for the electroless nickel plating process. The plating process is established in four steps: dissolve the nickel chloride ($NiCl_2 \cdot 6H_2O$) in water; add potassium sodium tartrate as a complex agent with vigorous stirring; add ammonium chloride; and adjust the pH to 11. The solution was heated to 94 °C, and sodium hypophosphite (a reducing agent) was added. The precipitated powder was filtrated and washed with distilled water several times until the pH value of the supernatant became neutral. The produced Ni powder was dried at 80 °C for 2 h. This process has previously been discussed in detail [48,53–55].

**Table 1.** Chemical composition of the bath for the electroless nickel plating process.

| Element | Concentration |
|---|---|
| $NiCl_2 \cdot 6H_2O$ | 100 g/L |
| $NH_4Cl$ | 50 g/L |
| Potassium sodium tartrate | 80 g/L |
| pH | ~11 |
| Temperature | ~94 |
| Sodium hypophosphite | 90 g/L |

Six nanocomposite samples, in which the Al with 5 wt% Ni was the reference sample (S1) and the Al-5 wt% Ni-0.2 GNs sample was the second one (S2), were prepared. Then, SiC was added at 5, 10, 15, and 20 wt% (S3–S6) to the Al-5 wt% Ni-0.2 GNs mixture (S2), as shown in Table 2. All samples were prepared by mixing for 10 h using a high-energy ball mill (type RETSCH PM 100). For mixing, alumina balls with a 5 mm diameter were used. The ratio of balls to powders was 5:1 at 300 rpm. Methyl alcohol was added (0.5% of the powder's mass) to the powder mixture as a process control agent to avoid agglomeration and prevent the deposition of powder on the walls of the vial and balls.

**Table 2.** Composition of the nanocomposite specimens.

| Sample No. | Composition |
|---|---|
| S1 | 95 wt.% Al + 5 wt.% Ni |
| S2 | 99.8 wt.% (95%Al + 5% Ni) + 0.2% GNPs Base sample |
| S3 | 95 wt.% Base sample/5% SiC |
| S4 | 90 wt.% Base sample/10% SiC |
| S5 | 85 wt.% Base sample/15% SiC |
| S6 | 80% Base sample/20% SiC |

The mixed nanocomposite powders were cold-compacted in a steel die with a 12 mm inner diameter and an 80 mm high under high pressure (800 MPa) for 30 s. After that, the die was heated with a heating rate of 7.5 °C/min in an electric furnace with no atmosphere control at two different temperatures (550 °C and 600 °C) for 30 min.

### 2.3. Characterization

Samples were characterized by studying their densification, phase composition, microstructure, hardness, compression strength, wear rate, and thermal expansion. For the microstructure characterization, the specimens were ground with 400, 800, 1000, and 2500 grit SiC paper then polished with 3 mm alumina paste to reduce the grinding scratches. The microstructural analysis was carried out using a field-emission scanning electron microscope (FE-SEM; model: QUANTAFEG250). X-ray diffraction (XRD) (model x, pert PRO PANalytical) with Cu ka radiation (k = 0.15406 nm) was used to investigate the phase structure of samples and to detect any new phases that formed during the consolidation process.

A macro-Vickers hardness tester (model: 5030 SKV England) was used to study the plastic deformation of the fabricated samples at room temperature. The measurements for each weight were taken on each specimen's surface with an interval of 3 mm to avoid any effects of neighboring indentations. The mean value was recorded as the Vickers hardness (HV) number.

The wear characteristics of the prepared nanocomposite samples were investigated using a pin-on-disc apparatus. The experiment was performed at room temperature. Pins with a 316 L wheel with a 110 mm diameter and a 20 mm thickness were used as rings. The test was carried out at different rotational speeds (300, 600, and 900 rpm) and under different loads (10 and 15 N) for a 10 min loading time. The specimens were cleaned with ethanol before and after each run of the wear test.

The compressive strength of all prepared samples was measured using a universal testing machine (Shimadzu AD-X plus) at room temperature and a strain rate of 0.001/s.

The coefficient of thermal expansion of the fabricated samples was investigated by using a digital indicator with a sensitivity of 0.001 mm and an electrical furnace at a 5 °C/min heating rate. The temperature range of the experimental test was 150–450 °C. The thermal strain was measured and the mean value of 3 readings was used in the equation to determine the CTE.

For the electrochemical investigation, Al-Ni-GNs doped with different SiC ratios (5, 10, 15, and 20 wt%) served as an electrode in a conventional three-electrode cell. The 5 Al-Ni-Gr-SiC samples were used as working electrodes with an area of 1 cm$^2$. The counter electrode was made of a platinum plate with an area of 1 cm$^2$. A saturated silver/silver chloride (Ag/AgCl) electrode was used as a reference electrode (HANNA Instruments, Rhode, Italy).

The electrodeposition bath contained 0.05 M $KNO_3$ and 0.1 M Zn $(NO_3)_2$. The distance between electrodes was fixed at 0.3 cm. All electrodeposition experiments occurred at a temperature of 25 °C. First, before the film deposition, all 5 samples were polished with emery paper followed by etching in 80 g/L of NaOH (at 60 °C) for 20 min to eliminate abrasions and the oxide layer and rinsing in distilled water. After that, all 5 Al electrodes were cleaned and sonicated in an ethyl alcohol bath followed by drying at 80 °C. This preparation method provides a chemically clean and smooth surface, which is important for electrochemical studies. Linear sweep voltammetry curves were measured by sweeping from 0 to −2000 mV vs. the Ag/AgCl electrode at a scan rate of 10 mV s$^{-1}$. The voltammetric data were obtained from a high-performance 20 V/1A potentiostate/galvanostate (model: Volta Lab 21 PGP 201) coupled to a computer to record currents and potentials. The resulting ZnO film deposited on the tested Al electrode was rinsed with bi-distilled water and ethyl alcohol and then dried under a vacuum in a desiccator for 60 min.

## 3. Results and Discussion

### 3.1. Powder Characterization

The particle shape and size of the raw Al, GNs, electroless-deposited nano-Ni, and milled SiC powders are shown in Figure 1. The images show that the Al has an irregular shape with an average particle size of 1–4 μm, while the graphene appears in the form of sheets or layers. The precipitated nickel particles look irregular and have a 40–86 nm particle size. Due to the milling of the SiC powders for 20 h, a reduction in the particle size to the nano scale was achieved and the particles appear to have an irregular shape. An image of the GNs-Ag is overlaid on the image of the GNs to show the change before and after the coating of graphene layers with Ag. As shown, layers of graphene were coated with Ag. An X-ray of the uncoated GNs and Ag-painted GNs is also overlaid on the same image to emphasize the chemical deposition of Ag. The Ag element peaks were detected, indicating that the coating process by electroless chemical deposition was successful.

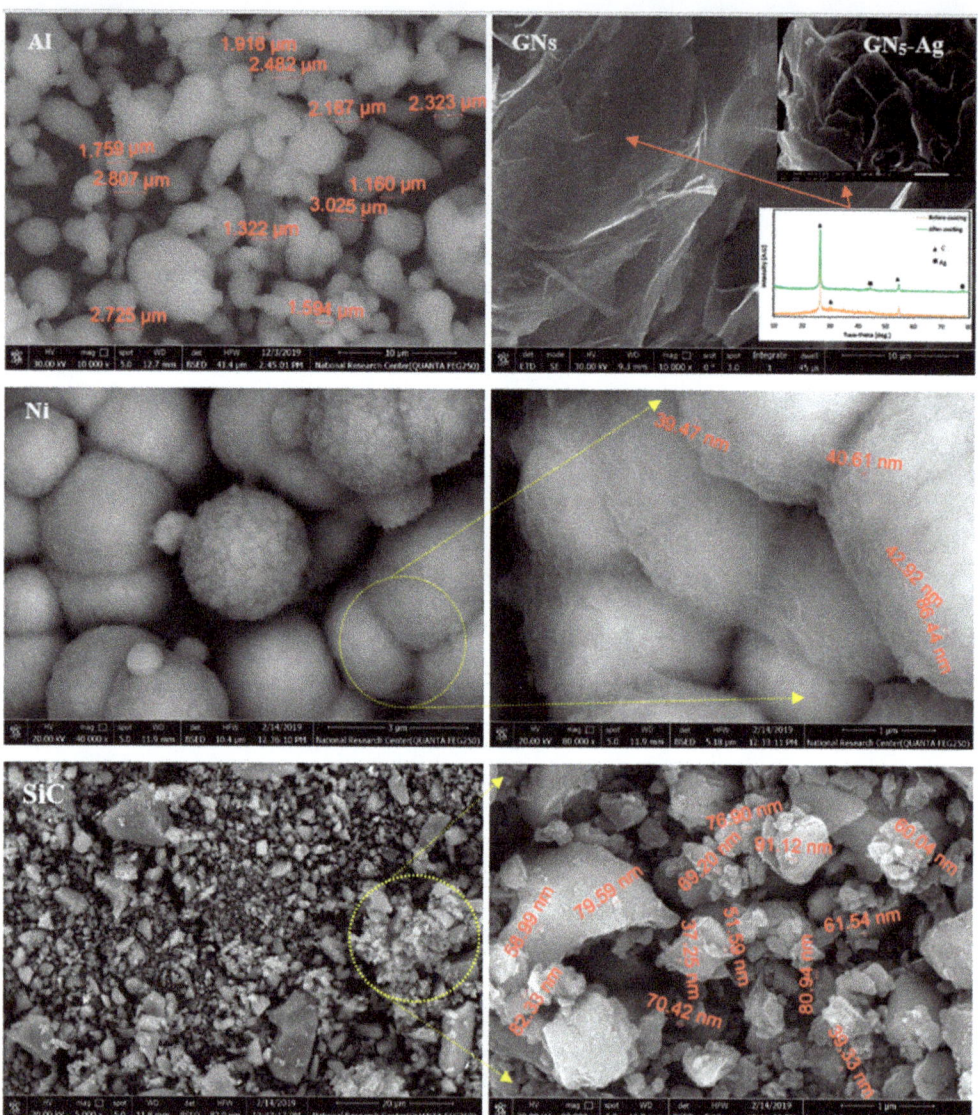

**Figure 1.** SEM (BSE) micrograph of pure Al, GNs, electroless-deposited Ni, and milled SiC powder.

*3.2. XRD*

The X-ray diffraction patterns the samples produced at 600 °C are shown in Figure 2. As shown in the figure, Al is a significant phase. The figure shows the mean peaks of the Al, SiC, and GNs. A new intermetallic compound was formed between the Al and the Ni, which is aluminum nickel ($Al_3N$) according to the phase diagram. Graphene reacted with the Ni and formed a nickel carbide ($Ni_3C$). The intensity of the carbon element increased as the SiC content increased, which was due to the interaction between GNs and SiC. The hot pressing technique not only improves the adhesion between the constituents of the composite but also increases the likelihood of interaction between the elements. Several aluminum and nickel compounds have been observed to form during various stages of the preparation of Al–Ni elemental powder mixtures [56–58]. It has been reported that solid-state reactions between elementary particles in the temperature interval of 550–600 °C result in Al-rich compounds, such as $Al_3Ni$ and $Al_3Ni_2$ [56–59]. According to the Al–Ni phase diagram, the combustion reaction occurs upon heating to about 640 °C, where eutectic melting is expected to happen [60].

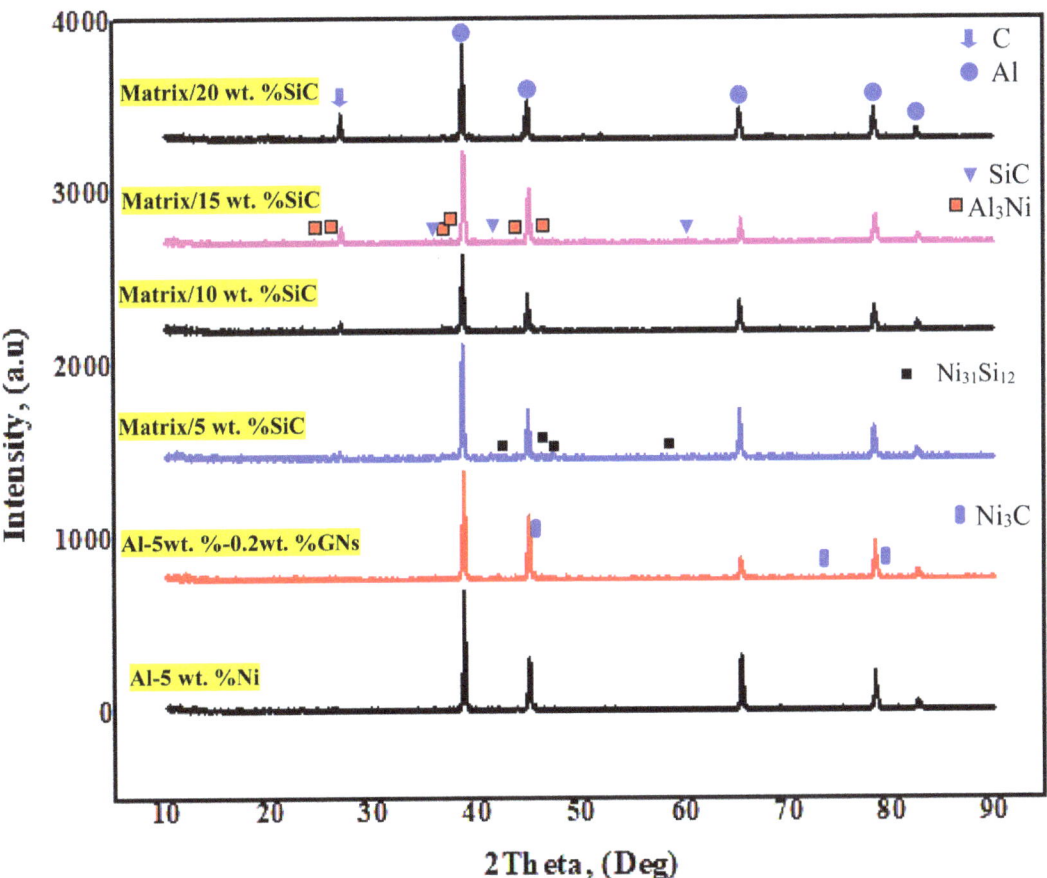

**Figure 2.** XRD patterns of consolidated samples at 600 °C.

### 3.3. Microstructure Investigation

The micrographs of samples produced by hot pressing at 600 °C for 30 min are shown in Figure 3 at a high magnification. The micrographs show four phases: the grey color represents the Al matrix; the white-gray color indicates the dispersed Ni phase; the dark grey spots refer to the SiC particles; and the dark spots refer to the GNs. A homogeneous distribution of Ni, GNs, and SiC particles and excellent adhesion can be observed. As a result of applying pressure before and after heating, enter-particles were established, and no grain boundaries between the Al matrix particles were observed. Coating SiC and GNs with nano Ag decreased the surface energy between the ceramic SiC and GNs particles and between the metallic Al and Ni particles. So, good wettability between the Al and SiC particles was observed. All constituents dispersed homogenously in the Al matrix. Due to increased SiC-Ag content, some accumulation on the aluminum grain boundaries occurred in the 20 wt% SiC sample.

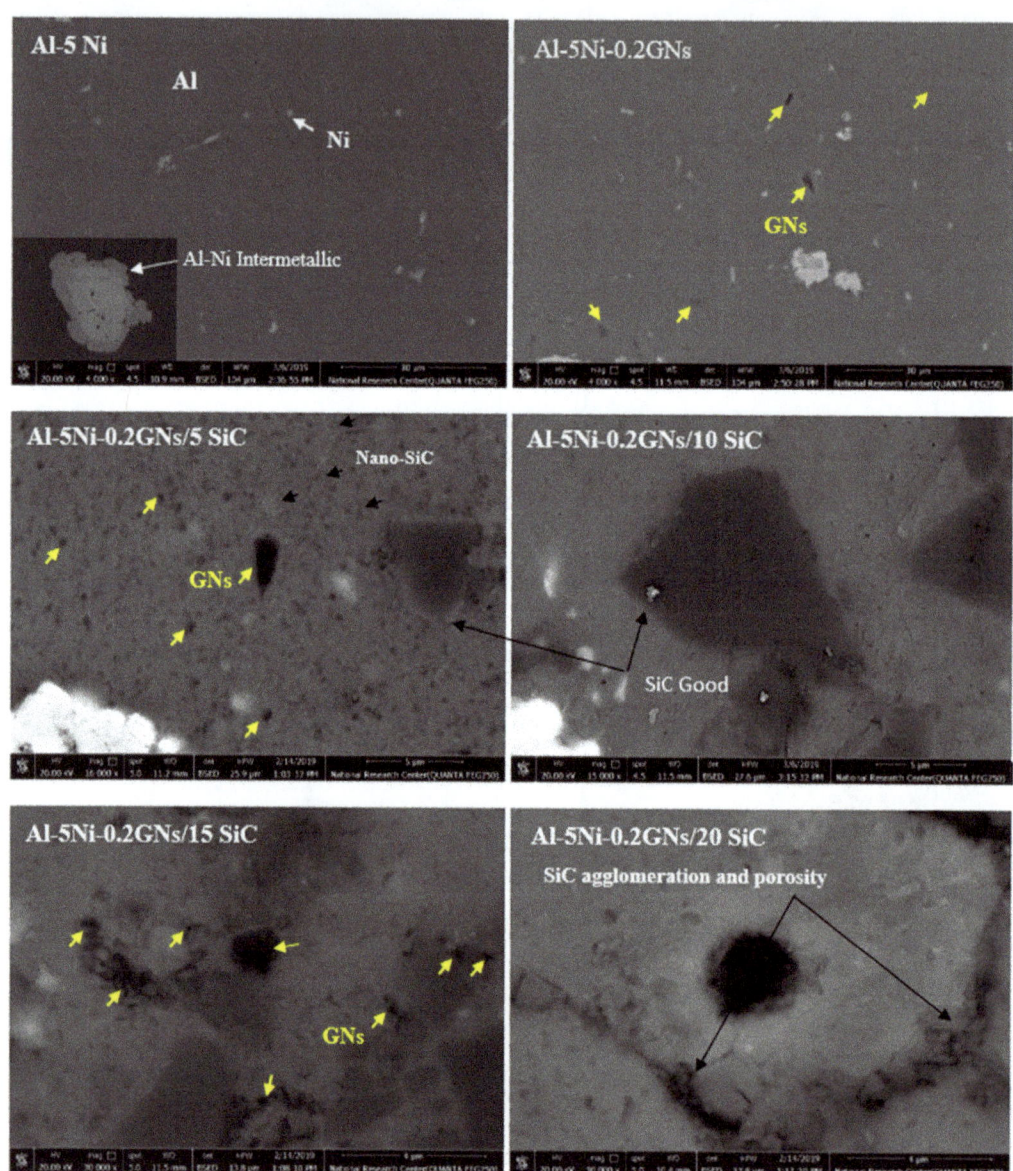

**Figure 3.** SEM (BSE) micrographs of fabricated samples.

*3.4. Density Measurement*

Figure 4 shows the variation in the theoretical and actual density of Al caused by adding Ni, GNs, and different weight percentages of SiC at consolidation temperatures of 550 °C and 600 °C. The theoretical density calculations show that the density of the aluminum increased with the addition of Ni and decreased with the addition of GNs.

Moreover, the density of Al-5Ni-0.2G increased gradually when increasing the content of the SiC-Ag particles up to 15 wt%. An increase or a decrease in theoretical density is related to the density values of the matrix and the reinforcement. When the reinforcement's density is higher than the matrix's density, the composite's density will increase. This is because the higher-density particles replace the lower-density ones. Three observations can be made from the figure. The first is that the density values of the samples consolidated at 600 °C are higher than those consolidated at 550 °C. The second is the gradual increase in density values when increasing the SiC content to 15 wt%, which then decrease at 20 wt%. The third observation is the slight increase in and the convergence of the density values. At 600 °C, the fabricated nanocomposites' actual density takes the same trend as the nanocomposite powders' theoretical density. The density increased when strengthening the aluminum with Ni, GNs, and up to 15 wt% SiC-Ag. Regardless of the decrease in the Al-5Ni nanocomposite matrix's theoretical density with the addition of 0.2 GNs, the actual density increased for the same sample. The increase in the Al-5Ni nanocomposite matrix's actual density may be due to the micro-pores being filled with the graphene layer during the hot pressing process. The increase in the Al-5Ni-0.2GNs nanocomposite's density produced by increasing the SiC-Ag content can be attributed to the efficient densification of the hot pressing process that uses high pressure during thermo-mechanical consolidation. Hot pressing helps to increase the number of interactions between particles, facilitating the high-temperature bonding of neighboring particles. In addition, it cracks any oxidation layer on the particles' surface that may disengage the excellent adhesion between them [47]. Applying pressure to the thermal load helps the aluminum flow and fill any gaps or voids and, consequently, to achieve the highest degree of densification. The decrease in the density of the 20 wt% SiC-Ag sample may be attributed to the production of agglomerations at this percentage that lead to the micro-pores shown in Figure 3. This participates in reducing the density of all the produced nanocomposites. Overall, the good densification results obtained for all compositions confirm that the selected temperature for the hot pressing process allowed for the plastic flow of aluminum under pressure.

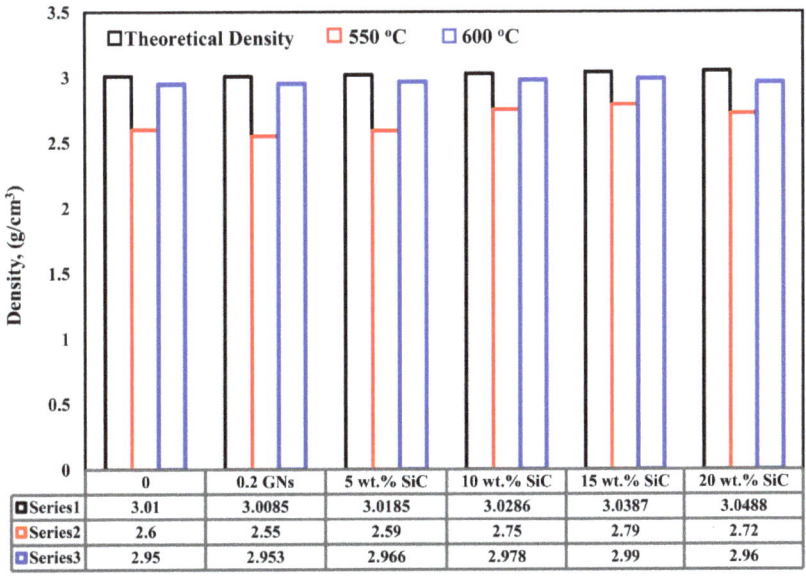

**Figure 4.** Density of the hybrid Al matrix reinforced with SiC-Ag.

## 3.5. Hardness Measurements

The hardness values of the investigated samples are presented in Figure 5. This test was performed to study the effect of Ni, GNs, and SiC-Ag wt.% on the plastic deformation of the aluminum matrix. The aluminum metal recorded a hardness of 31 HV [35]. Due to the addition of 5 wt% Ni, the hardness increased to 51.5 HV. This improvement may be due to the hardness of the Ni being higher than that of the Al and the production of $Al_3Ni$ and $Ni_3C$ intermetallic compounds. Yunya Zhang et al. [61] investigated the hardness behavior of Ni, Ni/$Ni_3C$ composites, and Ni-Ti-Al/$Ni_3C$ composites at different temperatures using a Vickers hardness tester. The result revealed that the formation of $Ni_3C$ enhanced the hardness of the Ni/$Ni_3C$ composite to 3.7 GPa compared with 2 GPa for the pure Ni at room temperature and the hardness decreased with increasing temperature. The addition of 0.2 wt% GNs to the Al/5Ni increased the hardness value to 77.8 HV (a 47.38% increment). Upon adding the SiC-Ag to the Al-5Ni-0.2 GNs matrix, the hardness increased gradually, and the 20 wt.% sample recorded a hardness of 121.6 HV (a 56.29% increment). The results reveal the importance of graphene as a strengthening material for the aluminum matrix, as 0.2 wt% graphene enhanced the hardness by 47.38%. This increment can be attributed to the characteristics of graphene. The graphene layer has a large surface area and nanoscale thickness. It represents a plane of atoms in the matrix. Because the high bonding strength between its atoms prevents the indenter from penetrating the matrix, the hardness increased. The addition of 20 wt% SiC enhanced the hardness by 56.29%. This improvement in the percentage is not high compared with the high content of SiC. This low degree of improvement may be due to the SiC's accumulation on the aluminum grain boundaries, which weakens the bonding between the particles. In addition to the effect of the SiC ceramic material, the X-ray analysis shows that new intermetallic compounds formed during the preparation process. Intermetallic compounds are characterized by high hardness, which also increases the aluminum matrix's hardness [47].

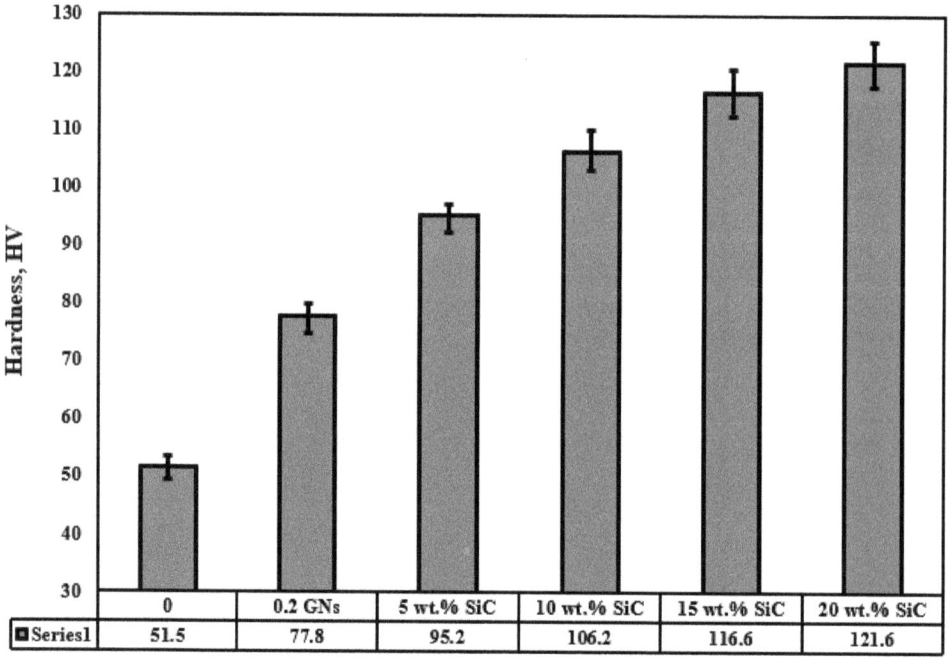

**Figure 5.** Hardness measurements of samples fabricated at 600 °C.

## 3.6. Compressive Strength

Figure 6 illustrates the stress–strain curves of the fabricated nanocomposites due to applying compression forces at room temperature. The maximum stress and total strain for each tested material are summarized in Table 3. As shown in the table, the compressive strength of the fabricated materials increased upon strengthening the aluminum matrix with 5 wt% Ni and 0.2 wt% GNs, respectively. Reinforcing the Al with SiC increased the maximum compressive strength gradually up to 15 wt% SiC, which then decreased for the 20 wt% SiC sample. According to the parameters that affect the powder metallurgy product, three main factors explain the gradual increase in compressive strength (Cs) with 5 wt% Ni, 0.2 wt% GNs, and up to 15 wt% SiC. The first is the strengthening effect of grain size according to the Hall–Petch equation, the second is the Orowan strengthening impact, which is related to the excellent distribution of the matrix's reinforcement, and the third is the strong cohesion force between the Al matrix and the ceramic reinforcement, which helps transfer the load from the ductile matrix to the rigid reinforcement. The $Al_3Ni$ and $Ni_3C$ intermetallic compounds on the grain boundaries of the Al particles and the GNs layers on the Al matrix act as load bearers and prevent cracks from appearing during the application of load, leading to an increase in the compressive strength of fabricated nanocomposites [52,62]. Yunya Zhang et al. [61] showed that the small grains and interstitial solution atoms of the $Ni_3C$ in a $Ni/Ni_3C$ composite prohibited the propagation of dislocations and enhanced the Ni matrix. In total, a 73% increase in strength with a 28% reduction in flexibility led to a 44% improvement in toughness. The decrease in the maximum compressive strength at 20 wt% SiC may be due to its accumulation at the Al matrix's grain boundaries. Regardless of the amount of SiC, the model (Al-5 wt% Ni-0.2 wt% GNs) was reinforced with, no cracks or fractures propagated during the compression test, indicating that the fabricated samples had high toughness. The recorded total strain values revealed that the real strain increased with the addition of Ni, GNs, and up to 15 wt% SiC.

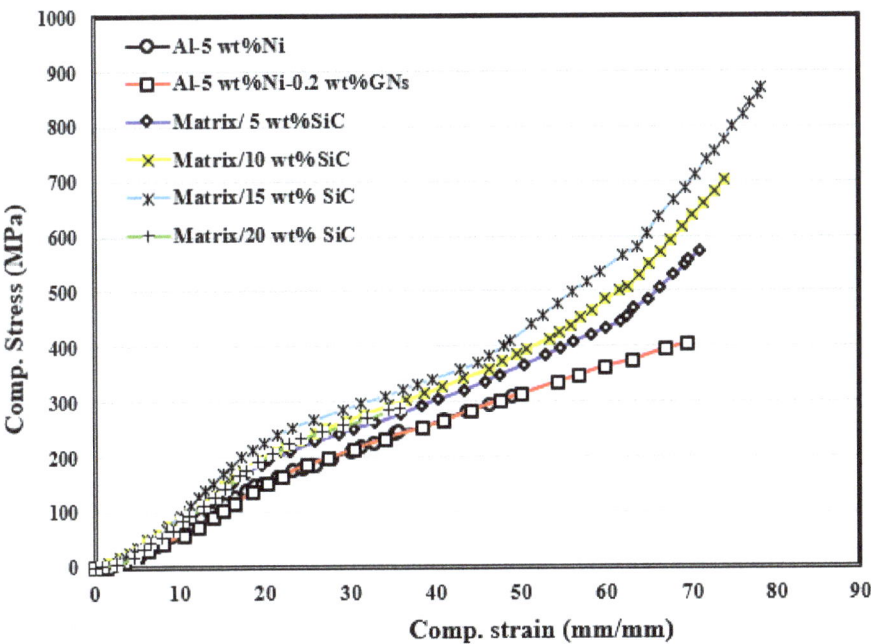

**Figure 6.** Stress–strain curves of the fabricated nanocomposites.

Table 3. Compressive stress and strain of the tested samples.

| Sample No. | Comp. Strength MPa | Strain |
|---|---|---|
| Al-5 wt% Ni | 311.82 | 48.86 |
| Al-5 wt% Ni-0.2 wt% GNs | 405.281 | 69.47 |
| Matrix/5 wt% SiC | 572.277 | 70.89 |
| Matrix/10 wt% SiC | 702.969 | 73.88 |
| Matrix/15 wt% SiC | 868.352 | 78.266 |
| Matrix/20 wt% SiC | 288.158 | 35.83 |

### 3.7. Wear Measurements

The influence of SiC content on the wear rate of the hybrid Al-5Ni-0.2GNs matrix at different sliding speeds and different loads for 10 min is shown in Figure 7. The results reveal that the wear rate decreased upon the addition of 5 wt% Ni, 0.2 wt% GNs, and up to 15 wt% SiC-Ag and then increased at 20 wt% SiC. Graphene nanosheets are hard ceramic materials with superior mechanical properties; reinforcing an Al matrix with them makes sliding more challenging, as shown in the hardness measurements (Figure 5), consequently increasing the wear resistance. Additionally, the agglomeration of GNs layers on the Al matrix's surface made it slide easily with a low wear rate. The SiC ceramic material helps improve the wear resistance as it confers strength on the ductile Al. The increase in the wear rate at 20 wt% SiC may be related to the agglomeration of the SiC at the grain boundaries of the Al matrix, which weakened the bonding strength between the particles and, consequently, increased the wear rate. The wear rate increased upon increasing the applied load. The application of load increases the contact area between the frictional parts and increases the wear rate. An essential phenomenon was also observed in which the wear rate increased as the sliding speed increased. As the sliding speed increased, the track of the sample on the frictional parts increased and, consequently, the wear rate increased [51].

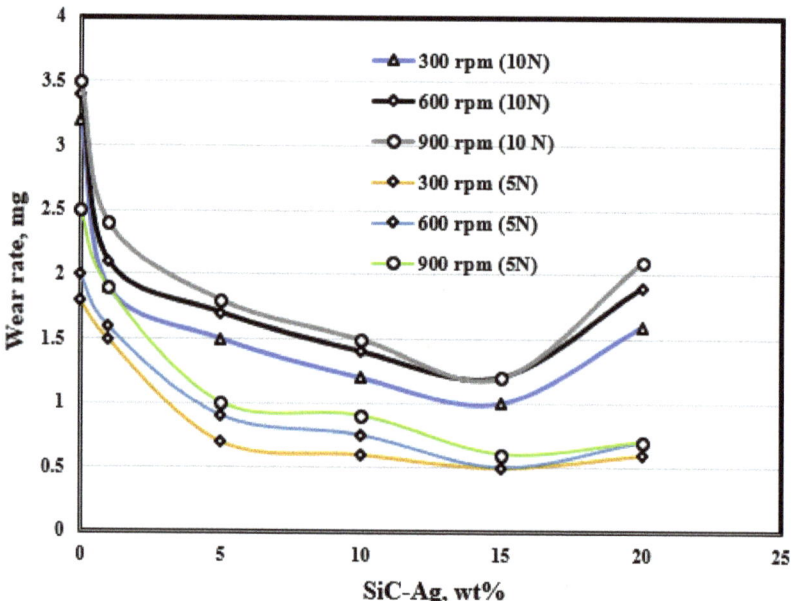

Figure 7. Wear rate (mg) of the fabricated materials at a load of 5 N and 10 N and different sliding speeds (600 rpm, 900 rpm, and 1200 rpm).

## 3.8. Thermal Expansion Estimation

The coefficient of thermal expansion (CTE) was determined in order to predict the fabricated samples' behavior under different applied thermal loads. It is expressed as the change in the material's dimensions as a function of temperature. The thermal expansion behavior depends on factors such as the type, microstructure, and percentage of the reinforcement, the morphology of the matrix, the thermal history, and the presence and number of pores. Additionally, the internal stress between the model and the support affects the thermal expansion behavior. Figure 8 illustrates the CTE of the fabricated samples at different heating temperatures. The CTE of the fabricated pieces changed upon the application of heat. The CTE increased upon increasing the heating temperature. As shown in the figure, the CTE of the Al-5Ni-0.2GNs sample decreased upon increasing the SiC content for reasons related to the low thermal expansion of the SiC and the porosity of the SiC at high contents. High amounts of SiC particles increase the restriction on the aluminum matrix and reduce the composites' expansion [35].

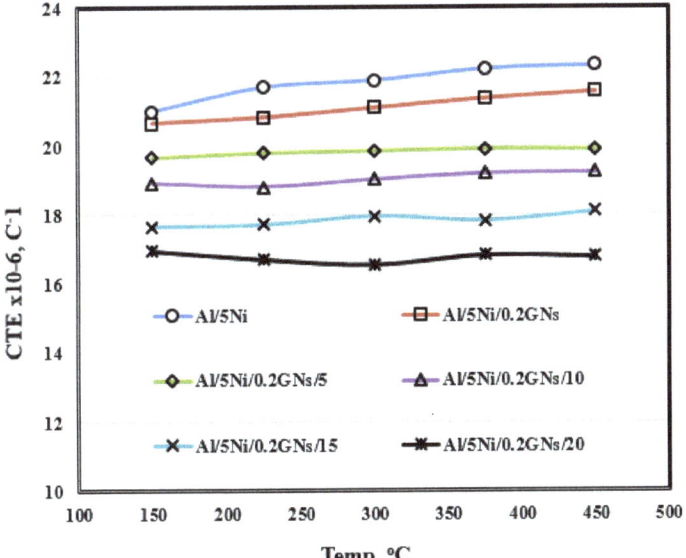

**Figure 8.** Coefficient of thermal expansion (CTE) of samples fabricated at 600 °C.

## 4. Electrochemical Investigation

### 4.1. Linear Sweep Voltammetric Study

We tested the synthesized Al-Ni-GNs substrates doped with different SiC ratios (5, 10, 15, and 20 wt%) by the electrodeposition of a ZnO layer on these substrates. We also studied factors affecting the deposition of ZnO by using linear sweep voltammetry and the potentiostatic technique.

Figure 9 shows the current–potential curves recorded with the Al-Ni-GNs-SiC substrates in a 0.1 M Zn (NO$_3$)$_2$ and 0.05 M KNO$_3$ solution. Linear sweep voltammetric studies were performed in the range of 0 to −2000 mV (versus the Ag/AgCl reference electrode) using a scan rate (SR) of 10 mV/s. During electrodeposition, nitrate ions are used as oxygen precursors. The cathodic deposition of ZnO film in a nitrate bath proceeds via the reduction of nitrate ions. The resulting Zn (OH)$_2$ ions react with Zn$^{+2}$ ions to form a hydroxide cathode. After that, Zn (OH)$_2$ is spontaneously dehydrated into ZnO [61]. The produced LSV curves for the pure Al-Ni and Al-Ni-GNs substrates are shown in Figure 1 (curves 1 and 2). These curves are similar and indicate similar behavior. The two curves are characterized by an initial weak cathodic deposition current at $-5$ mA cm$^{-2}$, associated with an applied potential of $-800$ mV.

On increasing the potential sweep, the cathode deposition current density begins to increase at a potential of $-1350$ mV. After that, at a potential of $-1350$ mV, the cathodic current sharply increases up to $-60$ mA cm$^{-2}$. This sharp increase in the cathodic current corresponds to nucleation (a crystallization step) and the electrodeposition of a thin ZnO layer on the tested Al-Ni and Al-Ni-GNs substrates. It is clear from all of the linear sweep curves that the tested Al-Ni and Al-Ni-GNs substrates have an identical ZnO deposition behavior. Curve 3 is associated with the presence of 5 wt% SiC in the Al-Ni-GNs electrode. On this curve, the deposition of ZnO film starts at a potential of $-750$ mV, earlier than on curves 1 and 2. After that, at a potential of $-1350$ mV, the cathodic current sharply increases up to $-80$ mA cm$^{-2}$. A similar behavior was observed with the presence of 10 wt% SiC (curve 4), but a sharp increase in the cathodic current up to $-85$ mA cm$^{-2}$ takes place. This indicates that the 5 wt% SiC in the Al-Ni-GNs composite (substrate) enhanced the cathodic deposition current by 20 mA/cm$^2$ and decreased the substrate's resistance to electrodeposition.

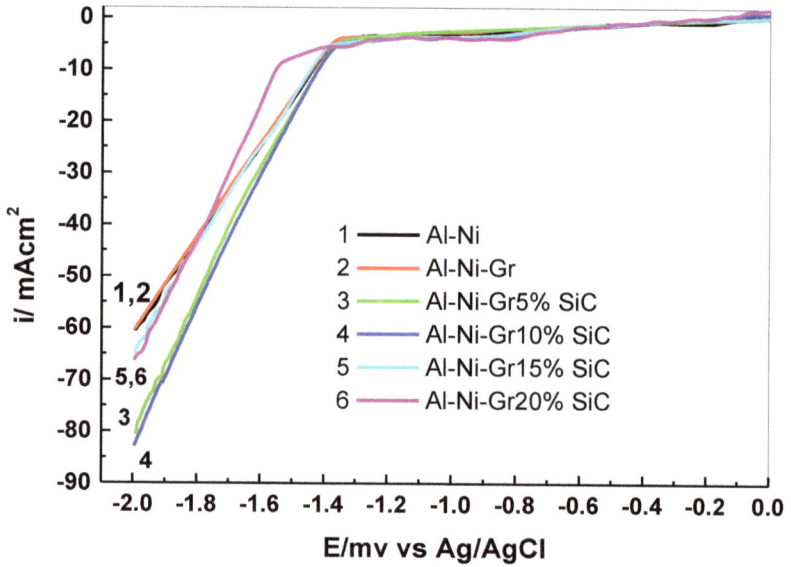

**Figure 9.** Current–potential curves recorded with various Al-Ni-Gr-SiC substrates.

Moreover, the 10 wt% SiC in the Al-Ni-GNs composite (substrate) improved the cathodic deposition current by 25 mA cm$^{-2}$ and decreased the resistance of the substrate to electrodeposition. The presence of the 10 wt% SiC in the Al-Ni-GNs composite improved the electrodeposition of the ZnO film to a greater degree than the presence of 5 wt% SiC. The presence of 10 wt% SiC recorded a higher deposition current ($-85$ mA cm$^{-2}$) and a higher rate of deposition than the $-80$ mA cm$^{-2}$ for 5 wt% SiC. On the other hand, with 15 wt% and 20 wt% SiC in the Al substrate (curves 5 and 6, respectively), the LSV curves are similar and indicate the same electrochemical deposition behavior. An initial deposition current of $-5$ mA cm$^{-2}$ characterizes the two curves and is associated with an applied potential of $-750$ mV. The deposition current density also begins to increase at a potential of $-1350$ mV, followed by a sharp increase in the cathodic current up to $-65$ mA cm$^{-2}$. In general, the presence of SiC in Al-Ni-GNs substrates causes a positive shift in the electrodeposition potential and shifts the film deposition current toward more negative values. This indicates that the presence of SiC in an Al-Ni-GNs substate improves the conductivity and decreases the resistance of the substrate to the electrochemical deposition process. Moreover, the 10 wt% SiC in the Al-Ni-GNs composite improved the electrodeposition of the ZnO film to a greater degree than the 15 and 20 wt% SiC. Additionally, the 5 wt% SiC in the Al-Ni-GNs composite enhanced the substrate's activity toward ZnO film deposition to a greater degree than the 15 and 20 wt% SiC. Each Al substrate has its own potential energy and conductivity. The conductivity is strongly related to the potential energy of the substrate. The ZnO crystal size is strongly affected by the conductivity of the substrate and its potential energy. The presence of 10 wt% SiC in the Al-Ni-GNs electrode improved the electrode's potential energy toward the ZnO deposition process and may have enhanced its conductivity.

*4.2. Potentiostatic Study*

Figure 10 presents the potentiostatic current versus time (*I-t*) transients for the growth and nucleation of ZnO with the Al-Ni-GNs substrate at potentials ranging between −600 and −1400 mV. The eight curves shown in the Figure were recorded with the Al-Ni-GNs substrate in a deposition solution of 0.1 M Zn $(NO_3)_2$ and 0.05 M $KNO_3$ under the same conditions.

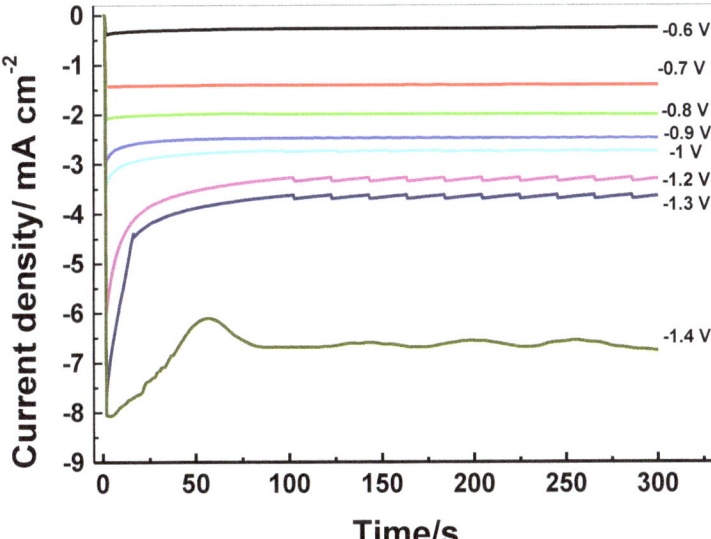

**Figure 10.** Current–time curves recorded with the Al-Ni-Gr cathode at a constant potential in an aqueous solution of 0.1 M Zn $(NO_3)_2$ and 0.05 M $KNO_3$.

The eight curves show that, immediately after the application of potential, the cathodic current rapidly increased because the nucleation of ZnO crystals started. Each crystal's three-dimensional growth rapidly increased the active surface area. The current transient is typical of a three-dimensional electro-crystallization growth process [63,64]. The transients were divided into three regions. The first region corresponds to short time periods ($t < 0.5$ s). In this region, the decrease in the cathodic current density was related to the charging of a double layer. The second region (1 to 50 s) relates to the crystal nucleation process and the crystals formed in the first region. The cathodic current densities of this region achieve their maximum value. The third region corresponds to long time periods (more than 50 s) and a decline in the current density, which represents the diffusion process [63,65]. A further increase in the deposition time (more than 60 s) causes a leveling out of the current density value (to almost a constant plateau). The current behavior for the eight curves in Figure 10 (the eight plateaus) becomes similar because the eight substrates are completely covered by the ZnO layer, which is especially visible. The figure shows the current plateau with a value of about $-0.28$ mA cm$^{-2}$ at a potential of $-600$ mV, a current plateau with a value of about $-1.4$ mA cm$^{-2}$ at a potential of $-700$ mV, a current plateau with a value of about $-1.94$ mA cm$^{-2}$ at a potential of $-800$ mV, a current plateau with a value of about $-2.5$ mA cm$^{-2}$ at a potential of $-900$ mV, a current plateau with a value of about $-2.7$ mA cm$^{-2}$ at a potential of $-1000$ mV, a current plateau with a value of about $-3.2$ mA cm$^{-2}$ at a potential of $-1200$ mV, a current plateau with a value of about 3.66 mA cm$^{-2}$ at a potential of $-1300$ mV, and a current plateau with a value of about $-6.6$ mA cm$^{-2}$ at a potential of $-1400$ mV.

Figure 11 depicts the current–time (I-t) transient curves for the growth of ZnO with the Al-Ni-Gr-10% SiC substrate at a constant potential ranging between $-600$ and $-1400$ mV. The figure displays a current plateau with a value of about $-1.68$ mA cm$^{-2}$ at a potential of $-600$ mV, a current plateau with a value of about $-2.38$ mA cm$^{-2}$ at a potential of $-700$ mV, a current plateau with a value of about $-3.19$ mA cm$^{-2}$ at a potential of $-800$ mV, a current plateau with a value of about $-4.14$ mA cm$^{-2}$ at a potential of $-900$ mV, a current plateau with a value of about $-4.5$ mA cm$^{-2}$ at a potential of $-1000$ mV, a current plateau with a value of about $-5.3$ mA cm$^{-2}$ at a potential of $-1200$ mV, a current plateau with a value of about $-7.15$ mA cm$^{-2}$ at a potential of $-1300$ mV, and a current plateau with a value of about $-8.33$ mA cm$^{-2}$ at a potential of $-1400$ mV. Figures 10 and 11 indicate that the increase in the current density as the potential increases could be related to the complete deposition of the ZnO film. Additionally, during the nucleation process, the electrochemical behavior was found to depend on the applied potential in both tested substrates (Al-Ni-Gr and Al-Ni-Gr-10% SiC). The presence of 10 wt% SiC in the Al-Ni-Gr substrate (Figure 11) is associated with higher current–time transient values (i-t plateaus) than the SiC-free substrate (Figure 10). This indicates that the Al-Ni-GN-10% SiC substrate is more conductive and less resistive than the Al-Ni-GN substrate.

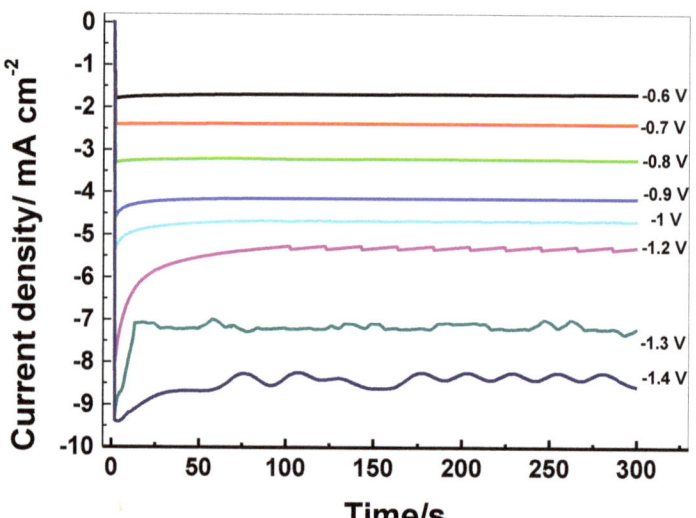

**Figure 11.** Current–time curves recorded with the Al-Ni-Gr-10% SiC cathode at a constant potential.

### 4.3. Ohmic Resistance Polarization Study

The corrosion behavior of six different Al-Ni-Gr substrates doped with different composite SiC percentages was tested using a 0.1 M Zn(NO$_3$)$_2$ + 0.05 M KNO$_3$ bath under the same conditions. The polarization resistance (Rp) is the transition resistance between the Al-Ni-Gr electrode with the composite SiC and the electrolyte. A high Rp value of a metal electrode implies high abrasion and erosion resistance, while a low Rp value implies low abrasion resistance. Thus, the polarization resistance is the ratio of the applied potential to the resulting current response. This resistance is inversely related to a uniform corrosion rate. The ohmic polarization resistance is vital to determining and testing the abrasion behavior of an Al substrate doped with SiC. The polarization resistance behaved like an abrasion resistor for the tested Al electrodes. Figure 12 shows that the ohmic resistance polarization values increase steeply with increasing SiC wt% in the Al samples from 0 to 5 wt% SiC, recording a value of 412 Ohms cm$^2$ with 5 wt% SiC. After that, the ohmic resistance polarization attained a maximum value of 469 ohms cm$^2$ with 10 wt% SiC.

On the other hand, a further increase in the SiC ratio in the Al-Ni-GNs substrate to 15% decreased the ohmic resistance polarization gradually to 428 Ohms cm$^2$. The ohmic resistance polarization gradually decreased to 424 ohms cm$^2$ upon increasing the SiC weight percentage to 20%. Finally, the results reveal that the Al-Ni-GNs sample with 10 wt% SiC showed a higher maximum ohmic resistance polarization value (469 ohms. cm$^2$) than the Al-Ni-GNs-SiC-free sample (400 ohms. cm$^2$).

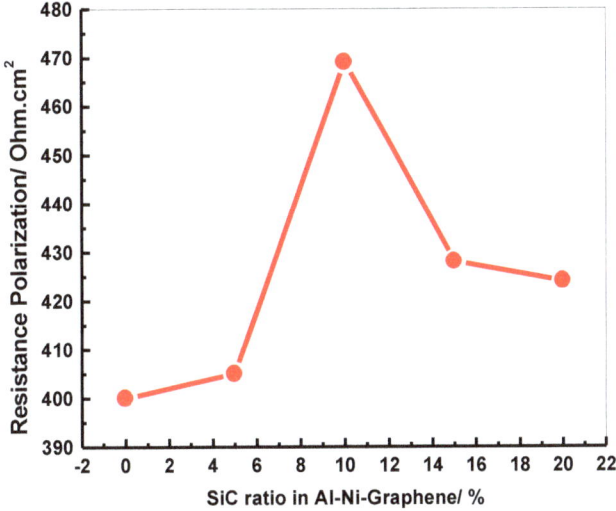

**Figure 12.** Variation in ohmic resistance polarization with different SiC weight percentages in the Al-Ni-Gr substrate for ZnO film.

## 5. Conclusions

In this work, an Al matrix was reinforced with Ni, GNPs, and different amounts of SiC using the powder technology–hot pressing technique. According to the obtained results, it can be concluded that:

1. Ni was successfully precipitated by the electroless chemical deposition technique on the nano scale;
2. All samples were successfully prepared by powder metallurgy–hot pressing at 600 °C for 30 min;
3. XRD revealed the formation of Al$_3$Ni and Ni$_3$C intermetallic compounds;
4. The density increased upon strengthening the aluminum with Ni, GNs, and up to 15 wt% SiC-Ag;
5. The Al-5Ni recorded a hardness of 51.5 HV. The addition of 0.2 wt% GNs increased the hardness value by 47.38%. The addition of SiC-Ag gradually increased the hardness. The 20 wt% SiC sample recorded a hardness of 121.6 HV with a 56.29% increment;
6. The results showed that the wear rate decreased upon the addition of 5 wt% Ni, 0.2 wt% GNs, and SiC-Ag content up to 15 wt% and then increased at 20 wt% SiC;

7. The high mechanical properties of GNs and SiC, the fair distribution, and the excellent adhesion to Al explained the gradual increases in compressive strength (Cs) with 5 wt% Ni, 0.2 wt% GNs, and up to 15 wt% SiC;
8. Due to the excellent adhesion between the Ni, GNs, SiC, and the Al matrix and their low thermal expansion, the CTE of the new nanocomposites decreased;
9. The results indicate that the presence of 10 wt% SiC in the Al-Ni-GNs substrate decreased the substrate's resistance to the electrochemical deposition process and consequently improved the electrodeposition of ZnO film to a greater degree than the 15 and 20 wt% SiC; and
10. The results suggest that the Al-Ni-GNs-10% SiC substrate achieved the maximum ohmic resistance polarization value (469 Ohms cm$^2$) compared with the 400 Ohms cm$^2$ of the Al-Ni-GNs-SiC-free sample.

**Author Contributions:** O.A.E. (Conceptualization, methodology, resources, project administration, and writing—original draft preparation), H.M.Y. (project administration, methodology, investigation, and writing—original draft preparation), A.A.I. (Investigation), A.M.E. (data curation, and writing—review and editing), E.M.E. (investigation, writing—review and editing), and A.A.M. (Investigation, and writing—review and editing). All authors have read and agreed to the published version of the manuscript.

**Funding:** This research received no external funding.

**Conflicts of Interest:** The authors declare no conflict of interest.

## References

1. Deaquino-Lara, R.; Gutierrez, E.; Estrada-Guel, I.; Hinojosa-Ruiz, G.; Sanchez, E.G.; Herrera-Ramirez, J.; Perez-Bustamante, R.; Sánchez, R.M. Structural characterization of aluminium alloy 7075–graphite composites fabricated by mechanical alloying and hot extrusion. *Mater. Des.* **2014**, *53*, 1104–1111. [CrossRef]
2. Baradeswaran, A.; Perumal, A.E. Wear and mechanical characteristics of Al 7075/graphite composites. *Compos. Part B Eng.* **2014**, *56*, 472–476. [CrossRef]
3. Ibrahim, I.A.; Mohamed, F.A.; Lavernia, E.J. Particulate reinforced metal matrix composites—A review. *J. Mater. Sci.* **1991**, *26*, 1137–1156. [CrossRef]
4. Barakat, W.S.; Elkady, O.; Abuoqail, A.; Yehya, H.; El-Nikhaily, A. Effect of Al$_2$O$_3$ coated Cu nanoparticles on properties of Al/Al$_2$O$_3$ composites. *J. Pet. Min. Eng.* **2020**, *22*, 53–60. [CrossRef]
5. Mansour, N.S.S.; Yehia, H.M.; Ali, A.I. Graphene reinforced copper matrix nano-composite for resistance seam welding electrode. *Bull. Tabbin Inst. Metall. Stud. (TIMS)* **2021**, *109*, 25–34. [CrossRef]
6. Li, Y.; Zhao, J.; Tang, C.; He, Y.; Wang, Y.; Chen, J.; Mao, J.; Zhou, Q.; Wang, B.; Wei, F.; et al. Highly Exfoliated Reduced Graphite Oxide Powders as Efficient Lubricant Oil Additives. *Adv. Mater. Interfaces* **2016**, *3*, 1–8. [CrossRef]
7. Zidan, H.M.; Hegazy, M.; Abd-Elwahed, A.; Yehia, H.M.; El Kady, O.A. Investigation of the Effectuation of Graphene Nanosheets (GNS) Addition on the Mechanical Properties and Microstructure of S390 HSS Using Powder Metallurgy Method. *Int. J. Mater. Technol. Innov.* **2021**, *1*, 52–57. [CrossRef]
8. Worsley, M.; Olson, T.Y.; Lee, J.; Willey, T.; Nielsen, M.H.; Roberts, S.K.; Pauzauskie, P.J.; Biener, J.; Satcher, J.H.; Baumann, T.F. High Surface Area, sp2-Cross-Linked Three-Dimensional Graphene Monoliths. *J. Phys. Chem. Lett.* **2011**, *2*, 921–925. [CrossRef] [PubMed]
9. Kausar, A.; Rafique, I.; Anwar, Z.; Muhammad, B. Perspectives of Epoxy/Graphene Oxide Composite: Significant Features and Technical Applications. *Polym. Technol. Eng.* **2015**, *55*, 704–722. [CrossRef]
10. Kumar, S.J.N.; Keshavamurthy, R.; Haseebuddin, M.R.; Koppad, P. Mechanical Properties of Aluminium-Graphene Composite Synthesized by Powder Metallurgy and Hot Extrusion. *Trans. Indian Inst. Met.* **2017**, *70*, 605–613. [CrossRef]
11. Geim, A.K. Graphene: Status and Prospects. *Science* **2009**, *324*, 1530–1534. [CrossRef] [PubMed]
12. Li, X.; Zhu, Y.; Cai, W.; Borysiak, M.; Han, B.; Chen, D.; Piner, R.D.; Colombo, L.; Ruoff, R.S. Transfer of Large-Area Graphene Films for High-Performance Transparent Conductive Electrodes. *Nano Lett.* **2009**, *9*, 4359–4363. [CrossRef] [PubMed]
13. Tang, C.; Wang, H.-F.; Huang, J.-Q.; Qian, W.; Wei, F.; Qiao, S.-Z.; Zhang, Q. 3D Hierarchical Porous Graphene-Based Energy Materials: Synthesis, Functionalization, and Application in Energy Storage and Conversion. *Electrochem. Energy Rev.* **2019**, *2*, 332–371. [CrossRef]
14. Jeon, C.-H.; Jeong, Y.-H.; Seo, J.-J.; Tien, H.N.; Hong, S.-T.; Yum, Y.-J.; Hur, S.-H.; Lee, K.-J. Material properties of graphene/aluminum metal matrix composites fabricated by friction stir processing. *Int. J. Precis. Eng. Manuf.* **2014**, *15*, 1235–1239. [CrossRef]
15. Stankovich, S.; Dikin, D.A.; Dommett, G.H.B.; Kohlhaas, K.M.; Zimney, E.J.; Stach, E.A.; Piner, R.D.; Nguyen, S.; Ruoff, R.S. Graphene-based composite materials. *Nature* **2006**, *442*, 282–286. [CrossRef]

16. Bastwros, M.; Kim, G.-Y.; Zhu, C.; Zhang, K.; Wang, S.; Tang, X.; Wang, X. Effect of ball milling on graphene reinforced Al6061 composite fabricated by semi-solid sintering. *Compos. Part B Eng.* **2014**, *60*, 111–118. [CrossRef]
17. Lei, Y.; Jiang, J.; Bi, T.; Du, J.; Pang, X. Tribological behavior of in situ fabricated graphene–nickel matrix composites. *RSC Adv.* **2018**, *8*, 22113–22121. [CrossRef]
18. Hossam, M.Y.; Mohamed, M.; Allam, S.; Saleh, K. Fabrication of Aluminum Matrix Nanocomposites by Hot Compaction. *J. Pet. Min. Eng.* **2020**, *22*, 16–20. [CrossRef]
19. Wang, J.; Li, Z.; Fan, G.; Pan, H.; Chen, Z.; Zhang, D. Reinforcement with graphene nanosheets in aluminum matrix composites. *Scr. Mater.* **2012**, *66*, 594–597. [CrossRef]
20. Hu, Z.; Tong, G.; Lin, D.; Chen, C.; Guo, H.; Xu, J.; Zhou, L. Graphene-reinforced metal matrix nanocomposites—A review. *Mater. Sci. Technol.* **2016**, *32*, 930–953. [CrossRef]
21. Haghighi, M.; Shaeri, M.H.; Sedghi, A.; Djavanroodi, F. Effect of Graphene Nanosheets Content on Microstructure and Mechanical Properties of Titanium Matrix Composite Produced by Cold Pressing and Sintering. *Nanomaterials* **2018**, *8*, 1024. [CrossRef] [PubMed]
22. Li, Z.; Fan, G.; Tan, Z.; Guo, Q.; Xiong, D.; Su, Y. Uniform dispersion of graphene oxide in aluminum powder by direct electrostatic adsorption for fabrication of graphene/aluminum composites. *Nanotechnology* **2014**, *25*, 325601. [CrossRef]
23. Zhang, J.; Chen, Z.; Zhao, J.; Jiang, Z.; Zhao, C.J.; Jiang, Z. Microstructure and mechanical properties of aluminium-graphene composite powders produced by mechanical milling. *Mech. Adv. Mater. Mod. Process.* **2018**, *4*, 4. [CrossRef]
24. Tabandeh-Khorshid, M.; Omrani, E.; Menezes, P.L.; Rohatgi, P.K. Tribological performance of self-lubricating aluminum matrix nanocomposites: Role of graphene nanoplatelets. *Eng. Sci. Technol. Int. J.* **2016**, *19*, 463–469. [CrossRef]
25. Zhang, W.L.; Gu, M.Y.; Wang, D.Z.; Yao, Z.K. Rolling and annealing textures of a SiCw/Al composite. *Mater. Lett.* **2004**, *58*, 3414–3418. [CrossRef]
26. Rashad, M.; Pan, F.; Tang, A.; Asif, M.; She, J.; Gou, J.; Mao, J.; Hu, H. Development of magnesium-graphene nanoplatelets composite. *J. Compos. Mater.* **2014**, *49*, 285–293. [CrossRef]
27. Peter, N.; Omayma, E.; Hossam, M.Y.; Atef, S.H.; Mohsen, A.H. Effect of Bimodal-Sized Hybrid TiC-CNT Reinforcement on the Mechanical Properties and Coefficient of Thermal Expansion of Aluminium Matrix Composites. *Met. Mater. Int.* **2021**, *27*, 753–766.
28. Nyanor, P.; El-Kady, O.; Yehia, H.M.; Hamada, A.S.; Nakamura, K.; Hassan, M.A. Effect of Carbon Nanotube (CNT) Content on the Hardness, Wear Resistance and Thermal Expansion of In-Situ Reduced Graphene Oxide (rGO)-Reinforced Aluminum Matrix Composites. *Met. Mater. Int.* **2019**, *27*, 1315–1326. [CrossRef]
29. Elasser, A.; Chow, T. Silicon carbide benefits and advantages for power electronics circuits and systems. *Proc. IEEE* **2002**, *90*, 969–986. [CrossRef]
30. Matsunami, H. Technological Breakthroughs in Growth Control of Silicon Carbide for High Power Electronic Devices. *Jpn. J. Appl. Phys.* **2004**, *43*, 6835–6847. [CrossRef]
31. Kretz, F.; Gácsi, Z.; Kovács, J.; Pieczonka, T. The electroless deposition of nickel on SiC particles for aluminum matrix composites. *Surf. Coatings Technol.* **2004**, *181*, 575–579. [CrossRef]
32. Fathy, A.; Sadoun, A.; Abdelhameed, M. Effect of matrix/reinforcement particle size ratio (PSR) on the mechanical properties of extruded Al-SiC composites. *Int. J. Adv. Manuf. Technol.* **2014**, *73*, 1049–1056. [CrossRef]
33. Zhang, S.; Han, K.; Cheng, L. The effect of SiC particles added in electroless Ni–P plating solution on the properties of composite coatings. *Surf. Coat. Technol.* **2008**, *202*, 2807–2812. [CrossRef]
34. Zhang, L.; He, X.; Qu, X.; Duan, B.; Lu, X.; Qin, M. Dry sliding wear properties of high volume fraction SiCp/Cu composites produced by pressureless infiltration. *Wear* **2008**, *265*, 1848–1856. [CrossRef]
35. Abolkassem, S.A.; Elkady, O.A.; Elsayed, A.H.; Hussein, W.A.; Yehya, H.M. Effect of consolidation techniques on the properties of Al matrix composite reinforced with nano Ni-coated SiC. *Results Phys.* **2018**, *9*, 1102–1111. [CrossRef]
36. Sikka, V.K.; Mavity, J.T.; Anderson, K. Processing of nickel aluminides and their industrial applications. In Proceedings of the Second International ASM Conference on High Temperature Aluminides and Intermetallics, San Diego, CA, USA, 16–19 September 1991; pp. 712–721.
37. Koch, C. Intermetallic matrix composites prepared by mechanical alloying—A review. *Mater. Sci. Eng. A* **1998**, *244*, 39–48. [CrossRef]
38. Carlson, T. Emerging applications of intermetallics N.S. *Trans. Am. Math. Soc.* **2016**, *369*, 2897–2916. [CrossRef]
39. Hodge, A.; Dunand, D. Synthesis of nickel–aluminide foams by pack-aluminization of nickel foams. *Intermetallics* **2001**, *9*, 581–589. [CrossRef]
40. Daoush, W.; Lim, B.K.; Mo, C.B.; Nam, D.H.; Hong, S.H. Electrical and mechanical properties of carbon nanotube reinforced copper nanocomposites fabricated by electroless deposition process. *Mater. Sci. Eng. A* **2009**, *513*, 247–253. [CrossRef]
41. Yehia, H.M.; Elkady, O.A.; Reda, Y.; Ashraf, K.E. Electrochemical Surface Modification of Aluminum Sheets Prepared by Powder Metallurgy and Casting Techniques for Printed Circuit Applications. *Trans. Indian Inst. Met.* **2018**, *72*, 85–92. [CrossRef]
42. Canakci, A.; Ozsahin, S.; Varol, T. Modeling the influence of a process control agent on the properties of metal matrix composite powders using artificial neural networks. *Powder Technol.* **2012**, *228*, 26–35. [CrossRef]

43. Yehia, H.M.; El-Tantawy, A.; Ghayad, I.; Eldesoky, A.S.; El-Kady, O. Effect of zirconia content and sintering temperature on the density, microstructure, corrosion, and biocompatibility of the Ti–12Mo matrix for dental applications. *J. Mater. Res. Technol.* **2020**, *9*, 8820–8833. [CrossRef]
44. Canakci, A.; Varol, T.; Erdemir, F. The Effect of Flake Powder Metallurgy on the Microstructure and Densification Behavior of B4C Nanoparticle-Reinforced Al–Cu–Mg Alloy Matrix Nanocomposites. *Arab. J. Sci. Eng.* **2015**, *41*, 1781–1796. [CrossRef]
45. Hassan, M.; Yehia, H.; Mohamed, A.; El-Nikhaily, A.; Elkady, O. Effect of Copper Addition on the AlCoCrFeNi High Entropy Alloys Properties via the Electroless Plating and Powder Metallurgy Technique. *Crystals* **2021**, *11*, 540. [CrossRef]
46. El-Tantawy, A.; El Kady, O.A.; Yehia, H.M.; Ghayad, I.M. Effect of Nano $ZrO_2$ Additions on the Mechanical Properties of $Ti_{12}Mo$ Composite by Powder Metallurgy Route. *Key Eng. Mater.* **2020**, *835*, 367–373. [CrossRef]
47. Yehia, H.M.; Allam, S. Hot Pressing of Al-10 wt% Cu-10 wt% Ni/x ($Al_2O_3$–Ag) Nanocomposites at Different Heating Temperatures. *Met. Mater. Int.* **2020**, *27*, 500–513. [CrossRef]
48. Yehia, H.M.; Daoush, W.; Mouez, F.A.; El-Sayed, M.H.; El-Nikhaily, A.E. Microstructure, Hardness, Wear, and Magnetic Properties of (Tantalum, Niobium) Carbide-Nickel–Sintered Composites Fabricated from Blended and Coated Particles. *Mater. Perform. Charact.* **2020**, *9*, 543–555. [CrossRef]
49. Yehia, H.M. Microstructure, physical and mechanical properties of the Cu/(WC-TiC-Co) nano-composites by the electro-less coating and powder metallurgy technique. *J. Compos. Mater.* **2019**, *53*, 1963–1971. [CrossRef]
50. El-Kady, O.; Yehia, H.M.; Nouh, F. Preparation and characterization of Cu/(WC-TiC-Co)/graphene nano-composites as a suitable material for heat sink by powder metallurgy method. *Int. J. Refract. Met. Hard Mater.* **2019**, *79*, 108–114. [CrossRef]
51. Yehia, H.M.; Abu-Oqail, A.; Elmaghraby, M.A.; Elkady, O.A. Microstructure, hardness, and tribology properties of the (Cu/MoS2)/graphene nanocomposite via the electroless deposition and powder metallurgy technique. *J. Compos. Mater.* **2020**, *54*, 3435–3446. [CrossRef]
52. Yehia, H.M.; Nouh, F.; El-Kady, O. Effect of graphene nano-sheets content and sintering time on the microstructure, coefficient of thermal expansion, and mechanical properties of (Cu/WC-TiC-Co) nano-composites. *J. Alloys Compd.* **2018**, *764*, 36–43. [CrossRef]
53. Yehia, H.M.; Daoush, W.M.; El-Nikhaily, A.E.; Yehia, H.M.; Daoush, W.M. Microstructure and physical properties of blended and coated (Ta, Nb) C/Ni cermets. *Powder Metall. Prog.* **2015**, *15*, 2.
54. Mordechay, S.; Milan, P. *Electroless Deposition of Nickel. Modern Electroplating*, 5th ed.; Wiley: Hoboken, NJ, USA, 2010; pp. 447–485.
55. Yehia, H.; El-Kady, O.; Abuoqail, A. Effect of diamond additions on the microstructure, physical and mechanical properties of WC-TiC-Co/Ni Nano-composite. *Int. J. Refract. Met. Hard Mater. Elsevier* **2018**, *71*, 198–205. [CrossRef]
56. Kim, D.W.; Kim, K.T.; Kwon, G.H.; Song, K.; Son, I. Self-propagating heat synthetic reactivity of fine aluminum particles via spontaneously coated nickel layer. *Sci. Rep.* **2019**, *9*, 1–8. [CrossRef]
57. Biswas, A.; Roy, S. Comparison between the microstructural evolutions of two modes of SHS of NiAl: Key to a common reaction mechanism. *Acta Mater.* **2003**, *52*, 257–270. [CrossRef]
58. Dyer, T.S.; Munir, Z.A. The synthesis of nickel aluminides by multilayer self-propagating combustion. *Met. Mater. Trans. A* **1995**, *26*, 603–610. [CrossRef]
59. Plazanet, L.; Nardou, F. Reaction process during relative sintering of NiAl. *J. Mater. Sci.* **1998**, *33*, 2129–2136. [CrossRef]
60. Sina, H.; Surreddi, K.B.; Iyengar, S. Phase evolution during the reactive sintering of ternary Al-Ni-Ti powder compacts. *J. Alloys Compd.* **2016**, *661*, 294–305. [CrossRef]
61. Zhang, Y.; Heim, F.M.; Bartlett, J.L.; Song, N.; Isheim, D.; Li, X. Bioinspired, graphene-enabled Ni composites with high strength and toughness. *Sci. Adv.* **2019**, *5*, eaav5577. [CrossRef] [PubMed]
62. Hernández-Méndez, F.; Altamirano-Torres, A.; Miranda-Hernández, J.G.; Térres-Rojas, E.; Rocha-Rangel, E. Effect of Nickel Addition on Microstructure and Mechanical Properties of Aluminum-Based Alloys. *Mater. Sci. Forum* **2011**, *691*, 10–14. [CrossRef]
63. Elsayed, E.; Harraz, F.A.; Saba, A. Nanocrystalline zinc oxide thin films prepared by electrochemical technique for advanced applications. *Int. J. Nanoparticles* **2012**, *5*, 136. [CrossRef]
64. Pauporté, T.; Lincot, D. Electrodeposition of semiconductors for optoelectronic devices: Results on zinc oxide. *Electrochim. Acta* **2000**, *45*, 3345–3353. [CrossRef]
65. Moharam, M.; Elsayed, E.; Nino, J.; Abou-Shahba, R.; Rashad, M. Potentiostatic deposition of $Cu_2O$ films as p-type transparent conductors at room temperature. *Thin Solid Film.* **2016**, *616*, 760–766. [CrossRef]

Article

# Extrusion Dwell Time and Its Effect on the Mechanical and Thermal Properties of Pitch/LLDPE Blend Fibres

Salem Mohammed Aldosari [1,2,*] and Sameer Rahatekar [1]

1. Enhanced Composite and Structures Centre, School of Aerospace, Transport, and Manufacturing, Cranfield University, Cranfield MK43 0AL, UK; S.S.Rahatekar@cranfield.ac.uk
2. National Center for Aviation Technology, King Abdulaziz City for Science and Technology (KACST), Riyadh 11442, Saudi Arabia
* Correspondence: S.M.aldosari@cranfield.ac.uk

**Abstract:** Mesophase pitch-based carbon fibres have excellent resistance to plastic deformation (up to 840 GPa); however, they have very low strain to failure (0.3) and are considered brittle. Hence, the development of pitch fibre precursors able to be plastically deformed without fracture is important. We have previously, successfully developed pitch-based precursor fibres with high ductility (low brittleness) by blending pitch and linear low-density polyethylene. Here, we extend our research to study how the extrusion dwell time (0, 6, 8, and 10 min) affects the physical properties (microstructure) of blend fibres. Scanning electron microscopy of the microstructure showed that by increasing the extrusion dwell from 0 to 10 min the pitch and polyethylene components were more uniformly dispersed. The tensile strength, modulus of elasticity, and strain at failure for the extruded fibres for different dwell times were measured. Increased dwell time resulted in an increase in strain to failure but reduced the ultimate tensile strength. Thermogravimetric analysis was used to investigate if increased dwell time improved the thermal stability of the samples. This study presents a useful guide to help with the selection of mixes of linear low-density polyethylene/pitch blend, with an appropriate extrusion dwell time to help develop a new generation of potential precursors for pitch-based carbon fibres.

**Keywords:** blend; extrusion; dwell time; morphology; carbon-fibres; mesophase pitch; polyethylene

Citation: Aldosari, S.M.; Rahatekar, S. Extrusion Dwell Time and Its Effect on the Mechanical and Thermal Properties of Pitch/LLDPE Blend Fibres. *Crystals* **2021**, *11*, 1520. https://doi.org/10.3390/cryst11121520

Academic Editors: Walid M. Daoush, Pavel Lukáč, Fawad Inam, Mostafa Ghasemi Baboli and Maha M. Khayyat

Received: 13 October 2021
Accepted: 30 November 2021
Published: 5 December 2021

**Publisher's Note:** MDPI stays neutral with regard to jurisdictional claims in published maps and institutional affiliations.

**Copyright:** © 2021 by the authors. Licensee MDPI, Basel, Switzerland. This article is an open access article distributed under the terms and conditions of the Creative Commons Attribution (CC BY) license (https://creativecommons.org/licenses/by/4.0/).

## 1. Introduction

Polymers are progressive substitutes for metal and wood, but their relatively poor performance regarding strength and stiffness, limits their ability to compete in many applications. Thus, in modern industries, from textiles to aerospace, there is an ongoing demand for improvements in the performance of polymer-based materials [1–3]. For example, there is great demand for carbon fibre (CF)-reinforced composite in the aerospace industry. Indeed, CFs using rayon as a precursor [4,5] were in widespread use in the early 1960s and, since 1963, pitch has successfully been used as the precursor for CFs with a superior elastic modulus [6].

The superior properties of CFs mean they are used in numerous applications from healthcare to space exploration [2]. Nevertheless, their industrial usefulness could be enormously improved if they were manufactured at lower cost and with enhanced mechanical properties [7]. Manufacturing CFs using petroleum derived pitch could reduce material costs for manufacturing CFs [8,9].

Previous reports [10] have shown that if the temperature is not controlled, the pitch filament could break in a brittle manner because the mesophase pitch fibre extrusion is sensitive to temperature changes. Researchers have also found that the fibres are also easily damaged during the spinning process and were difficult to handle before carbonization [11]. Other reports mention that decreasing the temperature by a few degrees significantly increased the viscosity of the mesophase pitch and increased the likelihood of brittle fracture during winding [12]. Lim and Yeo (2017) reported that it is very hard to wind pitch fibres because they are brittle and have low tensile strength [13].

Aldosari et al. (2020) [14] showed how to obtain ductile pitch based carbon fibre precursors using pitch and LLDPE blend fibres, which were shown to be an appropriate material for manufacturing carbon fibres [15,16]. However, during the mixing of different materials/polymers the extrusion dwell time will have an effect on the physical properties of the final product [17,18]. The aim of this work is to systematically study the effects of extrusion dwell time on the mechanical and thermal properties of pitch and polyethylene blend fibres.

As a part of the current research, we varied the extrusion dwell time of LLDPE and pitch blend from 0 min to 10 min. The effect of the extrusion dwell time on the tensile strength, tensile modulus and onset degradation temperature was then determined experimentally. Such a study will be a valuable aid in the selection of appropriate dwell times for manufacturing ductile pitch-based precursor fibres for the future development of low-cost carbon fibre manufacturing.

## 2. Experimental Procedures

*2.1. Materials*

Bonding Chemical supplied the mesophase pitch, mesophase content 92% and Sabic supplied the LLDPE. Table 1 shows relevant properties.

**Table 1.** Relevant properties of MP precursor and LLDPE.

| Relevant Property | MP Precursor | LLDPE |
|---|---|---|
| Softening Point | 268 °C | 99 °C |
| Melting point | 298 °C | 121 °C |
| Density | 1.425 g/cm$^3$ | 0.918 g/cm$^3$ |

*2.2. Materials Processing*

Figure 1 shows a schematic flow chart illustrating the different factors affecting the screw extrusion process, which is used to blend consistent, uniform mixes, often of several materials [19]. To successfully investigate how to improve the material properties of the fibres it is necessary to appreciate how the mixing ratios and extrusion process affect the morphology of the fibres.

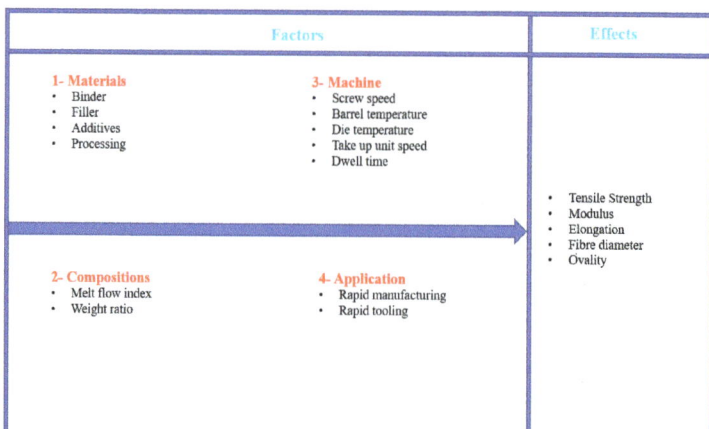

**Figure 1.** Effect of different factors on mixes of materials added to the extruder (adapted from Boparai et al. [19]).

A novel and valuable aspect of this work is that it completes the study of the interrelations between physical properties (both thermal and mechanical) and the microstructures of a linear low-density polyethylene/mesophase pitch (LLDPE/MP) blend over a range of extrusion dwell times.

"Mixing" is the processes whereby nonuniformity of concentrations are reduced. By definition, mixing increases the system's configurational entropy, which is a maximum when the locations of the constituents are random [20]. The time the polymer remains inside the extruder as a physically and chemically active hot melt will be termed the "effective dwell time". The distribution of pellets will determine their effective dwell time; the nearer the die the shorter the time, the nearer the hopper the longer the time Figure 2 [21].

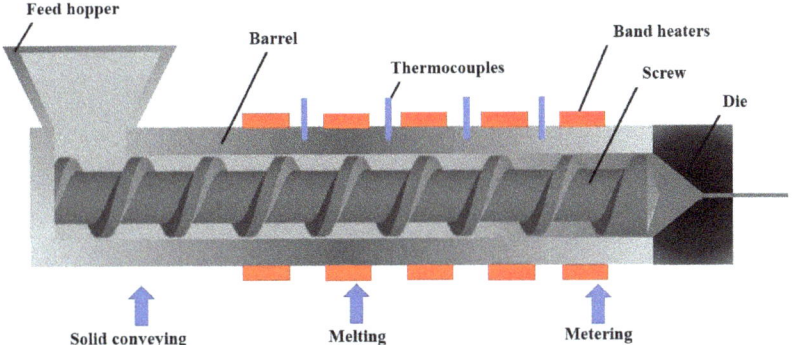

**Figure 2.** Extrusion of process of materials (adapted from Zhang et al. [22]). (Red blocks indicate band heaters).

The diameter of the nozzle of the Noztek extruder was 0.5 mm and its length was 1.5 mm, see Figure 2. The extrusion temperature is the temperature on the extreme right of Figure 2, at the entrance to the die was Figure 2. set to 315 °C with a volume rate of extrusion through the nozzle of close to 500 mm$^3$/min and a linear speed of 2.5 m/min, which was equal to the stretching speed. These values were maintained constant and the dwell time could be varied by changing the rate at which pitch and PE were fed into the extruder. The relative proportions to obtain the required fibre blend with 20 wt% LLDPE were, of course, constant. Figure 2 shows the stages within the extrusion process. The extrusion dwell times chosen to study the extrusion conditions for optimum thermal and mechanical properties of the CF were 0, 6, 8, and 10 min.

Note that a mix with 20 wt% LLDPE meant the mix contained 80 wt% mesophase pitch. All results presented below are for a blend of LLDPE/MP with 20 wt% of LLDPE unless otherwise stated.

### 3. Measurement Techniques

*3.1. Scanning Electron and Optical Microscopy*

The diameters of the fibres were measured using an optical microscope (a Nikon ECLIPSE ME600 at 20× magnification, Melville, NY, USA) with digital interference contrast microscopy to enhance sample contrast.

A Tescan VEGA3 SEM (Brno-Kohoutovice, Czech Republic) with Aztec software (Abingdon, UK) was used to investigate the specimens. They were prepared by being cut in 5.0 cm lengths, then having a thin layer of gold sputter-coated onto their end surfaces. For surface features, the specimens were placed lengthwise on horizontal aluminium stubs, and for examination of the end faces, were mounted end-on for analysis. For additional information, see [14].

## 3.2. Mechanical Tests

Tensile testing of the drawn and spun fibres were carried out according to ISO 11566-1996 [23], see Appendix A for Figure A1 and details of equipment used. A single filament was lightly stretched longitudinally across the centre of the elliptic slot, with the ends of the filament temporarily fixed using adhesive tape. With the filament in place, it was bonded to the mount using a single drop of Loctite 406 adhesive. To confirm repeatability and reproducibility, each fibre was the subject of six tests.

## 3.3. Differential Scanning Calorimetry and Thermogravimetric Analysis

The differential scanning calorimetry (DSC) measurements were made using a Mettler Toledo DSCQ2000 (New Castle, DE, USA) and thermogravimetric analysis (TA) using a Mettler Toledo Thermogravimetric Analyser TGAQ500 (New Castle, DE, USA). Both sets of measurements were performed in a nitrogen environment. The DSC samples were heated from 0 °C to 200 °C, then maintained at that temperature for 3 min and 20 s to remove all traces of any previous thermal events. The relevant thermal characteristics were found after carrying out cyclical heating and cooling of the specimens at a steady rate of 20 °C/min.

For the TA measurements, the samples were heated from 50 °C to 800 °C at 20 °C/min, then maintained at 800 °C for 5 min to remove any traces of prior thermal events.

# 4. Results and Discussion

## 4.1. Optical Microscopy

The morphology of the different blends of fibres was investigated using the Nikon system described above. As shown in Table 2, the diameter of the sample fibres increased with extrusion dwell time, see also Figure 3. Extrusion viscosity from the nozzle increases as temperature increases because the viscosity of the blend decreased. Our previous differential scanning calorimetry (DSC) data showed similar melting temperature for neat LLDPE and blend of pitch/LLDPE [14].

**Table 2.** Fibre diameter for LLDPE/MP as a function of extrusion dwell time.

| Extrusion Dwell Time (min) | Fibre Diameter, μm |
|---|---|
| 0 | 154 (±0.51) |
| 6 | 189 (±0.26) |
| 8 | 223 (±0.54) |
| 10 | 234 (±0.30) |

The amount of time that a randomly selected, small volume element in the feed stream spends inside the mixing equipment (extruder) is referred to as the residence or dwell time [24].

Shrinkage was observed in fibre diameter with an increase in LLDPE content for dwell time and all other factors kept constant. This was because fibre morphology changed as the LLPDE content increased, with the same extensive force producing a larger axial elongation [25]. Factors influencing the diameters of the extruded fibres included the viscosity of the polymer, die orifice diameter, and speed of extrusion.

## 4.2. SEM of LLDPE/MP Fibres

The images of the pitch fibres seen in Figure 4 show that microfibrils are present within fibres. In Figures 4a–h, and A2 the SEM images are for LLDPE/MP fibres with extrusion dwell times of 0 Figure 4a,b; 6 Figure 4c,d; 8 Figure 4e,f and 10 min Figure 4g,h.

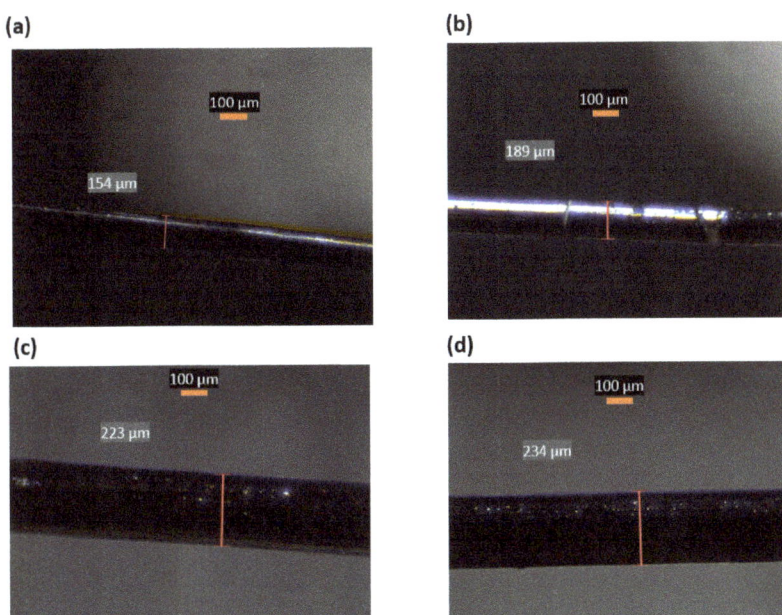

**Figure 3.** Images of LLDPE/MP fibres extrusion dwell times: (**a**) 0, (**b**) 6, (**c**) 8, and (**d**) 10 min.

There are two types of mixing: (1) distributive, or simple mixing, which refers to achieving a uniform spatial distribution of the different components, usually distributive mixing does not need high stresses, and (2) dispersive, or intensive mixing, which refers to achieving a fine level of dispersion and will often require reduction in the size of the components and so only occurs when the stress in the melt exceeds that necessary to rupture the component [21].

For zero minute dwell time the LLDPE pitch blend shows simple distributive mixing, which resulted in large mesophase pitch particles distributed in the LLDPE matrix Figure 4a,b.

However, with increases in dwell time from zero to 6, and then 8 min, distributive mixing of the pitch fibre in the LLDPE matrix was observed. By increasing the dwell time during the extrusion process from 0, to 6, to 8, and to 10 min the homogenous dispersion of pitch domain particles in the LLDPE matrix is increased Figure 4a–h. Thus, with a dwell time of 10 min, the matrix shows minimum number of voids; the pores trapped in the LLDPE/MP composite matrix had coalesced together. It may be, by increasing the extrusion dwell time, the diffusion bonding of LLDPE between the pitch domain is increased and reached its greatest value at 10 min. However, the 0, 6 and 8 min extrusion dwell times are not sufficient to form diffusion bonding between the fused LLDEP at the extrusion temperature of 315 °C.

The effect of the extrusion dwell time on the micrographs of the produced composite samples as shown in Figure 4 were determined by measuring the pitch domain diameters. Table 3 shows that the pitch domain size is reduced with increase in the extrusion dwell time. The MP molecules have a powerful inclination to align with the longitudinal axis of the fibre, a process which tends to start in the liquid crystalline phase [7].

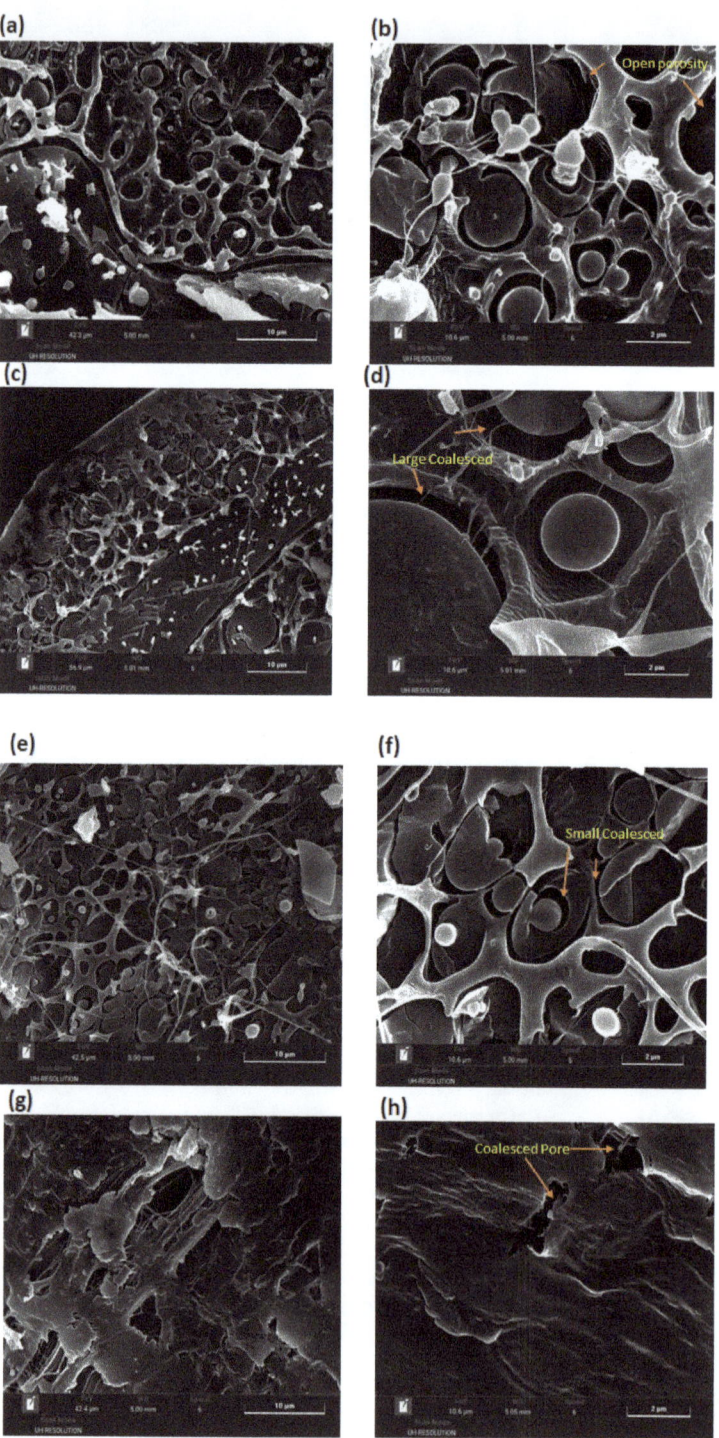

**Figure 4.** Morphology of LLDPE/MP fibres for different dwell times: zero (**a**,**b**), 6 (**c**,**d**), 8 (**e**,**f**), and 10 min (**g**,**h**).

**Table 3.** Particle size and air gap with extrusion dwell time.

| Extrusion Dwell Time Duration (min) | Average Pitch Domain Size (μm) |
|---|---|
| 0 | 1.38 (±0.2) |
| 6 | 1.16 (±0.4) |
| 8 | 1.08 (±0.3) |
| 10 | 0.92 (±0.2) |

*4.3. Effect of Extrusion Dwell Time on the Tensile Strength*

LLDPE is extensively used because it is inexpensive and versatile and has a tensile strength of about 6 MPa. This can be increased by adding fibre straws or multiple-walled carbon nanotubes to as high as 22 MPa [26–28], which would enable the uses of LLDPE to expand if it were to be used in combination with suitable additives.

The stress limit, the ultimate strength of extruded, neat LLDPE can be as high as 45 MPa, with strain of approximately 0.8, which implies that the LLDPE component within the fibres is critical for the load-bearing capacity of the LLDPE/MP fibres [14]. Figure 5 presents plots of stress against strain for LLDPE/MP fibres produced with different extrusion dwell times (Table 4). The samples showed high tensile strength (10.3 MPa) for zero dwell time but low strain to failure (0.23). However, samples with the highest extrusion dwell time (10 min) showed the lowest tensile strength (4.08 MPa) and highest strain to failure (0.60). We attribute this increase in strain to failure to the increased dwell time which allowed the LLDPE (the ductile polymer) to distribute more uniformly in the LLDPE/MP fibres and act as a plasticizer. Similar behaviour has been observed in epoxy rubber composites where the ductile rubber component helped to increase the strain to failure in the composite [29–33]. The samples with 6 min extrusion dwell time showed a good combination of high strength (8.36 MPa) and relatively high strain to failure (0.46). The results for an extrusion dwell time of 6 min appear anomalous because for this time we obtained moderate values of both ultimate strength and elongation, whereas for other durations we obtained either larger elongation values at the expense of ultimate strength, or higher ultimate strengths at the expense of elongation.

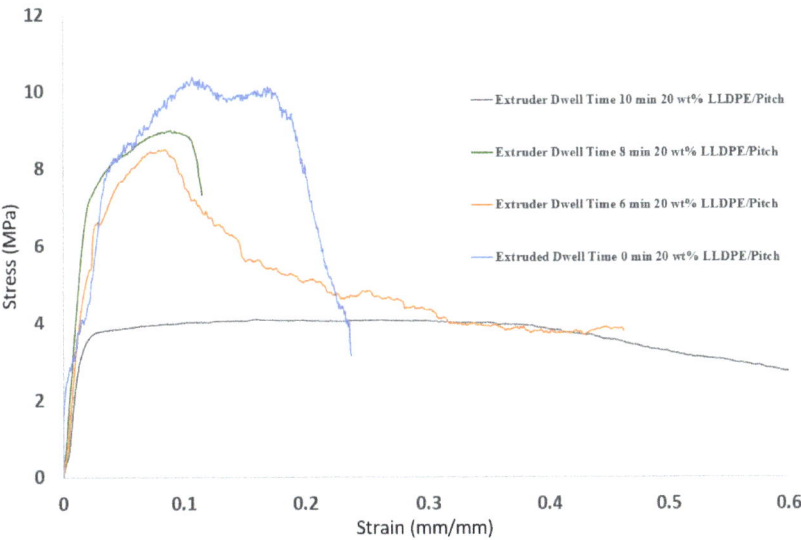

**Figure 5.** LLDPE/MP fibres, tensile strength of four different extrusion dwell times.

**Table 4.** Tensile strength, tensile modulus, and strain at failure of LLDPE/MP fibres for different extrusion dwell times. (Figures in brackets represent the standard deviation).

| Extrusion Dwell Time (Min) | Tensile Strength (MPa) | Tensile Modulus (MPa) | Strain at Failure |
|---|---|---|---|
| 0 | 10.3 (±0.87) | 763 (±5.3) | 0.23 (±0.025) |
| 6 | 8.36 (±0.83) | 823 (±4.5) | 0.46 (±0.022) |
| 8 | 8.98 (±0.57) | 842 (±3.8) | 0.11 (±0.036) |
| 10 | 4.08 (±0.65) | 857 (±5.6) | 0.60 (±0.028) |

*4.4. Differential Scanning Calorimetry for LLDPE/MP Blend*

As we have previously reported, Aldosari et al. [14] previously reported crystallization of different blends of LLDPE with LDPE and HDPE has been investigated using DSC. It was found that, typically, increasing the proportion of LLDPE, reduced crystallization temperature (usually ascribed to the LLDPE's higher molar mass), which increased chain entanglements, making crystallisation more difficult.

Analysis of the experimental results to obtain the required kinetic data revealed that the degradation rate of LLDPE was of relatively minor importance; the degradation started at 310 °C and 50% of the polymer degraded and was volatile at 430 °C [34]. The thermal properties of the given blend of LLDPE/MP with changes in extrusion dwell time between 0 and 10 min were investigated using DSC. Figure 6 shows the results of tests performed for a range of crystallization and melting temperatures.

For the given LLDPE/MP blend, Figure 6a,b presents well-ordered plots with clear crystallization peaks, showing that the more extended the extrusion dwell time the larger the peak value of the heat flow, except at 6 min. It is also observed, see Figure 6c,d, that for the given blend, increasing the extrusion dwell time increases the enthalpy of crystallization. Figure 6a,b also shows that the crystallisation temperature increased with increase in extrusion dwell time, confirming that the crystallisation temperature of the blend is affected by extrusion dwell time.

Figure 6c,d and Table 5 show that for the given LLDPE/MP blend, the melt temperature is a function of extrusion dwell time, as is the corresponding enthalpy of fusion. The melting or fusion temperature and maximum value of the enthalpy of fusion increased as extrusion dwell time increased. This is because the longer the extrusion dwell time the fewer crystalline spheres are present in the blend [14], confirming that the extrusion dwell time influenced both the fusion and crystallisation temperatures of the blends. The effect of extrusion dwell time on the fusion and crystallization temperature is negligible at all extrusion dwell times and the percentage difference in their values between maximum and minimum time is less than 1%. On the other hand, the extrusion dwell time has a more pronounced effect on the enthalpy of fusion compared with fusion and crystallization temperature. Therefore, the percentage difference in the enthalpy of fusion between 6 and 8 min is equal 37.5%. On other hand, the percentage difference between 6 and 10 min is equal 50% as shown in Table 5. This means that the effect of extrusion dwell temperature is more pronounced on the enthalpy of fusion compared to fusion and crystallization temperature by increasing the dwell time.

**Table 5.** Fusion and crystallisation temperatures of LLDPE/MP blend.

| Extrusion Dwell Time (min) | Fusion Temperature (°C) | Crystallization Temperature (°C) | Enthalpy of Fusion (kJ/mol) | Enthalpy of Crystallisation (kJ/mol) |
|---|---|---|---|---|
| 0 | 123.0 | 102.0 | 350 | 1750 |
| 6 | 122.8 | 101.6 | 271 | 1355 |
| 8 | 123.5 | 102.7 | 372 | 1860 |
| 10 | 123.6 | 102.9 | 406 | 2030 |

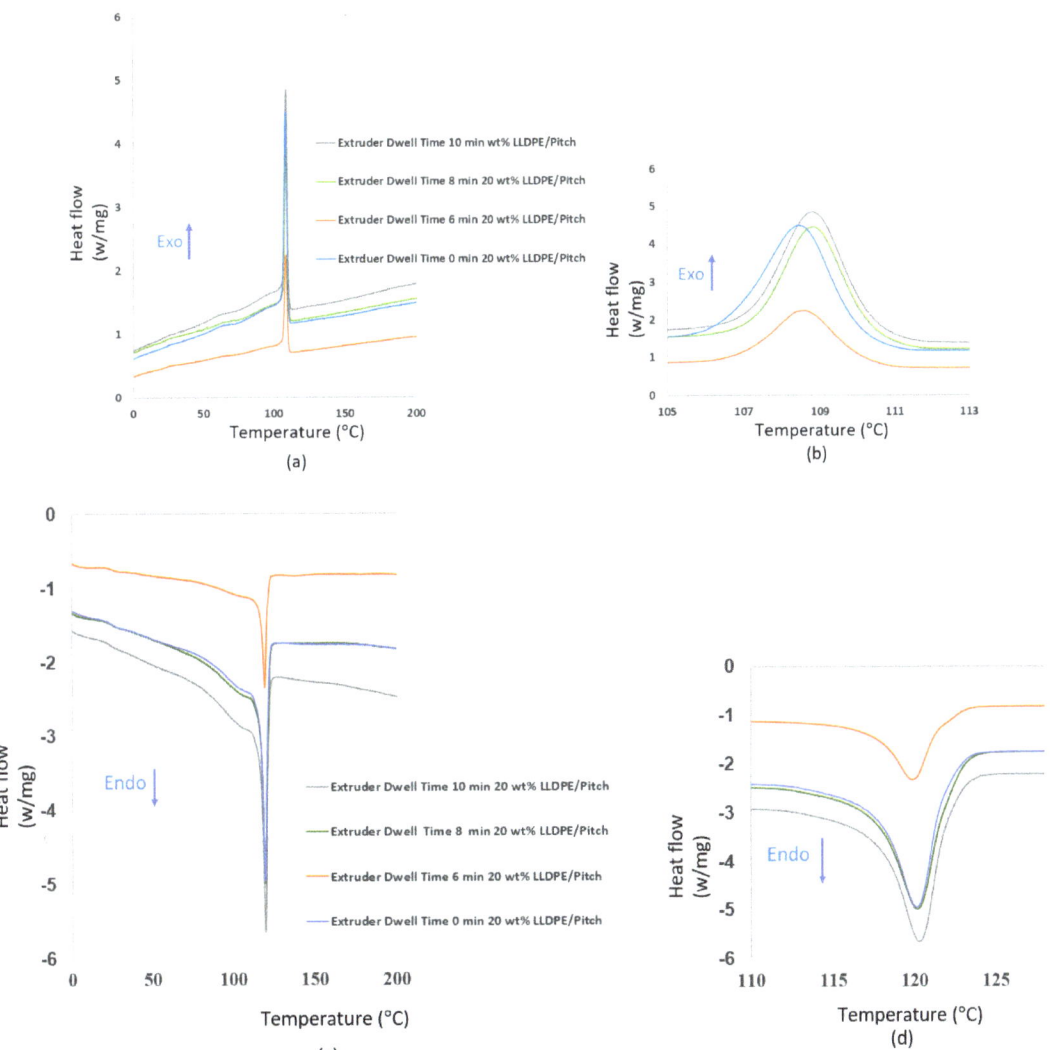

**Figure 6.** (**a**,**b**) Exothermal heat flow vs. temperature for crystallisation of LLDPE/MP sample and (**c**,**d**) endothermic heat flow for melting of LLDPE/MP sample. Four different extrusion dwell times. (Enthalpy of crystallisation, $\Delta H_c$, and enthalpy of fusion, $\Delta H_m$, are the areas under the respective curves).

DSC was used to assess the crystallization temperatures of the LDPE and LLDPE blend for temperatures between 110 °C and 120 °C, a temperature range in which pure LDPE does not melt. The population of crystallites melting increased with a 10 °C increase in temperature [35]. The longest extrusion dwell time (10 min) had the highest level of crystallization, possibly due to the influence of dwell time on crystallization during the extrusion process.

The enthalpy of fusion and melt temperature of the given LLDPE/MP blend changed with extrusion dwell time. It was observed that the longest dwell time (10 min) had the highest value for the melt temperature at 123 °C which may be due to the high crystallinity of the sample subject to the 10-min dwell time.

## 4.5. Thermogravimetric Analysis of the LLDPE/MP Blend

It is reported in the literature that higher decomposition temperatures of composite materials mean greater thermal stability [28,36,37]. Thermal decomposition for the given LLDPE/MP blend was investigated for different extrusion dwell times, and the resulting TGA curves, loss of mass vs. temperature, can be seen in Figure 7. The analysis was performed for temperatures from 100 °C to 800 °C.

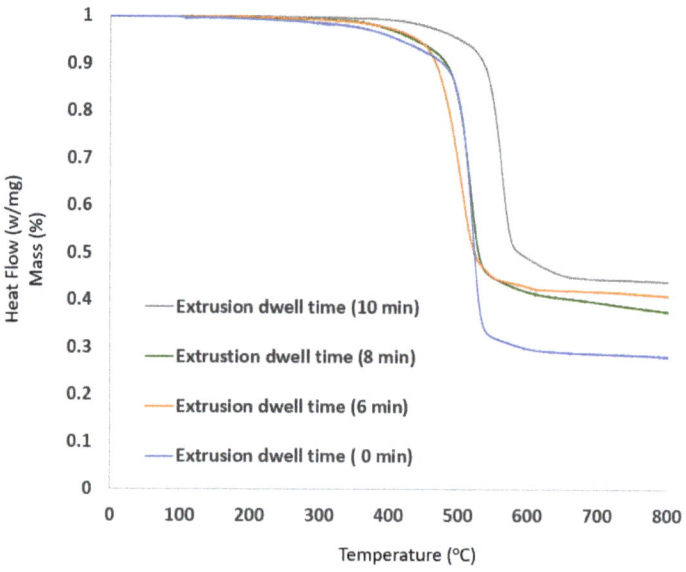

**Figure 7.** TGA curves for different extrusion dwell times for given LLDPE/MP blend.

Decomposition temperatures for various extrusion dwell times are shown in Figure 7 and Table 6. Those fibres subject to 10 min extrusion dwell time preserved more than 55% of the fibre mass until about 800 °C. For zero extrusion dwell time only about 30% of the mass was preserved, while for 8 min extrusion dwell time about 40% of mass was preserved. This confirms that, generally, the longer the extrusion dwell time the more thermally stable the blend, i.e., that the duration of the extrusion dwell time affects thermal stability. However, we also see that the mass preserved for 6 min extrusion dwell time can surpass that preserved for 8 min. With their decreased brittleness and relatively higher thermal stability, the 8 and 10 min extrusion dwell times showed higher onset degradation temperatures compared the other sample times and could provide acceptable precursors for the production of CFs. The onset points of decomposition that occur at which the sample shows a loss of 1 wt.% of its initial mass as shown in Table 6 due to the onset degradation temperature increased by increasing the extrusion dwell time.

**Table 6.** Onset and final degradation temperatures for LLDPE/MP with extrusion dwell time.

| Extrusion Dwell Time (min) | Onset Degradation Temperature * (°C) | Final Degradation Temperature (°C) | Final Residue (%) |
|---|---|---|---|
| 0  | 492.7 | 576.4 | 31 |
| 6  | 485.3 | 609.5 | 43 |
| 8  | 501.1 | 598.4 | 40 |
| 10 | 553.2 | 634.4 | 45 |

* Onset degradation temperature: temperature at which the sample has lost 1% of its initial mass.

## 5. Conclusions

This paper has shown that the extrusion dwell times for a LLDPE/MP blend (with 20 wt% LLDPE) can enhance the fibre morphology, mechanical and physical properties. For a range of extrusion dwell times from zero to 10 min, it has been shown that increasing the time significantly affects the morphology of the LLDPE/MP blend. It has also been demonstrated that the diameter of the fibres and the enthalpy of fusion both increased by increasing the extrusion dwell time due to die swell.

TGA results demonstrated that, generally, by increasing the extrusion dwell time the onset degradation temperature is increased and the final degradation temperature increased the residual mass (the mass remaining constant after 630 °C). Hence, optimisation of extrusion dwell time should be taken into account when considering the thermal stability and mechanical performance of LLDPE/MP blend fibres. An extension dwell time to 6 min showed a useful combination of high tensile strength, high strain to failure and relatively good thermal stability. Mechanical tests on the given LLDPE/MP blend showed that, generally, increasing the extrusion dwell time increased strain and enhancing failure.

It is expected that the findings reported in this work will be valuable to the development of a new generation of pitch-based carbon fibre precursors for end application in the aerospace industry.

**Author Contributions:** S.M.A. and S.R. conceptualized the theory and the method. S.M.A. performed the experimentation and the analysis. All authors have read and agreed to the published version of the manuscript.

**Funding:** This research received no external funding.

**Data Availability Statement:** Data can be available on a request from the corresponding author via email.

**Conflicts of Interest:** The authors declare no conflict of interest.

## Appendix A

The tensile test procedure was conducted according to the standard ISO 11566-1996.

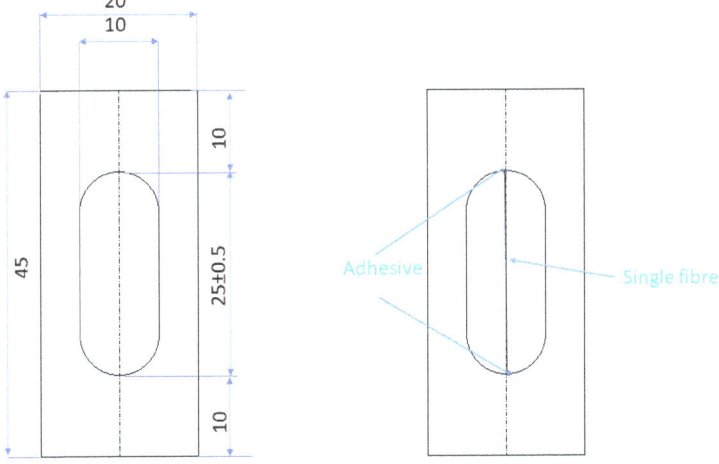

**Figure A1.** Tensile test setup for melt spun extrusion dwell time MP/PE fibre using a DEBEN Microtest fibre tensile tester and a Leica EC4 microscope.

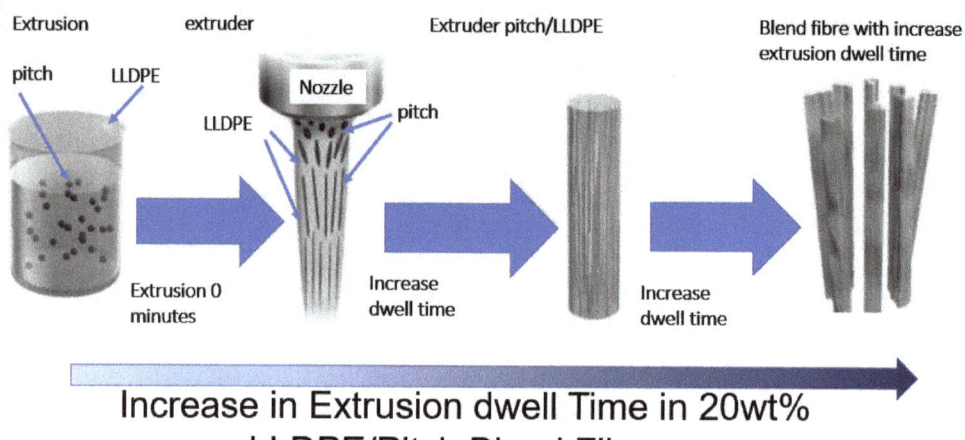

**Figure A2.** Increase in extrusion dwell time in 20 wt% LLDPE/pitch blend fibres.

## References

1. Toh, H.W.; Toong, D.W.Y.; Ng, J.C.K.; Ow, V.; Lu, S.; Tan, L.P.; En HouWong, P.; Venkatraman, S.; Huang, Y.; YingAng, H.; et al. Polymer blends and polymer composites for cardiovascular implants. *Eur. Polym. J.* **2021**, *146*, 1–15. [CrossRef]
2. Iqbal, A.; Saeed, A.; Ul-Hamid, A. A review featuring the fundamentals and advancements of polymer/CNT nanocomposite application in aerospace industry. *Polym. Bull.* **2020**, *78*, 539–557. [CrossRef]
3. Sharma, N. *Polymers and Textiles: You Are What You Wear*; American Institute of Chemical Engineers: New York, NY, USA, 2020.
4. Fitzer, E. Carbon Fibres Present State and Future Expectations. In *Carbon Fibers Filaments and Composites*; Figueiredo, J., Bernardo, C.A., Baker, R.T., Huttinger, K.J., Eds.; Kluwer Academic: Karlsruhe, Germany, 1990; pp. 3–41.
5. Liu, Y.; Kumar, S. Recent progress in fabrication, structure, and properties of carbon fibers. *Polym. Rev.* **2012**, *52*, 234–258. [CrossRef]
6. Park, S.J.; Lee, S.Y. History and Structure of Carbon Fibers. In *Carbon Fibers*; Springer: Incheon, Korea, 2015; pp. 1–30. [CrossRef]
7. Aldosari, S.M.; Khan, M.; Rahatekar, S. Manufacturing carbon fibres from pitch and polyethylene blend precursors: A review. *J. Mater. Res. Technol.* **2020**, *9*, 7786–7806. [CrossRef]
8. Ko, S.; Choi, J.E.; Lee, C.W.; Jeon, Y.P. Preparation of petroleum-based mesophase pitch toward cost-competitive high-performance carbon fibers. *Carbon Lett.* **2020**, *30*, 35–44. [CrossRef]
9. Singer, L.S.; Bacon, R. Carbon Fibres From Mesophase Pitch. *Fuel* **1979**, *60*, 839–847. [CrossRef]
10. Edie, D.D.; Dunham, M.G. Melt spinning pitch-based carbon fibers. *Carbon* **1989**, *27*, 647–655. [CrossRef]
11. Edie, D.D. The effect of processing on the structure and properties of carbon fibers. *Carbon* **1998**, *36*, 345–362. [CrossRef]
12. Gallego, N.C.; Edie, D.D. Structure-property relationships for high thermal conductivity carbon fibers. *Compos. Part A Appl. Sci. Manuf.* **2001**, *32*, 1031–1038. [CrossRef]
13. Lim, T.H.; Yeo, S.Y. Investigation of the degradation of pitch-based carbon fibers properties upon insufficient or excess thermal treatment. *Sci. Rep.* **2017**, *7*, 4733. [CrossRef] [PubMed]
14. Aldosari, S.; Khan, M.; Rahatekar, S. Manufacturing Pitch and Polyethylene Blends-Based Fibres as Potential Carbon Fibre Precursors. *Polymer* **2021**, *13*, 1445. [CrossRef]
15. Huang, X. Fabrication and properties of carbon fibers, Review. *Materials* **2009**, *2*, 2369–2403. [CrossRef]
16. Horikiri, S.; Amagasaki, J.; Minobe, M. Process for Production of Carbon Fiber. US4070446, 24 January 1978.
17. Buser, M.D.; Abbas, H.K. Effects of Extrusion Temperature and Dwell Time on Aflatoxin Levels in Cottonseed. *J. Agric. Food Chem.* **2002**, *50*, 2556–2559. [CrossRef] [PubMed]
18. Swolfs, Y.; Zhang, Q.; Baets, J.; Verpoest, I. The influence of process parameters on the properties of hot compacted self-reinforced polypropylene composites. *Compos. Part A Appl. Sci. Manuf.* **2014**, *65*, 38–46. [CrossRef]
19. Boparai, K.S.; Singh, R.; Singh, H. Modeling and optimization of extrusion process parameters for the development of Nylon6–Al–Al2O3 alternative FDM filament. *Prog. Addit. Manuf.* **2016**, *1*, 115–128. [CrossRef]
20. Baird, D.G. Polymer Processing. In *Encyclopedia of Physical Science and Technology*. Third; Meyers, R.A., Ed.; Academic Press: New York, NY, USA, 2003; pp. 611–643. [CrossRef]
21. Chung, C.I. *Physical Description of Single-Screw Extrusion*. *Extrusion of Polymers Theroy and Practice*, 2nd ed.; Hanser: Munich, Germany, 2019; pp. 14–57. [CrossRef]

22. Zhang, W.; Chen, J.; Zeng, H. Polymer processing and rheology. In *Polymer Science and Nanotechnology: Fundamentals and Applications*, 1st ed.; Narain, R., Ed.; Elsevier Inc.: Amsterdam, The Netherlands, 2020; pp. 149–178. [CrossRef]
23. British Standards. *BS ISO 11566: Carbon Fibre-Determination of the Tensile Properties of the Tensile-Filament Specimens*; British Standards Institute: London, UK, 1996.
24. Drobny, J.G. (Ed.) Processing Methods Applicable to Thermoplastic Elastomers. In *Handbook of Thermoplastic Elastomers*, 2nd ed.; William Andrew Publishing: Oxford, UK, 2014; pp. 33–173. [CrossRef]
25. Fakirov, S.; Bhattacharyya, D.; Lin, R.J.T.; Fuchs, C.; Friedrich, K. Contribution of Coalescence to Microfibril Formation in Polymer Blends during Cold Drawing. *J. Macromol. Sci. Part B* **2007**, *46*, 183–194. [CrossRef]
26. Zhang, L.; Xu, H.; Wang, W. Performance of Straw/Linear Low Density Polyethylene Composite Prepared with Film-Roll Hot Pressing. *Polymers* **2020**, *12*, 860. [CrossRef]
27. Durmus, A.; Kaşgöz, A.; Macosko, C.W. Mechanical properties of linear low-density polyethylene (LLDPE)/clay nanocomposites: Estimation of aspect ratio and interfacial strength by composite models. *J. Macromol. Sci. Part B Phys.* **2008**, *47*, 608–619. [CrossRef]
28. Tai, J.H.; Liu, G.Q.; Caiyi, H.; Shangguan, L.J. Mechanical properties and thermal behaviour of LLDPE/MWNTs nanocomposites. *Mater. Res.* **2012**, *15*, 1050–1056. [CrossRef]
29. Kargarzadeh, H.; Ahmad, I.; Abdullah, I. Mechanical Properties of Epoxy-Rubber Blends. In *Handbook of Epoxy Blends*; Springer International: Cham, Switzerland, 2015; pp. 1–36. [CrossRef]
30. Unnikrishnan, K.P.; Thachil, E.T. Toughening of epoxy resins. *Des. Monomers Polym.* **2006**, *9*, 129–152. [CrossRef]
31. Abadyan, M.; Bagheri, R.; Kouchakzadeh, M.A. Fracture Toughness of a Hybrid-Rubber-Modified Epoxy. I. Synergistic Toughening. *J. Appl. Polym. Sci.* **2012**, *125*, 2467–2475. [CrossRef]
32. Barcia, F.L.; Amaral, T.P.; Soares, B.G. Synthesis and properties of epoxy resin modified with epoxy-terminated liquid polybutadiene. *Polymer* **2003**, *44*, 5811–5819. [CrossRef]
33. Bucknall, C.B.; Yoshii, T. Relationship between structure and mechanical properties in Rubber-Toughened Epoxy Resins. *Br. Polym. J.* **1978**, *10*, 53–59. [CrossRef]
34. Bhardwaj, I.S.; Kumar, V.; Palanivelu, K. Thermal characterisation of LDPE and LLDPE blends. *Thermochim. Acta* **1988**, *131*, 241–246. [CrossRef]
35. Drummond, K.M.; Hopewell, J.L.; Shanks, R.A. Crystallization of low-density polyethylene- and linear low-density polyethylene-rich blends. *J. Appl. Polym. Sci.* **2000**, *78*, 1009–1016. [CrossRef]
36. Bottom, R. Thermogravimetric Analysis. In *Principles and Applications of Thermal Analysis*, 1st ed.; Gabbott, P., Ed.; Blackwell Publishing Ltd.: Oxford, UK, 2008; pp. 87–118. [CrossRef]
37. Wang, Z.; Cheng, Y.; Yang, M.; Huang, J.; Cao, D.; Chen, S.; Xie, Q.; Lou, W.; Wu, H. Dielectric properties and thermal conductivity of epoxy composites using core/shell structured Si/SiO$_2$/Polydopamine. *Compos. Part B Eng.* **2018**, *140*, 83–90. [CrossRef]

Article

# Influence of High-Concentration LLDPE on the Manufacturing Process and Morphology of Pitch/LLDPE Fibres

Salem Mohammed Aldosari [1,2,*], Muhammad A. Khan [3] and Sameer Rahatekar [1,*]

1. Enhanced Composite and Structures Centre School of Aerospace, Transport, and Manufacturing, Cranfield University, Cranfield MK43 0AL, UK
2. National Center for Aviation Technology, King Abdulaziz City for Science and Technology (KACST), Riyadh 11442, Saudi Arabia
3. Centre of Life-Cycle Engineering and Management School of Aerospace, Transport, and Manufacturing, Cranfield University, Cranfield MK43 0AL, UK; Muhammad.A.Khan@cranfield.ac.uk
* Correspondence: S.M.aldosari@cranfield.ac.uk (S.M.A.); S.S.Rahatekar@cranfield.ac.uk (S.R.)

Citation: Aldosari, S.M.; Khan, M.A.; Rahatekar, S. Influence of High-Concentration LLDPE on the Manufacturing Process and Morphology of Pitch/LLDPE Fibres. Crystals 2021, 11, 1099. https://doi.org/10.3390/cryst11091099

Academic Editors: Walid M. Daoush, Fawad Inam, Mostafa Ghasemi Baboli and Maha M. Khayyat

Received: 13 August 2021
Accepted: 5 September 2021
Published: 9 September 2021

**Publisher's Note:** MDPI stays neutral with regard to jurisdictional claims in published maps and institutional affiliations.

**Copyright:** © 2021 by the authors. Licensee MDPI, Basel, Switzerland. This article is an open access article distributed under the terms and conditions of the Creative Commons Attribution (CC BY) license (https://creativecommons.org/licenses/by/4.0/).

**Abstract:** A high modulus of elasticity is a distinctive feature of carbon fibres produced from mesophase pitch. In this work, we expand our previous study of pitch/linear low-density polyethylene blend fibres, increasing the concentration of the linear low-density polyethylene in the blend into the range of from 30 to 90 wt%. A scanning electron microscope study showed two distinct phases in the fibres: one linear low-density polyethylene, and the other pitch fibre. Unique morphologies of the blend were observed. They ranged from continuous microfibres of pitch embedded in linear low-density polyethylene (occurring at high concentrations of pitch) to a discontinuous region showing the presence of spherical pitch nodules (at high concentrations of linear low-density polyethylene). The corresponding mechanical properties—such as tensile strength, tensile modulus, and strain at failure—of different concentrations of linear low-density polyethylene in the pitch fibre were measured and are reported here. Thermogravimetric analysis was used to investigate how the increased linear low-density polyethylene content affected the thermal stability of linear low-density polyethylene/pitch fibres. It is shown that selecting appropriate linear low-density polyethylene concentrations is required, depending on the requirement of thermal stability and mechanical properties of the fibres. Our study offers new and useful guidance to the scientific community to help select the appropriate combinations of linear low-density polyethylene/pitch blend concentrations based on the required mechanical property and thermal stability of the fibres.

**Keywords:** mesophase-pitch; polyethylene; carbon-fibres; morphology; winder; blend

## 1. Introduction

There is an increasing need for polymer materials with improved properties to meet modern requirements since no single polymer has all desired properties [1]. Most polymer blends are prepared by melting [2] and polymer blending, with the fibres produced by a melt compound extruder. This is an economic and useful method of manufacturing new materials with advantageous properties [3–7]. To investigate how this can be improved, one must understand how the morphology of the obtained fibre is formed, changed, and controlled during the extrusion or mixing process, or with the help of mixing material ratios. An important and novel aspect of this research is that it completes the definition of the mechanical and thermal properties of LLDPE/MP blends for the complete range of mixes.

Carbon fibres (CFs) with manmade rayon as precursor had been created by 1960 [8,9], and by 1963, carbon fibres with a high elastic modulus were being manufactured from pitch [10]. Due to superior mechanical, electrical, and thermal features, CFs are extensively employed in numerous applications, ranging from space exploration to healthcare. However, their industrial application would greatly increase if they could be manufactured at a significantly lower cost with improved mechanical characteristics [11]. MP-based carbon fibre (MPCF) is one material that appears very promising due to how easily the mesophase appears [12,13].

CFs made from pitch precursors are classified into one of two types based on their characteristics and pitch precursor used: MP-based or derived from isotropic pitch. The former shows high resistance to longitudinal stress and, as used here, has a molar mass of roughly 2600 g/mol [14], while the latter possesses desirable mechanical properties. Controlled production of isotropic pitch is considered difficult because, above a specific temperature, mesophase spheres can suddenly appear [15]. Moreover, anisotropy can occur with the enhanced alignment of the fibres, which results in improved mechanical properties compared to CFs derived from isotropic pitch or polyacrylonitrile (PAN) [9]. However, while the precursor may be cheap, purification of the pitch is costly, and its widespread use in industry depends on the degree to which flaws are introduced in the course of manufacturing.

The manufacture of CFs using pitch and synthetic polymers that are easily available should lower the cost of materials, allowing CFs to be used more widely. Relative to expensive PAN-based fibres, the mechanical properties of relatively easily available, low-cost materials—such as organic polymers—have been found to be insufficient [16], whereas CFs have superior mechanical properties and are light weight [17]. Compared to PAN, textile-grade polyethylene (PE) is attractive as a precursor for CF synthesis for three reasons: It has a relatively high carbon content, a higher carbonation rate, and its use could substantially reduce the cost of production [18–21]. PE is cheap compared to PAN because between half and two-thirds of the production cost of PAN is incurred in synthesizing the precursor, which PE does not require. Fusion spinning is also a quicker and more environmentally friendly method of producing PE compared to PAN. However, MP has intrinsic shortcomings: MPCFs are brittle and that increases the difficulty of successfully spinning fibres [3,4]. This is a complex phenomenon that depends on stress level, intrinsic material properties (crystallinity and molecular orientation of the polymer), and other external parameters [22]. Oxidation of MPCF is the most critical step in the fabrication process, but it is also the most inefficient due to the time required [23].

As part of an extensive program to produce low-cost PE-based CFs with superior physical and other desirable characteristics, Huang et al., (2009) presented a new technique for the manufacture of a melt-spun, carbonised, and sulfonated PE precursor [24]. The same process could produce CFs with a 75% yield [18,20,23–31].

We blended LLDPE with MP to reduce fibre brittleness and increase fibre spinnability. LLDPE is very ductile, can be converted into CFs, and promises to be a superb blending material that can be used to produce less brittle MP carbon fibre precursors. Polymers are elements formed of a lengthy chain of molecules that repeat themselves. Low density, good strength-to-weight properties, resistance to corrosion, and low thermal conductivity are some of their advantages [32,33]. The molar mass of LLDPE is high (between 50,000 and 200,000 g/mol), higher than either HDPE or LDPE [34]. Our ultimate goal was to produce CF-based PE and MP, with a relatively low molar mass of between 400 and 800 g/mole, which could increase the proportion of carbon in the carbon fibre when it is added to PE [19,20,35,36]. Innovatively generating MP/PE-derived CFs at a reduced cost and combining enhanced ductility with stronger mechanical properties and fewer flaws can only improve the usability of both MPCF and PE carbon. Reinforcing using fibres is an important technological innovation in today's world, with applications extending from strengthening concrete [37] to improving agricultural mulching film [38]. Because LLDPE is so often a component of these advances, greater knowledge of its properties and behaviour when blended should be of wide interest.

Aldosari et al., (2020) have reviewed the possibility of achieving these desired objectives and showed that, despite relevant research being at an early stage, they could be achieved by using a PE/pitch blend to produce CF precursors with improved ductility [11]. In our previous work, we showed the feasibility of mixing LLDPE/pitch as a possible precursor for the manufacture of CFs [39]. That work was limited to a narrow LLDPE/MP range, from 0 to 20 wt% LLDPE, with a corresponding range from 100 to 80 wt% MP. In this paper, we have expanded our previous study into a more comprehensive investigation of LLDPE/MP blend fibres, extending the range investigated to between 30 and 90 wt% LLDPE in 10 wt% increments. This extended study allows us to better compare the variations introduced into the manufacturing process and the corresponding changes in mechanical properties and thermal stability of the fibres over the complete range of LLDPE in LLDPE/MP blend fibres from 0 to 90 wt%. This allows us to select the best possible LLDPE/MP blend. The optimum precursor fibre can potentially overcome the issue of the high level of the brittleness of pitch fibres for future high-performance CF manufacturing. As this research is a direct continuation of our previously published work, the same equipment was used and the reader may refer to [39] for details.

This work studies the effect of the LLDPE content of the blend on its morphology, ranging from continuous microfibres of pitch contained in LLDPE (at high concentrations of pitch) to discontinuous spheres of pitch (at high concentration of LLDPE) see Graphical Abstract. It considers factors affecting the chemical composition of the pitch mesophase dispersed in a matrix of LLDPE on the tensile strength and tensile modulus, and its correlation with the structural morphology of the pitch/LLDPE composite.

## 2. Experimental Methods

### 2.1. Materials

In this study, the MP precursor was purchased from Bonding Chemical. Its softening point was 268 °C, melting point 298 °C, and density 1.425 g/cm$^3$. The mesophase content was 92%. The LLDPE was bought from Sabic, Saudi Arabia; it had a softening point of 99 °C, a melting point of 121 °C, and a density of 0.918 g/cm$^3$.

### 2.2. Materials Processing

A Noztek Pro Filament single-screw melt fibre spinning extruder was used. Details of this and the fibre collection are given in our previous paper [35]. The nozzle was 0.5 mm in diameter and 1.5 mm in length. The extruder was set to 315 °C and at a 2.5 m/min extrusion speed, with the stretching speed set to 2.5 m/min. To obtain different fibre blends, pitch and PE were fed into the extruder in different ratios. The manufacturing processing stages are shown in Figure 1. The sample designation for the different proportions is LLDPE (x) wt% and MP (100-x) wt%. Figure 2 is a diagrammatic representation of cold stretching.

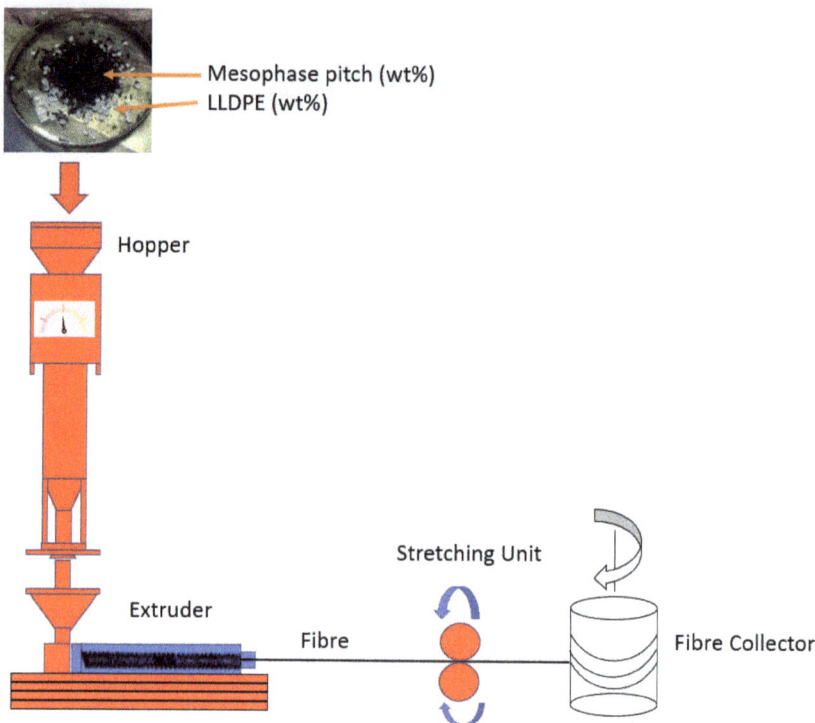

**Figure 1.** A flowchart of processing of materials.

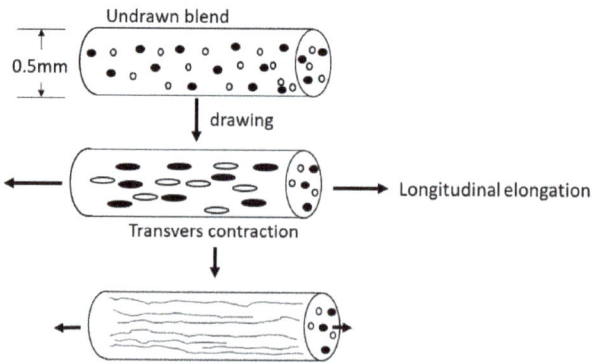

**Figure 2.** Melt blends cold drawn to make fibres; discontinuous pitch domains are stretched from spheres to ellipses.

## 3. Characterisation Methods

### 3.1. Microscopy: Optical and SEM

Optical microscopy: fibre diameter was measured via a Nikon ECLIPSE ME600 at 20× magnification. Nomarski microscopy was used to enhance sample contrast and determine the diameter of the fibres.

The SEM used to examine the prepared specimens was a Tescan VEGA3 and associated Aztec software. Before the measurements were made, a thin gold layer was sputter-coated onto the specimens. The specimens were cut into 50 mm lengths. The specimens were mounted end-on for the vertical image and for surface features laid horizontally on aluminium stubs, for more details see [39].

*3.2. Tensile Mechanical Test*

Tensile measurements on the spun MP/PE fibres were made in accordance with Standard ISO 11566-1996, Figure A1 [40] with a DEBEN Microtest fibre tensile tester connected to a Leica EC4 Microscope; see Figure A1 in Appendix A. We placed a single filament over the centre of the slot and temporarily fixed one end of specimen, mounting it with adhesive tape, and lightly stretched the specimen across the slot, then fixed the other end of the specimen to other ends of the mounting with adhesive tape, and then bound the specimen to the mounting by applying a drop of adhesive. Each sample was subjected to six tests to confirm repeatability.

*3.3. Differential Scanning Calorimetry (DSC)*

A Mettler Toledo DSCQ2000 was used for the DSC measurements. Specimens were heated from 0 to 200 °C at 20 °C/min in an inert environment. The sample was then kept at a higher temperature for 200 s to eliminate any prior thermal history. The non-isothermal behaviour and kinetics of the samples were studied by cooling them at 20 °C/min after completing the heating and cooling cycles.

*3.4. Thermogravimetric Analysis*

For the prepared specimens, a Mettler Toledo Thermogravimetric Analyser TGAQ500 was used for the TGA analysis. Under a nitrogen environment, the specimens were heated from 50 to 800 °C at 20 °C/min, after which the specimens were kept at 800 °C for 300 s to remove any preceding thermal history.

## 4. Results and Discussion

*4.1. Optical Microscopy*

Optical microscopy was used to study the morphology of the blend's fibres. The diameter of the sample fibres depended upon the wt% LLDPE, decreasing as the LLDPE content increased, see Table 1 and Figure 3.

All other factors remaining constant, we observed shrinkage in the diameter of the fibres with increased LLPDE wt% due to the greater axial elongation obtained by the same applied axial force; see Figure 2. The morphology of the blend changed with LLDPE wt% [41].

Many parameters influence the diameter of the extruded fibres, including the diameter of the die orifice, extrusion speed, and polymer viscosity. Figure 3 shows the optical images obtained for cold-drawn specimens using an optical microscope. Figure 3 also shows that different fibre diameters were obtained by reducing or increasing the LLDPE content.

**Table 1.** LLDPE/MP fibre diameter as a function of LLDPE wt%.

| Blend Designation | Fibre Diameter, µm |
|---|---|
| LLDPE (30 wt%)/MP | 152 (±0.54) |
| LLDPE (40 wt%)/MP | 149 (±0.30) |
| LLDPE (50 wt%)/MP | 146 (±0.36) |
| LLDPE (60 wt%)/MP | 143 (±0.38) |
| LLDPE (70 wt%)/MP | 140 (±0.51) |
| LLDPE (80 wt%)/MP | 138 (±0.53) |
| LLDPE (90 wt%)/MP | 135 (±0.24) |

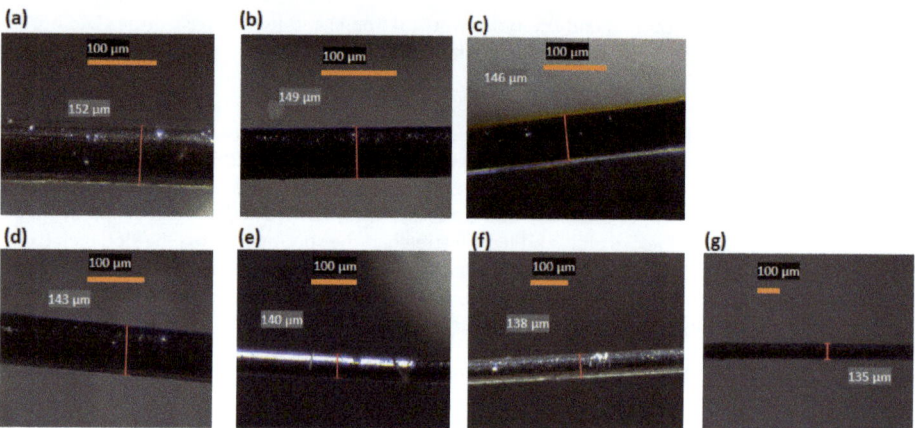

**Figure 3.** Images of LLDPE/MP fibres taken using a Nikon ECLIPSE ME600 optical microscope: (**a**) LLDPE 30 wt%, (**b**) LLDPE 40 wt%, (**c**) LLDPE 50 wt%, (**d**) LLDPE 60 wt%, (**e**) LLDPE 70 wt%, (**f**) LLDPE 80 wt%, and (**g**) LLDPE 90 wt%.

*4.2. SEM of Fibres*

Pitch fibres tend to generate microfibrils within their fibres, as shown in the SEM images [42]; also see Figure 4i–viii, and Figure A2 for LLDPE/MP blends in the range of between 30 and 100 wt% of LLDPE. The MP molecules show a strong preference to align themselves with the longitudinal fibre axis, and this process has a tendency to begin in the liquid crystalline phase [11]. This could be the reason for micro-fibre development in our samples. As the LLDPE wt% content of the LLDPE/MP fibres increased to about 60%, the development of microfibres inside the blend was significantly reduced; see Figure 4iii,iv.

It was observed from the results that the fibrous pitch/LLDPE domains were due to the continuously increasing elongational force along the spinning line, and they gradually extended from spheres into long continuous nanofibrillar formations (Figure 2). The creation of fibrillar structures is necessary to improve the mechanical properties of polymer blend fibres [43,44]. The PE does not form the liquid crystal phase, so when PE is added to the pitch, it reduces the microfibre content. With a high LLDPE content, e.g., 70 wt% LLDPE (Figure 4v,vi), we see discontinuous elliptical-shaped inclusions of pitch domains. At 80 wt% LLDPE (Figure 4vii,viii), no continuous microfibre formations in the LLDPE/pitch blend fibres are observed, and we see small spherical pitch domains dispersed in LLDPE. Pure LLDPE fibres (Figure 4viiii) do not show any micro-fibres.

With immiscible polymer blends, the morphology of the dispersed phase is crucial in determining its physical properties [45]. As a result, we deduced that increasing the wt% LLDPE in a blend substantially affects the morphology of the fibre. It was observed from the results that the extruder rotation rate affected morphology by changing the polymer dwell time in the extruder [46]. It is widely known that the performance and properties of polymer blends are directly related to the morphology of the pitch/LLDPE polymer blends [47], so the size, shape, and orientation of the phases can be used to characterise the morphology of pitch/LLDPE polymer blends. A major problem in the development of polymer blends is how to influence morphology. Here, we have shown that viscoelastic drop deformation is essentially a problem of coalescence and breakup. Many investigations on the blending of polymer process assumed Newtonian fluid systems using, for example, Taylor's minor deformation theory or Grace's breakup curve [46]. However, the Newtonian fluids theories were shown to be inapplicable for polymer blending, which is not surprising [48,49].

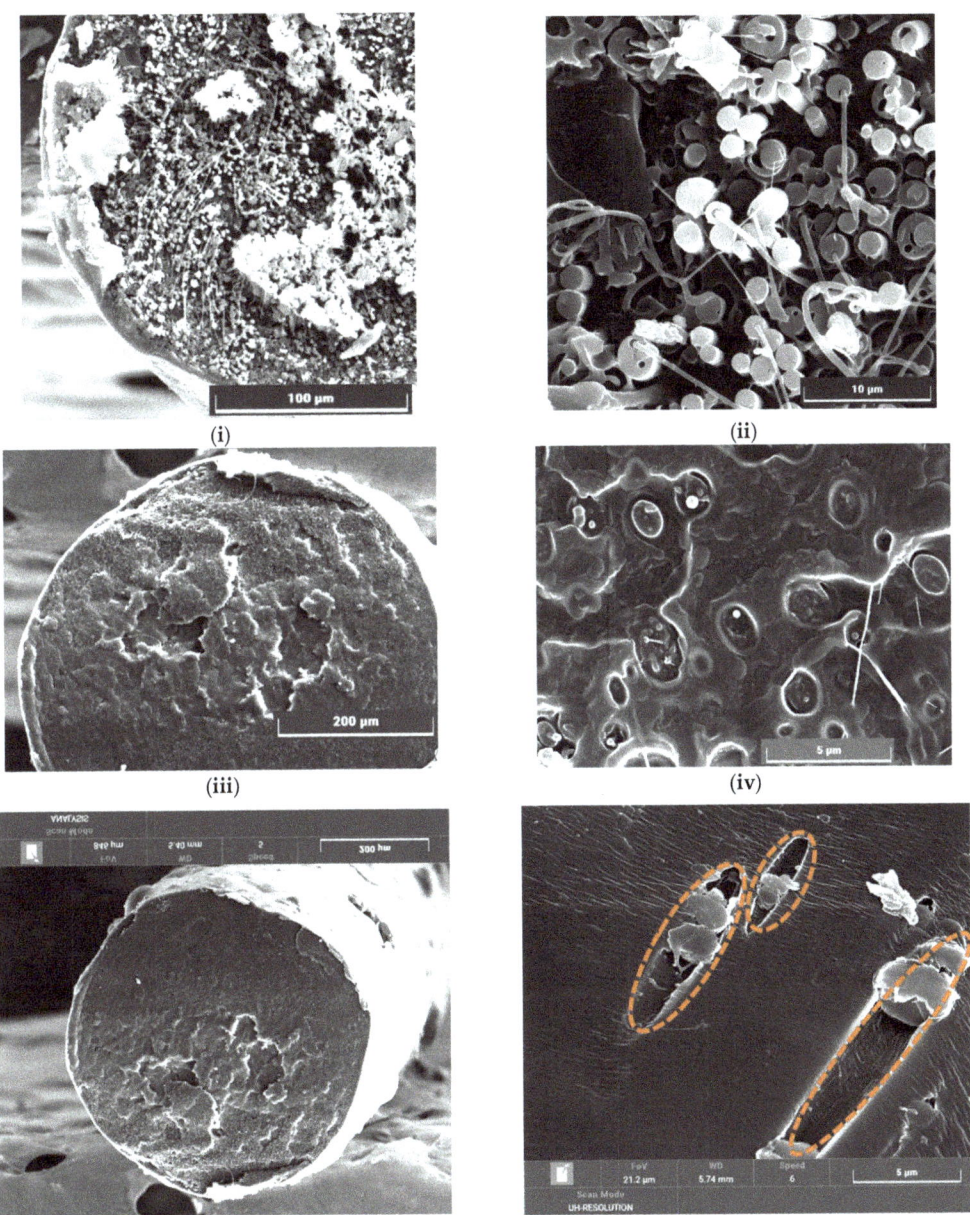

**Figure 4.** *Cont.*

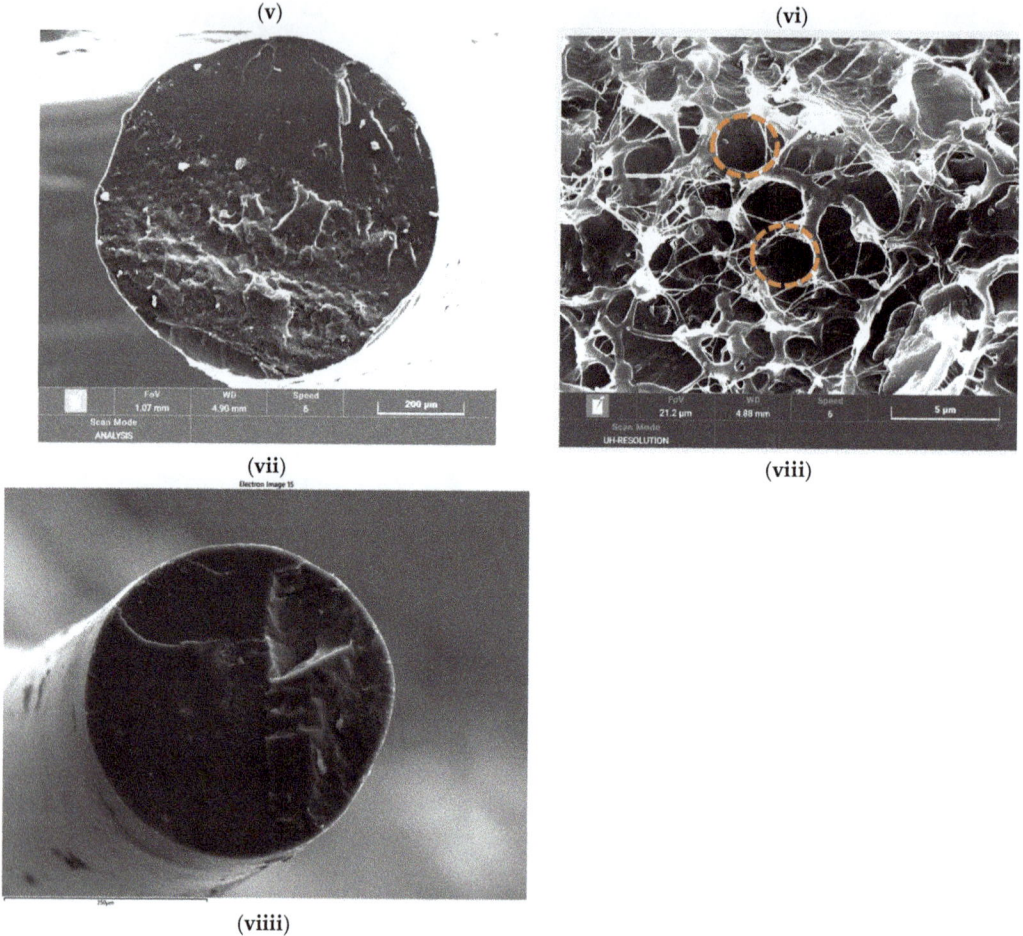

**Figure 4.** Morphology of LLDPE/MP fibres, 30 wt% LLDPE (**i,ii**); 60 wt% LLDPE (**iii,iv**); 70 wt% LLDPE (**v,vi**); 80 wt% LLDPE (**vii,viii**); and neat LLDPE (**viiii**).

### 4.3. Tensile Tests of Pitch Blends

LLDPE is widely used in many and various forms because it is cheap and flexible. One investigation has demonstrated that LLDPE's average tensile strength is around 6.1 MPa, and can be enhanced by the addition of fibre straws [50]. A second investigation claimed to have demonstrated that LLDPE can have a tensile strength as high as 9.9 MPa [51]. Others have claimed that, by adding multiple-walled carbon nanotubes, the tensile strength can rise to 22 MPa [52]. Enhanced values such as these suggest that the possible uses of LLDPE could be expanded if it is used in conjunction with other materials.

The stress limit of extruded neat LLDPE is typically about 40 MPa, and the typical strain will be about 0.8 which suggests that the LLDPE component in LLDPE/MP fibres has an important load-bearing role, and toughens the material [39]. Figure 5 shows that boosting the wt% LLDPE in relatively brittle LLDPE/MP fibres can be significant.

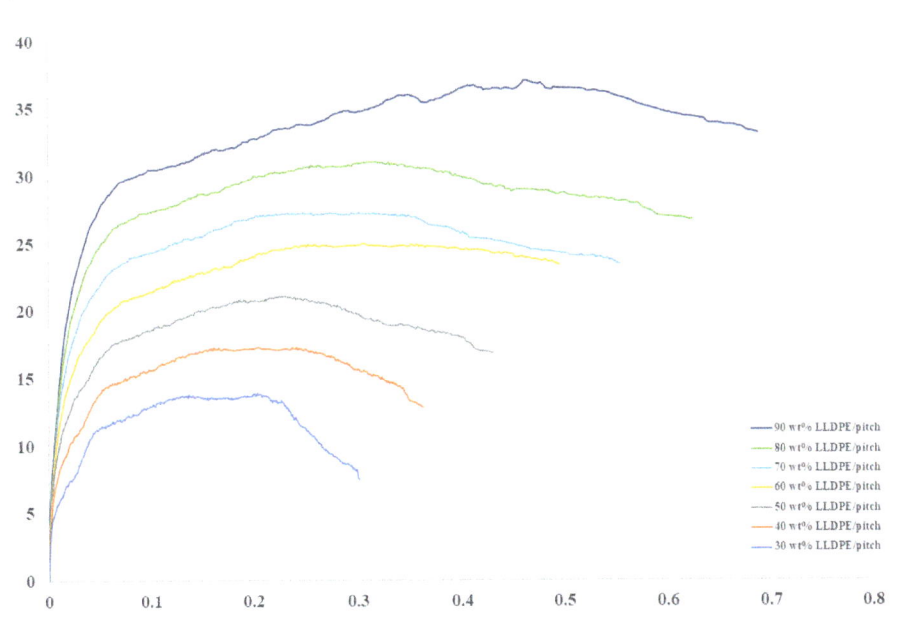

**Figure 5.** Tensile strength of LLDPE/MP fibres with differing wt% LLDPE content.

We also see in Figure 5 the stress vs. strain curves for LLDPE/MP blends for between 30 and 90 wt% LLDPE. Obviously, the stress vs. strain behaviour of the specimens is a function of wt% LLDPE. The maximum values of tensile strength and strain increase progressively with wt% LLDPE. The minimum tensile strength (13.84 MPa) and minimum strain (30%), occurred with an LLDPE content of 30 wt%, while the maximum tensile strength (36.97 MPa) and maximum strain (68%) occurred with an LLDPE content of 90 wt%.

As in our previous paper [39], we identified being brittle with low strain to failure, and the pitch fibre samples demonstrated such brittle behaviour. Increasing the wt% LLDPE in the blend increased the strain to failure, which was taken to mean that increasing the proportion of LLDPE in the fibres decreased brittleness. The data gathered on tensile strength and modulus, and strain to failure of the LLDPE/MP fibres, are presented in Table 2. We see that the greater the wt% LLDPE, the larger the tensile strength and modulus, and strain to failure. It has been reported in previous work that different morphologies have different properties that can be used to satisfy different requirements. Controlling the final morphology requires a good understanding of how the mechanism of morphology develops in polymer blends, and would be beneficial when designing the processes and equipment to improve the required properties in batch mixers or extruders [4,53,54]. However, in the current study, the time of the extrusion process and the length of extruder does not change throughout the fabrication process.

*4.4. DSC for LLDPE/Mesophase Pitch Blends*

As reported by Aldosari et al. [39] (2021), DSC has been utilized to evaluate the crystallization of various blends of low- and high-density polyethylene (LDPE and HDPE) with LLDPE. Typically, crystallization temperature is reduced with the addition of LLDPE. This decrease can be credited to the high molar mass of LLDPE, which increases chain entanglements, making crystallization more challenging.

**Table 2.** Tensile strength, modulus, and strain to failure of LLDPE/MP fibres with different wt% LLDPE (figures in brackets represent the standard deviation).

| Samples | Tensile Strength (MPa) | Tensile Modulus (MPa) | Strain at Failure |
|---|---|---|---|
| LLDPE (30 wt%)/MP | 13.84 (±0.33) | 798 (±4.3) | 0.30 (±0.031) |
| LLDPE (40 wt%)/MP | 17.53 (±0.44) | 817 (±4.5) | 0.36 (±0.038) |
| LLDPE (50 wt%)/MP | 21.47 (±0.34) | 838 (±3.8) | 0.43 (±0.041) |
| LLDPE (60 wt%)/MP | 25.06 (±0.85) | 852 (±5.6) | 0.48 (±0.028) |
| LLDPE (70 wt%)/MP | 28.91 (±0.52) | 873 (±6.9) | 0.56 (±0.021) |
| LLDPE (80 wt%)/MP | 33.06 (±0.48) | 944 (±6.1) | 0.62 (±0.029) |
| LLDPE (90 wt%)/MP | 36.97 (±0.43) | 989 (±3.4) | 0.68 (±0.035) |

Experimental data were analysed using least-squares and the rates of change of the thermograms calculated to provide the necessary kinetic data. The degradation rate of LDPE was found to be a second-order effect [55]. The thermal characteristics of LLDPE/MP blends within the range of between 30 and 90 wt% LLDPE, were explored using DSC. The tests were performed for a range of fusion and crystallization temperatures of the LLDPE/MP blends, as determined by the wt% of the LLDPE.

In Figure 6a and Table 3, we see that all the plots have clear and well-ordered crystallization peaks, and the greater the wt% of LLDPE, the higher the value of the peak heat flow. We also observe that increasing the LLDPE content of the LLDPE/MP blend increases the enthalpy of crystallization. Figure 6a also demonstrates that, as the wt% of the LLDPE in the LLDPE/MP blend increases, the crystallisation temperature decreases. This confirms the crystallisation temperature of the blends depends on LLDPE content.

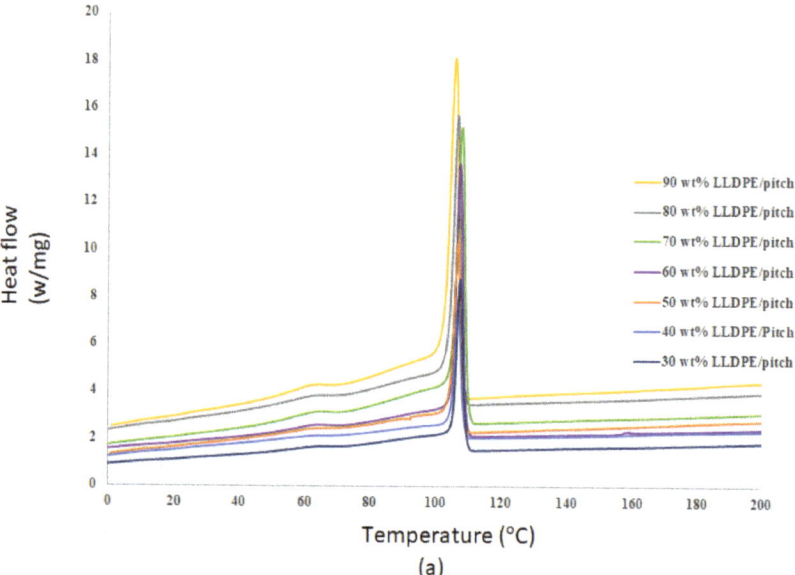

**Figure 6.** Cont.

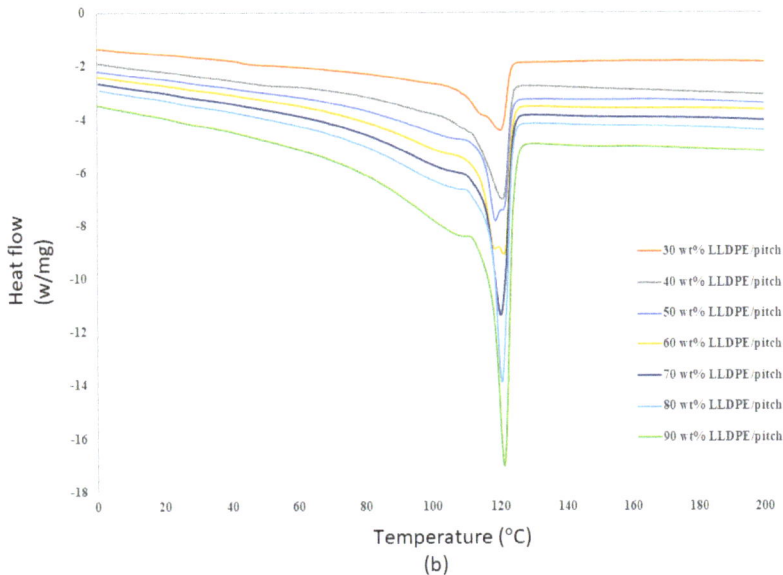

**Figure 6.** (a) Crystallisation (ΔH$_c$, enthalpy of crystallization, is the area under the exotherm curve) and (b) melting (ΔH$_m$, enthalpy of fusion, is the area under the endotherm curve) of LLDPE/mesophase pitch blends.

**Table 3.** Melting and crystallisation temperatures of LLDPE/mesophase pitch blends.

| Samples | Melting Temperature (°C) | Crystallization Temperature (°C) | Enthalpy of Fusion (J/g) Sample | Enthalpy of Fusion (J/g) LLDPE |
|---|---|---|---|---|
| LLDPE (30 wt%)/MP | 123.1 | 102.1 | 52 | 173 |
| LLDPE (40 wt%)/MP | 123.3 | 102.2 | 74 | 185 |
| LLDPE (50 wt%)/MP | 123.4 | 102.3 | 106 | 212 |
| LLDPE (60 wt%)/MP | 123.5 | 102.4 | 153 | 255 |
| LLDPE (70 wt%)/MP | 123.6 | 102.5 | 181 | 258 |
| LLDPE (80 wt%)/MP | 123.8 | 102.7 | 198 | 247 |
| LLDPE (90 wt%)/MP | 123.9 | 102.9 | 217 | 241 |

Figure 6b and Table 3 show that the temperature at which the blend melts, and the corresponding enthalpy of the test samples, is a function of the wt% LLDPE in the LLDPE/MP blend. The fusion temperature of the blend increases with LLDPE content. Thus, for the range considered here, the maximum melting temperature is at 90 wt% LLDPE. Correspondingly, the maximum enthalpy of fusion is also at 90 wt% LLDPE. This is because fewer crystalline spheres are present with the higher the percentage of LLDPE in the blend [39]. These results confirm the wt% of LLDPE in LLDPE/MP blends influences both the melting and crystallisation behaviour of the blends.

*4.5. Thermogravimetric Analysis of LLDPE/MP Blends*

TGA is commonly used to evaluate a composite material's thermal stability. Here, it was used to investigate the thermal decomposition of LLDPE/MP blends with different wt% of LLDPE burned in air. The higher the decomposition temperatures, the greater the thermal stability [52,56].

TGA measures the proportion and/or quantity of the mass of a material that is transformed, either as a function of temperature or isothermally, as a function of time. This is usually carried out in a regulated atmosphere [57]. TGA can be used to assess any substance that shows a change in weight when combusted, identifying phase changes during decomposition or oxidation [58]. Thermogravimetric analyses typically consist of a specimen pan supported by a precision scale [57]. These data are used to study the weight change, the material's chemical structure, and decomposition [59].

The TGA was performed with temperatures between 100 and 800 °C. Thermal decomposition for a range of LLDPE/MP blends was investigated, and the TGA curves obtained are presented in Figure 7. Figure 7 presents the loss of mass as a function of temperature for the test specimens. We see that the LLDPE/MP blend with 90 wt% LLDPE has suffered severe decomposition at 540 °C with over 90% mass loss. However, the blend with 30 wt% LLDPE shows only 60% mass loss at that temperature. This confirms that MP is more stable than LLDPE, and the wt% LLDPE in the blend influences its thermal stability. The decomposition temperatures of various LLDPE/MP blends are seen in Figure 7. Those fibres with 30 wt% and 40 wt% LLDPE could preserve more than 30% of the fibre mass until about 800 °C. With their decreased brittleness and relatively higher thermal stability, these two blends could provide acceptable precursors for the fabrication of CFs. We also studied the onset decomposition temperature of the samples. The onset degradation temperature is defined as the temperature at which the sample shows 5% initial mass loss. As seen in Table 4, onset degradation temperature also reduced with increasing LLDPE content in the LLDPE/pitch fibres. This observation further supports our initial finding that increased LLDPE content will reduce the thermal stability of the LLDPE/pitch blend.

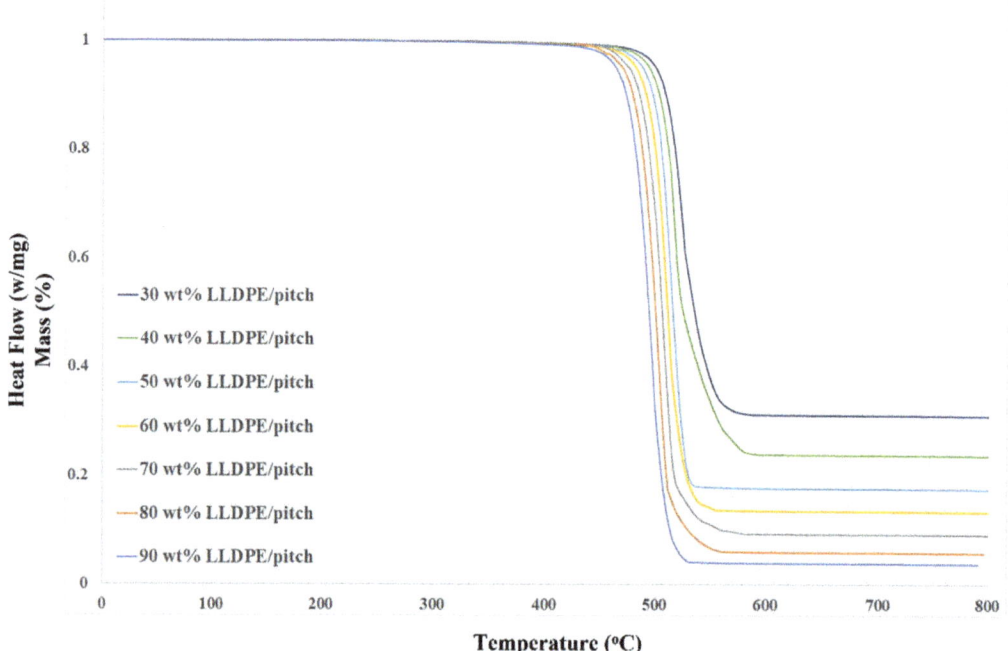

**Figure 7.** TGA curves for different wt% LLDPE in LLDPE/MP blends.

Table 4. Thermogravimetric analysis (TGA) result for 30 to 90 wt% LLDPE/MP.

| Samples | Onset Degradation Temperature * (°C) | Final Degradation Temperature (°C) | Final Residue (%) |
|---|---|---|---|
| LLDPE (30 wt%)/MP | 490.5 | 571.7 | 30 |
| LLDPE (40 wt%)/MP | 481.6 | 539.4 | 27 |
| LLDPE (50 wt%)/MP | 475.2 | 536.3 | 22 |
| LLDPE (60 wt%)/MP | 472.2 | 532.1 | 19 |
| LLDPE (70 wt%)/MP | 462.9 | 528.3 | 15 |
| LLDPE (80 wt%)/MP | 453.4 | 527.1 | 12 |
| LLDPE (90 wt%)/MP | 446.2 | 525.4 | 8 |

* Onset degradation temperature defined as temperature at which the sample shows 1% initial mass loss.

## 5. Conclusions

This paper shows how the concentration of LLDPE affects the morphology and physical properties of LLDPE/MP blend fibres over the range of between 30 and 90 wt% LLDPE. The SEM image analysis of higher LLDPE content blends showed significantly altered morphology from micro-fibres to non-microfibers. The fibre diameter also showed a marginal reduction with increased wt% LLDPE in the LLDPE/MP blend fibres. The DSC analysis reveals that the enthalpy of fusion increases with the increase in LLDPE content in the sample.

The micromechanical testing of the LLDPE/MP blends showed a clear increase in both tensile modulus and strength, as well as strain to failure with an increase in the LLDPE content. However, TGA revealed a reduction in the temperature of the onset of degradation and a reduction in the residual mass (after about 540 °C) with an increase in the wt% LLDPE in the LLDPE/MP fibres. Hence, selection of LLDPE/pitch blend fibres must be based on both mechanical performance and the thermal stability of fibres.

**Author Contributions:** S.M.A. and S.R. conceptualized the theory and the method. S.M.A. performed the experimentation and the analysis. M.A.K. enhanced the data analysis. S.M.A. worked on the original draft preparation. All authors have read and agreed to the published version of the manuscript.

**Funding:** This research received no external funding.

**Data Availability Statement:** Data can be available on a request from the corresponding author via email.

**Acknowledgments:** I would like to thank King Abdulaziz City for Science and Technology (KACST) for their generous funding throughout this project.

**Conflicts of Interest:** The authors declare no conflict of interest.

## Appendix A

The tensile test procedure was conducted according to the standard ISO 11566-1996.

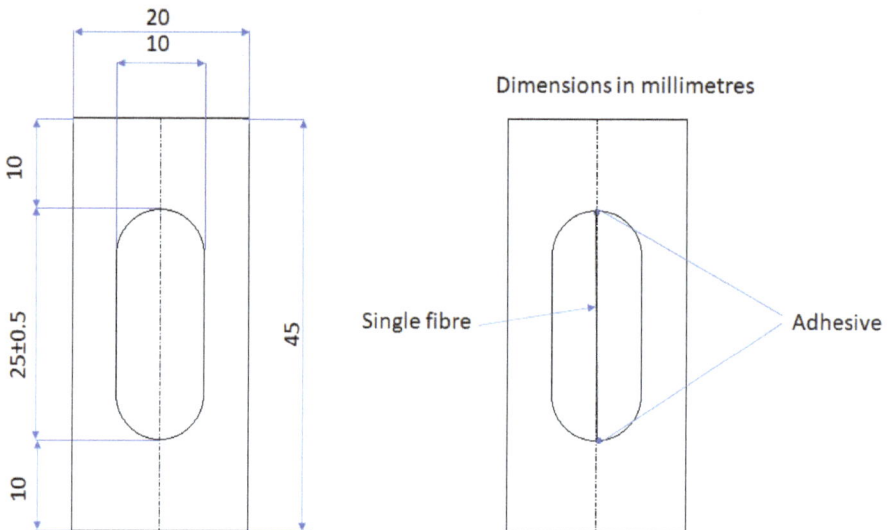

**Figure A1.** Tensile test setup for melt-spun MP/PE fibre.

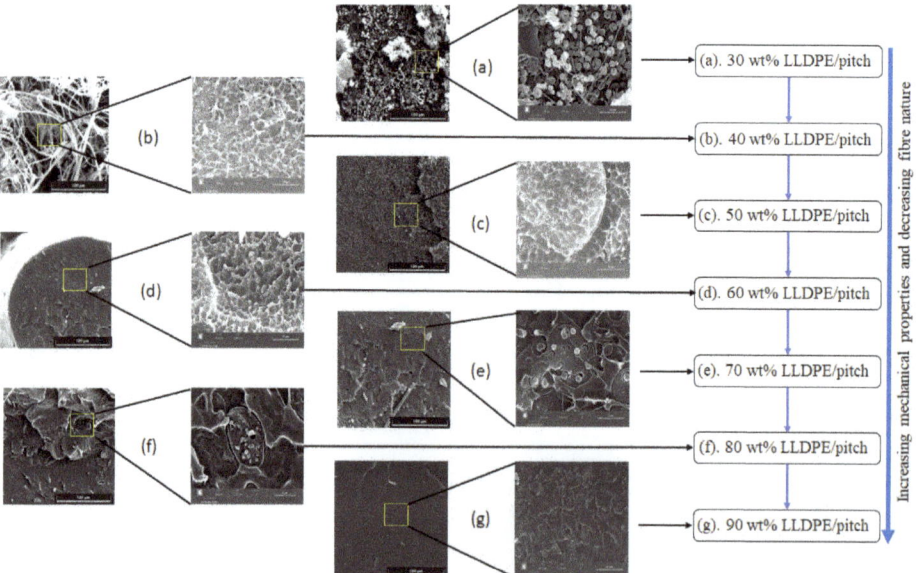

**Figure A2.** Morphology of fibers with LLDPE content in LLDPE/mesophase Pitch blend, (**a**) 30 wt% LLDPE/mesophase, (**b**) 40 wt% LLDPE/mesophase, (**c**) 50 wt% LLDPE/MP, (**d**) 60 wt% LLDPE/MP, (**e**) 70 wt% LLDPE/MP, (**f**) 80 wt% LLDPE/MP, (**g**) 90 wt% LLDPE/MP.

## References

1. Toh, H.W.; Toong, D.W.Y.; Ng, J.C.K.; Ow, V.; Lu, S.; Tan, L.P.; Wong, P.E.H.; Venkatraman, S.; Huang, Y.; Ang, H.Y. Polymer blends and polymer composites for cardiovascular implants. *Eur. Polym. J.* **2021**, *146*, 110249. [CrossRef]
2. Fortelný, I.; Jůza, J. Description of the Droplet Size Evolution in Flowing Immiscible Polymer Blends. *Polymers* **2019**, *11*, 761. [CrossRef]
3. Utracki, L.A. Economics of polymer blends. *Polym. Eng. Sci.* **1982**, *22*, 1166–1175. [CrossRef]

4. Macosko, C.W. Morphology Development and Control in Immiscible Polymer Blends. In *Macromolecular Symposia*; WILEY-VCH: Weinheim, Germany, 2000; Volume 149, pp. 171–184.
5. Włoch, M.; Datta, J. Rheology of polymer blends. In *Rheology of Polymer Blends and Nanocomposites: Theory, Modelling and Applications*, 1st ed.; Thomas, S.C.S., Chandran, N., Eds.; Elsevier: Amsterdam, The Netherlands, 2010; pp. 19–29. [CrossRef]
6. Utracki, L.A.; Wilkie, C.A. (Eds.) *Polymer Blends Handbook*, 2nd ed.; Springer: London, UK, 2014; pp. 1–2378. [CrossRef]
7. Chukov, N.A.; Ligidov, M.K.; Pakhomov, S.I.; Mikitaev, A.K. Polypropylene polymer blends. *Russ. J. Gen. Chem.* **2017**, *87*, 2238–2249. [CrossRef]
8. Fitzer, E. Carbon Fibres Present State and Future Expectations. In *Carbon Fibers Filaments and Composites*; Figueired, J.L., Bernardo, C.A., Baker, R.T., Huttinger, K.J., Eds.; Kluwer Academic: London, UK, 1990; pp. 3–41.
9. Liu, Y.; Kumar, S. Recent progress in fabrication, structure, and properties of carbon fibers. *Polym. Rev.* **2012**, *52*, 234–258. [CrossRef]
10. Park, S.J.; Lee, S.Y. History and Structure of Carbon Fibers. In *Carbon Fibers*; Springer: Incheon, Korea, 2015; pp. 1–30. [CrossRef]
11. Aldosari, S.M.; Khan, M.; Rahatekar, S. Manufacturing carbon fibres from pitch and polyethylene blend precursors: A review. *J. Mater. Res. Technol.* **2020**, *9*, 7786–7806. [CrossRef]
12. Fortin, F.; Yoon, S.H.; Korai, Y.; Mochida, I. Structure of round-shaped methylnaphthalene-derived mesophase pitch-based carbon fibres prepared by spinning through a Y-shaped die hole. *J. Mater. Sci.* **1995**, *30*, 4567–4583. [CrossRef]
13. Xiao, B.; Huang, Q.; Chen, H.; Chen, X.; Long, G. A fractal model for capillary flow through a single tortuous capillary with roughened surfaces in fibrous porous media. *Fractals* **2021**, *28*, 2150017. [CrossRef]
14. Yuan, G.; Cui, Z. Preparation, Characterization, and Applications of Carbonaceous Mesophase: A Review. In *Nematic Liquid Crystals*, 1st ed.; Carlescu, I., Ed.; IntechOpen: London, UK, 2019; pp. 1–20. [CrossRef]
15. Zeng, S.M.; Maeda, T.; Tokumitsu, K.; Mondori, J.; Mochida, I. Preparation of isotropic pitch precursors for general purpose carbon fibers (GPCF) by air blowing-II. Air blowing of coal tar, hydrogenated coal tar, and petroleum pitches. *Carbon* **1993**, *31*, 413–419. [CrossRef]
16. Gao, Z.; Zhu, J.; Rajabpour, S.; Joshi, K.; Kowalik, M.; Croom, B.; Schwab, Y.; Zhang, L.; Bumgardner, C.; Brown, K.R.; et al. Graphene reinforced carbon fibers. *Sci. Adv.* **2020**, *6*, eaaz4191. [CrossRef]
17. Tran, T.Q.; Lee, J.K.Y.; Chinnappan, A.; Loc, N.H.; Tran, L.T.; Ji, D.; Jayathilaka, D.; Kumar, V.V.; Ramakrishna, S. High-performance carbon fiber/gold/copper composite wires for lightweight electrical cables. *J. Mater. Sci. Technol.* **2020**, *42*, 46–53. [CrossRef]
18. De Palmenaer, A.; Wortberg, G.; Drissen, F.; Seide, G. Production of Polyethylene Based Carbon Fibres. *Chem. Eng. Trans.* **2015**, *43*, 1699–1704. [CrossRef]
19. Wortberg, G.; De Palmenaer, A.; Beckers, M.; Seide, G.; Gries, T. Polyethylene-Based Carbon Fibers by the Use of Sulphonation for Stabilization. *Fibers* **2015**, *3*, 373–379. [CrossRef]
20. Kim, K.-W.; Lee, H.-M.; Kim, B.S.; Hwang, S.-H.; Kwac, L.-K.; An, K.-H.; Kim, B.-J. Preparation and thermal properties of polyethylene-based carbonized fibers. *Carbon Lett.* **2015**, *16*, 62–66. [CrossRef]
21. Yang, K.S.; Kim, B.-H.; Yoon, S.-H. Pitch based carbon fibers for automotive body and electrodes. *Carbon Lett.* **2014**, *15*, 162–170. [CrossRef]
22. Pujadas, P.; Blanco, A.; Cavalaro, S.; de la Fuente, A.; Aguado, A. The need to consider flexural post-cracking creep behavior of macro-synthetic fiber reinforced concrete. *Constr. Build. Mater.* **2017**, *149*, 790–800. [CrossRef]
23. Mochida, I.; Toshima, H.; Korai, Y.; Takashi, H. Oxygen distribution in the mesophase pitch fibre after oxidative stabilization. *J. Mater. Sci.* **1989**, *24*, 389–394. [CrossRef]
24. Huang, X. Fabrication and properties of carbon fibers Review. *Materials* **2009**, *2*, 2369–2403. [CrossRef]
25. Zhang, D.; Bhat, G.S. Carbon Fibers from Polyethylene-Based Precursors. *Mater. Manuf. Process.* **1994**, *9*, 221–235. [CrossRef]
26. Barton, B.E.; Behr, M.J.; Patton, J.T.; Hukkanen, E.J.; Landes, B.G.; Wang, W.; Horstman, N.; Rix, J.E.; Keane, D.; Weigand, S.; et al. High-Modulus Low-Cost Carbon Fibers from Polyethylene Enabled by Boron Catalyzed Graphitization. *Small* **2017**, *13*, 1701926. [CrossRef]
27. Behr, M.J.; Landes, B.G.; Barton, B.E.; Bernius, M.T.; Billovits, G.F.; Hukkanen, E.J.; Patton, J.T.; Wang, W.; Wood, C.; Keane, D.T.; et al. Structure-property model for polyethylene-derived carbon fiber. *Carbon* **2016**, *107*, 525–535. [CrossRef]
28. Zhang, D. Carbon Fibers from Oriented Polyethylene Precursors. *J. Thermoplast. Compos. Mater.* **1993**, *6*, 38–48. [CrossRef]
29. Postema, A.R.; De Groot, H.; Pennings, A.J. Amorphous carbon fibres from linear low density polyethylene. *J. Mater. Sci.* **1990**, *25*, 4216–4222. [CrossRef]
30. Kim, J.W.; Lee, J.S. Preparation of carbon fibers from linear low density polyethylene. *Carbon* **2015**, *94*, 524–530. [CrossRef]
31. Kim, K.-W.; Lee, H.-M.; An, J.-H.; Kim, B.-S.; Min, B.-G.; Kang, S.-J.; An, K.-H.; Kim, B.-J. Effects of cross-linking methods for polyethylene-based carbon fibers: Review. *Carbon Lett.* **2015**, *16*, 147–170. [CrossRef]
32. Penning, J.P.; Lagcher, R.; Pennings, A.J. The effect of diameter on the mechanical properties of amorphous carbon fibres from linear low density polyethylene. *Polym. Bull.* **1991**, *25*, 405–412. [CrossRef]
33. Meza, A.; Pujadas, P.; López-Carreño, R.D.; Meza, L.M.; Pardo-Bosch, F. Mechanical Optimization of Concrete with Recycled PET Fibres Based on a Statistical-Experimental Study. *Materials* **2021**, *14*, 240. [CrossRef]
34. Boustead, I. *Eco-Profiles of the European Plastics Industry Linear Low Density Polyethylene (LLDPE)*; PlasticsEurope: Brussels, Belgium, 2005.

35. Bansal, R.C.; Donnet, J.B. Pyrolytic Formation of High-performance Carbon Fibres. In *Comprehensive Polymer Science and Supplements*; Bevington, G.A., Ed.; Elsevier: Amsterdam, The Netherlands, 1996; pp. 501–520. [CrossRef]
36. Kershaw, J.R.; Black, K.J.T. Structural Characterization of Coal-Tar and Petroleum Pitches. *Energy Fuels* **1993**, *7*, 420–425. [CrossRef]
37. Pujadas, P.; Blanco, A.; Cavalaro, S.; Aguado, A. Plastic fibres as the only reinforcement for flat suspended slabs: Experimental investigation and numerical simulation. *Constr. Build. Mater.* **2014**, *57*, 92–104. [CrossRef]
38. Kumar, R.P.; Wadgaonkar, K.; Mehta, L.; Jagtap, R. Enhancement of mechanical and barrier properties of LLDPE composite film via PET fiber incorporation for agricultural application. *Polym. Adv. Technol.* **2019**, *30*, 1251–1258. [CrossRef]
39. Aldosari, S.; Khan, M.; Rahatekar, S. Manufacturing Pitch and Polyethylene Blends-Based Fibres as Potential Carbon Fibre Precursors. *Polymer* **2021**, *13*, 1445. [CrossRef]
40. British Standards. *BS ISO 11566: Carbon Fibre-Determination of the Tensile Properties of the Tensile-Filament Specimens*; British Standards Institute: London, UK, 1996.
41. Fakirov, S.; Bhattacharyya, D.; Lin, R.J.T.; Fuchs, C.; Friedrich, K. Contribution of Coalescence to Microfibril Formation in Polymer Blends during Cold Drawing. *J. Macromol. Sci. Part B* **2007**, *46*, 183–194. [CrossRef]
42. Lu, S.; Blanco, C.; Rand, B. Large diameter carbon fibres from mesophase pitch. *Carbon* **2002**, *40*, 2109–2116. [CrossRef]
43. Grasser, W.; Schmidt, H.W.; Giesa, R. Fibers spun from poly(ethylene terephthalate) blended with a thermotropic liquid crystalline copolyester with non-coplanar biphenylene units. *Polymer* **2001**, *42*, 8517–8527. [CrossRef]
44. Chen, L.; Pan, D.; He, H. Morphology Development of Polymer Blend Fibers along Spinning Line. *Fibers* **2019**, *7*, 35. [CrossRef]
45. Gonzalez-Nunez, R.; Favis, B.D.; Carreau, P.J.; Lavallée, C. Factors influencing the formation of elongated morphologies in immiscible polymer blends during melt processing. *Polym. Eng. Sci.* **1993**, *33*, 851–859. [CrossRef]
46. Li, H.; Sundararaj, U. Morphology development of polymer blends in extruder: The effects of compatibilization and rotation rate. *Macromol. Chem. Phys.* **2009**, *210*, 852–863. [CrossRef]
47. Iii, C.L.T.; Moldenaers, P. Microstructural Evolution in Polymer Blends. *Annu. Rev. Fluild Mech.* **2002**, *34*, 177–210.
48. Taylor, G.I. The Formation of Emulsions in Definable Fields of Flow. *R. Soc.* **1934**, *146*, 501–523. [CrossRef]
49. Taylor, P.; Grace, H.P. Dispersion Phenomena in High Viscosity Immiscible Fluid Systems and Application of Static Mixers As Dispersion Dispersion Phenomena in High Viscosity Immiscible Fluid Systems and Application O F Static Mixers as Dispersion Devices in Such Systems. *Chem. Eng. Commun.* **1982**, *14*, 225–277. [CrossRef]
50. Zhang, L.; Xu, H.; Wang, W. Performance of Straw/Linear Low Density Polyethylene Composite Prepared with Film-Roll Hot Pressing. *Polymers* **2020**, *12*, 860. [CrossRef] [PubMed]
51. Durmus, A.; Kaşgöz, A.; Macosko, C.W. Mechanical properties of linear low-density polyethylene (LLDPE)/clay nanocomposites: Estimation of aspect ratio and interfacial strength by composite models. *J. Macromol. Sci. Part B Phys.* **2008**, *47*, 608–619. [CrossRef]
52. Tai, J.H.; Liu, G.Q.; Caiyi, H.; Shangguan, L.J. Mechanical properties and thermal behaviour of LLDPE/MWNTs nanocomposites. *Mater. Res.* **2012**, *15*, 1050–1056. [CrossRef]
53. Evstatiev, M.; Fakirov, S.; Bechtold, G.; Friedrich, K. Structure-Property Relationships of Injection- and Compression-Molded Microfibrillar-Reinforced PET/PA-6 Composites. *Adv. Polym. Technol.* **2000**, *19*, 249–259. [CrossRef]
54. Gonzalez-Montiel, A.; Keskkula, H.; Paul, D.R. Impact-modified nylon 6/polypropylene blends: 1. Morphology-property relationships. *Polymer* **1995**, *36*, 4587–4603. [CrossRef]
55. Bhardwaj, I.S.; Kumar, V.; Palanivelu, K. Thermal characterisation of LDPE and LLDPE blends. *Thermochim. Acta* **1988**, *131*, 241–246. [CrossRef]
56. Wang, Z.; Cheng, Y.; Yang, M.; Huang, J.; Cao, D.; Chen, S.; Xie, Q.; Lou, W.; Wu, H. Dielectric properties and thermal conductivity of epoxy composites using core/shell structured Si/SiO2/Polydopamine. *Compos. Part B Eng.* **2018**, *140*, 83–90. [CrossRef]
57. Cai, J.; Xu, D.; Dong, Z.; Yu, X.; Yang, Y.; Banks, S.W. Processing Thermogravimetric Analysis Data for Isoconversional Kinetic Analysis of Lignocellulosic Biomass Pyrolysis: Case Study of Corn Stalk. *Renew. Sustain. Energy Rev.* **2018**, *82*, 2705–2715. [CrossRef]
58. Bottom, R. Thermogravimetric Analysis. In *Principles and Applications of Thermal Analysis*, 1st ed.; Gabbott, P., Ed.; Blackwell Publishing Ltd.: Oxford, UK, 2008; pp. 87–118. [CrossRef]
59. Saddawi, A.; Jones, J.M.; Williams, A.; Wójtowicz, M.A. Kinetics of the Thermal Decomposition of Biomass. *Energy Fuels* **2010**, *24*, 1274–1282. [CrossRef]

Article

# Synthesis and Characterization of Antibacterial Carbopol/ZnO Hybrid Nanoparticles Gel

Sameh H. Ismail [1], Ahmed Hamdy [1,2], Tamer Ahmed Ismail [3], Heba H. Mahboub [4], Walaa H. Mahmoud [5] and Walid M. Daoush [6,7,*]

1. Faculty of Nanotechnology for Postgraduate Studies, Sheikh Zayed Campus, Cairo University, 6th October City, Giza 12588, Egypt; drsameheltayer@yahoo.com (S.H.I.); ahsadek@zewailcity.edu.eg (A.H.)
2. Environmental Engineering Program, Zewail City of Science, Technology and Innovation, 6th October City, Giza 12578, Egypt
3. Department of Clinical Laboratory Sciences, Turabah University College, Taif University, P.O. Box 11099, Taif 21944, Saudi Arabia; t.ismail@tu.edu.sa
4. Department of Fish Diseases and Management, Faculty of Veterinary Medicine, Zagazig University, P.O. Box 44511, Sharkia 44519, Egypt; Hebamahboub@zu.eg
5. Faculty of Science, Cairo University, Giza 12613, Egypt; wmahmoud@sci.cu.edu.eg
6. Department of Chemistry, College of Science, Imam Mohammad ibn Saud Islamic University (IMSIU), Othman ibn Affan St., P.O. Box 5701, Riyadh 11432, Saudi Arabia
7. Department of Production Technology, Faculty of Technology and Education, Helwan University, Saray–El Qoupa, El Sawah Street, Cairo 11281, Egypt
* Correspondence: wmdaoush@imamu.edu.sa

**Abstract:** This study recommends Carbopol/zinc oxide (ZnO) hybrid nanoparticles gel as an efficient antibacterial agent against different bacterial species. To this end, ZnO nanoparticles were synthesized using chemical precipitation derived from a zinc acetate solution with ammonium hydroxide as its precipitating agent under the effect of ultrasonic radiation. The synthesized ZnO nanoparticles were stabilized simultaneously in a freshly prepared Carbopol gel at a pH of 7. The chemical composition, phase identification, particle size and shape, surface charge, pore size distribution, and the BET surface area of the ZnO nanoparticles, as well as the Carbopol/ZnO hybrid Nanoparticles gel, were by XRD, SEM, TEM, AFM, DLS, Zeta potential and BET instruments. The results revealed that the synthesized ZnO nanoparticles were well-dispersed in the Carbopol gel network, and have a wurtzite-crystalline phase of spherical shape. Moreover, the Carbopol/ZnO hybrid nanoparticles gel exhibited a particle size distribution between ~9 and ~93 nm, and a surface area of 54.26 m$^2$/g. The synthesized Carbopol/ZnO hybrid nanoparticles gel underwent an antibacterial sensitivity test against gram-negative *K. pneumonia* (ATCC 13883), *Bacillus subtilis* (ATCC 6633), and gram-positive *Staphylococcus aureus* (ATCC 6538) bacterial strains, and were compared with ampicillin as a reference antibiotic agent. The obtained results demonstrated that the synthesized Carbopol/ZnO hybrid nanoparticles gel exhibited a compatible bioactivity against the different strains of bacteria.

**Keywords:** ZnO nanoparticles; hybrid materials; chemical precipitation; Carbopol; BET surface area; zeta-potential; antibacterial activity

## 1. Introduction

In recent decades, discussions focused on the application of biomedical nanomaterials have increased in the medical field due to their eminent biological properties. With their broad applications in various fields, metal oxide nanoparticles have shown far-reaching and promising prospects in the field of biomedicine, especially for drug delivery/antibacterial genes, biosensing, cytometry, cancer, and others [1–3].

Many synthetic techniques are used for the synthesis of zinc oxide nanoparticles (ZnO NPs). These methods could be classified into three main types: physical, chemical, and biological methods. Furthermore, chemical synthesis includes liquid-phase synthesis and gas-phase synthesis. Liquid phase synthesis involves some sub-methods, such as polyol [4], sonochemical [5], solvothermal [6], hydrothermal [7], water-oil microemulsions [8], sol-gel processing [9], co-precipitation [10], and precipitation methods [11], while methods such as inert gas condensation and pyrolysis fall within fabrication methods in the vapor phase [12]. ZnO NPs are classified as II–VI semiconductors and are characterized by a high excitation energy of 60 eV and wide band gap energy of 3.3 eV. Thus, they can tolerate large electric fields, high temperatures, and high power operations [13]. These properties make ZnO NPs highly applicable in chemical sensors, photocatalysis, and solar cells [14]. Additionally, the crystal structure of ZnO NPs significantly contributes to the emergence of their piezoelectric properties. Accordingly, this makes ZnO NPs suitable for acoustic wave resonators and acoustic-optic modulators. In addition, the Centro-symmetric structure of ZnO NPs made them the highest tensors among all semiconductors, providing a large electromechanical coupling [15]. Moreover, the GRAS substances (SCOGS) database allows access to opinions and conclusions from 115 SCOGS reports published between 1972–1980 on the safety of over 370 Generally Recognized as Safe (GRAS) food substances by the U.S. Food and Drug Administration (FDA). Accordingly, the Select Committee on GRAS Substances (SCOGS) opinion concluded that there is no evidence in the available information on zinc oxide that demonstrates, or suggests reasonable grounds to suspect, a hazard to the public [16].

Additionally, ZnO NPs are commonly used in various fields due to their distinct physical and chemical properties as one of the most important semiconductor metal oxide nanoparticles [17,18]. Because of the ability of ZnO to absorb UV radiation, it is increasingly used in personal care products, such as sunscreens and cosmetics [19]. ZnO NPs have also been applied in the rubber industry as they can provide the abrasion resistance for the rubber composite, as well as improve the high polymer performance in anti-aging, toughness and strength, and other functions [20]. In addition, ZnO NPs have excellent antibacterial and antimicrobial properties. Furthermore, when used in the textile industry, fabrics treated with the addition of ZnO NPs gain attractive properties, such as UV blocking, visible light resistance, deodorant, and antibacterial agents [21,22]. Zinc oxide can also be used in other branches of industry, including electronics, photocatalysis, concrete production, and other technologies [23,24].

These hydrogels are three-dimensional systems of hydrophilic polymers, which can swell when absorbing water; however, they do not dissolve. Moreover, these hydrogels can interact with the surrounding environments and simultaneously exhibit changes in both their chemical and physical properties [25,26]. Great efforts have been exerted on hydrogels containing pH and/or temperature-sensitive properties due to their extraordinary potential in bioengineering and biomedical uses, especially in cell culture, molecular separation, and drug release [27,28]. The term "nanogel" was introduced to define the cross-linked dual-functional networks of a polyion and a nonionic polymer for polynucleotide delivery [29,30]. Nanogels are composed of nanoparticles that are formed by physically or chemically cross-linked polymer networks that stabilize and swell in a fine solvent [31,32]. Breakthroughs in nanotechnology have generated the need to develop nanogel systems which demonstrate the ability to deliver drugs in controlled, stable, and targetable settings. With the promising field of polymer science, it is now possible to develop smart nanoscale systems which could provide effective treatments and diagnoses, as well as advance clinical trials [33,34]. Carbopol 940, or Carbomer 940, is a synthetic high-molecular-weight polymer of the acrylic acid monomer. They may be homopolymers of acrylic acid or cross-linked with an allyl ether of propylene, allyl ether of sucrose, or an allyl ether of pentaerythritol. Carbopol 940 contains not less than 56% and not more than 68% of carboxylic (–COOH) groups. The viscosity of a neutralized 0.5% aqueous dispersion of Carbopol 940 is between 40,000 and 60,000 centipoises [35]. In an aqueous solution of a neutral pH, Carbopol 940 is an anionic polymer, i.e., many of the side chains of Carbopol 940 will lose their protons and acquire a negative charge. This gives Carbopol 940 polyelectrolytes the ability to absorb and retain water and swell to many times their original volume [36]. Dry Carbopol 940 is produced as white, fluffy powders that are frequently used as gels in personal care and cosmetic products. Their role in cosmetics is to suspend solids in liquids, prevent emulsions from separating, and to control the consistency in the flow of cosmetics. Carbopol codes (910, 934, 940, 941, and 934P) are an indication of molecular weight and the specific components of the polymer [37]. For many applications, Carbopol 940 is used in form of alkali metal or ammonium salts, e.g., sodium polyacrylate. In the dry powder form, the positively charged sodium ions are bonded to the polyacrylate; however, in aqueous solutions, the sodium ions can be dissociated. Instead of the formation of an organized polymer chain, this leads to a swollen gel that can absorb a high amount of water [38].

The targets of synthesis and fabrication of any antimicrobial compound are to inhibit the causal microbe without any side effects on the patients. Besides, it is worthy to stress here the basic idea of applying any chemotherapeutic agent, which depends essentially on the specific control of at least one biological function while avoiding multiple ones. ZnO nanogel exhibits good and effective antimicrobial and antibacterial activity. It is highly selective in bactericidal activity and reveals biocompatibility as well as low effects in human cells [39–41]. ZnO nanogel delays the microbial growth of foodborne pathogens, such as *B. subtilis*, *E. coli*, *Pseudomonas* fluorescent, etc. In vitro culture, media studies showed that ZnO nanogel is very effective at killing microbes such as *S. enteritidis*, *E. coli*, *Listeria monocytogenes*, and others [42,43]. ZnO nanoparticles of about 10–25 nm particle size can penetrate deeply into the ventral cell and enhance its membrane permeability. They disintegrate and produce complete sets of bacterial and microbial cell membranes [18]. There are several types of mechanisms that explain the antimicrobial and antibacterial activities of ZnO nanogel. Complete cell lysis and an elevation in membrane permeability due to the production of hydrogen peroxide ($H_2O_2$) from the ZnO nanoscale surface, are some of the most recognized mechanisms within the scientific community [44,45]. The size of the ZnO nanogel is very important for its activity; it is shown that the smaller sizes of ZnO nanoparticles per unit volume in the aqueous medium increase the surface area and enhance the production of hydrogen particles [46,47]. Another study reported that the production of zinc particles causes severe damage to the bacterial cell membrane. This results in the formation of small pores on the cell surface and leads to leakage of cellular contents, causing bacterial cell death [48–50].

Furthermore, many researchers have used Carbopol-based inorganic metal/metal oxides nanoparticles and organic additive hybrid systems in different fields. For example, Jana et al. have used Carbopol gel containing chitosan-egg albumin nanoparticles for transdermal aceclofenac delivery [51], while Sareen et al. have formulated and evaluated the meloxicam Carbopol-based gels for drug release uses [52]. In addition, Bonacucina et al. have analyzed the thickening properties of Carbopol 974 and 971 in a 50:50 mixture of water/Silsense™ A-21 as a new cationic silicon miscible in any proportion with water. In addition, they have also evaluated the rheological properties of Silicon/Carbopol hydrophilic gel systems as a vehicle for the delivery of water-insoluble drugs [53].

In this study, ZnO nanoparticles were synthesized using chemical precipitation derived from a zinc acetate solution with ammonium hydroxide as its precipitating agent under the effect of ultrasonic radiation. The prepared ZnO NPs were stabilized using Carbopol gel to obtain a hybrid system of well-dispersed nanoparticles in the gel network. Similarly, the synthesized Carbopol/ZnO hybrid nanoparticles gel was also successfully prepared using a chemical precipitation reaction in a Carbopol stabilizing agent under the effect of ultrasonic irradiation. The chemical composition, phase identification, adsorption/desorption behavior, and pore size distribution properties of ZnO NPs and Carbopol/ZnO hybrid nanoparticles gel were evaluated by XRD and BET analyzers. In addition, topographical and morphological textures of the synthesized ZnO NPs, as well as the Carbopol/ZnO hybrid nanoparticles gel, have been investigated using AFM, SEM, and TEM microscopics. DLS and zeta-potential studies were also conducted to investigate the size distribution and charge measurements of the prepared ZnO NPs and Carbopol/ZnO hybrid nanoparticles gel, respectively. Finally, the synthesized Carbopol/ZnO hybrid nanoparticles gel was examined as antibacterial nanoparticles/gel hybrid system against gram-negative *K. pneumonia* (ATCC 13883) and gram-positive (*Bacillus subtilis* (ATCC 6633) and *Staphylococcus aureus* (ATCC 6538) bacterial strains.

## 2. Experimental Section

### 2.1. Preparation of ZnO NPs and Carbopol/ZnO Hybrid Nanoparticles Gel

ZnO NPs were synthesized using the chemical precipitation method under the effect of ultrasound irradiation. In a typical procedure, zinc acetate dihydrate ($Zn(CH_3COO)_2 \cdot 2H_2O$, Loba Chemie, Mumbai, India) as a precursor, and an ammonia solution of 30–33% ($NH_3$) in an aqueous solution ($NH_4OH$, Advent chembio, Mumbai, India) as a reducing agent, were used [54]. The ZnO nanoparticles were produced by dissolving the appropriate amount of zinc acetate in 100 mL of deionized water to produce 0.1 M of a zinc ions solution. Subsequently, the zinc ions solution was subjected to ultrasonic wave irradiation using a Hielscher UP400S (400 W, 24 kHz, Berlin, Germany) at an amplitude of 79% and a cycle of 0.76 for 5 min at a temperature of 40 °C. Then, the ammonia solution was added dropwise to the zinc ions solution under the effect of the ultrasonic waves. After few moments, the ZnO NPs began to precipitate and grow, and the ammonia solution was continuously added until the complete precipitation of ZnO NPs occurred.

The obtained ZnO NPs were washed using deionized water several times and were left out to settle down. Posteriorly, the obtained precipitate was dried at room temperature. To prepare the ZnO nanogel, the produced ZnO NPs were rinsed with double deionized water and were outfitted for the next step. On the other hand, 0.5 g of Carbopol 940 (Loba Chemie, Mumbai, India) was dissolved in 300 mL of doubled deionized water, followed by addition of the freshly washed ZnO NPs. Because Carbopol is naturally acidic [55], the solution needed to be neutral, otherwise it would not thicken. Thus, the mixture had undergone continuous sonication using an ultrasound prop (Hielscher, UP400S Berlin, Germany) with an amplitude of 95 and a cycle of 95% for 1 h. Then, 50 mL of trimethylamine (TEA) as a neutralizing agent (raise the pH to 7) was added dropwise under continuous sonication until the formation of the ZnO white gel occurred, and where the Carbopol would thicken when the pH was near to the neutral conditions [56]. Figure 1 shows a summarized schematic flowchart of the synthesis procedures for the ZnO NPs as well as the Carbopol/ZnO hybrid nanoparticles gel.

### 2.2. Characterization of Synthesized ZnO NPs and Carbopol/ZnO Hybrid Nanoparticles Gel

The synthesized ZnO NPs and Carbopol/ZnO hybrid nanoparticles gel were investigated for determining the chemical composition and the crystalline phase using an X-ray diffractometer (XRD, D8-Discover, Bruker, CuK$\alpha$ radiation, Madison, WI, USA) working at a current of 30 mA and voltage of 20 kV. The Raman shift spectrum of the synthesized ZnO NPs and Carbopol/ZnO hybrid nanoparticles gel were investigated using the Raman spectrometer of a model (Horiba labRAM HR evolution visible single spectrometer, Edison, NJ, USA). The measurement processes were performed at room temperature and the acquisition time was 20 seconds. The Raman spectroscopy was supplied with a He-Cd green LASER which provided a wavelength of 532 nm/edge, and a grating of 1800 (450–850 nm), supported with a 100% ND filter and an objective of X50/Vis. The scanning electron microscope (SEM, JSM-6701F Plus, JEOL, Peabody, MA, USA), and transmission electron microscope (JEOL, TEM-2100, Peabody, MA, USA) operated at a potential of 20 kV, and were used to investigate the morphology, shape, and size of the Carbopol/ZnO hybrid nanoparticles gel. In the investigations of the samples using TEM, a copper grid was prepared to support the NPs by sputtering them with gold. The ZnO NPs stabilized Carbopol gel sample was diluted with distilled water and sonicated with an ultrasonic cleaner (Elma, Singen, Germany) for 30 min. Then, a few drops of the Carbopol/ZnO hybrid nanoparticles gel sample were deposited onto the coated copper grid and allowed to dry at room temperature before the investigations were performed by the TEM microscopy. The specific surface area of the Carbopol/ZnO hybrid nanoparticles gel sample was determined by the $N_2$ adsorption/desorption isotherm using Brunauer–Emmett Teller (BET) analyzer (NOVA touch LX2, model; NT2LX-2, Quantachrome, FL, USA). An Atomic Force Microscope (AFM, 5600LS, Agilent, California, CA, USA) was used to provide 2D and 3D topographic images of the synthesized Carbopol/ZnO hybrid nanoparticles gel. Finally, the particle size and zeta potential of the prepared ZnO NPs and Carbopol/ZnO hybrid nanoparticles gel samples were measured using a DLS and zeta potential analyzer (Nano Sight NS500, Malvern Instruments Ltd., Kassel, Germany).

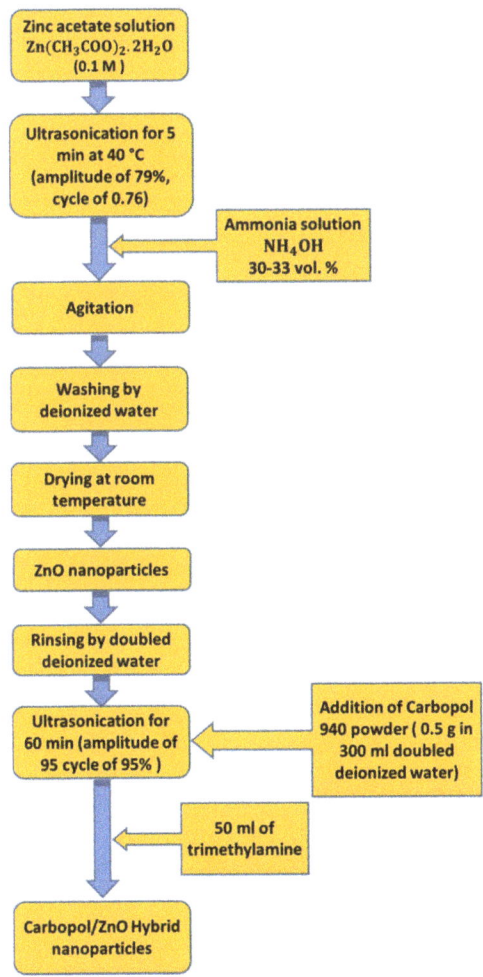

**Figure 1.** Schematic flowchart for the synthesis of ZnO NPs and Carbopol/ZnO hybrid nanoparticles gel.

*2.3. Antibacterial Sensitivity Test*

The antibacterial activity of the investigated Carbopol/ZnO hybrid nanoparticles gel was employed on Mueller–Hinton agar plates using an Agar well diffusion technique against gram-negative *K. pneumonia* (ATCC 13883) and gram-positive *Bacillus subtilis* (ATCC 6633), and *Staphylococcus aureus* (ATCC 6538) bacterial strains, and by applying ampicillin as a reference antibacterial agent. A stock solution was prepared by dissolving 10 mg of Carbopol/ZnO hybrid nanoparticles gel in 1 mL DMSO. The nutrient agar medium of the composition (0.5% Peptone, 0.1% Beef extract, 0.2% Yeast extract, 0.5% NaCl, and 1.5%

Agar-Agar) was prepared by heating the contents in a water bath, cooling them down to 47 °C, and seeding them with the investigated microorganisms. After the solidification of the Agar media, 5 mm diameter holes were punched aseptically and carefully using a sterile cork borer. The investigated Carbopol/ZnO hybrid nanoparticles gel was introduced in Petri-dishes (10 cm diameter) after their dissolving in DMSO to reach a $1.0 \times 10^{-3}$ M concentration. The prepared culture plates were then incubated at 37 °C for 20 h to enhance the growth of the bacteria. The activity was determined by measuring the diameter of the inhibition zone in mm. The plates were kept for incubation at 37 °C for 24 h and then the plates were investigated for a recording of the zone of inhibition in millimeters. Antimicrobial activities were performed in triplicate and the average was taken as the final reading.

## 3. Results and Discussion

### 3.1. Synthesis and Stabilization Mechanism of ZnO NPs by Carbopol Gel

The consequence chemical precipitation reactions of the synthesis method of ZnO NPs, as well as the Carbopol/ZnO hybrid nanoparticles gel, can be explained by the following chemical equations;

$$Zn(CH_3COO)_2 \cdot 2H_2O + 2NH_4OH \rightarrow Zn(OH)_2 + 2CH_3COONH_4 + 2H_2O \quad (1)$$

$$Zn(OH)_2 + 2H_2O \rightarrow Zn^{2+} + 2HO^- = [Zn(OH)_4]^{2-} \quad (2)$$

$$[Zn(OH)_4]^{2-} \leftrightarrow ZnO_2^{2-} + 2H_2O \quad (3)$$

$$ZnO_2^{2-} \xrightarrow{24\ kHz,\ 400\ W,\ \Delta} ZnO \downarrow + O_2 \uparrow \quad (4)$$

(5)

According to the chemical equations mentioned above, (Equation (1) to Equation (4), the zinc acetate reacted with an equivalent amount of ammonium hydroxide solution forming zinc hydroxide precipitate, which dissolved to form ammonium zincate in the presence of an excess amount of ammonium hydroxide solution. The formed ammonium zincate was converted into ZnO nanoparticles by heating under the effect of ultrasonic wave irradiation. The produced ZnO NPs were stabilized by the interaction with Carbopol according to the chemical Equation (5). The Carbopol gel is usually formed by adding the Carbopol powder to distilled water, which was previously neutralized to pH = 7 using few drops of inorganic bases, such as sodium hydroxide or potassium hydroxide, or low molecular weight amines and alkanolamines, which can provide satisfactory neutralization. Some of the amine bases that are effective as neutralizing agents for aqueous formulations include TEA (triethanolamine), AMP-95 (aminomethyl propanol), Tris Amino (tromethamine), and Neutrol TE (tetrakis-2-hydroxypropyl ethylenediamine) [57–59]. Then, the obtained mixture was subjected to continuous rigorous mixing for a few minutes to avoid agglomeration, and then continued stirring until viscosity built up before it turned to gel. In the current work, and according to Equation (5), trimethylamine was used as a neutralizing agent to produce a matrix gel between ZnO NPs and the Carbopol network. The Carbopol/ZnO hybrid nanoparticles gel is formed by the agitation of the ZnO NPs within Carbopol solution to achieve the homogeneity of mixture and to facilitate the polymer-solvent and polymer-polymer interactions giving rise to a better-network structure, typically of a gel-like system. Besides, adding the neutralizing agent induces the entanglement between the different polymer chains. The neutralization reaction ionizes the polymer and generates negative charges along the chain of the polymer. The repulsions between similar charges cause the uncoiling of the molecule into an extended strained structure. This reaction occurs rapidly and provides an instantaneous thickening and an emulsion formation/stabilization. Consequently, the mixture starts to convert to the Carbopol gel network and chelates with the ZnO NPs inside its chain structure. The thick structure and higher elastic character of the hydrated Carbopol/ZnO hybrid nanoparticles gel may be attributed to the electrostatic, Vander Walls, dipole-dipole, hydrophobic type interactions, and the formation of H-bonding, which can be established between ZnO NPs and the hydroxyl groups of Carbopol base-polymer due to the high electronegativity of the oxygen atom. Moreover, there is a possibility for H-bonding creation between ZnO NPs and the nitrogen atoms of trimethylamine agents, even though it has a lower electronegativity than that of the oxygen atom. Furthermore, ultrasonic wave irradiation causes particle-particle interaction by enhancing the molecular vibration of the constituent in the reaction mixture, which generally enhances the thickening process promoted by the polymer-solvent interactions, and aids in the solvation of Carbopol. In addition, ultrasound wave irradiation enhances the polymer–ZnO NPs interaction and improves the viscoelastic properties of the prepared Carbopol/ZnO hybrid nanoparticles gel. Accordingly, many studies have introduced similar interpretations [60–64].

Figure 2 shows the prepared ZnO NPs as well as the Carbopol/ZnO hybrid nanoparticles gel. The results indicate that the prepared ZnO NPs, in the absence of any stabilizing agent, were partially suspended and settled down in the solution within a short period. However, the Carbopol/ZnO hybrid nanoparticles gel was completely suspended and did not settle down for a long time, eventually forming a homogeneous gel. Therefore, one can see that Carbopol is a suitable stabilizing agent for the ZnO NPs in the solution.

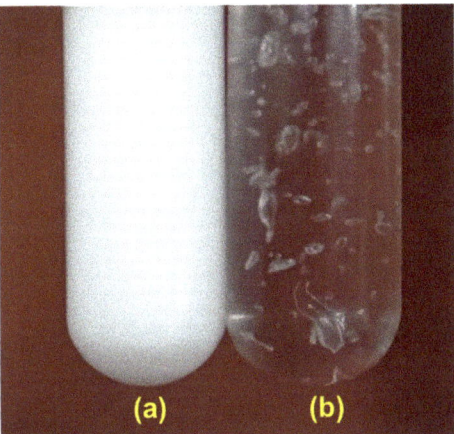

**Figure 2.** ZnO NPs synthesized by the chemical precipitation method under the effect of the ultrasound wave irradiation, where (**a**) is in the aqueous solution, and (**b**) is stabilized by Carbopol gel.

*3.2. Chemical Composition and Phase Identification of Synthesized ZnO NPs and Carbopol/ZnO Hybrid Nanoparticles Gel*

Figure 3 shows the XRD pattern of the synthesized ZnO NPs and Carbopol/ZnO hybrid nanoparticles gel. It can be observed from the results that the diffraction peaks have a high intensity, which implies an ideal crystalline structure within the synthesized ZnO NPs. The Carbopol gel is in an amorphous state as no peaks were recorded. Besides, the characteristic peaks of zinc oxide are prominently featured in the XRD pattern of the Carbopol/ZnO hybrid nanoparticles gel. In addition, the crystalline structure of the synthesized ZnO NPs demonstrates a hexagonal structure of the high-purity ZnO wurtzite phase according to the reference COD no. 2300113. Thus, the observed peaks corresponding to the (100), (002), (101), (102), (110), (103), (200), (112), (201), (004) and (202) planes. Furthermore, the Zn element represents 80.3%, while the oxygen element represents 19.7% of the sample. The average crystallite size (D) of ZnO NPs was estimated from the highly intense and sharp diffraction peak corresponding to the (101) plane using the Debye-Scherer formula [65] according to the following equation:

$$D = \frac{0.9\lambda}{\beta cos\theta} \quad (6)$$

where D is the average crystalline size (nm), λ the CuKα radiation wavelength, i.e., 1.54060 Å, β the full-width at half maximum in radians, and θ the scattering angle in degree. The average crystallite size of the synthesized ZnO nanoparticle was found to be 48.70 nm. Similar results were reported in the previous work [66]. The assessment of different diffraction peaks and the detailed XRD analysis of ZnO NPs and Carbopol/ZnO hybrid nanoparticles gel are listed in Table 1.

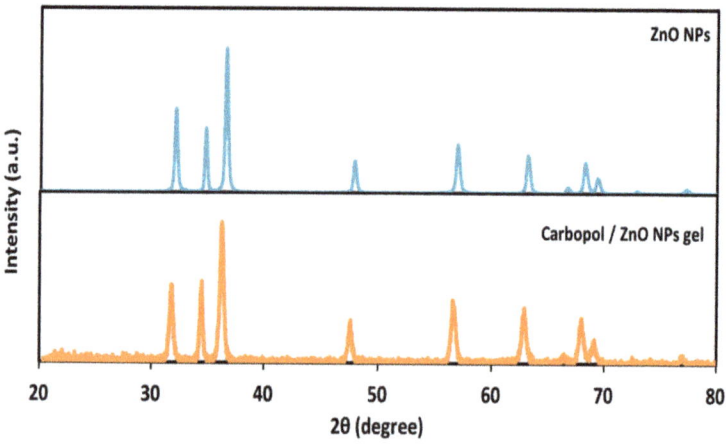

**Figure 3.** XRD patterns of the synthesized ZnO NPs and Carbopol/ZnO hybrid nanoparticles gel.

**Table 1.** Appraised parameters of the XRD analysis of synthesized ZnO NPs and Carbopol/ZnO hybrid nanoparticles gel.

| Index | ZnO NPs | | | Carbopol/ZnO Hybrid Nanoparticles Gel | | |
|---|---|---|---|---|---|---|
| | d-Value (Å) | 2θ (°) | hkl | d-Value (Å) | 2θ (°) | hkl |
| 1 | 2.8009 | 31.926 | 100 | 2.8141 | 31.773 | 100 |
| 2 | 2.5886 | 34.624 | 002 | 2.6027 | 34.430 | 002 |
| 3 | 2.4635 | 36.442 | 101 | 2.4755 | 36.259 | 101 |
| 4 | 1.901 | 47.808 | 102 | 1.9107 | 47.551 | 102 |
| 5 | 1.6171 | 56.894 | 210 | 1.6247 | 56.604 | 210 |
| 6 | 1.4692 | 63.242 | 103 | 1.4769 | 62.875 | 103 |
| 7 | 1.4004 | 66.742 | 200 | 1.4070 | 66.388 | 200 |
| 8 | 1.3715 | 68.34 | 212 | 1.3782 | 67.962 | 212 |
| 9 | 1.3519 | 69.471 | 201 | 1.3583 | 69.097 | 201 |
| 10 | 1.2943 | 73.046 | 004 | 1.3013 | 72.591 | 004 |
| 11 | 1.2317 | 77.422 | 202 | 1.2377 | 76.978 | 202 |

Based on the aforementioned results, and in comparison with previous studies, the existence of ZnO NPs in the hexagonal wurtzite structure demonstrated a high purity and an excellent crystallinity in the synthesized particles. Table 2 shows the crystallinity and phase structures of ZnO NPs prepared using different methods.

**Table 2.** Crystallinity of synthesized ZnO NPs based on the synthesis route and structure compared with those in the literature.

| Synthesis Route | Crystal Phase | Lattice Structure | Crystallite Size (nm) | References |
|---|---|---|---|---|
| Green synthesis by sheep and goat fecal matter | Hexagonal | Wurtzite | 28.50 | [67] |
| Biological synthesis of ZnO NPs using *C. albicans* | Hexagonal | Wurtzite | 25.00 | [68] |
| Biogenic synthesis of ZnO NPs using an aqueous extract of *Papaver somniferum* L | Hexagonal | Wurtzite | 48.00 | [69] |
| ZnO NPs synthesized via a solvothermal method in triethanolamine (TEA) media | Hexagonal | Wurtzite | 33.00 | [70] |
| Hydrothermal synthesis of highly crystalline ZnO NPs | Hexagonal | Wurtzite | 17.00 | [71] |
| Sol-gel synthesis of ZnO NPs at three different calcination temperatures | Hexagonal | Wurtzite | 30.00 | [72] |
| Chemical precipitation/ultrasonication synthesis of ZnO NPs | Hexagonal | Wurtzite | 48.70 | This study |

As shown in Table 2, most of the studies in the literature reported a similar crystallite size for ZnO NPs, although prepared by different methods, and confirmed the same lattice structure.

### 3.3. Raman Spectrum of the Synthesized ZnO NPs and Carbopol/ZnO Hybrid Nanoparticles Gel

The Raman spectrum is a fundamental and multilateral diagnostic technique that can be used to investigate the structural disorder, crystallization, and defects in the micro and nanostructures materials. The vibrational modes of the synthesized ZnO NPs and Carbopol/ZnO hybrid nanoparticles gel are studied. The ZnO NPs of hexagonal wurtzite crystal-type structures belong to the space group of $P63_{mc}$ with two formula units per primitive cell. Also, for pure ZnO NPs crystals, the optical phonons at $\Gamma_{opt}$ point of the Brillouin zone are included in a first-order Raman scattering. According to the classical group theory, the zone center optical phonons can be assorted according to the following irreducible equation:

$$\Gamma_{opt} = A_1 + 2B_2 + E_1 + 2E_2 \quad (7)$$

where both the $A_1$ and $E_1$ modes are two polar branches and both are Raman and infrared active. Due to the macroscopic electric fields associated with the LO phonons, the two modes are divided into transversal optical (TO) and longitudinal optical (LO) branches with various frequencies. Otherwise, the $A_1$, $E_1$, and $E_2$ modes ($E_2$ low and $E_2$ high) are non-polar and first-order Raman-active modes, where the $E_2$ modes are only active for Raman-shift. As shown in Figure 4, the main phonon scattering modes of the ZnO NPs have recorded two bands at 58.05 and 473.87 cm$^{-1}$ that attributed to the 2nd order Raman scattering arising from the $E_{2L}$ and $E_{2H}$ vibrational modes. The broad mode centered at 515.64 cm$^{-1}$ is assigned to the $E_1$(LO) mode, which is defects induced: oxygen vacancies and zinc interstitials. Whereas, the main phonon scattering modes of the Carbopol/ZnO hybrid nanoparticles gel were recorded in eight bands at 59.80, 553.93, 1097.21, 1455.96, 1638.38, 2411.25, 2543.44, and 2940.24 cm$^{-1}$, which corresponds to the $E_{2L}$, $E_{2H}$, $A_1$(LO)/$E_1$(LO), C-O stretch, O-C-O bend, CH$_3$ bend, CH$_3$ stretch, and NH stretch modes, respectively, while, the second-order phonon mode recorded at about 178.58 cm$^{-1}$ is represented as 2$E_{2L}$, and the $E_{2H}$ Raman mode observed at 553.93 cm$^{-1}$ is dominantly attributed to the oxygen vibrations. On the other hand, according to the Raman selection rule, the $B_1$ modes are usually inactive in Raman spectra and are identified as silent modes. The multi phonon scattering modes are recorded at 464.59, 769.87, and 953.89 cm$^{-1}$, which may be assigned to $3E_{2H} - E_{2L}$, $2(E_{2H} - E_{2L})$, and $A_1(TO) + E_1(TO) + E_{2L}$, respectively. In addition, an acoustic combination of $A_1$ and $E_2$ is recorded at around 953.89 cm$^{-1}$. Moreover, the Raman spectra can help in identifying the chemical forms of the Carbopol ligands, as the vibration frequencies of the functional groups are sensitive to the chemical environment. The other vibrational bands were observed in the Raman spectrum at 1455.96, 1638.38, 2411.25, and 2543.44 cm$^{-1}$, and could be originated from Carbopol ligand groups attached to the ZnO NPs' surfaces. However, the C-O stretch, O-C-O bend, $CH_3$ bend, $CH_3$ stretch, and NH stretch modes are much more sensitive to the surrounding environment. It could be stated the Raman shift spectrum for Carbopol gel exhibited the same vibrational modes of ZnO NPs besides the vibrational, rotational, and other low-frequency modes of the hybrid system, causing the multi-phonon to shift to higher wavenumbers. In addition, the structural fingerprint for both ZnO NPs and Carbopol gel were identified. Diallo et al. have mentioned similar vibrational properties of ZnO NPs synthesized by *Aspalathus linearis* [73]. Moreover, Muchuweni et al. have mentioned similar results for the Raman shift of ZnO nanowires prepared using a hydrothermal method [74]. In addition, Taziwa et al. have presented the Raman spectra for both unmodified ZnO and C:ZnO NPs synthesized using the PSP technique. They found that the C:ZnO NPs recorded a red-shift by 4 cm$^{-1}$ compared to that found in bulk ZnO samples. They attributed this shift either to the phonon confinement and tensile stress within the nanocrystal (quantum dots) boundaries, or the localization of phonons inside the C:ZnO NPs hybrid system, which has a more inherent defect when compared to pristine ZnO NPs [75]. On the contrary, in a study by Jayachandraiah and Krishnaiah, both reported that the main phonon scattering modes of Er-doped ZnO NPs systems dropped with different Er ratios without a significant shift [76].

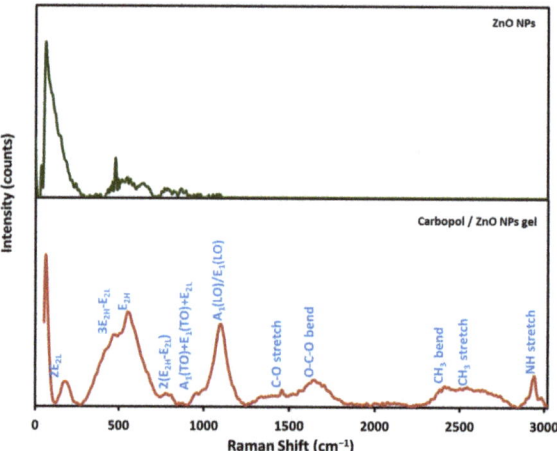

**Figure 4.** Raman shift spectrums of the synthesized ZnO NPs and Carbopol/ZnO hybrid nanoparticles gel.

*3.4. BET Surface Area and Pore Size Distribution Analysis of the Synthesized Carbopol/ZnO Hybrid Nanoparticles Gel*

The specific surface area of the synthesized Carbopol/ZnO hybrid nanoparticles gel was determined using the $N_2$ adsorption/desorption isotherm at a temperature of 77 K. Figure 5 shows the typical IV-type adsorption of the Carbopol/ZnO hybrid nanoparticles. Usually, the mesoporous adsorbents are exhibiting type IV isotherms (e.g., mesoporous molecular sieves, industrial adsorbents, and many oxide gels). In this case, the interactions between the molecules in the condensed state, and the adsorbent/adsorptive interactions, were designated to determine adsorption behavior in the mesoporous materials. Based on this reason, the initial monolayer-multilayer adsorption occurring on the mesoporous walls was followed by pore condensation. The phenomenon of pore condensation means that, at a pressure $p$ less than the saturation pressure $p_o$ of the bulk liquid, the gas condenses to a liquid-like phase in the pores [77]. According to the characteristics of the IV type, the capillary condensation taking place in mesoporous may be accompanied by hysteresis, which commonly occurs when the pore size exceeds a certain critical width. It could be said that the capillary condensation depends mainly on the adsorption system and temperature (for nitrogen adsorption in cylindrical pores at 77 K). Furthermore, as shown in the isotherm curve in Figure 4, the isotherm did not contain a hysteresis loop that starts to occur when the pores of the adsorbent are wider than ∼4 nm. Consequently, the specific surface area of Carbopol/ZnO hybrid nanoparticles gel was determined at 54.26 m$^2$/g, while the pore volume and the mean pore diameter were found to be 0.063 cm$^3$/g and 2.33 nm, respectively. The BET results of the Carbopol/ZnO hybrid nanoparticles gel are listed in Table 3. Kołodziejczak-Radzimska et al. reported similar results in their analysis [78].

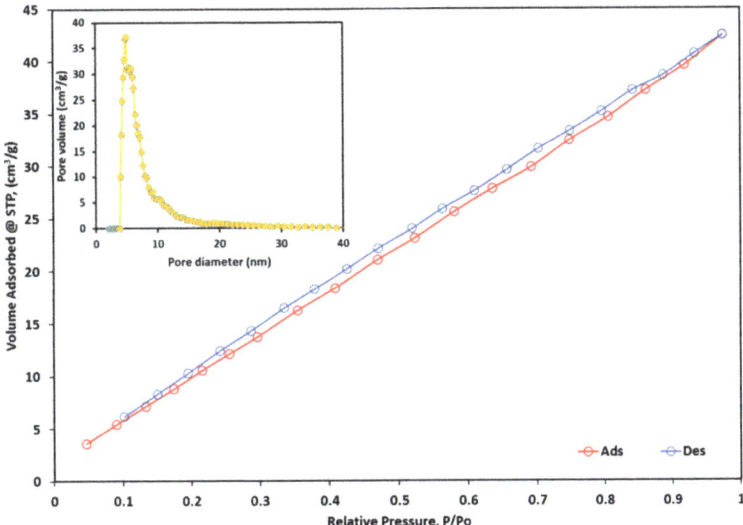

**Figure 5.** N$_2$ gas adsorption/desorption isotherm of the synthesized Carbopol/ZnO hybrid nanoparticles gel.

**Table 3.** BET parameters of the synthesized Carbopol/ZnO hybrid nanoparticles gel.

| Carbopol Stabilized ZnO NPs | |
|---|---|
| BET surface area, m$^2$/g | 54.26 |
| Average particle size, nm | 5.02 |
| Mean pore diameter, nm | 2.33 |
| Total pore volume, cm$^3$/g | 0.063 |

As shown in Table 4, the Carbopol/ZnO hybrid nanoparticles gel exhibited a relatively high surface area when compared with the bare ZnO NPs and some other hybrid systems. Accordingly, this demonstrates that the Carbopol gel provided an extra area for ZnO NPs to disperse, and subsequently contributed to increasing the overall surface area of the obtained sample. As presented in Table 3, and as stated by the different reports, all hybrid systems showed higher surface areas than those of bare metal oxides.

**Table 4.** Comparison of the obtained SBET of synthesized Carbopol/ZnO hybrid nanoparticles gel with those for bare ZnO NPs and the other hybrid systems in the literature.

| Materials | $S_{BET}$, m$^2$/g | References |
|---|---|---|
| Bare ZnO NPs | 3.29 | [79] |
| Pd–ZnO-EG | 4.93 | |
| ZnO precursor | 87.43 | |
| ZnO Xerogel nanostructers annealed at | | |
| 275 °C | 25.36 | [80] |
| 375 °C | 22.81 | |
| 475 °C | 15.16 | |
| 600 °C | 8.78 | |
| ZnO nanostructures modified chitosan and sodium chloroacetate with isopropyl alcohol | | |
| ZnO-CTS-450 | 23.76 | [81] |
| ZnO-CMC1-450 | 15.44 | |
| ZnO-CTS-650 | 11.92 | |
| ZnO-CMC1-650 | 5.88 | |
| Modified ZnO NPs | | |
| Ag/ZnO NPs | 7.75 | [82] |
| Cd/ZnO NPs | 10.6 | |
| Pb/ZnO NPs | 106.65 | |
| ZnO synthesized without any biotemplate | 5 | |
| ZnO synthesized at different volumes of palm olein | | |
| 1 mL PO | 10 | [83] |
| 2 mL PO | 13 | |
| Kaolin/ZnO nanocomposites | 31.8 | [84] |
| Gum arabic-crosslinked-poly(acrylamide)/zinc oxide hydrogels (GA-cl-PAM/ZnO hydrogel) | 39.0 | [85] |
| Flower-like ZnO NPs in a cellulose hydrogel microreactor | 39.18 | [86] |
| ZnO/PAAH hybrid nanomaterials (PAAH = polyacrylic acid) | | |
| ZnO/PAA2-350 °C | 51 | [87] |
| ZnO/PAA5-350 °C | 31 | |
| Carbopol/ZnO hybrid nanoparticles gel | 54.26 | This study |
| Alginate/Zn aerogel beads | 143 | [88] |
| Highly dispersed ZnO NPs supported on the silica gel matrix | 245 | [89] |
| Cellulose/ZnO hybrid aerogel (CA/ZnO) | 352.82 | [90] |
| Wheat gliadin/ZnO hybrid nanospheres | 523.88 | [91] |

*3.5. AFM Topographical Analysis of the Synthesized Carbopol/ZnO Hybrid Nanoparticles Gel*

Figure 6a–d shows the non-contact mode topographical AFM 2D images and the corresponding 3D images for the synthesized Carbopol/ZnO hybrid nanoparticles gel. The images demonstrated that the Carbopol/ZnO hybrid nanoparticles have a spherical particle shape. In addition, a good distribution and monodisperse of a large number of random nano pits are observed. Furthermore, the automated batch-mode particle-height functional analysis provided a corresponding particle size distribution histogram of the synthesized Carbopol/ZnO hybrid nanoparticles gel through the scanned area, whereas the synthesized Carbopol/ZnO hybrid nanoparticles gel exhibited a homogeneous particle size of normal distribution. Moreover, the maximum peak height was found to be 59.9 nm, which represents the average particle size of synthesized Carbopol/ZnO hybrid nanoparticles gel, as shown in Figure 6e,f. Furthermore, the corresponding particle volume distribution histogram showed that the volume of the Carbopol/ZnO hybrid nanoparticles gel was found to be 0.0113 (V.$\mu m^2/\mu m^2$), and the volume of the void per cross-sectional area was 0.206 (V.$\mu m^2/\mu m^2$) as presented in Figure 6g,h. Similar results were reported in previous studies [92]. Moreover, the results of the AFM are correlative with the results of the XRD.

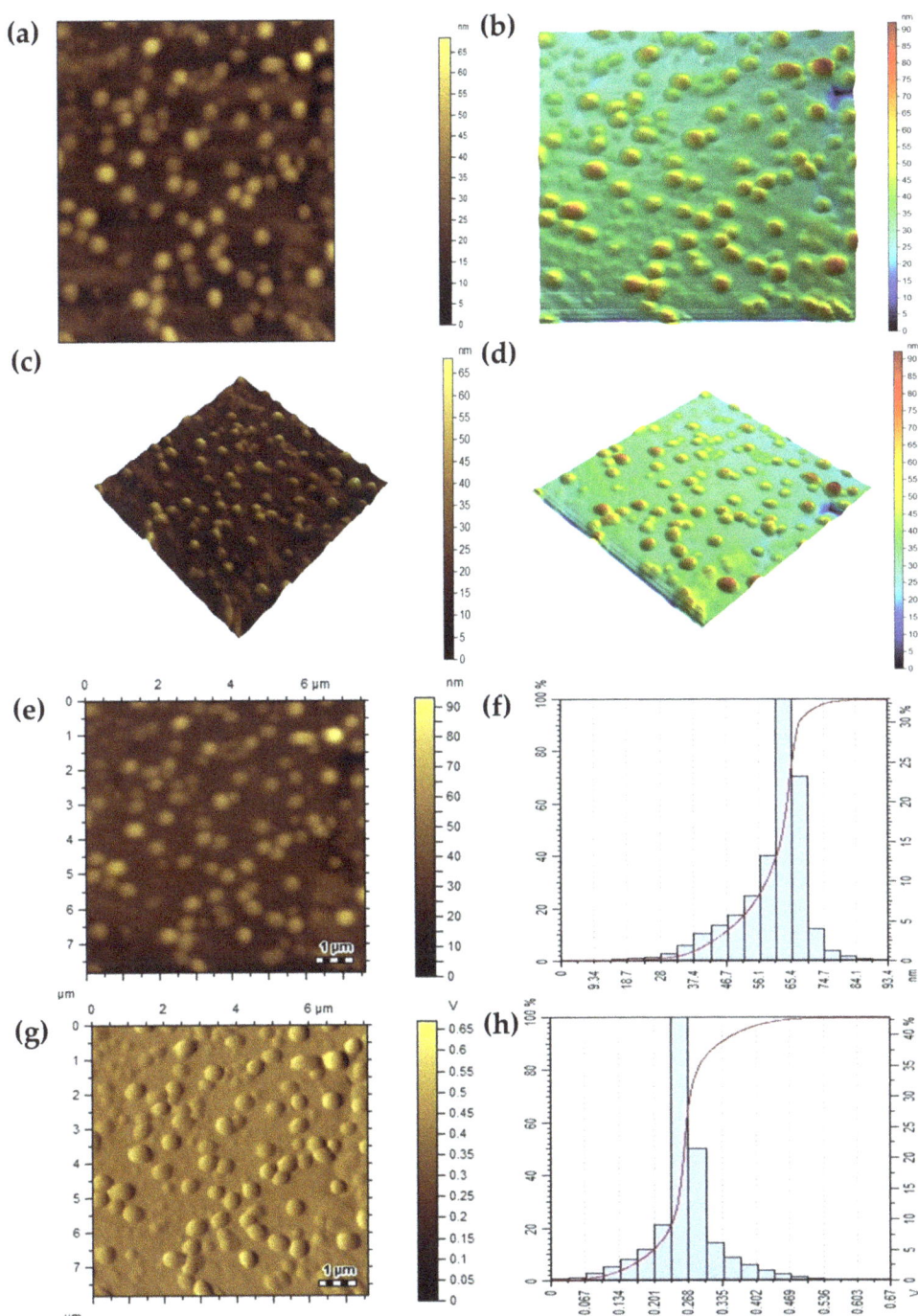

**Figure 6.** (**a**,**b**) Topographical 2D AFM images; (**c**,**d**) identical 3D AFM images; (**e**,**f**) typical histogram of particle size distribution; (**g**,**h**) typical histogram of particles volume distribution of the synthesized Carbopol/ZnO hybrid nanoparticles gel.

### 3.6. Particle Shape and Size Analysis of Synthesized ZnO NPs and Carbopol/ZnO Hybrid Nanoparticles Gel

The synthesized ZnO NPs and Carbopol/ZnO hybrid nanoparticles gel were investigated using a high-emission SEM microscope. Moreover, the obtained Carbopol/ZnO hybrid nanoparticles gel was further examined using a TEM microscope. The SEM and TEM results supported the results obtained by the AFM images. From Figure 7a, it could observed that the ZnO NPs were formed in nanospheres with uniform spherical shapes and sizes, as the majority of the particles lie in the nanodomain range. Moreover, the microstructure of nanocrystalline ZnO NPs has a skeletal form resulting from the aggregation process. These agglomerates have taken the shape of faceted crystals, which is characterized by a high porosity. Hutera et al. and Chai et al. reported similar results [93,94]. Figure 7b illustrates an SEM photograph of synthesized Carbopol/ZnO hybrid nanoparticles gel. It is clear that the Carbopol/ZnO hybrid nanoparticles gel have formed from Carbopol gel well-comprised by ZnO nanoclusters, and have created a network of ZnO NPs series connected to each other by the Carbopol gel. These observations were confirmed by the HR-TEM micrograph, as shown in Figure 8.

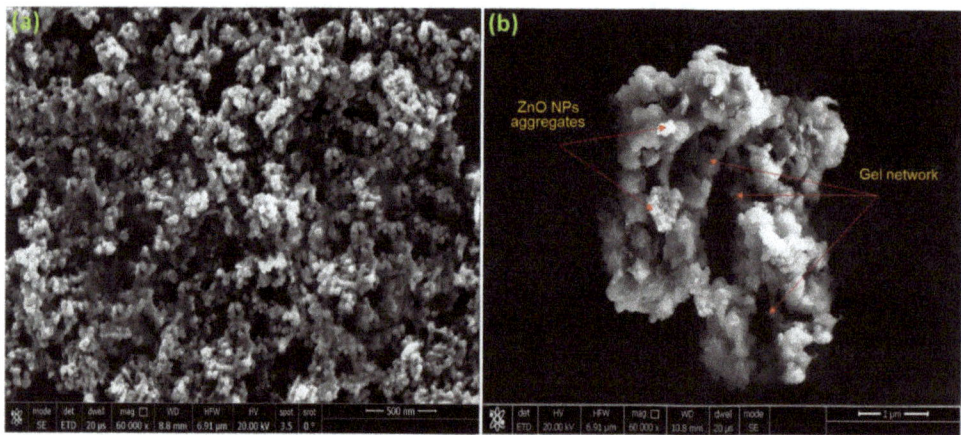

**Figure 7.** Field-Emission SEM image of (**a**) ZnO NPs and (**b**) Carbopol/ZnO hybrid nanoparticles gel.

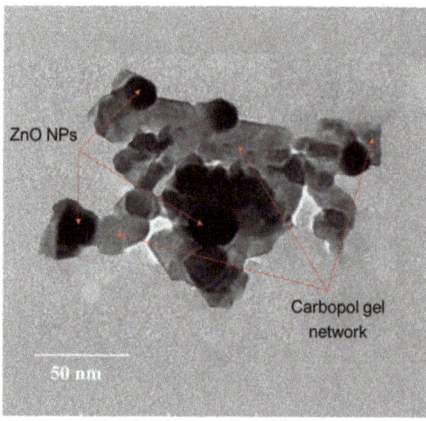

**Figure 8.** High-Resolution TEM image of the synthesized Carbopol/ZnO hybrid nanoparticles gel.

As shown in the TEM image, quazi-spherical ZnO NPs with sharp edges dispersed in the gel network. Moreover, the TEM image confirmed that the size of the Carbopol/ZnO hybrid nanoparticles was ~38 nm. Previous studies mention similar TEM morphologies. Mohammed et al. reported that the particle size investigations of the unloaded Carbopol AquaSF-1 nanogel and vancomycin-loaded Carbopol nanogel (VAC-AquaSF1) revealed that the prepared drug-loaded nanogel particles were almost spherical with a smooth morphology, appeared as black dots with bright surroundings, and were well dispersed and separated on the surface. The average particle size was found to be less than 115 nm and larger than unloaded nanogel, which was about lower than 100 nm [95]. In addition, Al-Awady et al. concluded that the TEM image of the dried-up suspension of collapsed Carbopol Aqua SF1 nanogel particles contained spherical particles of about 100 ± 20 nm, which matches with the average particle diameter measured by the Zetasizer analyzer [96].

*3.7. DLS and Zeta-Potential of the Synthesized ZnO NPs and Carbopol/ZnO Hybrid Nanoparticles Gel*

The synthesized ZnO NPs and Carbopol/ZnO hybrid nanoparticles gel were investigated using the dynamic laser scattering (DLS) technique and were subjected to zeta potential measurements. Figure 9 shows that the median particle size value of ZnO NPs was ~60 nm, and ~40nm for Carbopol/ZnO hybrid nanoparticles gel, which matched with the particle size estimated by the TEM image analysis. This difference may be attributed to the fact that the Carbopol gel improved the dispersity of the ZnO NPs and reduced the cluster aggregations.

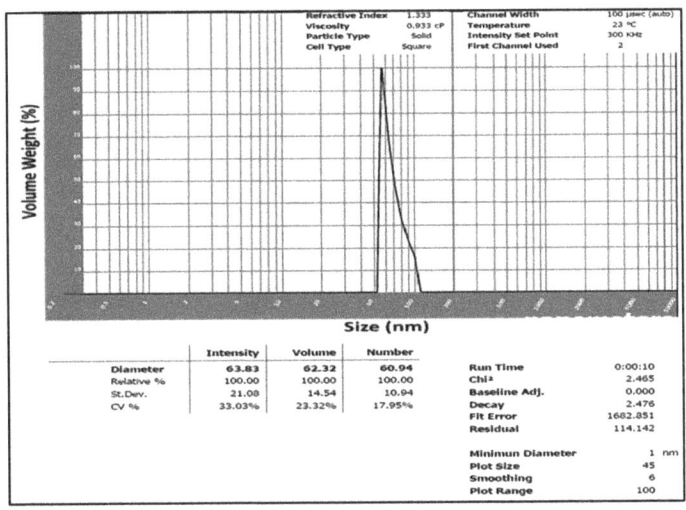

**Figure 9.** Particle size distribution of ZnO NPs measured by DLS.

Nanocrystals usually undergo agglomeration when dispersed in solutions, and this behavior has a major impact on the reactivity and responsiveness of nanomaterials when exposed to various cells or organisms. Therefore, the hydrodynamic sizes of the ZnO NPs samples suspended in Carbopol gel were measured at a neutral pH. According to the Derjaguin−Landau−Verwey−Overbeek (DLVO) model, the agglomeration of non-stabilized ZnO NPs nanocrystals depends on the repulsive interaction arising from an electrostatic force and the van der Waals force of attraction. Because the surface charges of nanocrystals influence the electrostatic repulsive force, nanoparticles with a larger zeta potential will generally reduce in hydrodynamic size. Punnoose et al. mentioned similar results [97].

The zeta-potential values of the ZnO NPs and Carbopol/ZnO hybrid nanoparticles gel were found at 5.82 mV (Figure 10), and −21.7 mV, respectively. Consequently, one can conclude from the results that the high value of zeta-potential might be interpreting the acquired high bioactivity of the Carbopol/ZnO hybrid nanoparticles gel against different types of bacterial species. It is clear from the results that the surface of the Carbopol/ZnO hybrid nanoparticles gel is negatively charged at a neutral pH. The high negative zeta potential observed for the Carbopol/ZnO hybrid nanoparticles gel signifies an efficient dispersion of the ZnO NPs powder in the neutral pH gel, where the well-dispersed powder provides a high effective surface for killing the bacteria [98]. The ZnO NPs stabilization by Carbopol gel is expected to modify the net surface charge density and inter-particle interactions. When the ZnO NPs are dispersed in a Carbopol gel, the nanocrystals undergo surface ionization and the adsorption of ions resulting in the generation of an enhanced surface charge. This surface charge leads to an electric potential between the ZnO NPs nanocrystals and the bulk of the dispersion medium (Carbopol gel), and this is measured as zeta potential in experiments that use electrophoretic/electrokinetic techniques. Besides, Weldrick et al. stated that the alcalase-coated clindamycin-loaded Carbopol Nanogels were narrow, although not monodisperse. The effect of pH on the nanogels not only causes swelling and de-swelling, but also changes the values of the zeta potential. There is an overall increase in the zeta potential according to the pH value of the solutions. The surface carboxylic groups of Carbopol gel that partially dissociate cause the appearance of a negative surface charge. Moreover, the nanogel particles are always negatively charged across the whole range of the pH, which allows for the cationic entities of materials [99].

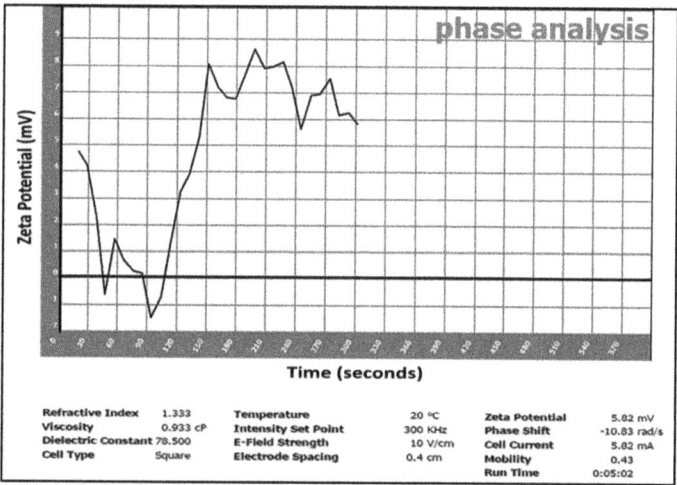

Figure 10. Zeta potential of ZnO NPs measured as a function of time.

A previous study by Mohammed, et al. reported that the particle size distribution and the zeta potential of bare Carbopol nanogel at pH 6.5 equaled 94.66 nm and −31.8 mV, respectively, while, the particle size distribution and the zeta potential of vancomycin-loaded Carbopol nanogel at pH 6.5 reached 391 nm and −38.5 mV, respectively [95]. Moreover, Al-Awady et al. reported that the average particle diameter of 0.05 wt% Carbopol Aqua SF1 was found to be approximately 100 nm, while the zeta potential was −44 mV, which indicates that the Carbopol Aqua SF1 nanogel particles have high stability when suspended in MilliQ water at pH 5.5. Furthermore, they measured the value of the average particle diameter of berberine-loaded Carbopol Aqua SF1 nanogel and found it to be about 135 nm at pH 5.5, while the berberine-loaded Carbopol Aqua SF1 nanogel suspension has a zeta potential value of −40 mV [96].

## 3.8. Bioactivity of Carbopol/ZnO Hybrid Nanoparticles Gel

Gram-positive bacteria (*Bacillus subtilis* and *Staphylococcus aureus*) and gram-negative bacteria (*K. pneumoniae*) are pathogenic bacteria, which is one of the ESKAPE pathogens with a multi-drug resistance. The emergence of resistance in *Bacillus subtilis*, *Staphylococcus aureus*, and *K. pneumoniae* causes high mortality and morbidity. Gram-positive and gram-negative bacteria have developed an ability to accumulate diverse resistance mechanisms. The rise in antibiotic resistance and emergence of antibiotic-resistant superbugs is stressing the need for innovative strategies to develop new antimicrobials [100]. In this study, two types of bacterial organisms, gram-positive bacteria (*Bacillus subtilis* and *Staphylococcus aureus*) and gram-negative bacteria (*K. pneumoniae*), were used to investigate the antibacterial activity of the prepared Carbopol/ZnO hybrid nanoparticles gel. The observed results revealed that the inhibition activity of the Carbopol/ZnO hybrid nanoparticles gel was more active than the reference sample of ampicillin (see Figure 11), where the inhibition zone diameters were found at $27 \pm 0.71$ and $33 \pm 0.62$ for *Bacillus subtilis* and *K. pneumoniae* (ATCC 13883), respectively. Thus, the obtained values of the inhibition zone can illustrate the great inhibition effect of Carbopol/ZnO hybrid nanoparticles gel against the different types of bacterial strains. However, the value of the inhibition zone of Carbopol/ZnO hybrid nanoparticles gel against *Staphylococcus aureus* reached $37 \pm 0.78$ mm with a high sensitivity without any contamination effect [101,102]. In the same context, the ampicillin showed an asymmetrical and moderate effect on the inhibition of bacterial growth for both *Bacillus subtilis* and *Staphylococcus aureus*, while also facing a drastic resistance from the *K. pneumonia*, and thus it did not record any activity.

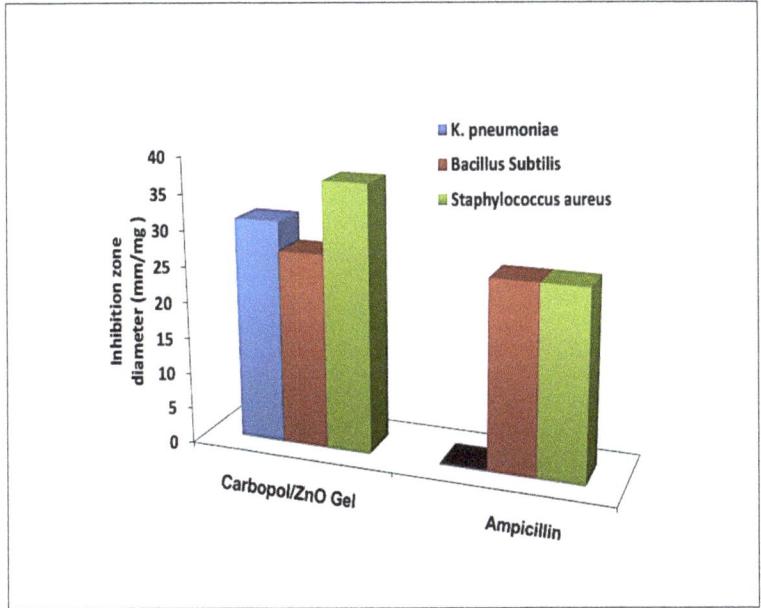

**Figure 11.** Antibacterial activity of the synthesized Carbopol/ZnO hybrid nanoparticles gel as compared with ampicillin standard samples.

Sadek et al. mentioned that the value of the inhibition zone of bare metal nanoparticles (e.g., nZVI) against *Staphylococcus aureus* reached 9 mm [103]. Moreover, many previous studies reported the efficient antibacterial activity of inorganic nanogels-based metals/metals oxides. Whaid et al. have investigated the antibacterial activity of β-chitin/ZnO nanocomposite hydrogels against *S. aureus* and *E. coli*, which exhibited good bactericidal properties, and mentioned that collagen-dextran-ZnO NPs nanocomposites could serve as a kind of promising wound dressing with sustained drug delivery properties [104].

Gokmen et al. have studied the synergetic effect of hydrogel, nanoTiO$_2$ doped hydrogel, and ZnO nanoflowers deposited nano TiO$_2$ doped in biodegradable hydrogel against *Staphylococcus aureus* and *Escherichia coli*. They reported that these composites have achieved antibacterial activity towards these two bacteria as follows: 79.45%, 90.37%, and 99.98% against *Staphylococcus aureus*; and 57.03%, 80.79%, and 97.46% against *Escherichia coli* [105]. In addition, Yang et al. have mentioned that due to the high hydrophilicity, unique three-dimensional network, fine biocompatibility, and cell adhesion, the hydrogels are one of the suitable biomaterials for drug delivery in antimicrobial areas. Furthermore, they also stated that the antimicrobial hydrogels loaded with metal nanoparticles are a potential method to solve antibiotic resistance [106]. Additionally, Scalzo et al. studied the interaction effects of polyacrylic acid polymers (Carbopol 940) on the antimicrobial activity of methyl parahydroxybenzoate against some gram-negative and gram-positive bacteria and yeast, and the results revealed that the hydrophilic polymer, widely employed in many formulations, exerts, on the contrary, an interesting synergism on the microbicidal activity of the preserving agent against *E. coli* and *P. Aeruginosa* [107].

It is known that gram-negative bacteria are surrounded by a thin peptidoglycan cell wall, which itself is surrounded by an outer membrane containing lipopolysaccharide, while gram-positive bacteria lack an outer membrane but are surrounded by layers of peptidoglycan many times thicker than in gram-negative bacteria. According to the previous studies, several mechanisms have been suggested to specify the antibacterial property of metals/metals oxides and hydrogels, including the penetration of the cell envelope by these metal oxides/hydrogels, and damaging the cell membrane due to the production of reactive oxygen species (ROS). The elevated ROS leads to producing effects on the bacteria, of which lipid peroxidation is one of them, which affects the bacterial membrane's integrity [51]. Furthermore, the elevated amounts of ROS cause oxidative stress within cells, eventually leading to cell death. ROS generation is known to be a prominent mechanism of cell death when cells are treated with nanoparticles [108,109]. It could assign the high antimicrobial activity of Carbopol/ZnO hybrid nanoparticles gel to the high stability of Carbopol/ZnO hybrid nanoparticles gel for a long time, which allows the included ZnO NPs within Carbopol gel to effectively produce more ROS. The Carbopol/ZnO hybrid nanoparticles gel has a distinctive antibacterial activity towards both the gram-positive and gram-negative bacteria. However, the significant activity of Carbopol/ZnO hybrid nanoparticles gel against *Staphylococcus aureus* may be attributed to the amount of ROS species that can easily penetrate the cell wall of gram-positive bacteria, due to its lack of an outer membrane resulting in high bacterial inhibition. On the contrary, the ROS species forfeits the ability to penetrate the thick cell wall of gram-negative bacteria, which makes it less effective [110].

## 4. Conclusions

High purity ZnO NPs, as well as Carbopol/ZnO hybrid nanoparticles gel, were synthesized using a wet chemical precipitation reaction with a high yield under the effect of the ultrasonic wave irradiation. The properties of the produced nanoparticles and the hybrid system were characterized using different characterization techniques. The obtained results revealed that the synthesized nanoparticles were composed mainly of ZnO wurtzite crystalline phases with a quazi-spherical shape and median particle sizes of 59.9 nm, a particle size distribution between ~9 and ~93 nm, and a surface area of 54.26 m$^2$/g. It could be stated that the high values of zeta-potential in the Carbopol/ZnO hybrid nanoparticles ($-21.7$ mV) might be interpreted as the reason for the high bioactivity of Carbopol stabilized ZnO NPs against different bacterial species. The antibacterial activity of Carbopol/ZnO hybrid nanoparticles gel was screened against different gram-positive and gram-negative bacterial strains. It was found that the inhibition activity of Carbopol/ZnO hybrid nanoparticles gel was higher than the reference sample of the ampicillin drug itself. This study has highlighted the properties of Carbopol/ZnO hybrid nanoparticles gel as a potent antibacterial hybrid system with regards to the growth inhibition of microorganisms. This makes Carbopol/ZnO hybrid nanoparticles gel an interest of note as an efficient bacterial inhibitor for many applications.

**Author Contributions:** Conceptualization S.H.I.; A.H.; T.A.I.; H.H.M.; W.H.M.; W.M.D.; Methodology, A.H.; W.M.D.; formal analysis, S.H.I., A.H.; W.H.M.; investigation, T.A.I.; H.H.M.; resources, S.H.I.; A.H.; H.H.M.; writing—original draft preparation, A.H.; writing—review and editing, W.H.M.; W.M.D.; All authors have read and agreed to the published version of the manuscript.

**Funding:** This research was funded by the Taif University Researchers Supporting Project number (TURSP-2020/134), Taif University, Taif, Saudi Arabia.

**Institutional Review Board Statement:** Not applicable.

**Informed Consent Statement:** Not applicable.

**Data Availability Statement:** All data are available in this manuscript.

**Acknowledgments:** The authors are grateful to Taif University Researchers Supporting Project number (TURSP-2020/134), Taif University, Taif, Saudi Arabia for their financial support. We gratefully acknowledge the faculty of nanotechnology for postgraduate studies, Cairo University (https://fnt.cu.edu.eg/en/) (accessed on 1 July 2021) for providing facilities.

**Conflicts of Interest:** The authors declare no conflict of interest.

# References

1. Jiang, J.; Pi, J.; Cai, J. The advancing of zinc oxide nanoparticles for biomedical applications. Available online: https://www.hindawi.com/journals/bca/2018/1062562/ (accessed on 7 July 2021).
2. Rasmussen, J.W.; Martinez, E.; Louka, P.; Wingett, D.G. Zinc oxide nanoparticles for selective destruction of tumor cells and potential for drug delivery applications. *Exp. Opin. Drug Deliv.* **2010**, *7*, 1063–1077. [CrossRef]
3. Hamdy, A.; Ismail, S.H.; Ebnalwaled, A.A.; Mohamed, G.G. Characterization of Superparamagnetic/Monodisperse PEG-Coated Magnetite Nanoparticles Sonochemically Prepared from the Hematite Ore for Cd(II) Removal from Aqueous Solutions. *J. Inorg. Organomet. Polym. Mater.* **2021**, *31*, 397–414. [CrossRef]
4. Chieng, B.W.; Loo, Y.Y. Synthesis of ZnO nanoparticles by modified polyol method. *Mater. Lett.* **2012**, *73*, 78–82. [CrossRef]
5. Banerjee, P.; Chakrabarti, S.; Maitra, S.; Dutta, B.K. Zinc oxide nano-particles–sonochemical synthesis, characterization and application for photo-remediation of heavy metal. *Ultrason. Sonochem.* **2012**, *19*, 85–93. [CrossRef] [PubMed]
6. Ghoshal, T.; Biswas, S.; Paul, M.; De, S.K. Synthesis of ZnO nanoparticles by solvothermal method and their ammonia sensing properties. *J. Nanosci. Nanotechnol.* **2009**, *9*, 5973–5980. [CrossRef] [PubMed]
7. Bharti, D.B.; Bharati, A. Synthesis of ZnO nanoparticles using a hydrothermal method and a study its optical activity. *Luminescence* **2017**, *32*, 317–320. [CrossRef]
8. Pineda-Reyes, A.M.; Olvera, M.d.l.L. Synthesis of ZnO nanoparticles from water-in-oil (w/o) microemulsions. *Mater. Chem. Phys.* **2018**, *203*, 141–147. [CrossRef]
9. Somoghi, R.; Purcar, V.; Alexandrescu, E.; Gifu, I.C.; Ninciuleanu, C.M.; Cotrut, C.M.; Oancea, F.; Stroescu, H. Synthesis of Zinc Oxide Nanomaterials via Sol-Gel Process with Anti-Corrosive Effect for Cu, Al and Zn Metallic Substrates. *Coatings* **2021**, *11*, 444. [CrossRef]
10. Adam, R.E.; Pozina, G.; Willander, M.; Nur, O. Synthesis of ZnO nanoparticles by co-precipitation method for solar driven photodegradation of Congo red dye at different pH. *Photonics Nanostructures Fundam. Appl.* **2018**, *32*, 11–18. [CrossRef]
11. Lijuan, A.N.; Jun, W.; Tiefeng, Z.; Hanlin, Y.; Zhihui, S. Synthesis of ZnO nanoparticles by direct precipitation method. *Adv. Mat. Res.* **2012**, *380*, 335–338.
12. Naveed Ul Haq, A.; Nadhman, A.; Ullah, I.; Mustafa, G.; Yasinzai, M.; Khan, I. Synthesis approaches of zinc oxide nanoparticles: The dilemma of ecotoxicity. *J. Nanomater.* **2017**, *2017*. [CrossRef]
13. Zhang, Y.; Ram, M.K.; Stefanakos, E.K.; Goswami, D.Y. Synthesis, characterization, and applications of ZnO nanowires. *J. Nanomater.* **2012**, *2012*, 1–22. [CrossRef]
14. Chaudhary, S.; Umar, A.; Bhasin, K.K.; Baskoutas, S. Chemical sensing applications of ZnO nanomaterials. *Materials* **2018**, *11*, 287. [CrossRef] [PubMed]
15. Chen, R.; Huo, L.; Shi, X.; Bai, R.; Zhang, Z.; Zhao, Y.; Chang, Y.; Chen, C. Endoplasmic reticulum stress induced by zinc oxide nanoparticles is an earlier biomarker for nanotoxicological evaluation. *ACS Nano* **2014**, *8*, 2562–2574. [CrossRef]
16. Food and Drug Administration. Select Committee on GRAS Substances (SCOGS) Opinion: Zinc Salts 2015. Washington, DC. 2015. Available online: https://www.accessdata.fda.gov/scripts/cdrh/cfdocs/cfcfr/CFRSearch.cfm (accessed on 1 April 2020).

17. Mirzaei, H.; Darroudi, M. Zinc oxide nanoparticles: Biological synthesis and biomedical applications. *Ceram. Int.* **2017**, *43*, 907–914. [CrossRef]
18. Siddiqi, K.S.; Rahman, A.; Husen, A. Properties of zinc oxide nanoparticles and their activity against microbes. *Nanoscale Res. Lett.* **2018**, *13*, 1–13. [CrossRef]
19. Smijs, T.G.; Pavel, S. Titanium dioxide and zinc oxide nanoparticles in sunscreens: Focus on their safety and effectiveness. *Nanotechnol. Sci. Appl.* **2011**, *4*, 95. [CrossRef]
20. Sahoo, S.; Maiti, M.; Ganguly, A.; George, J.J.; Bhowmick, A.K. Effect of zinc oxide nanoparticles as cure activator on the properties of natural rubber and nitrile rubber. *J. Appl. Polym. Sci.* **2007**, *105*, 2407–2415. [CrossRef]
21. Hatamie, A.; Khan, A.; Golabi, M.; Turner, A.P.F.; Beni, V.; Mak, W.C.; Sadollahkhani, A.; Alnoor, H.; Zargar, B.; Bano, S.; et al. Zinc oxide nanostructure-modified textile and its application to biosensing, photocatalysis, and as antibacterial material. *Langmuir* **2015**, *31*, 10913–10921. [CrossRef]
22. Sricharussin, W.; Threepopnatkul, P.; Neamjan, N. Effect of various shapes of zinc oxide nanoparticles on cotton fabric for UV-blocking and anti-bacterial properties. *Fibers Polym.* **2011**, *12*, 1037–1041. [CrossRef]
23. Kołodziejczak-Radzimska, A.; Jesionowski, T. Zinc oxide—from synthesis to application: A review. *Materials* **2014**, *7*, 2833–2881. [CrossRef]
24. Bica, B.O.; de Melo, J.V.S. Concrete blocks nano-modified with zinc oxide (ZnO) for photocatalytic paving: Performance comparison with titanium dioxide ($TiO_2$). *Constr. Build. Mater.* **2020**, *252*, 119120. [CrossRef]
25. Ahmed, E.M. Hydrogel: Preparation, characterization, and applications: A review. *J. Adv. Res.* **2015**, *6*, 105–121. [CrossRef]
26. Singh, S.K.; Dhyani, A.; Juyal, D. Hydrogel: Preparation, characterization and applications. *Pharma Innov.* **2017**, *6*, 25.
27. Chai, Q.; Jiao, Y.; Yu, X. Hydrogels for biomedical applications: Their characteristics and the mechanisms behind them. *Gels* **2017**, *3*, 6. [CrossRef] [PubMed]
28. Chaterji, S.; Kwon, I.K.; Park, K. Smart polymeric gels: Redefining the limits of biomedical devices. *Prog. Polym. Sci.* **2007**, *32*, 1083–1122. [CrossRef]
29. Sultana, F.; Manirujjaman; Imran-Ul-Haque, M.; Arafat, M.; Sharmin, S. An overview of nanogel drug delivery system. *J. Appl. Pharm. Sci.* **2013**, *3*, 95–105.
30. Oh, J.K.; Drumright, R.; Siegwart, D.J.; Matyjaszewski, K. The development of microgels/nanogels for drug delivery applications. *Prog. Polym. Sci.* **2008**, *33*, 448–477. [CrossRef]
31. Soni, K.S.; Desale, S.S.; Bronich, T.K. Nanogels: An overview of properties, biomedical applications and obstacles to clinical translation. *J. Controll. Release* **2016**, *240*, 109–126. [CrossRef]
32. Kabanov, A.V.; Vinogradov, S.V. Nanogels as pharmaceutical carriers: Finite networks of infinite capabilities. *Angew. Chem. Int. Ed.* **2009**, *48*, 5418–5429. [CrossRef]
33. Yin, Y.; Hu, B.; Yuan, X.; Cai, L.; Gao, H.; Yang, Q. Nanogel: A versatile nano-delivery system for biomedical applications. *Pharmaceutics* **2020**, *12*, 290. [CrossRef] [PubMed]
34. Patra, J.K.; Das, G.; Fraceto, L.F.; Campos, E.V.R.; Rodriguez-Torres, M.P.; Acosta-Torres, L.S.; Diaz-Torres, L.A.; Grillo, R.; Swamy, M.K.; Sharma, S.; et al. Nano based drug delivery systems: Recent developments and future prospects. *J. Nanobiotechnol.* **2018**, *16*, 1–33. [CrossRef] [PubMed]
35. Mohanambal, E. Formulation and Evaluation of pH Triggered in Situ Gelling System of Levofloxacin. M.Sc. Thesis, Madurai Medical College, Madurai, India, 2010.
36. Gutowski, I.A. The Effects of pH and Concentration on the Rheology of Carbopol Gels. M.Sc. Thesis, Department of Physics; Simon Fraser University, Burnaby, BC, Canada, 2010.
37. Alam, S.; Algahtani, M.S.; Ahmad, M.Z.; Ahmad, J. Investigation utilizing the HLB concept for the development of moisturizing cream and lotion: In-vitro characterization and stability evaluation. *Cosmetics* **2020**, *7*, 43. [CrossRef]
38. Lorca, S.; Santos, F.; Fernández Romero, A.J. A Review of the Use of GPEs in Zinc-Based Batteries. A Step Closer to Wearable Electronic Gadgets and Smart Textiles. *Polymers* **2020**, *12*, 2812. [CrossRef] [PubMed]
39. Sirelkhatim, A.; Mahmud, S.; Seeni, A.; Kaus, N.H.M.; Ann, L.C.; Bakhori, S.K.M.; Hasan, H.; Mohamad, D. Review on zinc oxide nanoparticles: Antibacterial activity and toxicity mechanism. *Nano-Micro Lett.* **2015**, *7*, 219–242. [CrossRef]
40. Wang, L.; Hu, C.; Shao, L. The antimicrobial activity of nanoparticles: Present situation and prospects for the future. *Int. J. Nanomed.* **2017**, *12*, 1227. [CrossRef]
41. Malmsten, M. Antimicrobial and antiviral hydrogels. *Soft Matter.* **2011**, *7*, 8725–8736. [CrossRef]
42. Shah, B.; Davidson, P.M.; Zhong, Q. Nanodispersed eugenol has improved antimicrobial activity against *Escherichia coli* O157: H7 and *Listeria monocytogenes* in bovine milk. *Int. J. Food Microbiol.* **2013**, *161*, 53–59. [CrossRef] [PubMed]
43. Kupnik, K.; Primožič, M.; Kokol, V.; Leitgeb, M. Nanocellulose in Drug Delivery and Antimicrobially Active Materials. *Polymers* **2020**, *12*, 2825. [CrossRef] [PubMed]
44. Da Silva, B.L.; Abuçafy, M.P.; Manaia, E.B.; Junior, J.A.O.; Chiari-Andréo, B.G.; Pietro, R.C.R.; Chiavacci, L.A. Relationship between structure and antimicrobial activity of zinc oxide nanoparticles: An overview. *Int. J. Nanomed.* **2019**, *14*, 9395. [CrossRef]
45. Diez-Pascual, A.M. Antibacterial nanocomposites based on thermosetting polymers derived from vegetable oils and metal oxide nanoparticles. *Polymers* **2019**, *11*, 1790. [CrossRef] [PubMed]
46. Nazoori, E.S.; Kariminik, A. In vitro evaluation of antibacterial properties of zinc oxide nanoparticles on pathogenic prokaryotes. *J. Appl. Biotechnol. Rep.* **2018**, *5*, 162–165. [CrossRef]

47. Alamdari, S.; Ghamsari, M.S.; Lee, C.; Han, W.; Park, H.; Tafreshi, M.J.; Afarideh, H.; Ara, M.H.M. Preparation and characterization of zinc oxide nanoparticles using leaf extract of *sambucus ebulus*. *Appl. Sci.* **2020**, *10*, 3620. [CrossRef]
48. Slavin, Y.N.; Asnis, J.; Häfeli, U.O.; Bach, H. Metal nanoparticles: Understanding the mechanisms behind antibacterial activity. *J. Nanobiotechnol.* **2017**, *15*, 1–20. [CrossRef] [PubMed]
49. Agarwal, H.; Menon, S.; Kumar, S.V.; Rajeshkumar, S. Mechanistic study on antibacterial action of zinc oxide nanoparticles synthesized using green route. *Chem. Biol. Interact.* **2018**, *286*, 60–70. [CrossRef] [PubMed]
50. Tiwari, V.; Mishra, N.; Gadani, K.; Solanki, P.S.; Shah, N.A.; Tiwari, M. Mechanism of anti-bacterial activity of zinc oxide nanoparticle against carbapenem-resistant *Acinetobacter baumannii*. *Front. Microbiol.* **2018**, *9*, 1218. [CrossRef]
51. Jana, S.; Manna, S.; Nayak, A.K.; Sen, K.K.; Basu, S.K. Carbopol gel containing chitosan-egg albumin nanoparticles for transdermal aceclofenac delivery. *Colloids Surf. B Biointerfaces* **2014**, *114*, 36–44. [CrossRef]
52. Sareen, R.; Kumar, S.; Gupta, G.D. Meloxicam carbopol-based gels: Characterization and evaluation. *Curr. Drug Deliv.* **2011**, *8*, 407–415. [CrossRef]
53. Bonacucina, G.; Cespi, M.; Misici-Falzi, M.; Palmieri, G.F. Rheological evaluation of silicon/carbopol hydrophilic gel systems as a vehicle for delivery of water insoluble drugs. *AAPS J.* **2008**, *10*, 84–91. [CrossRef]
54. Purwaningsih, S.Y.; Pratapa, S.; Triwikantoro; Darminto. Nano-sized ZnO powders prepared by co-precipitation method with various pH. In *Proceedings of the 3rd International Conference on Advanced Materials Science and Technology (ICAMST 2015), Proceedings of the AIP Conference 1725, 020063 (2016)*; Semarang, Indonesia, 6–7th October 2015; AIP Publishing LLC: Melville, NY, USA, 2016. [CrossRef]
55. Blanco-Fuente, H.; Esteban-Fernández, B.; Blanco-Méndez, J.; Otero-Espinar, F. Use of β-cyclodextrins to prevent modifications of the properties of carbopol hydrogels due to carbopol–drug interactions. *Chem. Pharm. Bull.* **2002**, *50*, 40–46. [CrossRef]
56. Suhail, M.; Wu, P.-C.; Minhas, M.U. Using carbomer-based hydrogels for control the release rate of diclofenac sodium: Preparation and in vitro evaluation. *Pharmaceuticals* **2020**, *13*, 399. [CrossRef]
57. Lamberti, G.; Caccavo, D.; Cascone, D.I.S. *Mathematical Description of Hydrogels' Behavior for Biomedical Applications*; Universita Degli Studi Di Salerno: Fisciano, Italy, 2013.
58. Viyoch, J.; Klinthong, N.; Siripaisal, W. Development of oil-in-water emulsion containing Tamarind fruit pulp extract I. Physical characteristics and stability of emulsion. *Naresuan Univer. J. Sci. Technol. NUJST* **2013**, *11*, 29–44.
59. Ethier, A.; Bansal, P.; Baxter, J.; Langley, N.; Richardson, N.; Patel, A.M. The role of excipients in the microstructure of Topical semisolid drug products. In *The Role of Microstructure in Topical Drug Product Development*; Springer: Amsterdam, The Netherlands, 2019; pp. 155–193.
60. Da Silva Ávila, D.M.; Zanatta, R.F.; Scaramucci, T.; Aoki, I.V.; Torres, C.R.G.; Borges, A.B. Randomized in situ trial on the efficacy of Carbopol in enhancing fluoride/stannous anti-erosive properties. *J. Dent.* **2020**, *101*, 103347. [CrossRef]
61. Abdullah, G.Z.; Abdulkarim, M.F.; Mallikarjun, C.; Mahdi, E.S.; Basri, M.; Abdul Sattar, M.; Noor, A.M. Carbopol 934, 940 and Ultrez 10 as viscosity modifiers of palm olein esters based nano-scaled emulsion containing ibuprofen. *Pak. J. Pharm. Sci.* **2013**, *26*, 75–83.
62. Fevola, M.J.; Walters, R.M.; LiBrizzi, J.J. A new approach to formulating mild cleansers: Hydrophobically-modified polymers for irritation mitigation. In *Polymeric Delivery of Therapeutics*; ACS Publications: Washington, DC, USA, 2010; pp. 221–242.
63. Varges, R.P.; Costa, C.M.; Fonseca, B.S.; Naccache, M.F.; Mendes, P.R.D. Rheological characterization of Carbopol® dispersions in water and in water/glycerol solutions. *Fluids* **2019**, *4*, 3. [CrossRef]
64. Panzade, P.; Puranik, P.K. Carbopol polymers: A versatile polymer for pharmaceutical applications. *Res. J. Pharm. Technol.* **2010**, *3*, 672–675.
65. Hamdy, A. Experimental Study of the Relationship Between Dissolved Iron, Turbidity, and Removal of Cu(II) Ion from Aqueous Solutions Using Zero-Valent Iron Nanoparticles. *Arab. J. Sci. Eng.* **2020**, *46*, 1–23. [CrossRef]
66. Suntako, R. Effect of zinc oxide nano-gel synthesized by a precipitation method on mechanical and morphological properties of the CR foam. *Bull. Mater. Sci.* **2015**, *38*, 1033–1038. [CrossRef]
67. Chikkanna, M.M.; Neelagund, S.E.; Rajashekarappa, K.K. Green synthesis of zinc oxide nanoparticles (ZnO NPs) and their biological activity. *SN Appl. Sci.* **2019**, *1*, 1–10. [CrossRef]
68. Mashrai, A.; Khanam, H.; Aljawfi, R.N. Biological synthesis of ZnO nanoparticles using C. albicans and studying their catalytic performance in the synthesis of steroidal pyrazolines. *Arab. J. Chem.* **2017**, *10*, S1530–S1536.
69. Muhammad, W.; Ullah, N.; Haroon, M.; Abbasi, B.H. Optical, morphological and biological analysis of zinc oxide nanoparticles (ZnO NPs) using *Papaver somniferum* L. *RSC Adv.* **2019**, *9*, 29541–29548. [CrossRef]
70. Zak, A.K.; Razali, R.; Abd Majid, W.H.; Darroudi, M. Synthesis and characterization of a narrow size distribution of zinc oxide nanoparticles. *Int. J. Nanomed.* **2011**, *6*, 1399.
71. Baruwati, B.; Kumar, D.K.; Manorama, S.V. Hydrothermal synthesis of highly crystalline ZnO nanoparticles: A competitive sensor for LPG and EtOH. *Sens. Actuators B Chem.* **2006**, *119*, 676–682. [CrossRef]
72. Acosta-Humánez, M.; Montes-Vides, L.; Almanza-Montero, O. Sol-gel synthesis of zinc oxide nanoparticle at three different temperatures and its characterization via XRD, IR and EPR. *Dyna* **2016**, *83*, 224–228. [CrossRef]
73. Diallo, A.; Ngom, B.D.; Park, E.; Maaza, M. Green synthesis of ZnO nanoparticles by *Aspalathus linearis*: Structural & optical properties. *J. All. Comp.* **2015**, *646*, 425–430.

74. Muchuweni, E.; Sathiaraj, T.; Nyakotyo, H. Hydrothermal synthesis of ZnO nanowires on rf sputtered Ga and Al co-doped ZnO thin films for solar cell application. *J. Alloy Comp.* **2017**, *721*, 45–54. [CrossRef]
75. Taziwa, R.; Meyer, E.; Katwire, D.; Ntozakhe, L. Influence of carbon modification on the morphological, structural, and optical properties of zinc oxide nanoparticles synthesized by pneumatic spray pyrolysis technique. *J. Nanomater.* **2017**, *2017*. [CrossRef]
76. Jayachandraiah, C.; Krishnaiah, G. Erbium induced raman studies and dielectric properties of Er-doped ZnO nanoparticles. *Adv. Mater. Lett.* **2015**, *6*, 743–748. [CrossRef]
77. Thommes, M.; Kaneko, K.; Neimark, A.V.; Olivier, J.P.; Rodriguez-Reinoso, F.; Rouquerol, J.; Sing, K.S. Physisorption of gases, with special reference to the evaluation of surface area and pore size distribution (IUPAC Technical Report). *Pure Appl. Chem.* **2015**, *87*, 1051–1069. [CrossRef]
78. Kołodziejczak-Radzimska, A.; Markiewicz, E.; Jesionowski, T. Structural characterisation of ZnO particles obtained by the emulsion precipitation method. *J. Nanomater.* **2012**, *2012*, 15. [CrossRef]
79. Umukoro, E.H.; Peleyejua, M.G.; Idrisa, A.O.; Ngilaab, J.C.; Mabubaab, N.; Rhymanac, L.; Ramasami, P.; Arotiba, O.A. Photo-electrocatalytic application of palladium decorated zinc oxide-expanded graphite electrode for the removal of 4-nitrophenol: Experimental and computational studies. *RSC Adv.* **2018**, *8*, 10255–10266. [CrossRef]
80. Ismail, M.A.; Taha, K.K.; Modwi, A.; Khezami, L. ZnO nanoparticles: Surface and X-ray profile analysis. *J. Ovonic Res.* **2018**, *14*, 381–393.
81. Thirumavalavan, M.; Huang, K.-L.; Lee, J.-F. Preparation and morphology studies of nano zinc oxide obtained using native and modified chitosans. *Materials* **2013**, *6*, 4198–4212. [CrossRef] [PubMed]
82. Nagaraju, P.; Puttaiah, S.H.; Wantala, K.; Shahmoradi, B. Preparation of modified ZnO nanoparticles for photocatalytic degradation of chlorobenzene. *Appl. Water Sci.* **2020**, *10*, 1–15. [CrossRef]
83. Ramimoghadam, D.; Hussein, M.Z.B.; Taufiq-Yap, Y.H. Synthesis and characterization of ZnO nanostructures using palm olein as biotemplate. *Chem. Central J.* **2013**, *7*, 1–10. [CrossRef]
84. Mustapha, S.; Tijani, J.O.; Ndamitso, M.M.; Abdulkareem, S.A.; Shuaib, D.T.; Mohammed, A.K.; Sumaila, A. The role of kaolin and kaolin/ZnO nanoadsorbents in adsorption studies for tannery wastewater treatment. *Sci. Rep.* **2020**, *10*, 1–22. [CrossRef]
85. Mittal, H.; Morajkar, P.P.; Al Alili, A.; Alhassan, S.M. In-situ synthesis of ZnO nanoparticles using gum arabic based hydrogels as a self-template for effective malachite green dye adsorption. *J. Polym. Environ.* **2020**, *28*, 1637–1653. [CrossRef]
86. Qin, C.; Li, S.; Jiang, G.; Cao, J.; Guo, Y.; Li, J.; Zhang, B.; Han, S. Preparation of flower-like ZnO nanoparticles in a cellulose hydrogel microreactor. *BioResources* **2017**, *12*, 3182–3191. [CrossRef]
87. Omar, L.; Perret, N.; Daniele, S. Self-Assembled Hybrid ZnO nanostructures as supports for copper-based catalysts in the hydrogenolysis of glycerol. *Catalysts* **2021**, *11*, 516. [CrossRef]
88. Abou Oualid, H.; Amadine, O.; Essamlali, Y.; Dânoumb, K.; Zahouily, M. Supercritical CO$_2$ drying of alginate/zinc hydrogels: A green and facile route to prepare ZnO foam structures and ZnO nanoparticles. *RSC Adv.* **2018**, *8*, 20737–20747. [CrossRef]
89. Safavinia, L.; Akhgar, M.R.; Tahamipour, B.; Ahmadi, S.A. Green Synthesis of highly dispersed zinc oxide nanoparticles supported on silica gel matrix by daphne oleoides extract and their antibacterial activity. *Iran. J. Biotechnol.* **2021**, *19*, 86–95.
90. Hasanpour, M.; Motahari, S.; Jing, D.; Hatami, M. Investigation of the different morphologies of zinc oxide (ZnO) in Cellulose/ZnO hybrid aerogel on the photocatalytic degradation efficiency of methyl orange. *Topics Catal.* **2021**, *2021*, 1–14.
91. Wang, Q.; Ji, P.; Yao, Y.; Liu, Y.; Zhang, Y.; Wang, X.; Wang, Y.; Wu, J. Gliadin-mediated green preparation of hybrid zinc oxide nanospheres with antibacterial activity and low toxicity. *Sci. Rep.* **2021**, *11*, 1–11.
92. Jassim, A.N.; Alwan, R.M.; Kadhim, Q.A.; Nsaif, A.A. Preparation and characterization of ZnO/polystyrene nanocomposite films using ultrasound irradiation. *Nanosci. Nanotechnol.* **2016**, *6*, 17–23.
93. Hutera, B.; Kmita, A.; Olejnik, E.; Tokarski, T. Synthesis of ZnO nanoparticles by thermal decomposition of basic zinc carbonate. *Arch. Metall. Mater.* **2013**, *58*, 489–491. [CrossRef]
94. Chai, M.H.H.; Amir, N.; Yahya, N.; Saaid, I.M. Characterization and colloidal stability of surface modified zinc oxide nanoparticle. *J. Phys. Conf. Ser.* **2018**, *1123*, 012007. [CrossRef]
95. Mohammed, W.H.; Ali, W.K.; Al-Awady, M.J. Evaluation of in vitro drug release kinetics and antibacterial activity of vancomycin HCl-loaded nanogel for topical application. *J. Pharm. Sci. Res.* **2018**, *10*, 2747–2756.
96. Al-Awady, M.J.; Fauchet, A.; Greenway, G.M.; Paunov, V.N. Enhanced antimicrobial effect of berberine in nanogel carriers with cationic surface functionality. *J. Mater. Chem. B* **2017**, *5*, 7885–7897. [CrossRef]
97. Punnoose, A.; Dodge, K.; Rasmussen, J.W.; Chess, J.; Wingett, D.; Anders, C. Cytotoxicity of ZnO nanoparticles can be tailored by modifying their surface structure: A green chemistry approach for safer nanomaterials. *ACS Sustain. Chem. Eng.* **2014**, *2*, 1666–1673. [CrossRef]
98. Sapkota, B.B.; Mishra, S.R. Preparation and Photocatalytic Activity Study of p-CuO/n-ZnO composites. *MRS Proceedings* **2012**, *1443*, 19–26. [CrossRef]
99. Weldrick, P.J.; San, S.; Paunov, V.N. Advanced Alcalase-Coated Clindamycin-Loaded Carbopol Nanogels for Removal of Persistent Bacterial Biofilms. *ACS Appl. Nano Mater.* **2021**, *4*, 1187–1201. [CrossRef]
100. Weldrick, P.J.; Iveson, S.; Hardman, M.J.; Paunov, V.N. Breathing new life into old antibiotics: Overcoming antibacterial resistance by antibiotic-loaded nanogel carriers with cationic surface functionality. *Nanoscale* **2019**, *11*, 10472–10485. [CrossRef]

101. Batool, M.; Khurshid, S.; Daoush, W.M.; Siddique, S.A.; Nadeem, T. Green synthesis and biomedical applications of ZnO nanoparticles: Role of PEGylated-ZnO nanoparticles as doxorubicin drug carrier against MDA-MB-231(TNBC) cells line. *Crystals* **2021**, *11*, 344. [CrossRef]
102. Batool, M.; Khurshid, S.; Qureshi, Z.; Daoush, W.M. Adsorption, antimicrobial and wound healing activities of biosynthesised zinc oxide nanoparticles. *Chem. Pap.* **2021**, *75*, 893–907. [CrossRef]
103. Sadek, A.H.; Asker, M.S.; Abdelhamid, S.A. Bacteriostatic impact of nanoscale zero-valent iron against pathogenic bacteria in the municipal wastewater. *Biologia* **2021**, *76*, 1–25. [CrossRef] [PubMed]
104. Wahid, F.; Zhong, C.; Wang, H.; Hu, X.; Chu, L. Recent advances in antimicrobial hydrogels containing metal ions and metals/metal oxide nanoparticles. *Polymers* **2017**, *9*, 636. [CrossRef]
105. Gokmen, F.O.; Temel, S.; Yaman, E. Enhanced antibacterial property by the synergetic effect of $TiO_2$ and ZnO nano-particles in biodegradable hydrogel. *Eur. Sci. J.* **2019**, *15*, 33. [CrossRef]
106. Yang, K.; Han, Q.; Chen, B.; Zheng, Y.; Zhang, K.; Li, Q.; Wang, J. Antimicrobial hydrogels: Promising materials for medical application. *Int. J. Nanomed.* **2018**, *13*, 2217. [CrossRef]
107. Scalzo, M.; Orlandi, C.; Simonetti, N.; Cerreto, F. Study of interaction effects of polyacrylic acid polymers (Carbopol 940) on antimicrobial activity of methyl parahydroxybenzoate against some gram-negative, gram-positive bacteria and yeast. *J. Pharm. Pharmacol.* **1996**, *48*, 1201–1205. [CrossRef]
108. Alpaslan, E.; Geilich, B.M.; Yazici, H.; Webster, T.J. pH-controlled cerium oxide nanoparticle inhibition of both gram-positive and gram-negative bacteria growth. *Sci. Rep.* **2017**, *7*, 1–12. [CrossRef]
109. Mahboub, H.H.; Shahin, K.; Zaglool, A.W.; Roushdy, E.M.; Ahmed, S.S.A. Efficacy of nano zinc oxide dietary supplements on growth performance, immunomodulation and disease resistance of African Catfish, Clarias gariepinus. *Dis Aquat Org.* **2020**, *142*, 147–160. [CrossRef] [PubMed]
110. Dakal, T.C.; Kumar, A.; Majumdar, R.S.; Yadav, V. Mechanistic basis of antimicrobial actions of silver nanoparticles. *Front. Microbiol.* **2016**, *7*, 1831. [CrossRef] [PubMed]

Article

# Green Synthesis and Biomedical Applications of ZnO Nanoparticles: Role of PEGylated-ZnO Nanoparticles as Doxorubicin Drug Carrier against MDA-MB-231(TNBC) Cells Line

Madiha Batool [1], Shazia Khurshid [1], Walid M. Daoush [2,3,*], Sabir Ali Siddique [4] and Tariq Nadeem [4]

1. Department of Chemistry, Government College University (GCU), Lahore 54000, Pakistan; tweetchem56@gmail.com (M.B.); shaziakhurshid@gcu.edu.pk (S.K.)
2. Department of Production Technology, Faculty of Technology and Education, Helwan University, Saray-El Qoupa, El Sawah Street, Cairo 11281, Egypt
3. Department of Chemistry, College of Science, Imam Mohammad ibn Saud Islamic University (IMSIU), Othman Ibn Affan Street, P.O. Box 5701, Riyadh 11432, Saudi Arabia
4. Institute of Chemistry, University of The Punjab, Lahore 54590, Pakistan; sabir.siddique@lums.edu.pk (S.A.S.); tariq.nadeem@cemb.edu.pk (T.N.)
* Correspondence: wmdaoush@imamu.edu.sa

Citation: Batool, M.; Khurshid, S.; Daoush, W.M.; Siddique, S.A.; Nadeem, T. Green Synthesis and Biomedical Applications of ZnO Nanoparticles: Role of PEGylated-ZnO Nanoparticles as Doxorubicin Drug Carrier against MDA-MB-231(TNBC) Cells Line. Crystals **2021**, 11, 344. https://doi.org/10.3390/cryst11040344

Academic Editor: Helmut Cölfen

Received: 5 January 2021
Accepted: 22 March 2021
Published: 28 March 2021

**Publisher's Note:** MDPI stays neutral with regard to jurisdictional claims in published maps and institutional affiliations.

**Copyright:** © 2021 by the authors. Licensee MDPI, Basel, Switzerland. This article is an open access article distributed under the terms and conditions of the Creative Commons Attribution (CC BY) license (https://creativecommons.org/licenses/by/4.0/).

**Abstract:** The present study aimed to develop the synthesis of zinc oxide nanoparticles (ZnO-NPs) using the green method, with *Aloe barbadensis* leaf extract as a stabilizing and capping agent. In vitro antitumor cytotoxic activity, as well as the surface-functionalization of ZnO-NPs and their drug loading capacity against doxorubicin (DOX) and gemcitabine (GEM) drugs, were also studied. Morphological and structural properties of the produced ZnO-NPs were characterized by scanning electron microscopy (SEM), transmission electron microscopy (TEM), energy dispersion X-ray diffraction (EDX), UV-Vis spectrophotometry, Fourier-transform infrared analysis (FTIR), and X-ray diffraction (XRD). The prepared ZnO-NPs had a hexagonal shape and average particle size of 20–40 nm, with an absorption peak at 325 nm. The weight and atomic percentages of zinc (50.58% and 28.13%) and oxygen (26.71% and 60.71%) were also determined by EDAX (energy dispersive x-ray analysis) compositional analysis. The appearance of the FTIR peak at 3420 m$^{-1}$ confirmed the synthesis of ZnO-NPs. The drug loading efficiency (LE) and loading capacity (LC) of unstabilized and PEGylated ZnO-NPs were determined by doxorubicin (DOX) and gemcitabine (GEM) drugs. DOX had superior LE 65% (650 mg/g) and higher LC 32% (320 mg/g) than GEM LE 30.5% (30 mg/g) and LC 16.25% (162 mg/g) on ZnO-NPs. Similar observation was observed in the case of PEG-ZnO-NPs, where DOX had enhanced LE 68% (680 mg/g) and LC 35% (350) mg/g in contrast to GEM, which had LE and LC values of 35% (350 mg/g) and 19% (190 mg/g), respectively. Therefore, DOX was chosen to encapsulate nanoparticles, along with the untreated nanoparticles, to check their in vitro antiproliferative potential against the triple-negative breast cancer (TNBC) cell line (MDA-MB-231) through the MTT (3-(4,5-Dimethylthiazol-2-Yl)-2,5-Diphenyltetrazolium Bromide) assay. This drug delivery strategy implies that the PEGylated biogenically synthesized ZnO-NPs occupy an important position in chemotherapeutic drug loading efficiency and can improve the therapeutic techniques of triple breast cancer.

**Keywords:** ZnO nanoparticles; green synthesis; cytotoxicity; anticancer activity; chemotherapeutic drugs; doxorubicin; gemcitabine; MDA-MB 231cell line; triple-negative breast cancer treatment

## 1. Introduction

A drug delivery system (DDS) refers to the engineered techniques for approaching, transporting, and formulating therapeutic agents for targeted release. Metal oxide nanoparticles are the residue of conventional drug delivery systems [1]. The major limitations in drug formulation system are poor drug loading capacity, low loading efficiency, and lowered ability to control the size distribution [2]. The defective drug loading efficiency is either due to the amount of the drug not approaching an equate pharmacological concentration in the targeted area, or the carrier substance amount exceeding optimal range associated with toxicity or uncertain side effects. Moreover, frequent release of the encapsulated drug results in a major fraction of the drug released before reaching its target site in the body, which causes inferior efficacy.

Consequently, there is a desperate need for better and more compatible materials for drug delivery systems, which can impart efficient loading and the controlled release of drugs [3]. The prospective application of MO-NPs (Metal oxide nano particles) in drug formulations holds the potential to overcome multidrug resistance (MDR) with improved biodistribution and the slow release of drugs in targeted body sites [4]. The DDS based on nanotechnology provides an approach to reduce MDR. An efficient drug release mechanism based on pH-triggered intracellular acidic environment offers efficient intracellular uptake, therapeutic effect, and cell reflux reduction in target areas of the body.

After completion of the history-making Human Genome Project in 2001, robust advances have been made in the field of cancer diagnostics and therapeutics through an elaborate understanding of novel signal transduction pathways orchestrating neoplastic diseases. A snowballing number of personalized treatment options are now available to manage various types of tumors, but gold-standard chemotherapy (cytotoxic agents that halt the cell cycle of rapidly dividing cells, preventing their division and initiating apoptosis) has always been a frontline strategy to treat various types of advance stage solid tumors [5]. On the other hand, long-term sequelae and side effects of cytostatic agents cannot be belittled and undervalued. Another key challenge is the emergence of multidrug-resistant tumors arising through increased efflux of the drug and decreased cellular uptake, altering epigenetic regulation and drug targets, drug metabolism, and enhanced DNA repair [6–8]. This is the reason why considerable attempts have been made for the efficient tumor-targeted delivery of cytotoxic drugs to mitigate toxic effects [9,10]. According to GLOBOCAN 2018 estimates, breast cancer is the second most commonly diagnosed cancer in the world (11.6% of the total cases) and the leading cause of cancer death among females [11]. The heterogeneous nature of this disease proves the bottleneck method to be effective in tumor management, taking into account clinical history (patient symptoms and past treatments), molecular characterization of tumor biology (estrogen receptor-ER positive, progesterone receptor-PR positive, and human epidermal growth factor receptor2-Her2 positive), and quality of life to narrow down therapeutic options [12,13]. Cytostatic chemotherapy (especially anthracyclines) is the solitary option in the case of advanced-stage metastatic breast cancer (MBC), triple-negative condition, or relapse after hormonal therapy [14,15]. Doxorubicin (DOX) is a cytotoxic antibiotic derived from *Streptomyces peucetius* belonging to an anthracycline class of chemotherapeutic drugs used to treat solid tumors such as breast, bladder and ovary cancer tumors. DOX impedes topoisomerase II, DNA, and RNA synthesis inhibition, causing breakage of DNA strands by intercalating within DNA base pairs [16,17]. Clinical evaluation of DOX has revealed that its therapeutic use is limited in the clinics due to severe side effects, including intrinsic myocardiotoxicity, emesis, alopecia, myelosuppression, and mucositis. Despite these implications, DOX is still under clinical use in combination with frontline chemotherapies [18–20].

Gemcitabine (GEM) is a pyrimidine nucleoside analogue antimetabolite chemotherapeutic agent that inhibits DNA synthesis and initiates apoptosis, being employed as a first-line treatment option against pancreatic cancer [21,22]. The drug vows enhanced efficacy in various types of tumors when used in different combinations. For example, anthracycline-resistant breast cancer patients signified susceptibility against gemcitabine/paclitaxel combination therapy [23,24]. However, such combination regimens also have the drawbacks of increased side effects, including diffused alveolar damage, nonspecific interstitial pneumonia, peripheral edema, and adverse hematological events such as thrombocytopenia [25,26]. As a result, efficient pharmaceutical formulations and drug delivery strategies need to be developed to alleviate toxicities of chemotherapeutic drugs and amplify their success to improve metastatic and triple-negative breast cancer management along with patient survival. Among various tumor-targeted drug delivery approaches, nano-size systems have been extensively investigated in a range of solid tumors due to their augmented efficacy and safety profile [27,28]. Some of the drug-encapsulated nano-formulations have been approved by the Food and Drug Administration (FDA-USA), best exemplified by Doxil (liposomal doxorubicin) in 1995 and Marqibo (liposomal Vincristine) in 2012, with reduced myocardiotoxicity and neuropathy, respectively [29]. Nanoparticles improve the pharmacokinetic properties of the anticancer drug and its half-life in plasma. Such drugs not only ameliorate adverse effects but also boost tumor localization through several mechanisms, including nanoparticles' passive tumor targeting ability to penetrate defective endothelial junctions due to the leakiness of tumor vasculature—a phenomenon known as th enhanced permeability and retention effect (EPR) [30,31]. Emerging drug delivery strategies hold promise to treat triple-negative breast cancer with reduced toxicity and enhanced efficacy. The green synthesis-based metal oxide nano-formulation drug delivery system modifies biodistribution and pharmacokinetics of encapsulated chemotherapeutic drugs, exposing the neoplastic tissue against the enhanced concentration of the drug released and reduced exposure to normal tissue. Active tumor targeting can be achieved by attaching ligands, aptamers, and antibodies at the surface of nanocarriers to bind their appropriate tumor-specific receptors [32,33].

For fabrication of nanocarriers, a broad range of organic (i.e., polyethylene glycol, chitosan, chondroitin sulphate, poly (lactic-co-glycolic acid), and hyaluronic acid) and inorganic (i.e., Zn, Mg, Mn, Cu, Ag, and Fe) matrices have been used to improve tumor targeting ability [34–36]. Among inorganic substances, zinc supplementation has been known as an effective therapy to treat age-related macular degeneration, depressive disorder, common cold, and sunburn [37]. Zinc oxide (source of zinc) has been classified as "GRAS" (generally recognized as safe) by Food and Drug Administration (FDA-USA) and reflects unique piezoelectric, catalytic, and optical properties [38]. ZnO-NPs readily dissolve in solution at low pH. Inflammatory and tumor tissues have significantly lower pH than their normal counterparts because of the comparatively high rate of glycolysis. Tumor-targeted increased intracellular drug concentration is facilitated through the fusion of endocytosed drug-loaded nanocarriers with lysosomes and the subsequent low pH, which triggers the dissolution of ZnO nanoparticles in the acidic environment [39]. Another unique anticancerous activity of ZnO nanoparticles is the generation of reactive oxygen species (ROS) responsible for cell death if ROS exceeds the cancerous cell antioxidative capacity [40,41]. In the current investigation, DOX with high loading capacity, loading efficiency, and drug efficiency was encapsulated on ZnO-NPs formulation to study its antitumor effect. In this study, the stealth drug delivery system was an established coating polymer corona, polyethylene glycol (PEG), on the surface of drug-loaded ZnO-NPs, which allows them to evade the immune system. Therefore, such stealth technology reduces protein adsorption and NPs uptake by the reticular endothelial system to prolong drugs' plasma circulation half-life. However, the presence of PEG may impede the release of a cytotoxic agent acting as a stumbling block between drug and the tumor cells, challenging future improvements to address this aspect [42]. Different synthetic models can be employed for the preparation of inorganic NPs, including physical (thermal ablation, laser ablation, evaporation-condensation) and chemical (reduction, coprecipitation, flow injection, electrochemical) ones. NPs produced by these methods reflect reproducible characters and uniformity in size distribution. These methods are relatively expensive and labor-intensive and generate toxic substances [43]. The emerging and promising green synthesis approach has been reported to be superior to other processes in which a variety of biological systems act as a biolaboratory for the synthesis of effective and safe metal oxide particles at the nanometer scale [44]. Several polyphenolic compounds have been reported in Aloe vera leaf extract that scavenges free radicals to prevent the onset and progression of metabolic diseases (tumors) [45].

Biosynthesis is a green and environmentally friendly technology of synthesizing nanoparticles from natural resources instead of toxic chemicals. Moreover, biogenic techniques offer safe processes for easy handling. Aloe Barbadensis Miller (Aloe Vera) belong to the Aloeaceae family and the genus Aloe vera, in which Barbadensis Miller, Aborescens, and Chinensis are familiar [46]. Additionally, Aloe vera plant leaf extract contains a viscous substance that carries vitamins A, C, beta-carotene antioxidants, choline, and folic acid, as well as contain calcium, copper, magnesium, potassium, and zinc, which are required for the functioning of enzymes in many metabolic pathways. These compounds are also antioxidants [47]. Aloe vera gel is used in food flavoring, cosmetic purposes, food supplements, and herbal remedies, which are enriched in cholesterol, campestral, steroids, sitosterol, and lupeol, having anti-inflammatory properties. Aloe vera gel is also comprised of salicylic acid, an anti-inflammatory agent, as well as the Carboxyl methyl and Sulphonyl groups [48]. These natural chemicals in plant-like salicylic acid and anthraquinones (aloin, aloetic acid, anthranol, cinnamic acid, anthracene) are responsible for the one-step reduction of metals in biogenic synthesis methods [49]. Polyphenols (aloin) in Aloe vera leaf extract can act as chelating agents, as well as capping and reducing agents for the biogenic formulation of metal nanoparticles. The leaf extract also has some chemicals, including polyphenols, acid, vitamins, and catechins (ECG), which are responsible for the reduction of metal [50].

In this study, a green process for the synthesis of ZnO-NPs using Aloe barbadensis (aloe vera) leaf extract was employed. The ZnO-NPs were investigated by UV (Ultra violet), SEM (Scanning electron microscope, TEM (transmission electron microscopy), FT-IR (Fourier transform) and XRD techniques. The obtained ZnO-NPs were used for the preparation of doxorubicin-encapsulated ZnO-nanoparticles (DOX-ZnO-NPs) and doxorubicin-encapsulated PEGylated nanoparticles (DOX-GEM-ZnO-PEGNPs). The anticancer activities of the prepared nanoparticles were investigated. In vitro antiproliferative potential against triple-negative breast cancer cell line (MDA-MB-231) through MTT (3-(4,5-Dimethylthiazol-2-Yl)-2,5-Diphenyltetrazolium Bromide) assay was also employed.

## 2. Materials and Methods

*2.1. Materials*

An Aloe barbadensis plant was purchased from a herbal clinic in Lahore, Pakistan, and authenticated by the Department of Botany, G.C. University Lahore, Pakistan. Doxorubicin HCl (DOX) was provided from Pfizer Laboratories (USA), whereas gemcitabine (GEM) was purchased from Novartis Pharma (Switzerland). Zinc nitrate was provided by the Department of Chemistry, University of Punjab (Pakistan). Phosphate buffered saline (PBS), fetal bovine serum (FBS), Dulbecco's modified Eagle's medium (DMEM), streptomycin, and penicillin were purchased from Gibco Life Technologies, Inc. (Grand Island, NY, USA). The MTT assay kit was provided from Sigma Aldrich (St-Lou, MO, USA).

*2.2. Green Synthesis of ZnO-NPs*

The Aloe barbadensis plant was rinsed 5 times with distilled water to remove any impurities. Then, 2 g of the cleaned leaves were collected, dried, and grinded into a fine powder. The obtained powder was dispersed in 100 mL of deionized distilled water with magnetic stirring and then boiled for 10 min at 100 °C. After allowing the extract to cool down at room temperature, it was filtered through a muslin cloth to collect the clear extract. An Erlenmeyer flask with 50 mL volume of 0.1 M zinc nitrate solution was prepared to react with 5 mL of aloe vera leaves extract through continuous magnetic stirring at 60 °C for 20 min. The color of the obtained solution changed from green to yellow, which confirmed the formation of ZnO-NPs stabilized by the aloe vera leaf extract [47]. The reaction mixture was centrifuged at 4000 rpm for 20 min followed by removal of the supernatant. ZnO nanoparticles were washed 3 times with distilled water and dried in an oven at 60 °C. Finally, the obtained powder was calcined at 500 °C for 1 h, and the formed black particles were collected for downstream processing. The obtained stabilized ZnO-NPs were allowed to interact with polyethylene glycol for 24 h with subsequent purification by dialysis against water, as depicted in Figure 1.

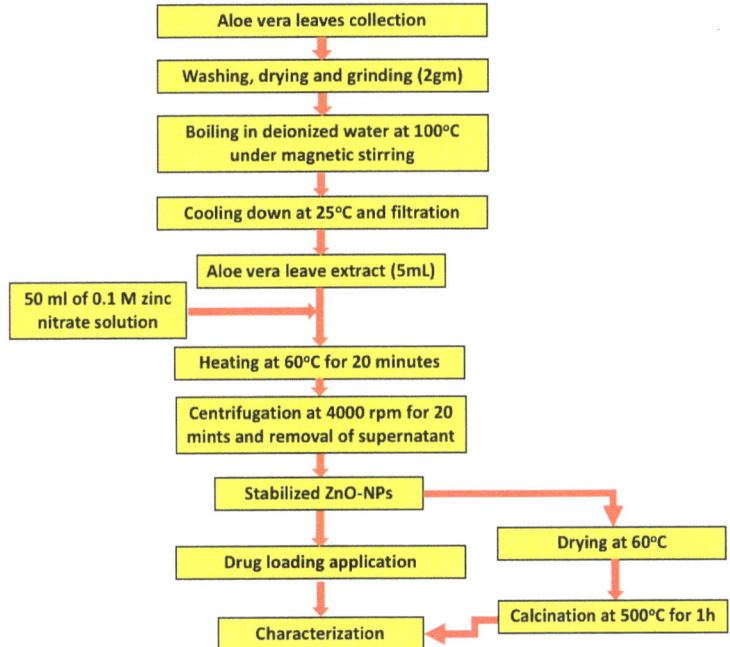

**Figure 1.** Schematic flowchart of the synthesis process of zinc oxide nanoparticles (ZnO-NPs) using Aloe barbadensis leaf extract.

*2.3. Preparation of Doxorubicin and Gemcitabine Non-PEGylated and PEGylated ZnO-NPs*

Doxorubicin (5 mg) was dissolved in 50 mL deionized distilled water to allow subsequent reaction with dispersed ZnO-NPs (10 mg/50 mL) in distilled water of 1:1 volume ratio. The solution was then incubated for 90 min at room temperature. After centrifugation, the supernatant was discarded, washed with ethanol, and dried in a vacuum oven overnight [48]. The same procedure was carried out for the preparation of GEM-ZnO NPs with ZnO-NPs.

The produced ZnO-NPs were modified using polymer polyethene glycol (PEG). Polyethene glycol (Sigma Aldrich) 400 g/mol were dissolved in anhydrous chloroform. The concentration of the hydrophobic chain was adjusted to be at least 0.04 M and was mixed with drug-loaded ZnO-NPs to make 1%weight/volume of the polymer [50]. The ZnO-NPs suspension and polymer solution were then merged in a round flask and heated at 40 °C under N$_2$ gas for 20 min. Finally, the obtained solution was stirred in the evaporator for 5 min and sonicated at 19 w and 40 kHz for 20 min.

*2.4. Drug Loading Analysis*

The investigated drugs were analyzed by UV-Vis spectrometry. Then, the drug solution was mixed with the lead extract, and the readings were recorded after 30 min, 60 min, 120 min, and 240 min intervals. The drug loading efficiency and loading capacity (LC) of ZnO-NPs can be calculated according to the following equations:

$$\text{LC}\left(\frac{\text{mg}}{\text{mg}}\right) = \frac{(\text{Drug}_i) - (\text{Drug}_f)}{(\text{Drug}_c)} \quad (1)$$

$$\%LC = [\text{entrapped drug/NP weight}] \times 100$$

$$LE = \frac{(\text{Drug}_i) - (\text{Drug}_f)}{(\text{Drug}_i)} \times 100 \qquad (2)$$

$\text{Drug}_i$ = Drug initial quantity (mg), $\text{Drug}_f$ = Free drug (mg) in the supernatant
$\text{Drug}_c$ = Carrier (ZnO-NPs drug cargo)

*2.5. Characterization of Nanoparticles*

Quantitative determination of ZnO-NPs was carried out by UV-Vis spectrophotometry (Lamda 25-Perkin Elmer, Waltham, MA, USA) within the 200–800 nm wavelength range. FT-IR spectrum was generated using (PerkinElmer Inc. Buckinghamshire, UK) within a range of 400–4000 cm$^{-1}$. X-ray diffraction analysis of the obtained nanoparticles was carried out using (XRD-6000, Shimadzu Corporation, Kyoto, Japan) at 40 kV with a copper source in a range of diffraction angles (10° to 80°). In the case of ZnO-NPs, the Joint Committee on Powder Diffraction Standards Database (JCPDS) was considered for analyzing the XRD reference patterns. The particle size and distribution, shape, and morphology of the prepared nanoparticles were analyzed by transmission electron microscope with the model (JOEL, JEM-2100) and scanning electron microscope with the model (JSM-6480LM, JEOL Ltd, Tokyo, Japan).

*2.6. In Vitro Anticancer Activity and MTT Assay on MDA-MB-231 Cell Line*

The breast cancer cell line (MDA-MB-231) was obtained from the National Centre of Excellence in Molecular Biology (University of The Punjab, Lahore, Pakistan). Cells were cultured in DMEM medium supplemented with 10% fetal bovine serum and 1% antibiotic/antimycotic. The cell was subcultured at 5000 cells/m$^2$ and extensive washing was performed with 1% FBS before the loading of the drug.

In vitro antiproliferative assessment was carried out when MDA-MB-231 cultured cells accomplished 70–80% confluency. Cultured cells were seeded onto a 96-well plate at a density of $1 \times 10^5$ cell into each well and incubated for 24 h at 37 °C. The cancer cells were then treated with various concentrations (3 µg/mL to 200 µg/mL) of DOX-PEG-ZnO-NPs and DOX-ZnO-NPs dispersions and incubated for 72 h at 37 °C under 5% $CO_2$. After removing the dissolved medium, 25 µL of the 5.5 mg/mL MTT solution (3-(4,5-dimethylthiazol-2-yl)-2,5-diphenyl tetrazolium bromide) was used to treat cells in each well and was incubated in the aluminum-foiled plate for a further 4 h in the dark. After immediate aspiration of the MTT solution, 100 µL DMSO was added in each well to dissolve the formazan compounds. In each well, absorbance was detected at 570 nm using Spectra Max Plus 384 UV-Vis plate reader. Cell viability was derived by comparison with the untreated group, which was referred to as a control.

For the assessment of cytotoxicity of ZnO-NPs and the modified doxorubicin-ZnO-NPs drug in the cell culture, an MTT assay was performed. MDA-MB 231 cells (10,000) were suspended in 100 µL of RPMI-1640 medium, which was added to each well of the 96-well plates. After 24 h, media was removed, and the wells were washed with $1 \times$ PBS. Then, the treatment of 7 doses (3.12 ug/mL to 200 ug/mL) of the test compound were given to the cells in triplicate, taking each untreated cell as Blank and DMSO as solvent control (0.2%). Media was removed after 24 h, and the wells were washed with $1 \times$ PBS. Then, 20 µL of MTT (5 mg/mL in $1 \times$ PBS) was added to each well. Plates were covered with aluminum foil and incubated at 37 °C with 5% $CO_2$ for 4 h. After incubation for 4 h, MTT was removed carefully and replaced with 100 µL DMSO in each well to dissolve the formazan products formed in the wells. Purple-colored formazan compounds were formed in response to the MTT reaction with live cells. DMSO was used to dissolve the formazan compounds. The appearance of less-intense purple color indicated high cell viability. The plates were kept at room temperature for 20 min. Then, the measured OD (Optical density) at 570 nm and 650 nm was determined, and the cell viability was calculated according to the following equation:

$$\text{Cell viability (\%)} = (\text{Test OD}_{570} - \text{OD}_{650} / \text{Blank OD}_{570} - \text{OD}_{650}) \times 100\% \quad (3)$$

Also, the IC50 (concentration inhibited 50% cell growth) was calculated by the formula:

$$\text{Inhibition growth} = [(\text{OD}_{control} - \text{OD}_{test}) \times 100]/\text{OD control} \quad (4)$$

The results expressed as the ± SD mean of the experiment in triplicate representation experiments.

*2.7. Statistical Analysis*

The MTT assay was performed in triplicate. The results were expressed as the mean standard error. Statistical analysis of all the data was performed using Origin 2019 software. Observed probability values less than 0.05 were considered to be statistically significant.

## 3. Results and Discussion

*3.1. Characterization of the ZnO-NPs*

The green synthesis of zinc oxide NPs could be demonstrated on the basis of the plant's ability to bioaccumulate metal ions involving the active phytoconstituents as stabilizers and bioreductants. The change in color correlates with the reduction of zinc nitrate to zinc oxide nanoparticles and may be attributed to the surface plasmon resonance phenomenon of nanoparticles [51].

3.1.1. SEM/EDAX Investigations

The average particle size of zinc oxide nanoparticles was analyzed by scanning and transmission electron microscopy. The SEM and the high-resolution TEM investigations of the ZnO-NPs, as show in Figure 2a,b, revealed that the ZnO-NPs were discrete, polydispersed, and hexagonal in shape. This structure depicts more iconicity and, consequently, enhanced catalytic activity of ZnO nanoparticles among their three 2D structures. The results are in agreement with the previously reported data in the literature. However, in some studies, oval- and spherical-shaped ZnO nanoparticles have also been reported using different plants [52–54]. The presence of dispersion and a few clusters of ZnO-NPs demonstrate the stability of the nanoparticles in aloe vera leaf extract due to the interaction with the flavonoids and phenols. Such compounds play a pivotal role in chelating nanoparticles to ligands [55,56]. The formation of ZnO-NPs was further confirmed by EDAX (energy dispersive x-ray analysis) compositional analysis. As shown in Figure 2c, significant peaks of zinc and oxygen appeared, confirming the formation of ZnO nanoparticles. The weight and atomic percentage of zinc (50.58% and 28.13%) and oxygen (26.71% and 60.71%) were determined and validated the formation of ZnO-NPs. The appearance of foreign peaks other than zinc and oxygen suggests the presence of various elements, including chlorine, calcium, potassium, phosphorous, aluminum, silicon, and mercury, which composed the capping agent that stabilized the ZnO-NPs. The source of these elements is the Aloe vera leaf extract, which acted as a reducing, capping, and stabilizing agent. The appearance of the peak of mercury suggests the accumulation of heavy metals inside the Aloe vera plant leaf tissues, which is related to the composition of the soil [57,58]. Figure 2d shows the particle size distribution of the prepared ZnO-NPs. It was observed from the results that the particle size ranged between 40–60 nm, with a mean particle size of 50 nm. It is well documented that larger surface-to-volume ratio makes the nanoparticles potent anticancer agents, which is subsequently linked with their decrease in size [59].

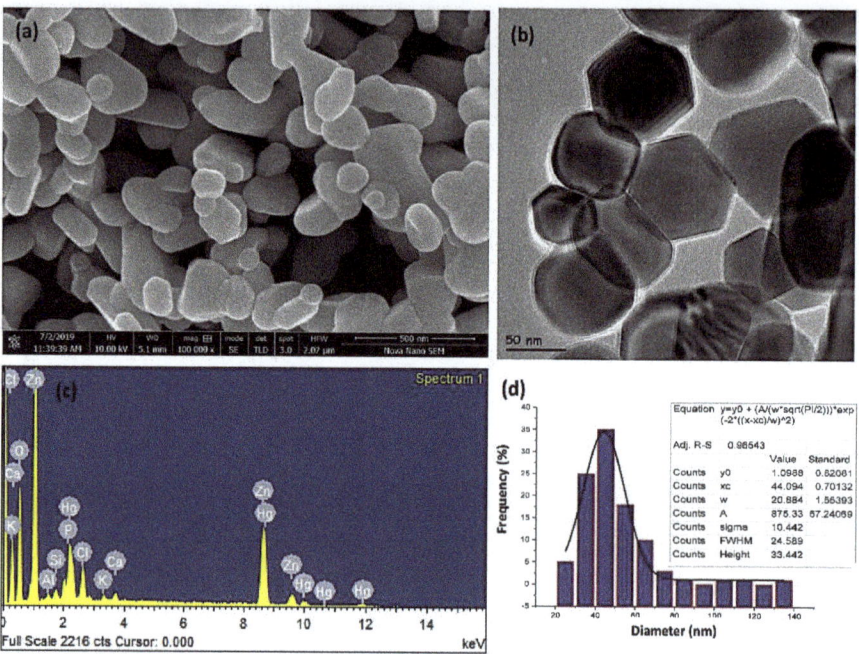

**Figure 2.** (a) Scanning electron microscopy (SEM) images, (b) high-resolution transmission electron microscopy (TEM) image, (c) EDAX (energy dispersive X-ray analysis) semi-quantitate compositional analusis, and (d) the particle size distribution of the investigated ZnO-NPs.

### 3.1.2. Phase Analysis and X-ray Diffraction

Figure 3a, b shows the X-ray diffraction (XRD) patterns of the as-prepared and calcined ZnO-NPs at 500 °C for 1 h, respectively. The results confirm that the peaks of $2\theta$ correspond to the patterns of the JCPD-00-005-0664 card from high-score software of the as-prepared ZnO-NPs, with miller indices lattices values of a (3.23 Å), b (3.243 Å), c (5.17 Å), and the JCPD-001-1136 of the calcined ZnO-NPs, respectively. The X-ray diffraction pattern further confirms that the ZnO-NPs are highly crystalline, with hexagonal-shaped crystal structure. The XRD patterns further confirm that the ZnO-NPs are of space group (P), with a value of 186, with a calculated density of 5.68 g/cm$^3$ at a temperature of 26 °C. The calculated values of the alpha, beta and gamma angles were 90.0, 90.0, and 120.0, with a Z value of 2. The graphical analysis confirmed the presence of (100), (002), (101), (102), (110), and (201) lattice plans at $2\theta$ angles 34.3, 36.4, 35.4, 47.5, 55.46, and 61.3, respectively (particle size = $K\lambda/\beta\cos\theta$) [56]. K reflected the wavelength of X-ray source used (1.541 Å), and $\beta$ was the full-width-at-half-maximum of the diffraction peak. It was observed from the results that the particle size was in the range of 20–40 nm [60].

The average crystalline size was calculated by the Scherrer equation (crystallite size = $K\lambda/\beta\cos\theta$, where K is a constant, $\lambda$ is the wavelength of the used X-ray source of 1.541 Å, $\theta$ is the angle of diffraction, and $\beta$ is the peak broadening at full width of max intensity FWHM ($\beta$) value of 0.75). A simple estimation of crystallite size from the breadths of a diffraction peak was obtained. It was observed from the results that the average crystallite size was about 37.86 nm [60].

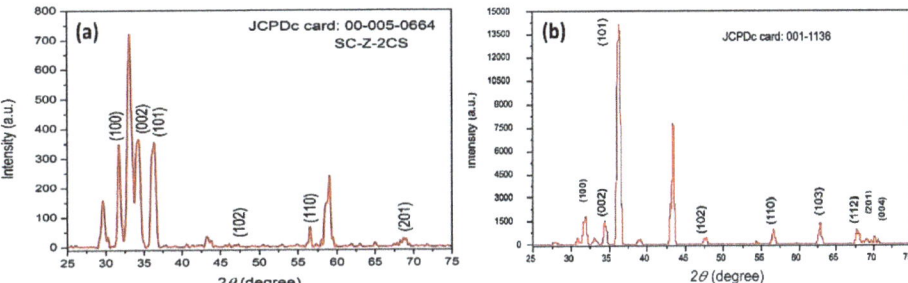

**Figure 3.** X-ray diffraction patterns of the investigated nanoparticles, where (**a**) the as prepared ZnO-NPs and (**b**) the ZnO-NPs after calcination at 500 °C.

### 3.1.3. Fourier-Transform Infrared (FTIR) and UV-Visible Spectrophotometry

The functional groups of the obtained stabilized ZnO-NPs by the Aloe vera leaf extract were detected by the FTIR spectral band identifications. The obtained data of the spectrum are listed in Table 1, with different bands stretching at 457 cm$^{-1}$ and 545 cm$^{-1}$ of the ZnO-NPs predicting the presence of the divalent metal oxide bond. Several other bands were observed at 3215.08 cm$^{-1}$, 2964.17 cm$^{-1}$, 1636.68 cm$^{-1}$, and 1541.82 cm$^{-1}$, which represent the presence of the –OH (hydroxyl), aromatic ring (C=C), amine (NH), and phenyl (C–H) groups, respectively. Such spectral data justify the presence of phenol, polyphenol, flavanol, and primary amine compounds in Aloe vera leaf extract and may be responsible for the stabilizing and capping of the ZnO-NPs. The presence of weak and broad bands at around 3200 cm$^{-1}$ was expected due to the formation of the hydroxyl group on the surface of ZnO-NPs (Figure 4). A decrease in the intensity of the hydroxyl bands after the reduction of $Zn^{2+}$ ions suggests the involvement of phenolic and flavanol sites in the binding of ZnO-NPs [61]. It is assumed that the possible mechanism of interaction between plant phytochemicals and $Zn^{2+}$ is the repetitive redox reaction, orchestrating the conversion of carbohydrates to energy during glycolysis process, coupled with the opulent hydrogen ions and ATP production. Thus, repetition of this redox reaction corroborates the conversion of zinc ions to $Zn^0$. Our findings are in agreement with previously reported studies addressing characterization of green-synthesized ZnO-NPs [52,62].

**Table 1.** Values of the FTIR spectrum bands of the investigated ZnO-NPs.

| Group | Bonding Vibration Mode | Detection Range |
|---|---|---|
| Zn–O | Stretch | 457–545 cm$^{-1}$ |
| C=C | Stretch | 1636.68 cm$^{-1}$ |
| N–H | Bend | 1541.82 cm$^{-1}$ |
| O–H | Stretch | 3215.08 cm$^{-1}$ |
| C–H | Bend | 2964.17 cm$^{-1}$ |

UV-Visible absorption spectrum of the obtained stabilized ZnO-NPs showed a strong peak at ~325 nm, detecting the formation of the stabilized ZnO-NPs which validated the formation of the ZnO-NPs with a characteristic of broad and continuous absorption spectrum. Electron transitions resulting from the valence band to the conduction band (O-2p to Zn-3d) may be attributed to the native band-gap absorption of ZnO-NPs. The catalytic activity and band gap of metal oxide NPs play a pivotal role in their cytotoxic response against biological systems [62–64] (Figure 5).

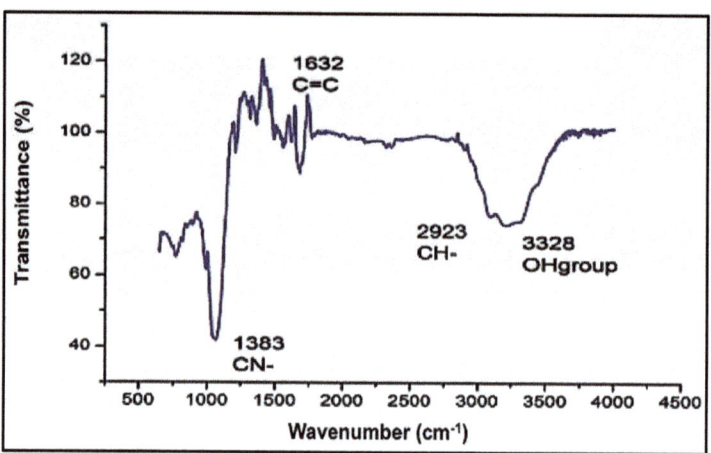

**Figure 4.** Fourier-transform infrared spectrum (FTIR) of the investigated ZnO-NPs.

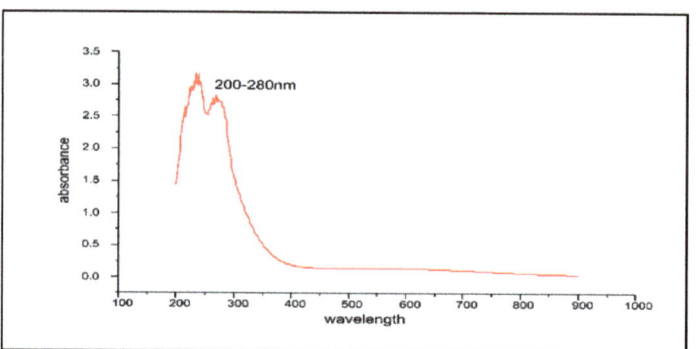

**Figure 5.** The UV-Visible spectrum of the ZnO-NPs synthesized by Aloe barbadensis leaf extract at a maximum wavelength of 200–280 nm.

*3.2. Anticancer Drug Loading Capacity and Efficiency*

UV-Vis spectrophotometry of DOX (Figure 6a) and GEM (Figure 6b) showed absorbance peaks at 480 nm and 283 nm, respectively. As mentioned in the previous section, the ZnO-NPs can be detected at a wavelength of 325 nm. Subsequent absorbance readings of solution which contained Aloe vera DOX and ZnO-NPs were recorded at discrete time intervals ranging from 0 min to 320 min. The lowering of the absorbance peak of DOX over time indicates the loading of the drug on the ZnO-NPs (Figure 7a). Likewise, the UV-Vis absorption spectrum of the solution containing GEM and ZnO-NPs was also recorded as shown in Figure 7b. It was observed from the results that GEM was successfully loaded on the surface of ZnO-NPs, as indicated by a lowering in the intensity of the absorbance peak (Figure 7b). Similarly, UV-Vis spectrophotometric measurements were observed during the PEGylation of drugs along with the ZnO-NPs. A decrease in the intensity of the absorbance peaks of DOX and GEM guaranteed the generation of DOX-PEG-ZnO-NPs (Figure 7c) and GEM-PEG-ZnO-NPs (Figure 7d), respectively. In the case of DOX-ZnO-NPs, the ZnO-NPs transferred their energy to DOX, which resulted in the reduction of the the corresponding absorption peak due to the $\pi$–$\pi$ interaction phenomenon among the DOX molecules. It was also assumed that a similar phenomenon can take place in the case of GEM [39].

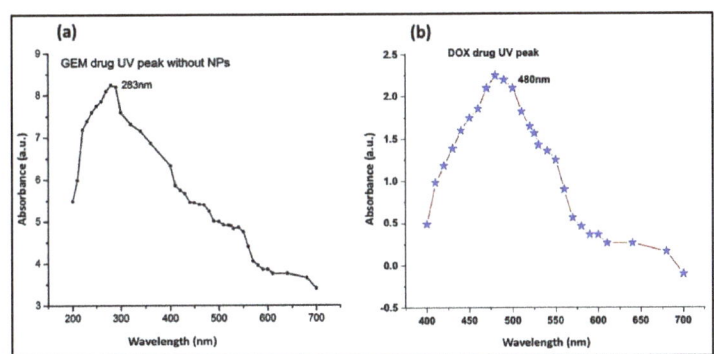

**Figure 6.** Absorption spectrum of (**a**) doxorubicin (DOX) and (**b**) gemcitabine (GEM).

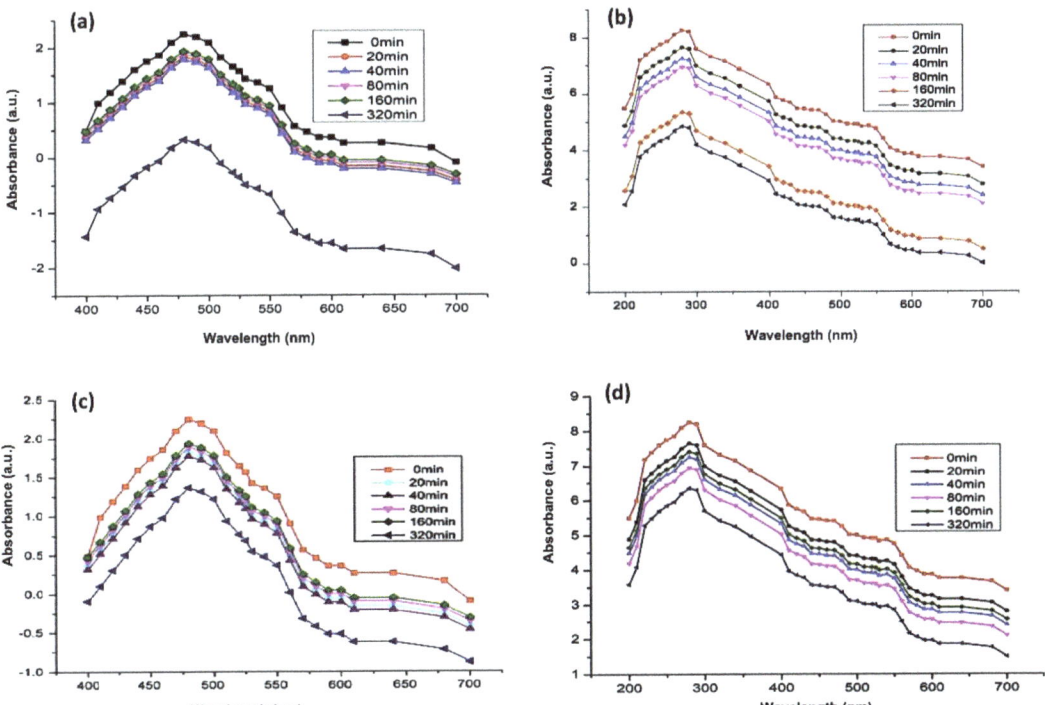

**Figure 7.** Absorption spectrum of (**a**) DOX-encapsulated ZnO-nanoparticles (DOX-ZnO-NPs), (**b**) GEM-encapsulated ZnO-nanoparticles (GEM-ZnO-NPs), (**c**) doxorubicin-encapsulated PEGylated nanoparticles (DOX-PEG-ZnO-NPs), and (**d**) GEM-encapsulated PEGylated nanoparticles (DOX-GEM-ZnO-PEGNPs (GEM-PEG-ZnO-NPs).

Figure 8 shows the change in the absorption of DOX-PEG-ZnO-NPs and GEM-PEG-ZnO-NPs by time. It was observed from the results that the DOX-loaded ZnO-NPs had a low absorption of nearly 35%. In contrast, GEM had an absorption of about 68% after a time interval of 320 min in the solution. This trend was replicated, with intensified behavior in the case of PEG-ZnO-NPs, which had 30% and 63% absorption, respectively. DOX revealed a high loading efficiency (LE) of 65% (650 mg/g) and a loading capacity (LC) of 32% (320 mg/g). Meanwhile, GEM-loaded ZnO-NPs had an LE of 30.5% (30 mg/g) and LC of 16.25% (162 mg/g). In the case of PEG-ZnO-NPs, DOX exhibited a high LE of 68%

(680 mg/g) and LC of 35% (350 mg/g) in contrast to GEM, which exhibited a LE of 35% (350 mg/g) and LC of 19% (190 mg/g) (Figure 9a,b). The main purpose behind this part of analysis is to determine the LC and LE of both investigated drugs on the PEGylated and the non-PEGylated-ZnO-NPs separately, as thse values depend on various factors such as electrostatic stabilization, Van der Waals forces, hydrogen bonding, static repulsions, size of the molecular core, specific surface area-to-volume ratio, molecular weight, and pH of the solution. The superior LE and concentration of DOX in comparison to GEM on ZnO-NPs as confirmed by UV spectrophotometry can be explained by the noncovalent electrostatic interactions between the positively charged DOX and the negatively charged ZnO-NPs. On the other hand, the PEG-coated ZnO-NPs seemed to upload a higher concentration of both drugs. Polyethene glycol is a polymer that can provide a larger molecular core, which makes it possible to upload a higher amount of drug, therefore enhancing the drug LE [65–68]. It was observed from the results that the DOX loaded more drug as compared to GEM due to its high stability. PEGylated ZnO-NPs seemed to load a higher concentration of both of the investigated anticancer drugs. Different functionalization factors account for drug loading capacity. The main analysis in this study was the LE of the different anticancer drugs on the coated and noncoated ZnO-NPs. The highly efficient loading may be due to different factors such as the electrostatic stabilization, Van der Waals forces, hydrogen bonding, and static repulsions. Drug loading and releasing can also be affected by experimental conditions such as pH.

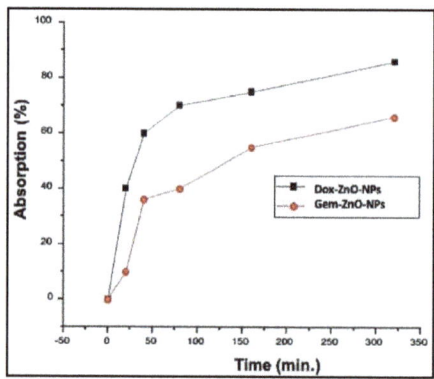

**Figure 8.** UV absorption spectra of DOX-ZnO-NPs and GEM-ZnO-NPs after 320 min.

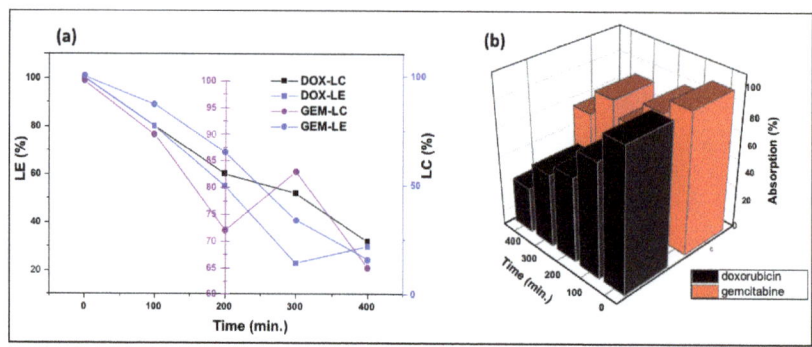

**Figure 9.** (a) Loading capacity (LC), loading efficiency (LE), and (b) relative absorption of gemcitabine and doxorubicin PEGylated ZnO-NPs.

### 3.3. In Vitro Cytotoxic Activity and MTT Assay on MDA-MB-231 Cell Line

Cancer is considered to be the second leading cause of mortality across the globe resulting from uncontrolled proliferation of cells. According to global cancer 2018 statistics, breast cancer is the second most prevalent malignancy among all cancer types [69]. Developing anticancer drugs with enhanced efficacy and minimal side effects poses dynamically a challenging task for the scientific community [70]. Patients suffering from triple-negative breast cancer show poor prognosis and a devious metastatic molecular mechanism that is hardly recognizable [71].

Many plant extracts and green synthesized NPs have been reported to reflect anticancer and antioxidant properties [72]. Furthermore, Aloe vera leaf extract-based ZnO-NPs have been reported to be associated with the inhibition of cellular malignancy [73,74]. Green-synthesized NPs could exert their antitumor potential through the production of reactive oxygen species involved in intracellular signaling, the regulation of cell proliferation, and phagocytosis [75]. The antiproliferative activity of green-synthesized ZnO-NPs and DOX-PEG-ZnO-NPs was determined in vitro on triple-negative breast cancer cell line MDA-MB-231 using an MTT assay. However, the cytotoxic assay of GEM-loaded formulations was not performed due to its low LC and LE. Also, the cells treated by incubation with different samples of ZnO-NPs at various concentrations exhibited a significant level of cytotoxicity. Similar behavior was observed in the case of DOX-ZnO-NPs (Figure 10). However, DOX-PEG-ZnO-NPs exhibited the best results. These results revealed that ZnO-NPs have anticancer potential that can be enhanced by anticancer drug loading with the subsequent PEGylation. Our findings validate the results of previous studies addressing the strong preferential cytotoxicity of green-synthesized ZnO-NPs against cancer cell lines [76].

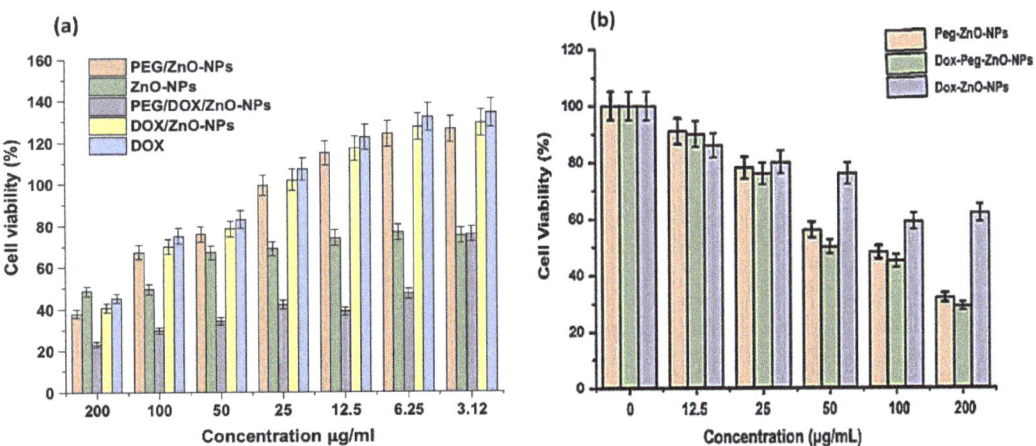

**Figure 10.** Anticancer activity of ZnO-NPs as a percentage of apoptosis in cell viability on MDA-MB-231 cancer cell line by MTT assay. (**a**) Comparison of DOX, ZnO-NPs, PEG-ZnO-NPs, DOX-ZnO-NPs, and DOX-PEG-ZnO-NPs at different concentration of nanoparticles. (**b**) Comparison of PEG-ZnO-NPs, DOX-ZnO-NPs, and DOX-PEG-ZnO-NPs at different concentration of nanoparticles. The results expressed as the ± SD mean subjected to one-way SPSS version. Experiments were performed in triplicate and repeated for three times and the calculated value of $p < 0.05$.

The ZnO-NPs generate reactive oxygen species (ROS), which are responsible for the cytotoxic activity exceeding a higher value than the antioxidant potential of the tumor cells [77]. Flavonoid and phenolic compounds, which are encapsulated on the surfaces of the stabilized ZnO-NPs by Aloe vera leaf extract, also exert anticancer effects through the suppression of metastasis and impairment of tumor angiogenesis, thus inhibiting proliferation, inactivating carcinogens, inducing cell arrest, triggering apoptosis, and promoting differentiation and oxidative destruction. Flavonoids also interact with the estrogen binding sites, downregulate kinase signal transduction pathways, and alter gene expression. Aloe vera leaf extract generates small-sized ZnO-NPs that exert potent toxicity effects [78–80]. The percentage cell viability of Dox-PEG-ZnO-NPs on MDA-MB-231 at 50 µg/mL was 40% and was reduced to 30% at 200 µg/mL. The Dox-ZnO-NPs showed the highest percentage viability (~60%) at a concentration of 200 µg/mL concentration. The MDA-MB-231 cell killing by Dox-PEG-ZnO-NPs was observed of ~70% at 200 µg/mL as shown in Figure 10. Enhancement of the activity of the nanoparticles is possible due to the dissociation of the encapsulated ZnO-NPs in the acidic tumor microenvironment, which releases $Zn^{2+}$ ions, enhancing the cellular uptake and retention of DOX. Also, the fairly high, effective, and powerful toxic activity of DOX-PEG-ZnO-NPs against cancer cells may be due to the high loading capacity and efficiency of polyethylene glycol [42,81,82].

Different concentrations of PEG-ZnO-NPs samples loaded with the anticancer drug treated against the MDA-MB-231 cancer cell line showed an IC50 value of 6.35 µg/mL in the MTT assay, as shown in Figure 10. Comparative study of PEG-ZnO-NPs and non-PEGylated ZnO-NPs showed cytotoxicity in a concentration-dependent manner. When the MDA-MB-231 cell line was treated without drug-loaded PEGylated-ZnO-NPs samples, viability was reduced to 50%, while viability was reduced to 35% with doxorubicin drug-loaded PEGylated ZnO-NPs. These findings indicate that the anticancer drug-loaded PEG-ZnO-NPs have high cytotoxicity. The relative cytotoxic effect on cancer cells between (0–200 µg/mL) concentrations was also examined. It was clear from the results, as shown in Figure 10 and listed in Table 2, that the PEGylated drug-loaded samples showed greater anticancer activity than the ZnO-NPs of calculated IC50 6.35 µg/mL as mentioned and compared with previous related works of the MTT assay of the DOX drug in the literature as listed in Table 2 and presented in Figure 11. It was mentioned in the literature that the size and dose of metal oxide nanoparticles are crucial factors which determine its cytotoxicity. As far as ZnO-NPs antitumor potential is concerned, the intracellular release of dissociated zinc ions with subsequent ROS induction is considered the underlying mechanism even though its precise cytotoxic mode of action is also under debate [83,84]. The IC50 value in the MTT assay describes the concentration of the drug at half inhibition. The IC50 profile of the MDA-MB-231 (TNBC) cancer cell line was estimated by plotting a graph between the concentrations of Dox-PEG-ZnO-NPs against the percentage of cell viability. Table 2 also shows the MTT assay comparison from the literature on different cancer cell lines and half-maximum inhibitory concentration measurements and the potency of biochemical function inhibition. According to the data listed in Table 2, MDA-MB-231, the triple-negative breast cancer cell line used in the present study, showed an IC50 value of 6.35 ± 0.5, which indicates the presence of the cancer cell. The percentage cell viability of PEGylated-ZnO-NPs with doxorubicin drug-loaded MDA-MB-231 at different concentrations shows that it is extremely toxic to cancer cells at 0–200 µg/mL concentration. These interested results need more detailed investigation in future. In our latest study, we proved that PEGylated-ZnO-NPs loaded with doxorubicin have slightly greater toxin then non-PEGylated drug-loaded nanoparticles and PEGylated-ZnO-NPs. The percentage cell viability of the effect of Dox-PEG-ZnO-NPs on MDA-MB-231 at 50 µg/mL was determined as 40% and was reduced to 30% at 200 µg/mL. The Dox-ZnO-NPs showed the highest percentage viability (~60%) at a concentration of 200 µg/mL. The MDA-MB-231 cell killing by Dox-PEG-ZnO-NPs was observed to be 70% at 200 µg/mL as shown in Figure 11.

Table 2. Comparative analysis of (IC50) value in MTT assay on MDA-MB231 cancer cell line (different cancer cell line values are in µg/mL).

| Cancer Cell Line | Assay | IC50 µg/mL | Reference |
|---|---|---|---|
| MDA-MB-231 | | 6.35 ± 0.5 | Present Study |
| MCF-7 | | 17.4 ± 4.2 | [67] |
| Hela | | 1.4 ± 0.1 | [68] |
| HCT-116 | MTT assay (Doxorubicin anticancer drug) | 19 ± 0.6 | [69] |
| HMEC | | 10.3 ± 4.6 | [70] |
| MCF-10A | | 30 ± 1.2 | [71] |
| MDA-MB-453 | | 55 ± 1.7 | [72] |
| Hs578T | | 2.9 ± 1.4 | [73] |
| MCF-7 | | 43.7 ± 07 | [74] |

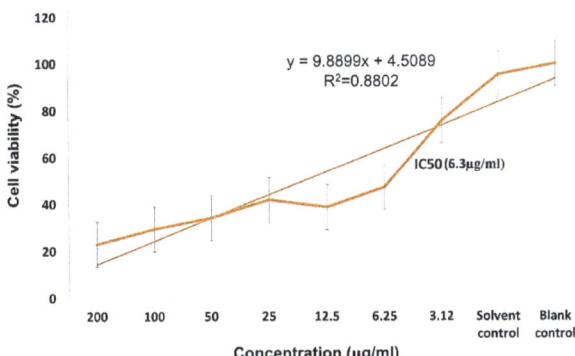

Figure 11. DOX/PEG-ZnO-NPs sample of metal oxide % cell viability and IC 50(6.3 µg/mL) analysis with a concentration range of 0–200 µg/mL.

## 4. Conclusions

The biogenic green synthesis process was employed to generate ZnO-NPs using Aloe vera plant extract as a stabilizing and capping agent in the medium. The optimum reaction conditions, easy and cost-effective procedures, and medicinal applications of the green synthesis ZnO-NPs-based drug delivery system are highlighted in our research findings. Literary contributions in the field of medical oncology for effective clinical cancer therapy were also studied. The synthesized ZnO-NPs had particle sizes in the range of 20–40 nm, which is considered ideal for the investigated drug carrier system. The DOX had better loading capacity and loading efficiency than GEM against PEGylated and non-PEGylated ZnO-NPs. Therefore, the DOX-loaded nanoparticles (PEGylated and non-PEGylated), along with the untreated ZnO-NPs, were shortlisted for in vitro analysis, which exhibited potent cytotoxicity against the investigated breast cancer. The antiproliferative activity of DOX/PEG-ZnO-NPs was determined in vitro on triple-negative breast cancer cell line MDA-MB-231 using the MTT assay. On the other hand, the cytotoxic assay of GEM-loaded formulations was not performed due to the its low loading capacity and loading efficiency. Cells incubated with ZnO-NPs at various concentrations exhibited a significant effect of cytotoxicity. A similar trend was observed in the case of the DOX-ZnO-NPs sample. These findings show that ZnO-NPs have anticancer potential that can be enhanced by subsequent PEGylated-ZnO-NPs, with biocompatibility, surface cancer cell targeting, and drug delivery capacity, and can be explored for cancer treatment. Doxorubicin drug-loaded PEGylated-ZnO-NPs exhibited the highest cytotoxicity, with a low concentration threshold for anticancer activity.

**Author Contributions:** Conceptualization, M.B; methodology, M.B; software W.M.D.; validation, W.M.D.; formal analysis, S.K.; investigation, M.B; resources, M.B; data curation, S.K.; writing—original draft preparation, M.B.; writing—review and editing, W.M.D.; visualization, T.N.; supervision, S.K.; project administration, S.A.S.; funding acquisition, W.M.D. All authors have read and agreed to the published version of the manuscript.

**Funding:** This research was supported by the CEMB (Center of excellence in molecular biology) Punjab University Lahore, Pakistan, grant No. HEC NRPU 7657.

**Institutional Review Board Statement:** Not applicable.

**Informed Consent Statement:** Informed consent was obtained from all subjects involved in the study.

**Data Availability Statement:** Not applicable.

**Acknowledgments:** The authors thank the CEMB institute and GCU Lahore for financial supports. The COMSAT Institute is acknowledged for the characterization of prepared samples, grant No. HEC NRPU 7657.

**Conflicts of Interest:** The authors declare no conflict of interest.

## References

1. Lunshof, J.E.; Bobe, J.; Aach, J.; Angrist, M.; Thakuria, J.V.; Vorhaus, D.B.; Hoehe, M.R.; Church, G.M. Personal genomes in progress: From the human genome project to the Personal Genome Project. *Dialogues Clin. Neurosci.* **2010**, *12*, 44–57. Available online: www.personalgenomes.org/mission.html (accessed on 5 January 2021).
2. Jackson, S.E.; Chester, J.D. Personalised cancer medicine. *Int. J. Cancer* **2015**, *137*, 262–266. [CrossRef] [PubMed]
3. Nagasaka, M.; Gadgeel, S.M. Role of chemotherapy and targeted therapy in early-stage non-small cell lung cancer. *Expert Rev. Anticancer. Ther.* **2018**, *18*, 63–70. [CrossRef]
4. Nevala-Plagemann, C.; Hidalgo, M.; Garrido-Laguna, I. From state-of-the-art treatments to novel therapies for advanced-stage pancreatic cancer. *Nat. Rev. Clin. Oncol.* **2020**, *17*, 108–123. [CrossRef] [PubMed]
5. Tokunaga, M.; Sato, Y.; Nakagawa, M.; Aburatani, T.; Matsuyama, T.; Nakajima, Y.; Kinugasa, Y. Correction to: Perioperative chemotherapy for locally advanced gastric cancer in Japan: Current and future perspectives. *Surg. Today* **2020**, *50*, 30–37, doi:10.1007/s00595-019-01896-5; Erratum in **2020**, *50*, 424, doi:10.1007/s00595-019-01950-2.
6. Mansoori, B.; Mohammadi, A.; Davudian, S.; Shirjang, S.; Baradaran, B. The different mechanisms of cancer drug resistance: A brief review. *Adv. Pharm. Bull.* **2017**, *7*, 339–348. [CrossRef] [PubMed]
7. Nurgali, K.; Jagoe, R.T.; Abalo, R. Editorial: Adverse effects of cancer chemotherapy: Anything new to improve tolerance and reduce sequelae? *Front. Pharmacol.* **2018**, *9*, 245. [CrossRef] [PubMed]
8. Ye, Q.; Liu, K.; Shen, Q.; Li, Q.; Hao, J.; Han, F.; Jiang, R.-W. Reversal of multidrug resistance in cancer by multi-functional flavonoids. *Front. Oncol.* **2019**, *9*, 487. [CrossRef]
9. Yang, Z.; Ma, Y.; Zhao, H.; Yuan, Y.; Kim, B.Y.S. Nanotechnology platforms for cancer immunotherapy. *Wiley Interdiscip. Rev. Nanomed. Nanobiotechnol.* **2020**, *12*, e1590. [CrossRef]
10. Shim, H. Bispecific antibodies and antibody-drug conjugates for cancer therapy: Technological considerations. *Biomolecules* **2020**, *10*, 360. [CrossRef]
11. Ferlay, J.; Colombet, M.; Soerjomataram, I.; Mathers, C.; Parkin, D.M.; Pineros, M.; Znaor, A.; Bray, F. Estimating the global cancer incidence and mortality in 2018: GLOBOCAN sources and methods. *Int. J. Cancer* **2019**, *144*, 1941–1953. [CrossRef]
12. Sauter, E.R. Cancer prevention and treatment using combination therapy with natural compounds. *Expert Rev. Clin. Pharmacol.* **2020**, *13*, 265–285. [CrossRef]
13. Anastasiadi, Z.; Lianos, G.D.; Ignatiadou, E.; Harissis, H.V.; Mitsis, M. Breast cancer in young women: An overview. *Updates Surg.* **2017**, *69*, 313–317. [CrossRef]
14. Subramani, R.; Lakshmanaswamy, R. Complementary and alternative medicine and breast cancer. *Prog. Mol. Biol. Transl. Sci.* **2017**, *151*, 231–274. [CrossRef]
15. Kolak, A.; Kamińska, M.; Sygit, K.; Budny, A.; Surdyka, D.; Kukiełka-Budny, B.; Burdan, F. Primary and secondary prevention of breast cancer. *Ann. Agric. Environ. Med.* **2017**, *24*, 549–553. [CrossRef]
16. Silva, E.F.; Bazoni, R.F.; Ramos, E.B.; Rocha, M.S. DNA-doxorubicin interaction: New insights and peculiarities. *Biopolymers* **2016**, *107*, e22998. [CrossRef] [PubMed]
17. Schwarzenbach, H.; Gahan, P.B. Predictive value of exosomes and their cargo in drug response/resistance of breast cancer patients. *Cancer Drug Resist.* **2020**. [CrossRef]
18. Benjanuwattra, J.; Siri-Angkul, N.; Chattipakorn, S.C.; Hattipakorn, N.C. Doxorubicin and its proarrhythmic effects: A comprehensive review of the evidence from experimental and clinical studies. *Pharmacol. Res.* **2020**, *151*, 104542. [CrossRef]
19. Najafi, M.; Shayesteh, M.R.H.; Mortezaee, K.; Farhood, B.; Haghi-Aminjan, H. The role of melatonin on doxorubicin-induced cardiotoxicity: A systematic review. *Life Sci.* **2020**, *241*, 117173. [CrossRef] [PubMed]

20. Xinyong, C.; Zhiyi, Z.; Lang, H.; Peng, Y.; Xiaocheng, W.; Ping, Z.; Liang, S. The role of toll-like receptors in myocardial toxicity induced by doxorubicin. *Immunol. Lett.* **2020**, *217*, 56–64. [CrossRef] [PubMed]
21. Stone, L. Gemcitabine reduces recurrence. *Nat. Rev. Urol.* **2018**, *15*, 466. [CrossRef]
22. Smith, S.D. Gemcitabine: End of a chemotherapy's era? *Acta Haematol.* **2019**, *141*, 91–92. [CrossRef]
23. Zhang, X.; Wang, S.; Cheng, G.; Yu, P.; Chang, J.; Chen, X. Cascade Drug-Release Strategy for Enhanced Anticancer Therapy. *Matter* **2021**, *4*, 26–53. [CrossRef]
24. Zhao, Y.; Lv, F.; Chen, S.; Wang, Z.; Zhang, J.; Zhang, S.; Cao, J.; Wang, L.; Cao, E.; Wang, B.; et al. Caveolin-1 expression predicts efficacy of weekly nab-paclitaxel plus gemcitabine for metastatic breast cancer in the phase II clinical trial. *BMC Cancer* **2018**, *18*, 1019. [CrossRef] [PubMed]
25. Hryciuk, B.; Szymanowski, B.; Romanowska, A.; Salt, E.; Wasąg, B.; Grala, B.; Jassem, J.; Duchnowska, R. Severe acute toxicity following gemcitabine administration: A report of four cases with cytidine deaminase polymorphisms evaluation. *Oncol. Lett.* **2018**, *15*, 1912–1916. [CrossRef]
26. Turco, C.; Jary, M.; Kim, S.; Moltenis, M.; Degano, B.; Manzoni, P.; Nguyen, T.; Genet, B.; Rabier, M.-B.V.; Heyd, B.; et al. Gemcitabine-induced pulmonary toxicity: A case report of pulmonary veno-occlusive disease. *Clin. Med. Insights* **2015**, *9*, 75–79. [CrossRef]
27. Yhee, J.; Son, S.; Lee, H.; Kim, K. Nanoparticle-based combination therapy for cancer treatment. *Curr. Pharm. Des.* **2015**, *21*, 3158–3166. [CrossRef]
28. Ashfaq, U.A.; Riaz, M.; Yasmeen, E.; Yousaf, M.Z. Recent advances in nanoparticle-based targeted drug-delivery systems against cancer and role of tumor microenvironment. *Crit. Rev. Ther. Drug Carr. Syst.* **2017**, *34*, 317–353. [CrossRef] [PubMed]
29. Bregoli, L.; Movia, D.; Gavigan-Imedio, J.D.; Lysaght, J.; Reynolds, J.; Prina-Mello, A. Nanomedicine applied to translational oncology: A future perspective on cancer treatment. *Nanomed. Nanotechnol. Biol. Med.* **2016**, *12*, 81–103. [CrossRef] [PubMed]
30. Pradhan, M.; Alexander, A.; Singh, M.R.; Singh, D.; Saraf, S.; Saraf, S.; Ajazuddin. Understanding the prospective of nano-formulations towards the treatment of psoriasis. *Biomed. Pharmacother.* **2018**, *107*, 447–463. [CrossRef]
31. Muhamad, N.; Plengsuriyakarn, T.; Na-Bangchang, K. Application of active targeting nanoparticle delivery system for chemotherapeutic drugs and traditional/herbal medicines in cancer therapy: A systematic review. *Int. J. Nanomed.* **2018**, *13*, 3921–3935. [CrossRef]
32. Alavi, M.; Hamidi, M. Passive and active targeting in cancer therapy by liposomes and lipid nanoparticles. *Drug Metab. Pers. Ther.* **2019**, *34*. [CrossRef]
33. Lin, G.; Chen, S.; Mi, P. Nanoparticles targeting and remodeling tumor microenvironment for cancer theranostics. *J. Biomed. Nanotechnol.* **2018**, *14*, 1189–1207. [CrossRef]
34. Jo, A.; Zhang, R.; Allen, I.C.; Riffle, J.S.; Davis, R.M. Design and fabrication of streptavidin-functionalized, fluorescently labeled polymeric nanocarriers. *Langmuir* **2018**, *34*, 15783–15794. [CrossRef]
35. Núñez, C.; Estévez, S.V.; Chantada, M.D.P. Inorganic nanoparticles in diagnosis and treatment of breast cancer. *J. Biol. Inorg. Chem.* **2018**, *23*, 331–345. [CrossRef] [PubMed]
36. Jiao, M.; Zhang, P.; Meng, J.; Li, Y.; Liu, C.; Luo, X.; Gao, M. Recent advancements in biocompatible inorganic nanoparticles towards biomedical applications. *Biomater. Sci.* **2018**, *6*, 726–745. [CrossRef]
37. Huang, L.; Drake, V.J.; Ho, E. Zinc. *Adv. Nutr.* **2015**, *6*, 224–226. [CrossRef] [PubMed]
38. Wieringa, F.T.; Dijkhuizen, M.A.; West, C.E. Iron and zinc interactions. *Am. J. Clin. Nutr.* **2004**, *80*, 787–788. [CrossRef] [PubMed]
39. Liu, J.; Ma, X.; Jin, S.; Xue, X.; Zhang, C.; Wei, T.; Guo, W.; Liang, X.-J. Zinc oxide nanoparticles as adjuvant to facilitate doxorubicin intracellular accumulation and visualize ph-responsive release for overcoming drug resistance. *Mol. Pharm.* **2016**, *13*, 1723–1730. [CrossRef]
40. Racca, L.; Cauda, V. Remotely Activated Nanoparticles for Anticancer Therapy. *Nano-Micro Lett.* **2021**, *13*, 1–34. [CrossRef]
41. Kim, S.; Lee, S.Y.; Cho, H.-J. Doxorubicin-wrapped zinc oxide nanoclusters for the therapy of colorectal adenocarcinoma. *Nanomaterials* **2017**, *7*, 354. [CrossRef] [PubMed]
42. Suk, J.S.; Xu, Q.; Kim, N.; Hanes, J.; Ensign, L.M. PEGylation as a strategy for improving nanoparticle-based drug and gene delivery. *Adv. Drug Deliv. Rev.* **2016**, *99*, 28–51. [CrossRef]
43. Li, C.; Wang, J.; Wang, Y.; Gao, H.; Wei, G.; Huang, Y.; Yu, H.; Gan, Y.; Wang, Y.; Mei, L.; et al. Recent progress in drug delivery. *Acta Pharm. Sin. B* **2019**, *9*, 1145–1162. [CrossRef]
44. Unni, M.; Uhl, A.M.; Savliwala, S.; Savitzky, B.H.; Dhavalikar, R.; Garraud, N.; Arnold, D.P.; Kourkoutis, L.F.; Andrew, J.S.; Rinaldi, C. Thermal decomposition synthesis of iron oxide nanoparticles with diminished magnetic dead layer by controlled addition of oxygen. *ACS Nano* **2017**, *11*, 2284–2303. [CrossRef]
45. Mikhailov, O.V.; Mikhailova, E.O. Elemental silver nanoparticles: Biosynthesis and bio applications. *Materials* **2019**, *12*, 3177. [CrossRef]
46. Sanaeimehr, Z.; Javadi, I.; Namvar, F. Antiangiogenic and antiapoptotic effects of green-synthesized zinc oxide nanoparticles using Sargassum muticum algae extraction. *Cancer Nanotechnol.* **2018**, *9*, 1–16. [CrossRef]
47. Singh, J.; Dutta, T.; Kim, K.-H.; Rawat, M.; Samddar, P.; Kumar, P. 'Green' synthesis of metals and their oxide nanoparticles: Applications for environmental remediation. *J. Nanobiotechnol.* **2018**, *16*, 1–24. [CrossRef]
48. Christy, S.R.; Priya, L.S.; Durka, M.; Dinesh, A.; Babitha, N.; Arunadevi, S. Simple combustion synthesis, structural, morphological, optical and catalytic properties of ZnO nanoparticles. *J. Nanosci. Nanotechnol.* **2019**, *19*, 3564–3570. [CrossRef] [PubMed]

49. Mozar, F.S.; Chowdhury, E.H. Impact of PEGylated nanoparticles on tumor targeted drug delivery. *Curr. Pharm. Des.* **2018**, *24*, 3283–3296. [CrossRef] [PubMed]
50. Manikandan, A.; Durka, M.; Selvi, M.A.; Antony, S.A. Aloe vera plant extracted green synthesis, structural and opto-magnetic characterizations of spinel CoxZn1-xAl$_2$O$_4$ nano-catalysts. *J. Nanosci. Nanotechnol.* **2016**, *16*, 357–373. [CrossRef]
51. Hassan, S.S.M.; Azab, W.I.M.E.; Ali, H.R.; Mansour, M.S.M. Green synthesis and characterization of ZnO nanoparticles for photocatalytic degradation of anthracene. *Adv. Nat. Sci. Nanosci. Nanotechnol.* **2015**, *6*, 045012. [CrossRef]
52. Zare, E.; Pourseyedi, S.; Khatami, M.; Darezereshki, E. Simple biosynthesis of zinc oxide nanoparticles using nature's source, and it's in vitro bio-activity. *J. Mol. Struct.* **2017**, *1146*, 96–103. [CrossRef]
53. Karnan, T.; Selvakumar, S.A.S. Biosynthesis of ZnO nanoparticles using rambutan (*Nephelium lappaceum* L.) peel extract and their photocatalytic activity on methyl orange dye. *J. Mol. Struct.* **2016**, *1125*, 358–365. [CrossRef]
54. Bhuyan, T.; Mishra, K.; Khanuja, M.; Prasad, R.; Varma, A. Biosynthesis of zinc oxide nanoparticles from Azadirachta indica for antibacterial and photocatalytic applications. *Mater. Sci. Semicond. Process.* **2015**, *32*, 55–61. [CrossRef]
55. Rasli, N.I.; Basri, H.; Harun, Z. Zinc oxide from aloe vera extract: Two-level factorial screening of biosynthesis parameters. *Heliyon* **2020**, *6*, e03156. [CrossRef]
56. Chabala, L.F.G.; Cuartas, C.E.E.; López, M.E.L. Release behavior and antibacterial activity of chitosan/alginate blends with aloe vera and silver nanoparticles. *Mar. Drugs* **2017**, *15*, 328. [CrossRef] [PubMed]
57. Singh, R.; Gautam, N.; Mishra, A.; Gupta, R. Heavy metals and living systems: An overview. *Indian J. Pharmacol.* **2011**, *43*, 246–253. [CrossRef]
58. Keshtkar, M.; Dobaradaran, S.; Soleimani, F.; NorooziKarbasdehi, V.; Mohammadi, M.; Mirahmadi, R.; FarajiGhasemi, F. Data on heavy metals and selected anions in the Persian popular herbal distillates. *Data Brief* **2016**, *8*, 2352–3409. [CrossRef]
59. Azizi, S.; Mohamad, R.; Shahri, M.M. Green microwave-assisted combustion synthesis of zinc oxide nanoparticles with *Citrullus colocynthis* (L.) schrad: Characterization and biomedical applications. *Molecules* **2017**, *22*, 301. [CrossRef] [PubMed]
60. Muniz, F.T.L.; Miranda, M.A.R.; Dos Santos, C.M.; Sasaki, J.M. The Scherrer equation and the dynamical theory of X-ray diffraction. *Acta Crystallogr. Sect. A Found. Adv.* **2016**, *72*, 385–390. [CrossRef]
61. Barad, S.; Roudbary, M.; Omran, A.N.; Daryasari, M.P. Preparation and characterization of ZnO nanoparticles coated by chitosan-linoleic acid; fungal growth and biofilm assay. *Bratisl. Med. J.* **2017**, *118*, 169–174. [CrossRef]
62. Ezealisiji, K.M.; Siwe-Noundou, X.; Maduelosi, B.; Nwachukwu, N.; Krause, R.W.M. Green synthesis of zinc oxide nanoparticles using Solanum torvum (L) leaf extract and evaluation of the toxicological profile of the ZnO nanoparticles–hydrogel composite in Wistar albino rats. *Int. Nano Lett.* **2019**, *9*, 99–107. [CrossRef]
63. Zhang, H.; Ji, Z.; Xia, T.; Meng, H.; Low-Kam, C.; Liu, R.; Pokhrel, S.; Lin, S.; Wang, X.; Liao, Y.-P.; et al. Use of metal oxide nanoparticle band gap to develop a predictive paradigm for oxidative stress and acute pulmonary inflammation. *ACS Nano* **2012**, *6*, 4349–4368. [CrossRef] [PubMed]
64. Zak, A.K.; Majid, W.A.; Mahmoudian, M.; Darroudi, M.; Yousefi, R. Starch-stabilized synthesis of ZnO nanopowders at low temperature and optical properties study. *Adv. Powder Technol.* **2013**, *24*, 618–624. [CrossRef]
65. Punnoose, A.; Dodge, K.; Rasmussen, J.W.; Chess, J.; Wingett, D.; Anders, C. Cytotoxicity of ZnO nanoparticles can be tailored by modifying their surface structure: A green chemistry approach for safer nanomaterials. *ACS Sustain. Chem. Eng.* **2014**, *2*, 1666–1673. [CrossRef]
66. Kc, B.; Paudel, S.N.; Rayamajhi, S.; Karna, D.; Adhikari, S.; Shrestha, B.G.; Bisht, G. Enhanced preferential cytotoxicity through surface modification: Synthesis, characterization and comparative in vitro evaluation of TritonX-100 modified and unmodified zinc oxide nanoparticles in human breast cancer cell (MDA-MB-231). *Chem. Cent. J.* **2016**, *10*, 16. [CrossRef]
67. Kenechukwu, F.C.; Attama, A.A.; Ibezim, E.C.; Nnamani, P.O.; Umeyor, C.E.; Uronnachi, E.M.; Momoh, M.A.; Akpa, P.A.; Ozioko, A.C. Novel intravaginal drug delivery system based on molecularly PEGylated lipid matrices for improved antifungal activity of miconazole nitrate. *BioMed Res. Int.* **2018**, *2018*, 3714329. [CrossRef] [PubMed]
68. Kumari, P.; Ghosh, B.; Biswas, S. Nanocarriers for cancer-targeted drug delivery. *J. Drug Target.* **2016**, *24*, 179–191. [CrossRef]
69. Bray, F.; Ferlay, J.; Soerjomataram, I.; Siegel, R.L.; Torre, L.A.; Jemal, A. Global cancer statistics 2018: GLOBOCAN estimates of incidence and mortality worldwide for 36 cancers in 185 countries. *CA Cancer J. Clin.* **2018**, *68*, 394–424. [CrossRef]
70. Ali, R.; Mirza, Z.; Ashraf, G.M.D.; Kamal, M.A.; Ansari, S.A.; Damanhouri, G.A.; Abuzenadah, A.M.; Chaudhary, A.G.; Sheikh, I.A. New anticancer agents: Recent developments in tumor therapy. *Anticancer. Res.* **2012**, *32*, 2999–3005.
71. Davion, S.M.; Siziopikou, K.P.; Sullivan, M.E. Cytokeratin 7: A re-evaluation of the 'tried and true' in triple-negative breast cancers. *Histopathology* **2012**, *61*, 660–666. [CrossRef]
72. Huffman, M.A. Animal self-medication and ethno-medicine: Exploration and exploitation of the medicinal properties of plants. *Proc. Nutr. Soc.* **2003**, *62*, 371–381. [CrossRef]
73. Jiang, J.; Pi, J.; Cai, J. The advancing of zinc oxide nanoparticles for biomedical applications. *Bioinorg. Chem. Appl.* **2018**, *2018*, 1062562. [CrossRef] [PubMed]
74. Ali, K.; Dwivedi, S.; Azam, A.; Saquib, Q.; Al-Said, M.S.; Alkhedhairy, A.A.; Musarrat, J. Aloe vera extract functionalized zinc oxide nanoparticles as nanoantibiotics against multi-drug resistant clinical bacterial isolates. *J. Colloid Interface Sci.* **2016**, *472*, 145–156. [CrossRef] [PubMed]
75. Vinardell, M.P.; Mitjans, M. Antitumor activities of metal oxide nanoparticles. *Nanomaterials* **2015**, *5*, 1004–1021. [CrossRef] [PubMed]

76. Selim, Y.A.; Azb, M.A.; Ragab, I.; El-Azim, M.H.M.A. Green synthesis of zinc oxide nanoparticles using aqueous extract of deverra tortuosa and their cytotoxic activities. *Sci. Rep.* **2020**, *10*, 3445. [CrossRef]
77. El-Shorbagy, H.M.; Eissa, S.M.; Sabet, S.; El-Ghor, A.A. Apoptosis and oxidative stress as relevant mechanisms of antitumor activity and genotoxicity of ZnO-NPs alone and in combination with N-acetyl cysteine in tumor-bearing mice. *Int. J. Nanomed.* **2019**, *14*, 3911–3928. [CrossRef]
78. Radha, M.H.; Laxmipriya, N.P. Evaluation of biological properties and clinical effectiveness of Aloe vera: A systematic review. *J. Tradit. Complement. Med.* **2015**, *5*, 21–26. [CrossRef]
79. Kumar, R.; Singh, A.K.; Gupta, A.; Bishayee, A.; Pandey, A.K. Therapeutic potential of Aloe vera—A miracle gift of nature. *Phytomedicine* **2019**, *60*, 152996. [CrossRef]
80. Gao, Y.; Kuok, K.I.; Jin, Y.; Wang, R. Biomedical applications of Aloe vera. *Crit. Rev. Food Sci. Nutr.* **2019**, *59*, S244–S256. [CrossRef]
81. Wongpinyochit, T.; Uhlmann, P.; Urquhart, A.J.; Seib, F.P. PEGylated silk nanoparticles for anticancer drug delivery. *Biomacromolecules* **2015**, *16*, 3712–3722. [CrossRef] [PubMed]
82. Chung, B.H.; Chung, S.J. Recent advances in pH-sensitive polymeric nanoparticles for smart drug delivery in cancer therapy. *Curr. Drug Targets* **2018**, *19*, 300–317. [CrossRef]
83. Bisht, G.; Rayamajhi, S. ZnO nanoparticles: A promising anticancer agent. *Nanobiomedicine* **2016**, *3*, 9. [CrossRef] [PubMed]
84. Rasmussen, J.W.; Martinez, E.; Louka, P.; Wingett, D.G. Zinc oxide nanoparticles for selective destruction of tumor cells and potential for drug delivery applications. *Expert Opin. Drug Deliv.* **2010**, *7*, 1063–1077. [CrossRef]

# Silica Microspheres for Economical Advanced Solar Applications

Maha M. Khayyat

Materials Science Research Institute, King Abdullaziz City for Science and Technology (KACST), P.O. Box 6086, Riyadh 11442, Saudi Arabia; mkhayyat@kacst.edu.sa

**Abstract:** Solar cells made of silicon nanowires (Si-NWs) have several potential benefits over conventional bulk Si ones or thin-film devices related primarily to light absorption and cost reduction. Controlling the position of Si-NWs without lithography using silica microspheres is indeed an economical approach. Moreover, replacing the glass sheets with polycarbonates is an added advantage. This study employed the Nanoscale Chemical Templating (NCT) technique in growing Si-NWs seeded with Al. The growth was undertaken at the Chemical Vapor Deposition (CVD) reactor via the original growth process of vapor–liquid–solid (VLS). The bottom-up grown nanowires were doped with aluminum (Al) throughout the growth process, and then the p-n junctions were formed with descent efficiency. Further work is required to optimize the growth of Si-NWs between the spun microspheres based on the growth parameters including etching time, which should lead to more efficient PV cells.

**Keywords:** chemical vapor deposition (CVD); silicon (Si); nanowires (NWs); silica microspheres; nanoscale chemical templating (NCT); photovoltaic (PV) cells

## 1. Introduction

Conventional 3D crystal growth requires only two phases; however, nanowires grow in three-phase systems. The vapor–liquid–solid (VLS) mechanism is based on systems that combine three phases: a vapor that supplies the materials for crystal growth, a liquid droplet to seed the growth, and the solid crystal. One-dimensional crystal (NWs) growth occurs when the growth rate at the interface between the liquid phase and the solid crystal is higher than the growth rate at the interface between the vapor/solid phase boundary.

Vertical NWs made of silicon substrate are of great interest because they would allow for ultimate light trapping and distinguished charge carriers' separation for solar cell applications. They could therefore achieve, in principle, better efficiency than thin-film planar cells, with the added merits of minimal use of materials and much lower process cost [1]. Nanowire solar cells have some potential benefits over traditional wafer-based or thin-film devices related to optical, electrical, and strain relaxation effects, charge separation mechanisms, and cost reduction. Ordered arrays of vertical nanowires with radial junctions take advantage of all these effects, as explained in some detail by Wacaser et al. [2]. Controlling the position of nanowires is another important topic for research. A promising emerging field for future low-cost, decent efficiency solar cell devices is the use of vapor–liquid–solid (VLS)-grown Si-NWs. The bottom-up approach of Si-NW growth via the VLS mechanism has a key advantage for device applications, since it is possible to template the position of the NWs by controlling the placement of the initial metal seed or catalyst particle. This templating then allows the integration of NWs with other parts of the structure, as required for many of these applications [3–6]. The technique of microsphere lithography has been proposed to produce regular hexagonal arrays of Al-seeded nanowires, and the fabrication process continues to fabricate solar cells [5–8]. Templating the Si-NWs' growth has been explained by several research groups [1,2,6,9–11], using wide schemes of techniques including electron beam lithography. Employing microspheres to template the growth of NWs will reduce the production cost of device fabrication; however, this area of research requires further investigations. The current study is a direct application of the concept of Nanoscale Chemical Templating (NCT) [6]. There are several advantages of using microspheres at the level of templating the growth, as it is an economical approach that reduces the number of the required steps of lithography. Moreover, using polycarbonate sheets in solar cells is a potential application. According to the SABIC report, there is a 61% $CO_2$ footprint reduction for each Kg of polycarbonate based on certified renewable feedstock [9]. Polycarbonate plastic materials are transparent amorphous, although they are made commercially available in a variety of colors, and the raw material allows for the internal transmission of light nearly in the same capacity as glass. Polycarbonate polymers are used to produce a variety of materials and are particularly useful when impact resistance or transparency is a product requirement. Polycarbonate is a relatively hard, lightweight material. These properties make polycarbonates suitable for products intended for long open-air operation, such as PV cells and modules.

## 2. Experiments

A Chemical Vapor Deposition (CVD) reactor has been used to grow Si-NWs via the VLS process, which includes a home-built Al evaporator operated under Ultra High Vacuum (UHVCVD). Si-NWs, as previously observed [5,7], grow preferentially 45° tilted on Si (100), and perpendicularly in the (111) direction, so Si (111) substrates were primarily used. The detailed general growth conditions along with the pre-growth steps were described in previous studies [5–8].

The thickness of the Al catalyst layer is of critical importance to the grown NWs. As the thickness of the seed Al increases from 2 nm to 5 nm, the fidelity of the growth increases, then after reaching 5 nm Al thickness the growth fidelity decreases quite significantly. Based on these measurements, it has been decided to choose 5 nm of Al thickness. The samples were then transferred into the hot growth chamber for a pre-growth anneal at 600 °C, which is above the Al/Si bulk eutectic temperature of 577 °C. This anneal was for 20 min under full pumping and was intended to allow the Al film to agglomerate into small islands on the surface and ball up, forming liquid droplets. The furnace was then cooled to the desired growth temperature (usually 520 °C), and pure silane ($SiH_4$) was introduced at a fixed flow, 20 sccm, using a mass flow controller (MFC). The morphology of grown Si-NWs was examined with an environmental scanning electron microscope (SEM) FEI Co., Eindhoven, The Netherlands, model XL 30.

Silica microspheres are commercially available with narrow size distributions and have an average diameter of less than 1 μm (approximately 900 nm). When silica microspheres are dispersed properly using a spinner on the Si substrate surface, they form a close-packed array that forms gaps through which a material can be deposited. Si (111) of p-type substrates were cleaned using standard techniques [5–11], which leave a thin oxide on the Si surface. A drop of 1 μm diameter polystyrene microspheres in solution taken directly from the commercially provided stock solution was dispersed onto the substrate.

Reflectance measurements were undertaken using the F10-RT reflectometer, which captures reflectance and transmittance simultaneously (sample thickness range 15–1000 μm, wavelength range 200–900 nm). I-V characteristics were carried out using "keithley 2400" source meter.

## 3. Results and Discussion

The current experimental work is focused on using the economic approach using microspheres for templating the growth of nanowires (Section 3.1.1.). We assessed the light reflectance and I-V characteristics of Si-NWs (Section 3.1.2).

*3.1. Controlling the Growth Position of Si-NWs*

Controlling the growth position of Si-NWs, catalyzed with Al, can be undertaken using e-beam lithography [6,7].

### 3.1.1. Templating Si-NWs Using Polystyrene Microspheres and SiO$_2$ Thin-Layer Lithography

The concept of the NCT technique has been explained and employed, as shown in Figure 1. A schematic illustration of the NCT process in seven main steps is presented in Figure 1a as follows: Step (1) starts off with Si (111), where nanowires grow perpendicularly, which includes a thin layer of SiO$_2$ (approximately 50 nm), where polystyrene microspheres are spun on it. Step (2): In oxygen plasma, Rapid Thermal Annealing (RTA) is performed. Step (3) involves etching using BOE, and then the removal of polystyrene microspheres occurs in step (4), or the process could continue without the removal of the microspheres. Step (5) involves the deposition of a 5 nm thick layer of Al on the surface covering the top surfaces of the microspheres and bare Si surface. Step (6) is where annealing is undertaken, showing that Al on the top of SiO$_2$ oxidizes, forming Al$_2$O$_3$, or on the top of the microspheres (if they have not been removed), and the Al in the opening balls up. Then, in step 7, the growth of Si-NWs occurs after allowing SiH$_4$ to flow, providing the required medium of the VLS system. The epitaxial perpendicular Si-NW growth of approximately 2 µm long is accompanied by the planar deposition of Si of a few nanometers between Si-NWs (see Figure 1b) [12].

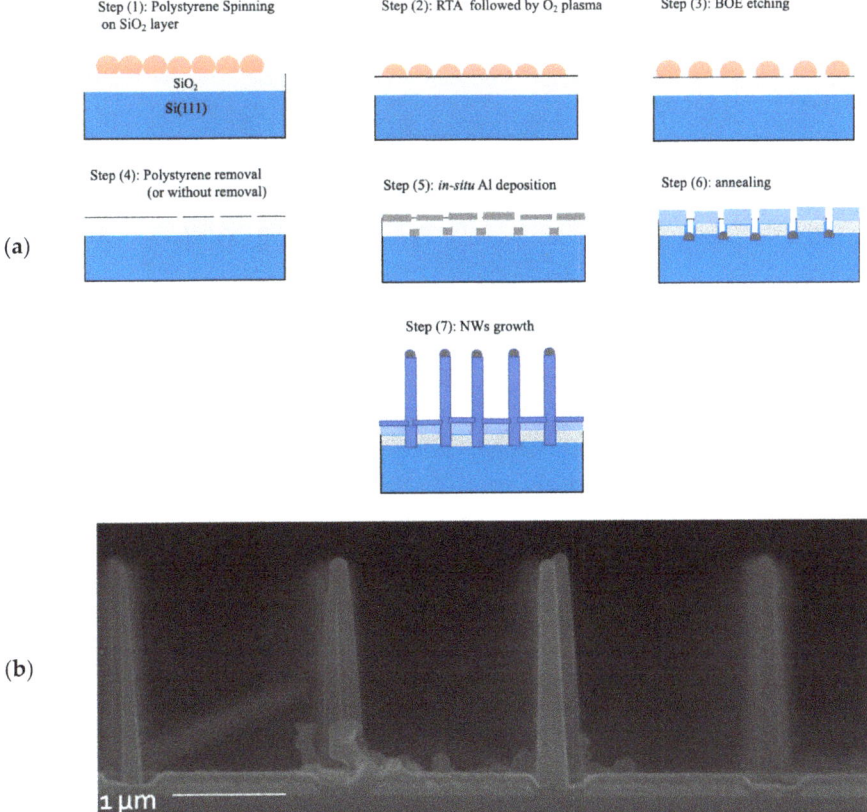

**Figure 1.** (a) A schematic presentation of the original NCT technique, showing that microspheres (MS) can be used for templating in 6 main steps. (b) SEM micrograph of grown Si (NWs) using NCT. See the layers between Si-NWs—the bright layer is SiO$_2$, which is the top surface of it that is oxidized, forming Al$_2$O$_3$, covered by the dark layer of planar growth of Si.

### 3.1.2. Templating Si-NWs Using Silica Microsphere Lithography

The concept of the original process of the NCT technique explained schematically in Figure 1 can be employed in fewer steps, as presented in Figure 2. After careful spinning of silica microspheres that are commercially available with narrow size distributions (step 1), in step 2, the oxide is etched with buffered oxide etch solution (BOE 9:1 Seidler Chemical Company; 9 parts 40% NH4F in water to 1 part 49% HF in water), typically for a few seconds. Then, annealing (step 3) is followed by NW growth, as explained earlier [5,7].

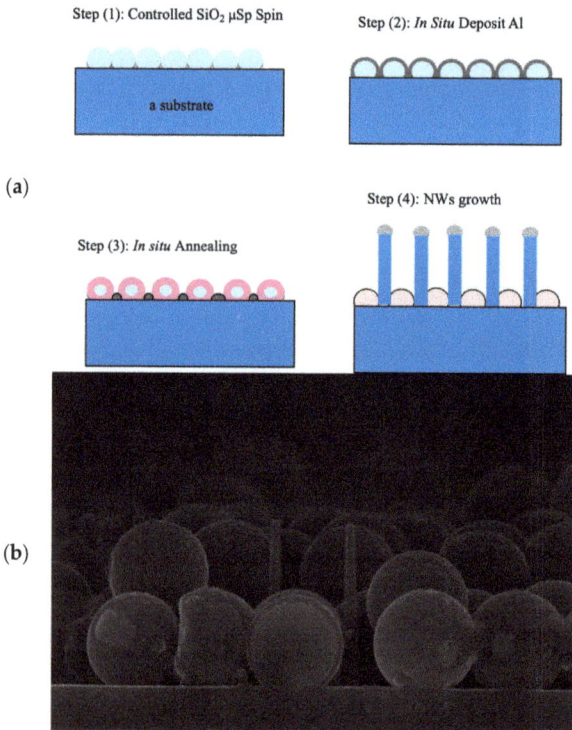

**Figure 2.** (**a**) A schematic representation showing that microspheres (MS) of silica were used for templating in four main steps, along with the side view SEM micrograph of grown Si-NWs between the MS (**b**). A thin unseeded planar Si layer forms on the surface of the oxidized Al on the outer surface of microspheres (just as for the standard lithography case). The gaps between microspheres could be controlled by varying the BOE time (scale bar is 500 nm).

The described concept in Figure 2 has been examined and it showed encouraging results; however, further optimization of the Si-NWs' growth has to be carried out. It can be shown that the outer shell of the silica microspheres appears in a different color, confirming that the evaporated Al on the outer surfaces of the silica microsphere has been oxidized, forming $Al_2O_3$ on the top of it. A thin unseeded planar Si layer forms on the surface of the microspheres, just as for the standard lithography case. We found that the gaps between microspheres could be controlled to some extent by varying the BOE time. If etch times exceeding 10 s were used, the gaps became large enough for NWs to be seeded in the openings. However, etch times longer than 45 s resulted in some of the microspheres lifting off the Si surface to leave large gaps where multiple nanowires could grow. This process has not been optimized for yield.

Si (NWs) will not grow on these microspheres due to the oxidation of the Al catalysts. We show the results of this process, where Si-NWs can be seen growing in the gaps between the microsphere array, where the Al catalyst is on the Si substrate. The microspheres of silica exhibit two main contents of distinct colors. The core, which represents more than 90% of the microsphere, remains as it is, while on the very outer exposed part of the shell of oxidized Al, a thin unseeded planar Si layer forms on the surface of the microspheres, just as for the standard lithography case. We found that the gaps between microspheres could be controlled to some extent by varying the BOE time. If etch times longer than 10 s were used, the gaps became large enough for NWs to be seeded in the openings. However, etch times longer than 45 s resulted in some of the microspheres lifting off the Si surface to leave large gaps where multiple nanowires could grow. This process has not been optimized for yield. Khan et al., in 2015 and 2016 [13,14], respectively, reported that liquid nanodroplets Indium Nanodroplets (In NDs) act as facilitator sites for Si deposition. These nano-sized droplets can be considered as the growth catalyst and it is part of the produced device, just like Al in our work. However, our work used silica microspheres, which can be removed after templating the growth or could stay there between the grown nanowires as an electrical insulator for several applications such as the MOSFETs and photovoltaic (PV) cells.

### 3.2. Light Absorption of Si-NWs and PV Applications

Nanowires have a very high surface to volume ratio, making them ideal components for light absorption or any related interface phenomena. The optical characteristics of Si-NWs exhibit excellent light harvesting characteristics (see Figure 3). The reflectance of light from the Si substrate covered with NWs is almost zero, in comparison to the pristine Si substrate, as shown in Figure 3a. The schematic illustration is presented in Figure 3b, describing how the light rays are reflected between the NWs, increasing the chance of absorption of the incident light rays [15–17].

Growing semiconductor nanowires can be catalyzed by chemically active materials using an efficient and economic approach. It has been well established in the literature that nanowires can be catalyzed using gold (Au). However, gold negatively affects the performance of semiconductor devices, as it acts as a deep-level trapping charge carrier. On the other hand, Al is a part of the semiconductor industry, as it is used as p-type dopants of silicon. However, Al is a chemically active element; as we deposit aluminum on the semiconductor substrate, we should not expose the samples to air, so the NCT technique can be employed. It is possible to form the p–n core–shell junctions in high-density arrays, which have the benefit of decoupling the absorption of light from charge transport by allowing lateral diffusion of minority carriers to the p–n junction, which is at most 50–500 nm away rather than many microns away as in Si bulk solar cells [18–20]. Based on this, the potential cost benefits come from lowering the purity standard and the amount of semiconductor material needed to obtain sophisticated efficiencies, increasing the defect tolerance, and lattice-matched substrates. The concept of NW-based PV cells has attracted the scientific community's attention because of their potential benefits in carrier transport, charge separation, and light absorption [21]. The Lieber [11,19] and Atwater [9,10] and other groups [10,22–24] have developed core–shell growth and contact strategies for their silicon p–n nanowire solar cells, with sophisticated efficiencies. After all, Si-NWs possess the combined attributes of cost effectiveness and mature manufacturing infrastructures for further advanced applications. Increasing the growth temperature increases the active doping of Al-catalyzed NWs.

Thicker lightly doped NWs and doping the outer areas by diffusion are shown in Figure 4a. The width of the junction can be engineered, if required, by controlling the width of the intrinsic layer, forming a p–i–n junction [22–24].

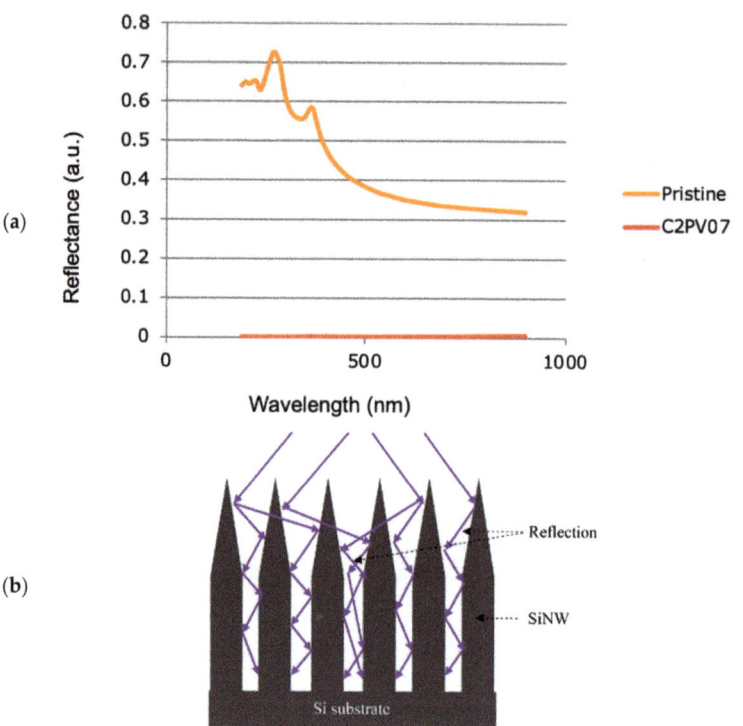

**Figure 3.** Si-NWs are excellent light absorbers. (**a**) The reflectance of Si-NWs is almost zero in comparison to pristine Si. (**b**) Schematic representation shows how the light rays bounce between the NWs to be absorbed eventually.

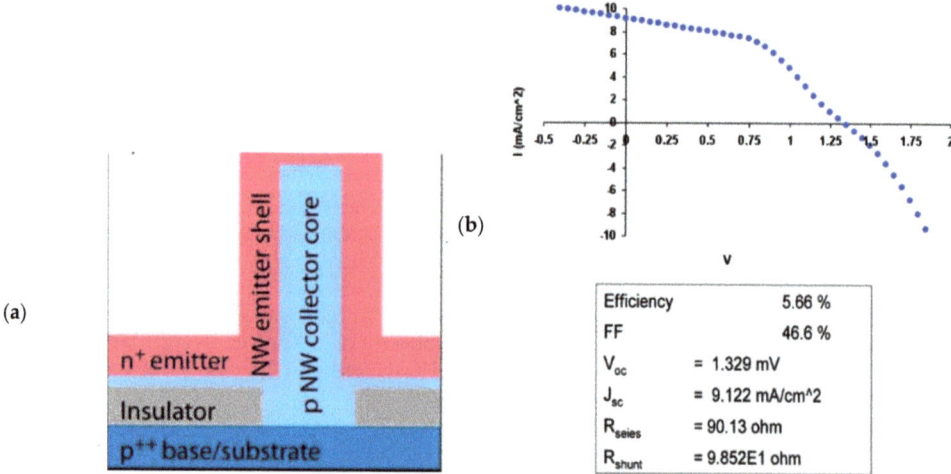

**Figure 4.** (**a**) Schematic illustration of p_n junction of Si-NWs, (**b**) I-V characteristics of the fabricated PV cell of Si-NWs.

The obtained electrical efficiency of the fabricated PV showed encouraging results of 5.66%, along with the other related parameters, as shown in Figure 4b. Clearly, much work is required to improve the performance of the PV cell made of Si-NWs. Forming an ideal p–n junction for core–shell NWs by CVD is not a trivial undertaking. Typically, when switching precursors in CVD, much care is needed to keep from forming defects at the interfaces. This problem is only compounded by the NWs because they have a three-dimensional shape with different facet orientations and characteristics. The engineering of this interface is an ongoing research topic that will need to be resolved to make core–shell grown nanowires competitive for large-scale PV [23–25].

## 4. Conclusions

The combination of the NCT technique along with silica microspheres is an economic approach, as it requires fewer steps compared to conventional patterning approaches, not requiring lift off of a metal layer or the removal of the mask. The encapsulation of silicon solar panels with PC sheets reduces the weight of the solar panels and opens up more applications of solar panels in harsh environments considering the mechanical properties of PC. The methodology used in achieving the controlled placement of single NWs seeded with Al is an example of proving the concept of the NCT technique; however, more detailed work is required to optimize the growth process and increase the NWs' growth fidelity, leading to more efficient PV cells. More oxygen-reactive materials such as Sn, Sb, In, Ga, and Ti templated by NCT will have more advanced applications in nanodevice fabrications.

**Funding:** This research has been conducted as part of the employment at KACST.

**Institutional Review Board Statement:** Not applicable.

**Informed Consent Statement:** Not applicable.

**Data Availability Statement:** The data used to support the findings of this study are included within the article.

**Acknowledgments:** The author would like to thank T. C. Chen of the T. J. Watson Research Center, IBM, Yorktown Heights, NY, for his scientific insights throughout working on this project.

**Conflicts of Interest:** The author declares that there is no conflict of interest regarding the publication of this paper.

## References

1. Kim, J.; Hong, A.J.; Nah, J.-W.; Shin, B.; Ross, F.; Sadana, D.K. Three-Dimensional a-Si:H Solar Cells on Glass Nanocone Arrays Patterned by Self-Assembled Sn Nanospheres. *ACS Nano* **2011**, *6*, 265–271. [CrossRef]
2. Wacaser, B.; Khayyat, M.; Reuter, M.; Sadana, D.; Ross, F. Technical advantages and challenges for core-shell micro/nanowire large area PV devices. In Proceedings of the 2010 35th IEEE Photovoltaic Specialists Conference, Honolulu, HI, USA, 20–25 June 2010.
3. Huang, B.; Lin, T.; Hung, W.; Sun, F. Performance evaluation of solar photovoltaic/thermal systems. *Sol. Energy* **2001**, *70*, 443–448. [CrossRef]
4. Gong, L.; Lu, J.; Ye, Z. Transparent and conductive Ga-doped ZnO films grown by RF magnetron sputtering on polycarbonate substrates. *Sol. Energy Mater. Sol. Cells* **2010**, *94*, 937–941. [CrossRef]
5. Wacaser, B.; Reuter, M.C.; Khayyat, M.; Wen, C.-Y.; Haight, R.; Guha, S.; Ross, F. Growth System, Structure, and Doping of Aluminum-Seeded Epitaxial Silicon Nanowires. *Nano Lett.* **2009**, *9*, 3296–3301. [CrossRef] [PubMed]
6. Khayyat, M.; Wacaser, B.; Reuter, M.; Ross, F.; Sadana, D.; Chin, T. Nanoscale chemical templating of Si-NWs seeded with Al. *Nanotechnology* **2013**, *24*, 235301. [CrossRef]
7. Khayyat, M.; Wacaser, B.; Sadana, D. Nanoscale Chemical Templating with Oxygen Reactive Materials. U.S. Patent Number 8349715, 2013.
8. Fan, H.J.; Werner, P.; Zacharias, M. Semiconductor Nanowires: From Self-Organization to Patterned Growth. *Small* **2006**, *2*, 700–717. [CrossRef]
9. Kayes, B.M.; Atwater, H.A.; Lewis, N.S. Comparison of the device physics principles of planar and radial p-n junction nanorod solar cells. *J. Appl. Phys.* **2005**, *97*, 114302. [CrossRef]
10. Kayes, B.M.; Filler, M.A.; Putnam, M.C.; Kelzenberg, M.D.; Lewis, N.S.; Atwater, H.A. Growth of vertically aligned Si wire arrays over large areas (>1 cm$^2$) with Au and Cu catalysts. *Appl. Phys. Lett.* **2007**, *91*, 103110. [CrossRef]
11. Huang, Y.; Lieber, C.M. Integrated nanoscale electronics and optoelectronics: Exploring nanoscale science and technology through semiconductor nanowires. *Pure Appl. Chem.* **2004**, *76*, 2051–2068. [CrossRef]
12. Wagner, R.S.; Ellis, W.C. Vapor-liquid-solid mechanism of single crystal growth. *Appl. Phys. Lett.* **1964**, *4*, 89–91. [CrossRef]

13. Khan, M.A.; Ishikawa, Y.; Kita, I.; Fukunaga, K.; Fuyuki, T.; Konagai, M. Control of verticality and (111) orientation of In-catalyzed silicon nanowires grown in the vapour–liquid–solid mode for nanoscale device applications. *J. Mater. Chem. C* **2015**, *3*, 11577–11580. [CrossRef]
14. Khan, M.A.; Ishikawa, Y.; Kita, I.; Tani, A.; Yano, H.; Fuyuki, T.; Konagai, M. Investigation of crystallinity and planar defects in the Si nanowires grown by vapor–liquid–solid mode using indium catalyst for solar cell applications. *Jpn. J. Appl. Phys.* **2015**, *55*, 01AE03. [CrossRef]
15. Kim, H.; Bae, H.; Chang, T.-Y.; Huffaker, D.L. III–V nanowires on silicon (100) as plasmonic-photonic hybrid meta-absorber. *Sci. Rep.* **2021**, *11*, 13813. [CrossRef]
16. Zeng, W.R.; Li, S.F.; Chow, W.K. Preliminary Studies on Burning Behavior of Polymethylmethacrylate (PMMA). *J. Fire Sci.* **2002**, *20*, 297–317. [CrossRef]
17. Wacaser, B.; Reuter, M.; Khayyat, M.; Haight, R.; Guha, S.; Ross, F. The Role of Microanalysis in Micro/Nanowire-Based Future Generation Photovoltaic Devices. *Microsc. Microanal.* **2010**, *16* (Suppl. S2), 1368–1369. [CrossRef]
18. Khayyat, M.; Wacaser, B.; Reuter, M.; Sadana, D. Templating silicon nanowires seeded with oxygen reactive materials. In Proceedings of the 2011 Saudi International Electronics, Communications and Photonics Conference (SIECPC), Riyadh, Saudi Arabia, 24–26 April 2011.
19. Law, M.; Goldberger, J.; Yang, P. Semiconductor nanowires and nanotubes. *Annu. Rev. Mater. Res.* **2004**, *34*, 83–122. [CrossRef]
20. Samuelson, L. Semiconductor nanowires for 0D and 1D physics and applications. *Physica E* **2004**, *25*, 313–318. [CrossRef]
21. Givargizov, E. Fundamental VLS growth. *J. Cryst. Growth* **1975**, *31*, 20–30. [CrossRef]
22. Wacaser, B.; Dick, K.; Johansson, J.; Borgstrom, M.; Deppert, K.; Samuelson, L. Preferential interface nucleation: An expansion of the VLS growth mechanism for nanowires. *Adv. Mater.* **2009**, *21*, 153–165. [CrossRef]
23. Westwater, J.; Gosain, D.; Pand, S. Control of the size and position of silicon nanowires grown via the vapor–liquid–solid technique. *Jpn. J. Appl. Phys.* **1997**, *136*, 6204–6209. [CrossRef]
24. Harper, C.A. *Handbook of Plastic Processes*; John Wiley & Sons: Hoboken, NJ, USA, 2005.
25. Lee, C.-L.; Goh, W.-S.; Chee, S.-Y.; Yik, L.-K. Enhancement of light harvesting efficiency of silicon solar cell utilizing arrays of poly(methyl methacrylate-co-acrylic acid) nano-spheres and nano-spheres with embedded silver nano-particles. *Photonics Nanostructures-Fundam. Appl.* **2017**, *23*, 36–44. [CrossRef]

Article

# Crystalline Silicon Spalling as a Direct Application of Temperature Effect on Semiconductors' Indentation

Maha M. Khayyat

Nanotechnology and Semiconductors Center, Materials Science Research Institute, King Abdulaziz City for Science and Technology (KACST), P.O. Box 6086, Riyadh 11442, Saudi Arabia; mkhayyat@kacst.edu.sa

**Abstract:** Kerf-less removal of surface layers of photovoltaic materials including silicon is an emerging technology by controlled spalling technology. The method is extremely simple, versatile, and applicable to a wide range of substrates. Controlled spalling technology requires a stressor layer, such as Ni, to be deposited on the surface of a brittle material; then, the controlled removal of a continuous surface layer can be performed at a predetermined depth by manipulating the thickness and stress of the Ni layer, introducing a crack near the edge of the substrate, and mechanically guiding the crack as a single fracture front across the surface. However, spalling Si(100) at 300 K (room temperature RT) introduced many cracks and rough regions within the spalled layer. These mechanical issues make it difficult to process these layers of Si(100) for PV, and in other advanced applications, Si does not undergo phase transformations at 77 K (Liquid Nitrogen Temperature, LNT); based on this fact, spalling of Si(100) has been carried out. Spalling of Si(100) at LNT improved material quality for further designed applications. Mechanical flexibility is achieved by employing controlled spalling technology, enabling the large-area transfer of ultrathin body silicon devices to a plastic substrate at room temperature.

**Keywords:** indentation; room temperature; liquid nitrogen temperature; spalling; Si-NWs; nanoscale chemical templating (NCT); PV

**Citation:** Khayyat, M.M. Crystalline Silicon Spalling as a Direct Application of Temperature Effect on Semiconductors' Indentation. *Crystals* 2021, 11, 1020. https://doi.org/10.3390/cryst11091020

**Academic Editor:** Bo Chen

Received: 1 August 2021
Accepted: 24 August 2021
Published: 25 August 2021

**Publisher's Note:** MDPI stays neutral with regard to jurisdictional claims in published maps and institutional affiliations.

**Copyright:** © 2021 by the author. Licensee MDPI, Basel, Switzerland. This article is an open access article distributed under the terms and conditions of the Creative Commons Attribution (CC BY) license (https://creativecommons.org/licenses/by/4.0/).

## 1. Introduction

The mechanical deformation of crystalline silicon induced by micro-indentation has been studied [1–4]; when crystalline silicon Si(100) is hydrostatically compressed at room temperature, to pressures in the range of 11–15 GPa, it transforms from face-centered cubic (diamond structure) phase silicon to a body-centered tetragonal phase (Si-II), which is metallic [1,2]. This transformation is not reversible, as when the hydrostatic pressure is released, the Si-II phase transforms into the body-centered cubic phase (Si-III), which is also metastable [3–6]. On the one hand, it has been well established that when Si single crystals undergo indentation at 300 K [6–9], their crystalline structure transforms mainly to Si-III at lower pressures, as indentation includes an element of shear stresses in addition to hydrostatic pressures [1,4]. On the other hand, it has been proven experimentally that Si(100) does not transform at 77 K using Raman spectroscopy [4], along with electrical characterizations [6–8]. The current work is based on a detailed study undertaken in 2007 by Khayyat et al. [6] at sample temperatures higher than 77 K but lower than 300 K, where the temperature range could be determined below which Vickers indentation-induced phase transitions in single crystals of silicon would not occur. Both in situ electrical resistance measurements and ex situ Raman spectroscopy of indentations were employed for these investigations, and it has been found that the sample temperature indeed has a very significant influence on the occurrence, or otherwise, of the indentation-induced phase transition from Si-I (face-centered cubic structure) to Si-II (body-centered tetragonal structure) [10,11].

Thin-film electronic materials have been extensively studied for the realization of a wide range of mechanically flexible electronic devices such as light-emitting diodes, thin-film transistors, photovoltaic solar cells, and sensors. So far, mainstream flexible electronics have been based on thin-film organic and amorphous semiconductors that allow direct device fabrication on a flexible substrate at relatively low temperatures (≤300 °C). The salient feature of this processing scheme is the ability to achieve very-large-area flexible electronics at a relatively low processing cost. However, the inherently defective and highly disordered crystalline structure in such materials severely limits overall device performance and reliability when the device dimensions are scaled down.

The aim of this study is to introduce additional control into the material spalling process, thus improving both crack initiation and propagation, and increasing the range of selectable spalling depths that are provided. The method includes providing a stressor layer on the surface of a base substrate at an initial temperature, which is room temperature. Next, the base substrate, including the stressor layer, is brought to a second temperature, which is lower than room temperature. The base substrate is spalled at the second temperature to form a spalled material layer. Thereafter, the spalled material layer is returned to room temperature, i.e., the first temperature. This investigation describes in some detail how indentations at 77 K have led to an important application in producing thin, flexible Si films for advanced applications [12,13].

## 2. Materials and Methods

The indentation experiments were conducted on the following sample: silicon single crystals Si(100). The single crystal specimens were of the dimensions 10 mm, 10 mm, 0.38 mm, all of which had been cut from a 50 mm diameter Si(100) wafer supplied by Wacker–Chemitronic GMBH (Munich, Germany). The wafer was n-type and the dopant was phosphorous; its resistivity was 50 Ω cm and carrier concentration was ~$10^{14}$ cm$^{-3}$. The temperature of the sample was measured with a thermocouple junction placed on the ceramic header package. In these investigations, the sample temperature could be varied in a controlled manner to an accuracy of ±5 K, as the experimental set-up allows $N_2$ (g), at temperatures as close as possible to 77 K, to follow above the sample to prevent air vapor condensation and temperature fluctuations in the range of 150 to 300 K (see Figure 1). A silica tube containing nitrogen gas, which was supplied from a metallic cylinder, was passed through a dewar cooler to bring it as close as possible to liquid nitrogen temperature. The cooled nitrogen gas passed through a cooling apparatus. Consequently, the sample was cooled by being mounted on the cooling apparatus. A thermocouple (type T) was attached to the sample and then, the temperature was measured using a FLUKE 54 II thermometer. The cooling apparatus consisted of a box of brass; a tube of brass was built inside the box with two openings to inlet and outlet nitrogen gas. The sample was mounted on the surface of the brass box. The sample was equilibrated at the temperature of interest for around 10 min, with fluctuations in temperature of ±5 °C or below for most of the time. This was achieved by controlling the nitrogen gas flow through the cooling apparatus.

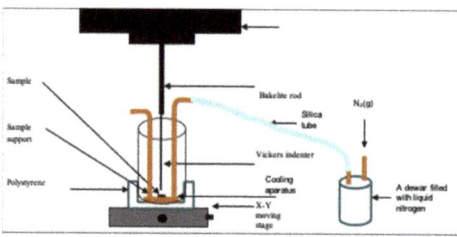

Figure 1. Set-up for indentation in an Instron machine at different temperatures.

In order to make relatively high load indentations in the silicon samples that were cooled to down temperatures in the range of 150–300 K and in a moisture-free atmosphere, which was suitable for Raman studies, we carried out another series of experiments. In this series, a single crystal Si(100) sample of the dimensions 10 mm, 10 mm, 0.38 mm was mounted onto the copper block (described above), and a thermocouple junction was stuck onto the sample with a piece of adhesive tape [11]. This assembly was then placed inside a 500 mL Pyrex glass beaker. The cooling of the copper block was carried out in the same manner, as described above. However, the cold, dry nitrogen gas exiting from the copper block was allowed to fill the Pyrex glass beaker. A Vickers diamond was used as the indenter, which was mounted on a 20 N load cell, that, in turn, was screwed onto the cross beam of an Instron Model 1122 mechanical testing machine. To make an indentation on the test sample, the indenter was loaded at a speed of 0.05 mm/min$^{-1}$. After reaching the desired indenter load, it was held constant for 15 s, and then, the indenter was unloaded at the same speed. Another site on the sample was then brought under the indenter, the sample cooled down to another desired temperature, and an indentation was made at another preselected load. In this manner, several indentations were made at various temperatures and under different indenter loads. After having made all the necessary indentations, the sample was allowed to gradually warm up to room temperature, making sure that, at no stage, did any water condensation occur on the sample. The indented samples were then stored in a desiccator. Some of the residual Vickers indentations made at 300 and 77 K were examined with an FEI Co., environmental scanning electron microscope (Eindhoven, The Netherlands, model XL 30). These samples were not coated, as surface charging did not take place inside the experimental chamber of this microscope.

## 3. Results and Discussion

### 3.1. Indentations at 300 K and 77 K

Environmental scanning electron micrographs of residual indentations made at 300 and 77 K are shown in Figure 2, respectively. It can be seen from Figure 2a that there is clear evidence for material extrusion from the indentation. This extrusion has generally been accepted as evidence for Si-I to Si-II phase transition. On the other hand, it can be seen from the micrograph of the residual indentation made at 77 K (see Figure 2b) that there is no extrusion of material, and instead, shear lines within the indentation are quite clearly visible. These observations also suggest that in indentations made at 77 K, there is no Si-I to Si-II phase transition.

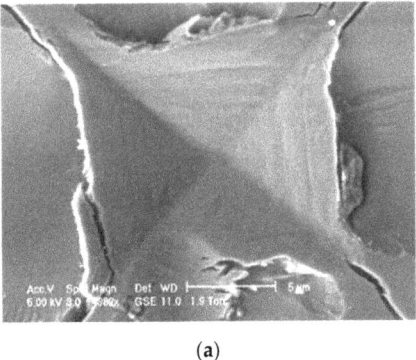 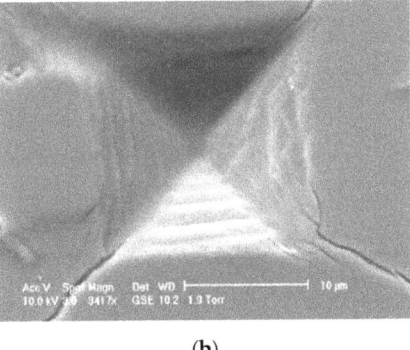

(a) (b)

**Figure 2.** (a) 300 K Vickers indentation in Si(100) under a load of 2.63 N; the arrow highlights the extrusion. (b) 77 K Vickers indentation in Si(100) under a load of 5.21 N; the arrow points at a shear line.

Using two complementary techniques, namely, in situ four-terminal dc electrical resistance measurements of the bare silicon when indented with a Vickers diamond indenter [10,11], and Raman spectroscopy [10] of the residual indentations, combined with environmental scanning electron microscopy of residual indentations made at different temperatures [11], it has been shown that indentation-induced phase transformation of a silicon crystal is significantly affected by its temperature. Whereas indentations at room temperature caused Si-I to transform to Si-II within the plastically deformed zone around the indentation, no such phase transition occurred when the sample had been cooled down to 200 K or lower.

During compression to ~11 GPa, Si-I (f.c.c) transforms into Si-II (body-centered tetragonal, also known as beta-tin structure); Si-II has low resistivity, which is similar to that of copper. In contrast, at decompression at 300 K and in the pressure range of 8 to 2 GPa, Si-II transforms to Si-XII, which is rhombohedral. From 2 GPa down to one atmosphere, Si-III forms, which is body-centered cubic or bc8; at these pressures, Si-XII is only a tiny fraction of the recovered silicon [2–5].

It has been assumed that there are temperature rises during indentation which assist phase transformation [11]. Figure 3 shows a schematic representation of the indentation-induced phase transition model, where the phase-transformed zone is just underneath the indenter and embedded within the plastic zone. It has been assumed that there is heat elevation in the indented zone of the material. An estimated local adiabatic temperature rises is generated during the indentation process, as explained in Equation (1) [12].

$$\Delta T = \frac{Y\varepsilon}{\rho c} = 460 \text{ K} \tag{1}$$

where $\Delta T$ is the maximum temperature rise, $Y$ is the uniaxial yield stress, $\varepsilon$ is the maximum plastic strain around a Vickers indentation, $\rho$ is the density, and $c$ is the heat capacity. The shear yield stress $\tau$ of silicon at room temperature has been given as 1 GPa, which gives $Y = 2\tau = 2$ GPa using the Tresca criterion [13]. The density $\rho$ is $2.33 \times 10^3$ kg m$^{-3}$ and heat capacity c of silicon is $0.67 \times 10^3$ J kg$^{-1}$ K$^{-1}$ [14]. Therefore, the estimated maximum temperature rise T during a Vickers indentation in silicon would be ~460 K. This temperature rise provides the required energy to rebuild the crystal structure-producing phase transitions. Cooling the sample down to 77 K will suppress this rebuilding of the crystal structure-producing structural phase-transformed zone within the plastically deformed area.

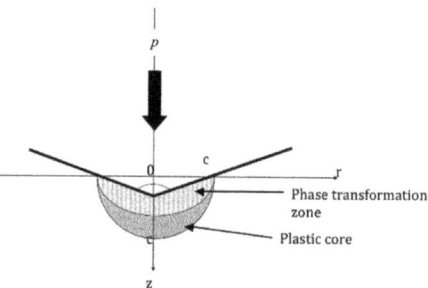

**Figure 3.** Schematic representation of the indentation-induced phase transition model; the phase-transformed zone is embedded within the plastic zone.

### 3.2. Si Crystals Spalling at 300 K and 77 K

Kerf-less removal of surface layers of materials, including silicon, is demonstrated by controlled spalling technology. The method is extremely simple, versatile, and applicable to a wide range of substrates. Controlled spalling technology, as has been described schematically in Figure 3, requires a stressor layer (Figure 4a) to be deposited on the surface of a brittle material, and the controlled removal and placement of a tape such as Kapton tape on the top of the stressor layer (Figure 4b); then, spalling of the surface layer is undertaken at a predetermined depth (Figure 4c). The stress layer (Ni) thickness affects the depth of the spalling, as strain distribution due to lattice-mismatch between these two layers increases with the increase in the thickness of the tensile layer, and consequently, increases the spalled layer thickness [15].

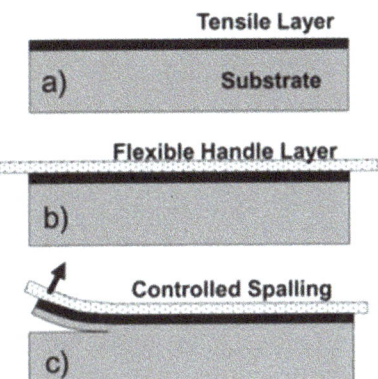

**Figure 4.** Schematic representation describes the main steps (**a**–**c**) of spalling [12,13].

Silicon (Si(100)) spalling at 77 K (liquid nitrogen temperature) is described schematically in Figure 4. Moreover, Figure 5 shows successful spalling of Si(111) wafer from the ingot.

**Figure 5.** Spalling of Si(111), wafer <111> from the ingot of a diameter of 100 mm.

By manipulating the thickness and stress of the Ni layer, a crack is introduced near the edge of the substrate and mechanically guided as a single fracture front across the surface at room temperature (300 K). Spalling from an ingot Si(111) is presented in Figure 5. However, there are many issues with spalling Si(100) at room temperature, such as cracks or irregularities in thickness. Based on previous knowledge of indentations at 77 K [11], where phase transformations and cracks disappear in comparison to indentation at 300 K, it has been suggested that spalling be undertaken at 77 K (Figure 6).

Figure 7 shows in detail the difference in appearance (optical microscope and SEM images) between Si(100) spalled at 300 K, and that spalled at 77 K, where the spalling at room temperature is performed first mechanically by introducing a crack as explained earlier, then by exposing the same sample to liquid nitrogen vapor at a fresh region, where it is spalled spontaneously. The resulted spalled layer at room temperature looks different to that spalled at liquid nitrogen temperature, as the outer surface looks rough (see Figure 7a, where this observation has been confirmed by SEM micrographs (Figure 7b,c)).

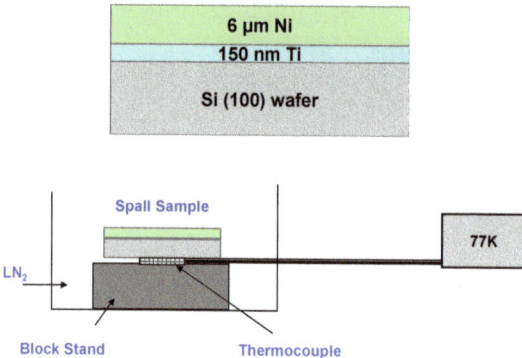

**Figure 6.** Liquid nitrogen temperature Si(100) spalling can be carried out as follows: pieces of 100 semiconductor materials (nominally 1.5, 3 inch pieces) are HF-dipped until hydrophobic (optional), $N_2$-dried, immediately placed into a sputter system for metal layer deposition, and then, Kapton tape is placed onto the Ni surface and cooled using liquid nitrogen.

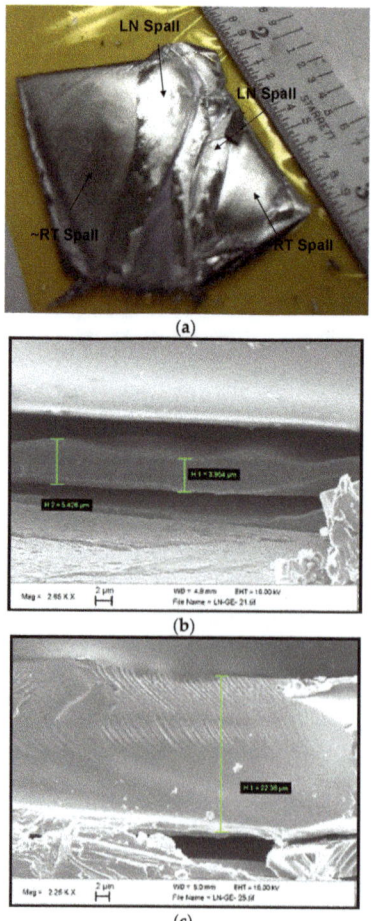

**Figure 7.** (a) Optical image of Si(100) spalled at 300 K and at 77 K in different areas of the same piece

of wafer. (**b**) SEM micrograph of the side view of Si(100) spalled at 300 K. The variation in spalled layer thickness, as shown by green lines, is between 3.91 to 5.43 µm. (**c**) Side view SEM micrograph of Si(100) spalled at 77 K, where the thickness of the spalled film is almost constant at around 22.38 µm.

When we spall Si(100) at room temperature, the spalled layer, as shown on the micrograph (Figure 7), has many cracks and rough regions, which makes it difficult to process for PV applications. As it has been shown previously that Si does not undergo phase transformations at 77 K, we cooled down the Si(100) (which has a stressor layer of Ni on its surface) as close as possible to 77 K. Then, when we carried out further examinations on the spalled samples, such as SEM micrographs, the spalled Si(100) samples at low temperatures showed less rough areas, and could be processed further for PV applications. Clearly, it is advantageous to carry out spalling, particularly for Si(100), at 300 K. Figure 8 shows a free standing chip of Si(100) spalled at 77 K (Figure 8a), along with side views of the spalled layers (Figure 8b–d).

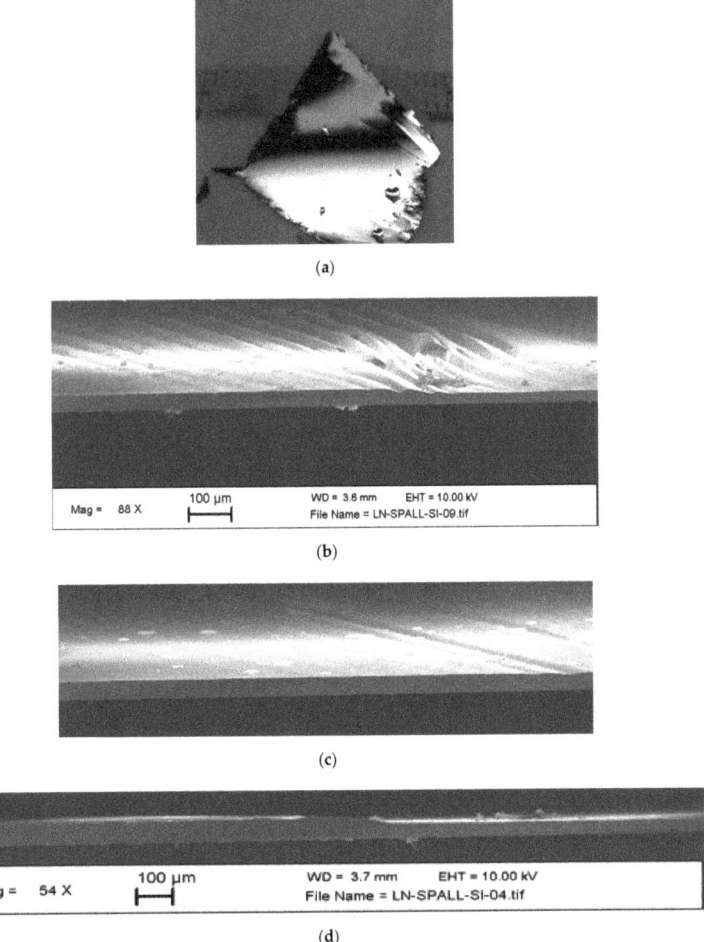

**Figure 8.** Cross-Section or side views of Si(100) spalled at 77 K. (**a**) Rough regions can be found. (**b**) Significant surface area is very smooth. (**c**) Spall material has thickness of 40–85 µm. (**d**) A broad area of the spalled layer of a length more than 1 mm shows the even spalled surface.

A detailed study of spalled Si(100) is presented in Figure 9. The thickness of the spalled Si(100) decreases from above 60 μm to below 50 μm, with the temperature increasing from 77 K to less than 200 K.

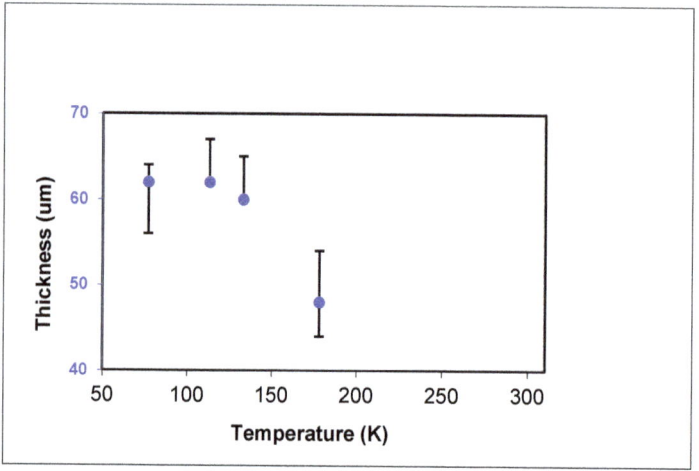

**Figure 9.** Temperature-dependence of spalling for Si(100). Spalled at various low temperature points. Data points represent thickness in smooth areas. Si error bars denote thickness variation in rough areas (see SEM images on Figure 8).

### 3.3. Applications on Spalling Technique

A stressor layer is formed atop a base substrate at a first temperature, which induces an initial tensile stress in the base substrate that is below its fracture toughness. The base substrate and stressor layer are then brought to a second temperature, which is lower than the first temperature. The second temperature induces a second tensile stress in the stressor layer which is greater than the first tensile stress, and which is sufficient to allow for spalling mode fracturing to occur within the base substrate. The base substrate is spalled at the second temperature to form a spalled material layer. Spalling occurs at a fracture depth, which is dependent upon the fracture toughness and stress level of the base substrate, and the second tensile stress of the stressor layer induced at the second temperature.

Spalling includes depositing a stressor layer on a substrate, placing an optional handle substrate on the stressor layer, and inducing a crack and its propagation below the substrate/stressor interface. This process, which is performed at room temperature, removes a thin layer of the base substrate below the stressor layer. By thin, it is meant that the layer thickness is typically less than 100 microns, with a layer thickness of less than 50 microns being more typical.

The ultimate goal of spalling is to produce thin films for advanced application of electronic device fabrications; Figure 10 describes the main steps involved in this process [14]. Si nanowires are an emerging PV technology [15,16]; nanoscale chemical templating (NCT) for the controlled growth of Si nanowires catalyzed by Al has shown good progress with regard to PV technology (see Figure 11), proving the principle of the NCT technique.

The fact that controlled spalling is able to remove layers of arbitrary size and shape allows one to design circuits and subsystems at the wafer scale and selectively remove them by selected deposition of the stressor layer on these regions [17,18]. The side view of previously grown nanowires [17,19], using controlled growth of Si nanowires, demonstrated a novel nanoscale chemical templating method, achieving controlled spatial placement of Si NWs by using patterned SiO$_2$ as a mask and Al as the seed material. The main advantage of this method lies in its suitability for the oxygen-reactive seed materials, which are of great interest for electronic applications. The NCT method can also have fewer steps compared to conventional patterning approaches, not requiring lift-off of a metal layer or removal of the mask. The method is also flexible, as it is amenable to both standard lithography techniques and self-assembled patterning techniques such as microsphere lithography. Patterning and growth parameters can be chosen to achieve high selectivity, growth yield, and fidelity; where no NWs grow between openings, most openings are occupied by one or more NWs and the majority are occupied by a single vertical NW. NCT will have several applications in nanotechnology research such as solar cells.

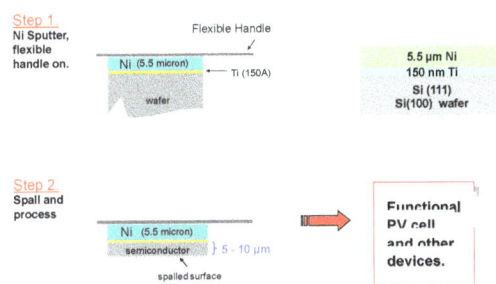

**Figure 10.** Handling and processing of spalled films for further applications such as solar cells (PV applications).

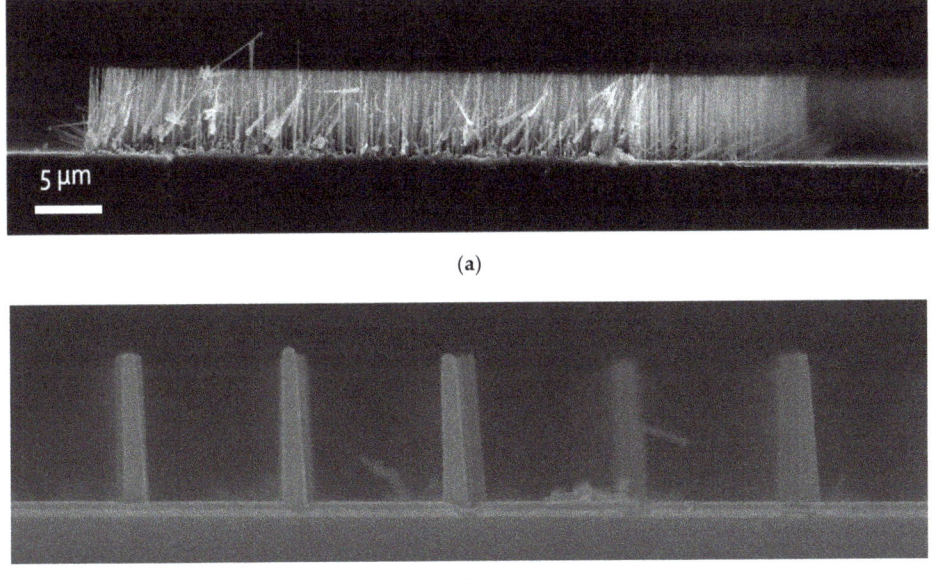

**Figure 11.** SEM Side views. (a) Nanoscale chemical templating (NCT), showing controlled growth of a group of Si nanowires. (b) Controlled growth of single Si nanowires at each opening; arrow 1 shows the thin oxidized layer of Al ($Al_2O_3$), and arrow 2 points to the planer growth of Si; the average nanowire length is 5 µm.

A promising field for future low-cost, medium-efficiency solar cell devices is the use of vapor–liquid–solid (VLS)-grown nanowires (NWs) as the active region of large scale (greater than 1 mm$^2$ area) photovoltaic devices. There are several advantages of using NWs. NWs can be doped as they are grown, helping with the formation of PV structures. NW-based PV structures require shorter carrier diffusion distances than are needed for a similarly thick planar absorber layer. At the same time, due to scattering and other optical phenomena, the NW structure is able to trap more light and improve overall light absorption. This, combined with the ability to grow nanowires on cheap substrates or reuse the growth substrate multiple times using spalling, makes NWs promising for future-generation PV devices.

## 4. Conclusions

Indeed, spalled single crystals are of immense technological importance. However, spalling Si(100) at room temperature exhibits several mechanical issues and corresponding technical limitations. It has been shown that when Si(100) single crystals are indented at liquid nitrogen temperature (77 K), there is no phase transformation. Owing to the presence of high shear stresses during indenter loading, most of the original Si remains unaltered in structure. It has been suggested that the difference in phase transition at the two temperatures can be explained by the following hypothesis: a temperature rise, which alongside high hydrostatic and shear stresses assists phase transition, occurs during indentation at room temperature.

Based on the fact that spalling of Si(100) at temperatures close to 77 K produces undeformed layers of relatively controlled thickness for further applications such as PV technologies, devices of flexible Si—if the spalled layer is thin enough—have many advantages over their bulk counterparts. By virtue of less material being used, thin-film devices ameliorate the material cost associated with device production and lower device weight, both of which are important in the semiconductor industries for a wide range of efficient thin materials applications—particularly, if this technology of bulk Si spalling is combined with controlled growth of Si-NWs, as has been proposed for PV devices. Furthermore, if a device layer is removed from a substrate that can be reused, additional fabrication cost reduction can be achieved.

**Funding:** This research has been carried out under the employment of KACST.

**Data Availability Statement:** The data used to support the findings of this study are included within the article.

**Acknowledgments:** The author would like to thank M. Munawar Chaudhri of Cavendish lab, University of Cambridge for his scientific insights throughout working on this project, and Gang Chen of T. J. Watson Research Center, IBM for valuable discussions.

**Conflicts of Interest:** The author declares that there is no conflict of interest regarding the publication of this paper.

## References

1. Hu, J.Z.; Merkle, L.D.; Menoni, C.S.; Spain, I.L. Crystal data for high-pressure phases of silicon. *Phys. Rev. B* **1986**, *34*, 4679. [CrossRef] [PubMed]
2. Crain, J.; Piltz, R.O.; Ackland, G.J.; Clark, S.J.; Payne, M.C.; Milman, V.; Lin, J.S.; Hatton, P.D.; Nam, Y.H. Erratum: Tetrahedral structures and phase transitions in III-V semiconductors [Phys. Rev. B 50, 8389 (1994)]. *Phys. Rev. B* **1995**, *52*, 16936. [CrossRef] [PubMed]
3. Kiran, M.S.R.N.; Tran, T.T.; Smillie, L.A.; Haberl, B.; Subianto, D.; Williams, J.S.; Bradby, J.E. Temperature-dependent mechanical deformation of silicon at the nanoscale: Phase transformation versus defect propagation. *J. Appl. Phys.* **2015**, *117*, 205901. [CrossRef]
4. Khayyat, M.M.; Hasko, D.G.; Chaudhri, M.M. Raman spectroscopy investigations and electrical characterisations of indentation induced phase transformations of Si. *Mater. Sci. Forum* **2005**, *480–481*, 225–230. [CrossRef]
5. Chaudhri, M.M.; Hasko, D.G.; Khayyat, M.M. Comment on "Phase transformations induced in relaxed amorphous silicon by indentation at room temperature". *Appl. Phys. Lett.* **2005**, *87*, 5559.
6. Khayyat, M.M.O.; Hasko, D.G.; Chaudhri, M.M. Effect of sample temperature on indentation-induced phase transitions in crystalline silicon. *J. Appl. Phys.* **2007**, *101*, 083515. [CrossRef]
7. Chaudhri, M.M.; Khayyat, M.; Hasko, D.G. Investigations of the indentation-induced crystallographic phase changes in silicon using raman spectroscopy. *Surf. Rev. Lett.* **2007**, *14*, 719–723. [CrossRef]
8. Hong, Y.; Zhang, N.; Xiong, L. Nanoscale plastic deformation mechanisms of single crystalline silicon under compression, tension and indentation. *J. Micromech. Mol. Phys.* **2016**, *1*, 1640007. [CrossRef]

9. Khayyat, M.M.; Banini, G.K.; Hasko, D.G.; Chaudhri, M.M. Raman microscopy investigations of structural phase transformations of crystalline and amorphous Si at room temperature and at 77 K. *J. Phys. D: Appl. Phys.* **2003**, *36*, 1300. [CrossRef]
10. Mujica, A.; Rubio, A.; Muñoz, A.; Needs, R.J. High-pressure phases of group-IV, III–V, and II–VI compounds. *Rev. Mod. Phys.* **2003**, *75*, 863–912. [CrossRef]
11. Ge, D.; Domnich, V.; Gogotsi, Y. Thermal stability of metastable silicon phases produced by nanoindentation. *J. Appl. Phys.* **2004**, *95*, 2725–2731. [CrossRef]
12. Chaudhri, M.M. Symposium (international) on detonation. In *Proceedings of the Ninth Symposium (International) on Detonation, OCNR 113291-7, Portland, OR, USA, 28 August–1 September 1989*; Office of Chief of Naval Research: Arlington, VA, USA, 1989; p. 857.
13. Rabier, J.; Demenet, J.L. Low temperature, high stress plastic deformation of semiconductors: The silicon case. *Phys. Status Solidi B* **2000**, *222*, 63–74. [CrossRef]
14. Cottrell, A.H. *The Mechanical Properties of Matter*; Wiley: New York, NY, USA, 1964; p. 5.
15. Zhang, Y.; Zhao, Y.P. Applicability range of Stoney's formula and modified formulas for a film/substrate bilayer. *J. Appl. Phys.* **2006**, *99*, 053513. [CrossRef]
16. Khayyat, M.M.; Cortes, N.E.S.; Saenger, K.L.; Bedell, S.W.; Sadana, D.K. Low-Temperature Methods for Spontaneous Material Spalling. USPTO Application Number 13/150,813, 1 June 2011.
17. Khayyat, M.M.; Cortes, N.E.S.; Bedell, S.W.; Fogel, K.E.; Sadana, D.K. Temperature-Controlled Depth of Release Layer. USPTO Application Number 13/448,939, 17 April 2012.
18. Shahrjerdi, D.; Bedell, S.W. Extremely flexible nanoscale ultrathin body silicon integrated circuits on plastic. *Nano Lett.* **2013**, *13*, 315–320. [CrossRef] [PubMed]
19. Wacaser, B.; Khayyat, M.; Reuter, M.C.; Sadana, D.K.; Ross, F.M. Technical advantages and challenges for core-shell micro/nanowire large area PV devices. In Proceedings of the 2010 35th IEEE Photovoltaic Specialists Conference, Honolulu, HI, USA, 20–25 June 2010.

# Article
# Development of Red Clay Ultrafiltration Membranes for Oil-Water Separation

Saad A. Aljlil

National Center for Water Treatment and Desalination Technology, King Abdulaziz City for Science and Technology, P.O. Box 6086, Riyadh 11442, Saudi Arabia; saljlil@kacst.edu.sa

**Abstract:** In this study, a red clay/nano-activated carbon membrane was investigated for the removal of oil from industrial wastewater. The sintering temperature was minimized using $CaF_2$ powder as a binder. The fabricated membrane was characterized by its mechanical properties, average pore size, and hydrophilicity. A contact angle of 67.3° and membrane spore size of 95.46 nm were obtained. The prepared membrane was tested by a cross-flow filtration process using an oil-water emulsion, and showed a promising permeate flux and oil rejection results. During the separation of oil from water, the flux increased from 191.38 to 284.99 L/m² on increasing the applied pressure from 3 to 6 bar. In addition, high water permeability was obtained for the fabricated membrane at low operating pressure. However, the membrane flux decreased from 490.28 to 367.32 L/m²·h due to oil deposition on the membrane surface; regardless, the maximum oil rejection was 99.96% at an oil concentration of 80 NTU and a pressure of 5 bar. The fabricated membrane was negatively charged, as were the oil droplets, thereby facilitating membrane purification through backwashing. The obtained ceramic membrane functioned well as a hydrophilic membrane and showed potential for use in oil wastewater treatment.

**Keywords:** ultrafiltration; red clay; calcium fluoride powder; wastewater; oil separation

## 1. Introduction

The daily generation of approximately 210 million barrels of water contaminated with oil can incur costs of $45 B in water purification [1]. Recently the membrane technology has been applied to separate mixtures of oil and water, showing greater efficacy than traditional technologies [2,3]. It is well-known that membranes synthesized from polymers are unstable compared to ceramic membranes [4]. Ceramic membranes are often used in industrial applications, such as wastewater treatment, despite the high cost compared to those of polymeric membranes [5], owing to their extraordinary chemical, mechanical, and thermal properties [6]. They have the capability of backwashing, high flux, good toughness, resistance to bacterial growth, and thermal stability [4,7]. Materials like alumina, zirconia, titanium oxide and zeolite materials have been used in ceramic membranes that resist high pH and pressure to separate oil-water mixtures [4,8].

Additionally, microfiltration (MF) and ultrafiltration (UF) have been used as pressure-driven membrane methods [3]. UF is recognized as the most efficient in oil-water separation with significant advantages compared to conventional separation techniques: it requires no additional chemicals and its energy consumption is low [3]. High-flux MF membranes have also been used for oil-water separation; however, they carry the risk of oil penetration [3]. NaA zeolite was deposited on an α-$Al_2O_3$ MF membrane by Cui et al. [9] and used in the separation of oil from water. The fabricated membrane had a pore diameter of 1.2 μm and showed 99% oil separation at a flow rate of 85 L/m²·h and pressure of 50 kPa. A porous MF aluminum ceramic membrane was applied by Liu et al. for oil-water separation [10]; they reported 99.98% removal of emulsified oil.

A UF membrane was used to separate oil from water in an oilfield [3] with more than 96% oil rejection. Depositing $TiO_2$ on the surface of a UF $ZrO_2$ membrane to separate oil and water [11] revealed that the cohesion between the oil droplets and the membrane surface decreased. This was because less oil was adsorbed on the membrane surface, thus decreasing surface fouling. UF ceramic membranes were used [12] to separate an oil-water–anionic surfactant emulsion to investigate the effect of pH and flow rate. They reported an occurrence of concentration polarization phenomena at low flow rates. In addition, the membrane surface was positively charged at low pH and attracted the anionic surfactant in the oil-water emulsion, causing a reduction in the water flux rate. Luo et al. [13] used a UF hollow-fiber membrane for oil rejection from water, which yielded highly pure water with very low turbidity. Tubular UF membranes with a pore size of 10 nm were fabricated [14] and showed a flux rate of 200 L/m$^2$·h for wastewater purification. Issaoui and Lionel [15] have reported the fabrication of commercialized ceramic membranes based on expensive materials such as cordierite, titania, zirconia, and silicon carbide for various industrial applications.

Despite their several advantages and promising results, the high cost of ceramic components in manufacturing is a major disadvantage of ceramic membranes [4]. Therefore, experts have focused on developing innovative, low-cost, effective, and secure materials that can be used to prepare ceramic membranes, while still exhibiting the qualities required for wastewater purification, such as chemical stability, fouling resistance, and good mechanical attributes [2,3]. Efforts have been dedicated toward utilizing low-cost materials such as natural clay, apatite powder, dolomite, kaolin, bauxite, and mineral coal fly ash for the fabrication of ceramic membranes exhibiting good performance for wastewater and pollutant treatment. An MF tubular ceramic membrane was fabricated from low-cost clay and kaolin [16] to separate oil-water mixtures and showed 93% rejection of the oil emulsion. Zhu et al. [17] fabricated a ceramic membrane from fly ash and $TiO_2$ that showed 97% oil rejection. Liu et al [18] fabricated a bentonite clay membrane for use in the separation of oil from saltwater, but it was unsuitable for the high-salinity environment. Furthermore, kaolin, quartz, feldspar, sodium carbonate as low-cost MF membrane materials were tested by Nandi et al. [19]. They noted that the fabricated membranes showed good oil removal efficiency over a 60-min experiment, with 98.8% oil rejection at the flux of $5.36 \times 10^{-6}$ m$^3$/m$^2$·s and applied pressures in the range 68.95–275.8 kPa. Kakali and Pugazhenthi [20] focused on using lithium aluminosilicate for the fabrication of a ceramic membrane via the slip-casting method, using starch as the pore-forming material for removing bacteria and oil from wastewater; they achieved good performance in terms of removal of bacteria and oil from wastewater. Ben Amar and Oun [21] used Tunisian mud as a low-cost material for the fabrication of tubular ceramic ultrafiltration membranes for removing pollutants from wastewater. These membranes were obtained via the slip-casting technique followed by sintering at 650 °C. With a pore size of 11 nm, the membranes demonstrated a permeability of 90 L/h·m$^2$·bar and oil pollutant removal of 90%. Hubadillah et al. [22] used kaolin as a low-cost ceramic raw material for the fabrication of a ceramic hollow fiber membrane and achieved improved mechanical strength and excellent performance in terms of pollutant removal from wastewater.

Our goal is to fabricate a UF ceramic membrane for oil removal from water using red clay as a low-cost material, nano-activated carbon powder for pore-forming, and $CaF_2$ as a binder and nucleating agent to minimize the sintering temperature. The aforementioned low-cost materials were processed using an extrusion method, and the sintering was performed by carbon pyrolysis. To the best of our knowledge, all prior researchers have fabricated MF membranes as substrates and used interlayers and filtration layers to obtain UF membranes. Additionally, the use of nano-activated carbon as a pore-forming material and $CaF_2$ as a binder, rather than the conventional carboxymethyl cellulose is also novel. This study demonstrates the effective preparation of a novel tubular ceramic membrane for oil-water separation by using low-cost and locally sourced materials like red clay.

## 2. Methodology

### 2.1. Raw Materials

Table 1 shows that the red clay used in fabricating the membrane contains higher amounts of $SiO_2$ (47.63%), $Al_2O_3$ (24.03%), and $Fe_2O_3$ (9.57%) with low amounts of $K_2O$ (2.08%) and $Na_2O$ (1.29%). The red clay was collected from the Biadh plant in Riyadh, Saudi Arabia. The role of $CaF_2$ (Sigma Aldrich, St. Louis, MO, USA) is to minimize the sintering temperature of the red clay membrane and improve the mechanical strength.

**Table 1.** Chemical composition of Saudi red clay (wt%).

| Oxides | $SiO_2$ | $Al_2O_3$ | $Fe_2O_3$ | $Na_2O$ | $K_2O$ | CaO | $TiO_2$ | Loss on Ignition |
|---|---|---|---|---|---|---|---|---|
| % | 47.63 | 24.03 | 9.57 | 0.23 | 2.08 | 1.29 | 1.28 | 11.55 |

Activated carbon (diameter = 4 mm; purity 98%) was obtained from Zhengzhou Company (Henan, China). Powdered activated carbon (65 µm) was obtained using a planetary ball mill at 300 rpm for 4 h, and was blended with water for 72 h and then treated ultrasonically (50 min, 540 W) using an SFX550 (Sonifier, Suwanee, GA, USA). Then, the collected suspension was centrifuged at 3500 rpm for 15 min to yield nano-activated carbon of size 91.6 nm (SEM, Figure 1). It is worth mentioning that activated carbon was converted to nanoactivated carbon for the sole purpose of a pore-forming material, i.e., to create nanopores in the body of the ceramic membrane. After sintering at temperatures above 500 °C, all of the activated carbon was burned from the ceramic membrane, which resulted in pore formation in the ceramic membrane (TGA, Figure 2).

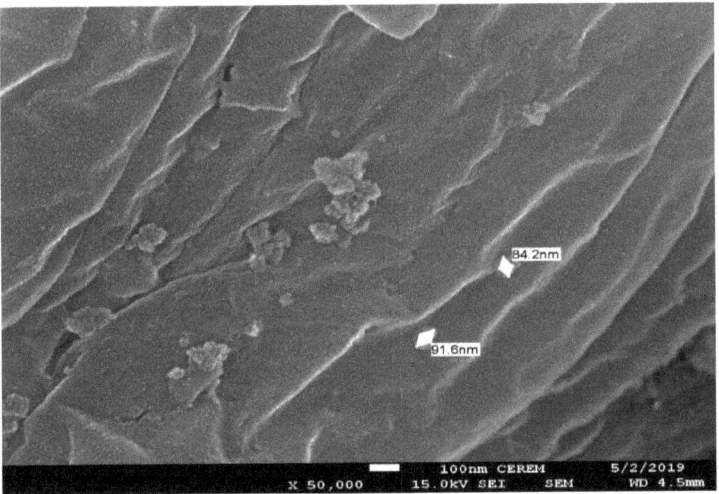

**Figure 1.** SEM of the nano-activated carbon prepared with 50000× magnification.

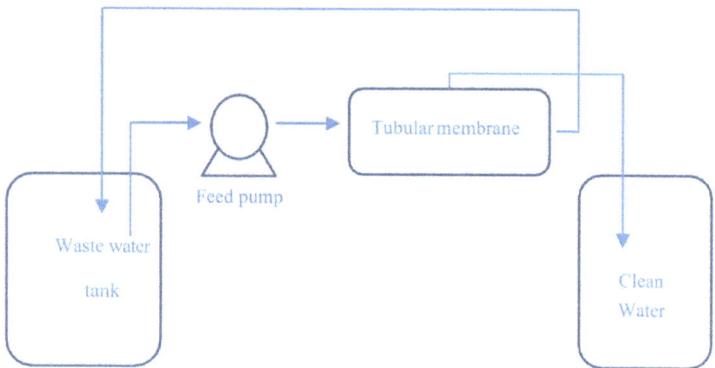

**Figure 2.** Sketch the experimental filtration process.

## 2.2. Membrane Fabrication

The fabricated raw material for the membrane comprised of a powder blend of 90 wt.% red clay, 5 wt.% $CaF_2$ as a binder, and 5 wt.% nano-activated carbon powder as a pore former. The red clay was ground at 250 rpm for 3 h using a planetary ball mill to achieve the particle size of 100 μm.

Az-mixer was used to combine the dry raw materials for 4 h. To this mixture, 400 mL water was slowly added. Through wet mixing, we obtained a paste with satisfactory plasticity, which was then fed into the extruder (Length = 200 cm, Width = 50 cm, Die diameter = 20 cm). Two sintering stages were applied: first, sintering was performed from 30 to 500 °C at an average heating rate of 1.5 °C/min to burn the organic material, thereby creating pores in the prepared membrane. Figure 2 presents the TGA data in an air of the nano-activated carbon powder, clarifying the sintering operation. It is clear from Figure 2 that complete burning of the nano-activated carbon powder was obtained at 450 °C. Subsequently, the produced membrane was densified by sintering in a furnace at temperatures from 400 °C to 1000 °C at a rate of 2 °C/min for 4 h.

## 2.3. Characterization of Ceramic Membrane

### 2.3.1. Scanning Electron Microscopy (SEM)

Scanning electron microscopy (SEM) (model NNL-200, Philips, 1-nm resolution) was used for morphological characterization and microstructural analysis of the fabricated sintered membrane. Samples smaller than 10 mm were obtained by cutting the sintered membrane samples by a cutting machine with a diamond cutting disc. The membrane sample was dried for 24 h in a vacuum oven, and then etched by 1% HF + 1% $HNO_3$ solution for 30 s. Finally, the membrane sample was coated by a thin gold layer using a sputter coater (SPI Inc., Lakewood, WA, USA) to increase the conductivity of the membrane and to obtain a clear image. SEM images of the membrane sample were obtained by scanning it with a focused beam of electrons, which interact with the electrons in the membrane sample, producing various detectable signals containing information about the sample's surface topography.

### 2.3.2. Apparent Porosity

The apparent porosity of the ceramic structures was determined by the standard test method (ISO EN 993-1) for ceramic structures using the Archimedes buoyancy technique with dry weights, soaked weights, and immersed weights in water. The membrane sample was dried in an oven at 105 °C for 24 h to eliminate the absorbed water. The dried membrane sample was weighed by the balance and the weight is recorded as $M_d$. Then the membrane sample was placed in a water-filled container for 24 h at room temperature. After that, the membrane sample has weighed and the weight recorded as $M_w$. In addition, the membrane sample was weighed inside water and the weight recorded as $M_a$ (suspension weight). Finally, the apparent porosity can be calculated from equation 1:

$$Apparen\ porosity = 100 \times \left(\frac{M_w - M_d}{M_w - M_a}\right) \quad (1)$$

### 2.3.3. Contact Angle Measurements

The capillary force liquid weight gain, which occurs when wetting the membrane sample, was precisely measured using a K100 force tensiometer (Kruss, Wissenschaftliche Laborgeräte, Borsteler Chaussee 85, Germany) with a particularly high resolution to obtain reliable and accurate contact angle (θ) data of the membrane sample. Before measuring, the membrane sample was dried in a vacuum oven for 24 h at 100 °C. The main procedure to measure the membrane contact angle using Kruss K100 involves sealing off the open ends of the tubular membrane sample with epoxy resins, hanging the sample on the microbalance in the K100 force tensiometer (Kruss, Wissenschaftliche Laborgeräte, Borsteler Chaussee 85, Germany)), and immersing the sample gradually into deionized water. The rate of immersion has to be adjusted to ~6 mm/min. Finally, the contact angle will be calculated from the forces acting on the membrane surface.

### 2.3.4. Mechanical Test

The mechanical properties of the ceramic membrane were determined by the three-point bending strength test using a Shimadzu-Universal testing machine (AGS-X, Riverwood Drive Columbia, MD 21046, USA) with a capacity of 5 kN, a total grip distance of 690 mm, a crosshead speed of 0.5 mm/min, a potential of 200 V, and power of 60 Hz. The stress-strain relationship of the membrane was found to be linear. Membrane samples with a length of at least 4 cm were obtained by cutting the samples by a cutting machine with an artificial diamond disc, the result is obtained by using the equation

$$\delta_f = \frac{FL}{\pi(d_2 - d_1)^3} \quad (2)$$

where "$\delta_f$" is the bending strength "F" is the flexural load in newton and "L", "$d_2$" and "$d_1$" is the span length, outer diameter, and inner diameter respectively.

### 2.3.5. X-Ray Diffraction

The X-ray diffraction technique (diffractometer used: model D8AD VANCE, BRUKER, Billerica, MA, USA) was used to identify the crystalline phases of the membrane sample. In this technique, the scattered intensity of an X-ray beam, generated upon hitting the membrane sample, is measured as a function of incident angle. The membrane sample for XRD analysis was dried for 24 h in a vacuum oven at 105 °C to eliminate any moisture present in the material. Then, the membrane sample was powdered and spread on the glass holder with a gap of 0.5 mm. The holder with the sample was then placed in the X-ray chamber and scanned at a constant temperature and a speed of 2°/min using CuKα radiation, over a diffraction angle (2θ) range from 10° to 80°, with a step size of 10°. The Joint Committee on Powder Diffraction Standards (JCPDS) diffraction file cards (2001) are used as reference for interpretation of the X-ray patterns obtained in the experiment.

### 2.3.6. Pore size Distribution Measurements

The pore size distribution of the membrane was determined using a constant-pressure fluid-fluid porometer (IFTS advanced fluid-fluid porometer, Institut de la Filtration et des Techniques Séparatives, Rue Marcel Pagnol, Foulayronnes, France).

### 2.4. Oil Emulsion Characterization

The performance of the prepared ceramic membrane was characterized by implementing it in the separation of oil from an oil-water mixture, where ultra-pure paraffin oil was used as a synthetic oil in the absence of a surfactant. The emulsion was vigorously and continuously mixed by using an agitator for 50 min at 1350 rpm and remained stable for several days unaffected by gravitational forces. The initial concentration of the oil emulsion was measured in terms of turbidity at 64 NTU. In addition, an oil-water mixture was obtained from an Aramco oilfield. Furthermore, a Zetasizer was used to obtain the zeta potential curve and to determine the oil droplet charge. To define the efficiency of the prepared membrane in oil-water separation, a turbidity meter was used to determine the oil turbidity.

*2.5. Ultrafiltration Testing*

Figure 2 shows the cross-flow filtration set-up, which consists of a feed tank, pump, pressure gauges, agitator, and a tubular-type ceramic membrane. In the flow circuit, feeds with a volumetric flow rate of 55 L/h at 25 °C and a specific pressure of a known and constant composition (water and oil emulsion) was pumped continuously through the cross-flow ultra-filtration membrane (diameter = 0.5 cm, length = 20 cm) at a specified cross flow pressure of 5 bar. In addition, a secondary agitator was used to provide mixing effects and to ensure emulsion stability. A water tank was used to collect the permeated water, and the weight was determined using a balance to calculate the flux rate of the clean water.

The water flux, $J$ (L/m$^2$·h), was determined from Equation (2):

$$J = V/A * t \qquad (3)$$

where V is the water volume collected through the pores of the membrane, A is the membrane surface area, and t is the time.

The turbidities of the water permeate and water in the feed tank were measured using a turbidimeter; the oil rejection (R%) was obtained from Equation (3):

$$R\% = [1 - (C_c/C_i)] \times 100, \qquad (4)$$

where $C_i$ is the raw oil turbidity and $C_c$ is the turbidity after filtration.

The membrane permeability ($P$) was obtained by Equation (4):

$$J = K_c * \Delta P \qquad (5)$$

where $K_c$ is the membrane permeability and $\Delta P$ is the applied pressure.

## 3. Results and Discussion

*3.1. Characterizations*

The zeta potential curves of the oil emulsion and membrane are presented in Figure 3. The isoelectric point of the oil droplets emulsion appears at pH 1, where the zeta potential was negative. The isoelectric point of the membrane was located at a slightly higher pH of 1.7, where the zeta potential was also negative. Since identical charges are known to experience electrostatic repulsion, it can be predicted that our fabricated membrane would prevent fouling during backwashing because of the electrostatic repulsion between the oil emulsion and the membrane.

Mechanical testing of the ceramic membranes with and without CaF$_2$ was performed using the three-point bending technique with a crosshead speed of 0.5 mm/min. The membrane without CaF$_2$ showed a lower bending strength of 49.53 MPa than the membrane with CaF$_2$ (54.13 MPa); the stress-strain relationship of both the membranes was linear. Further, a contact angle of 67.3° indicates that the membrane is hydrophilic. The membrane pore size distribution (Figure 4) indicated that a UF membrane was created without using a coating layer. The pore size of this membrane ranged from 40 nm to 110 nm with an average pore size of 96 nm, and 88% of the total pores were smaller than 96 nm. In addition, the measured porosity of the membrane was 32.56%. SEM was used to determine the morphology of the fabricated membrane (Figure 5). It was noted that the absence of cracks in the fabricated membrane was indicative of its high quality and good material properties in agreement with the results of the three-point bending strength test.

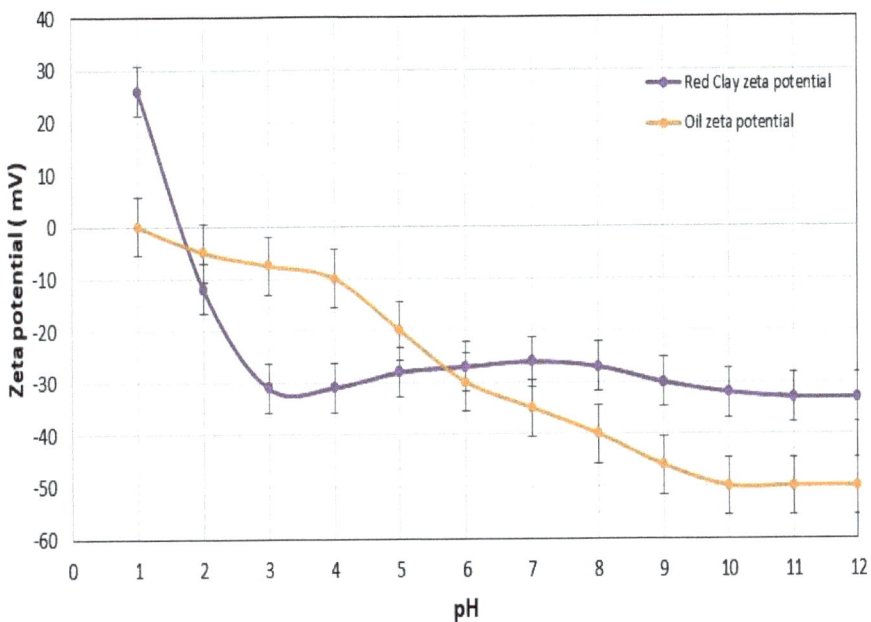

Figure 3. Zeta potential of oil droplets and the red clay membrane as a function of pH.

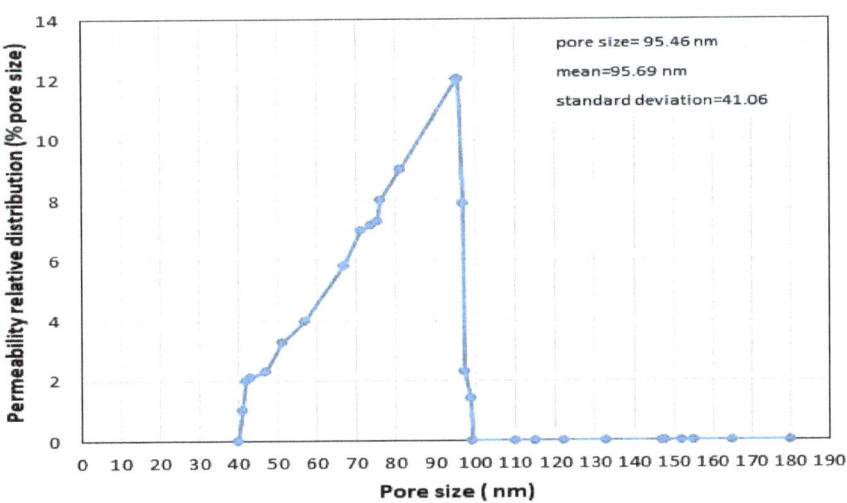

Figure 4. The pore size distribution of the fabricated membrane.

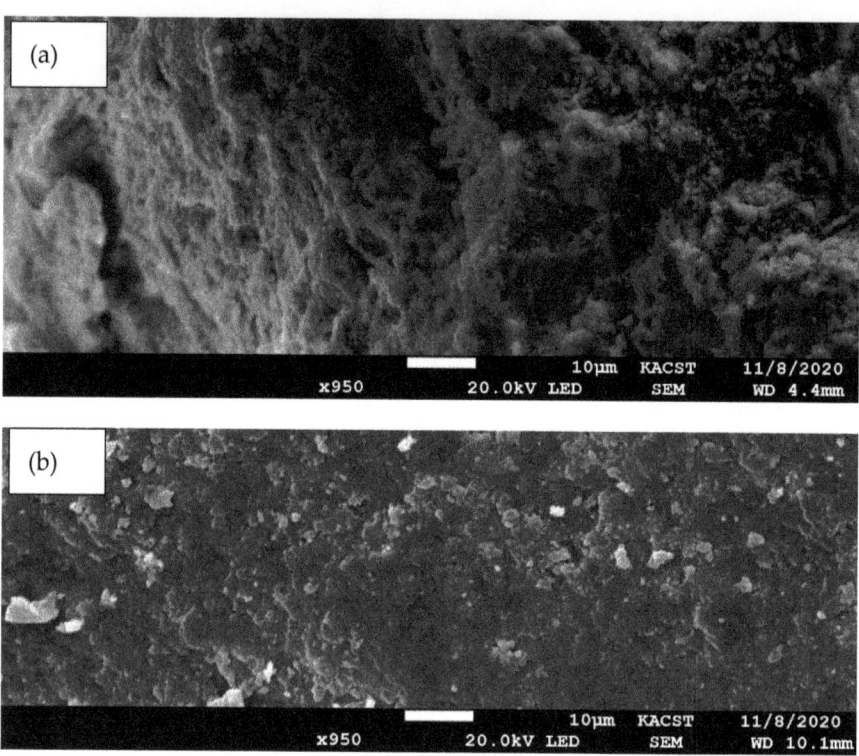

**Figure 5.** SEM of the tubular ceramic membrane: (**a**) cross-section of the membrane and (**b**) the top view of the membrane.

Figure 6 also presents the TGA data for the red clay paste, with a residual weight of 85 wt.% at 1000 °C. The holding time of the sintering procedure after treatment at 1000 °C was 2 h. $CaF_2$ as a binder and nucleating agent to minimize the sintering temperature was used. The sintering temperature of the ceramic membranes with and without $CaF_2$ was performed. The membrane with $CaF_2$ showed a lower sintering temperature of 1000 °C than the membrane without $CaF_2$ (1150 °C). The TGA data for $CaF_2$ (Figure 2) showed that the residual weight was ~94 wt.%. Crystal water decomposition was found to occur at 500–600 °C, and the standard decomposition or recrystallization possible during heat treatment occurs at 600–1100 °C.

The XRD pattern of the fabricated membrane (Figure 7), which contains red clay and $CaF_2$, shows that red clay is a highly illitic kaolinite-type clay. It includes illite, kaolinite, and hematite, which gives it a red color. Additionally, the clay contains some amount of free quartz. $CaF_2$ could be seen in the XRD.

Sintering at 1000 °C led to the decomposition of kaolinite and illite (clay minerals). Some amount of $CaF_2$ was also decomposed, while the remaining was observed in the membrane structure by XRD analysis. The calcium released from the $CaF_2$ decomposition and the aluminum silicate from the clay minerals were reacted to create an anorthite phase. During sintering, the free quartz and hematite remained stable, as evidenced by the XRD pattern of the sintered membrane. Most of the decomposed kaolinite was transformed to mullite. In addition, large amounts of amorphous phase were present in the sintered membrane, which was determined by the broad peak between $2\theta = 20°$ and $2\theta = 40°$.

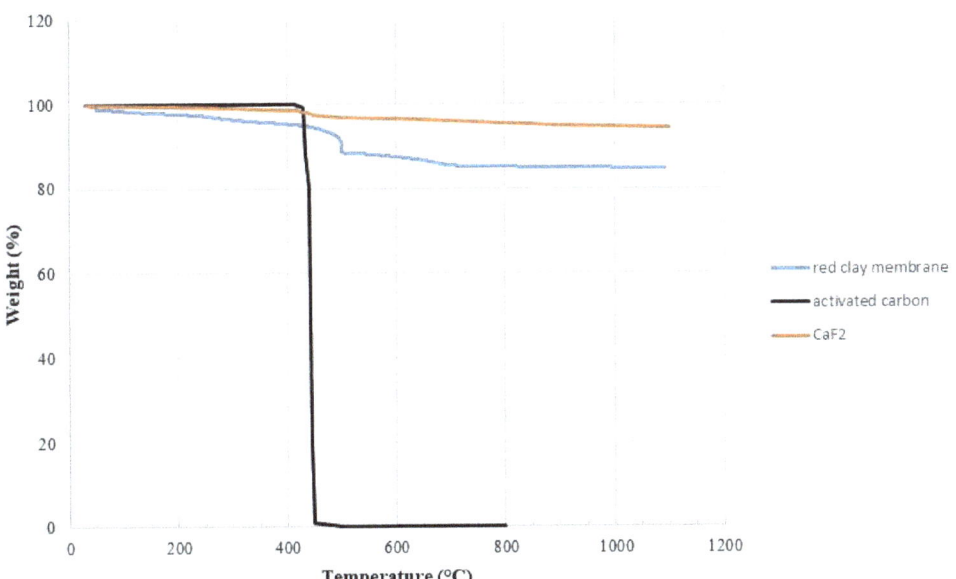

**Figure 6.** Thermogravimetric analysis of red clay membrane, activated carbon, and CaF$_2$ under air.

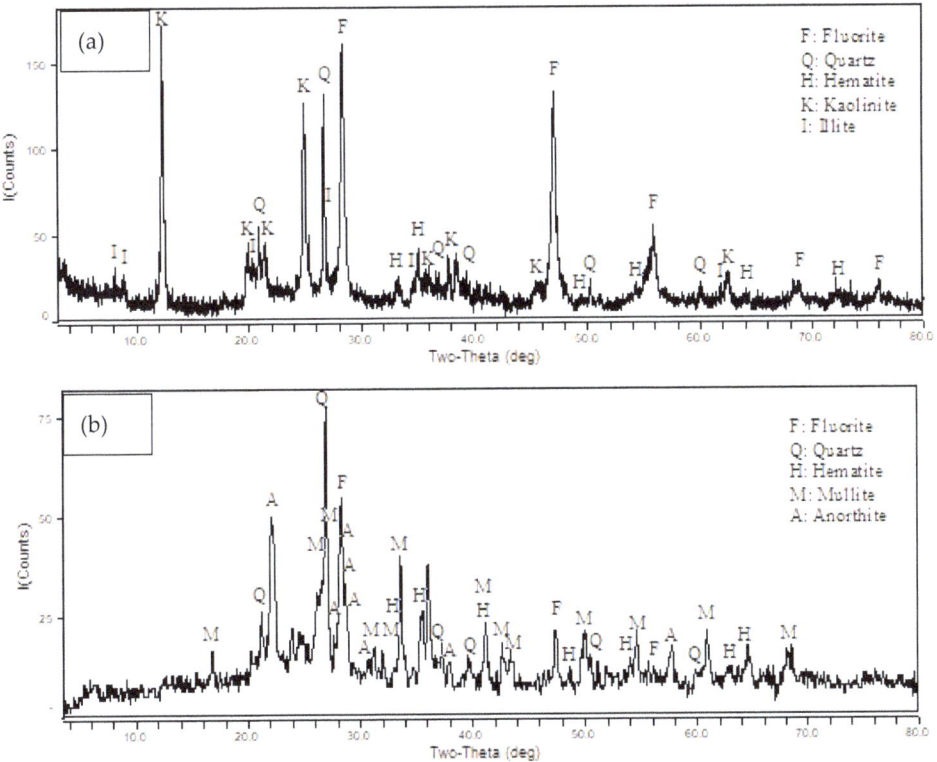

**Figure 7.** XRD for fabricated membrane: (**a**) without sintering operation and (**b**) with sintering operation at 1000 °C.

## 3.2. Evaluation of Fabricated Membrane

### 3.2.1. Evaluation with Synthetic Oil Emulsion

The water flux was obtained and the water permeate was collected for 4 h at different operating pressures (3, 4, 5, and 6 bar, corresponding to 300, 400, 500, and 600 kPa) and using ultra-pure paraffin oil as a feed at 64 NTU concentration. The water flux rate was calculated from Equation (1) and the membrane permeability was determined from Equation (3) (Figure 8).

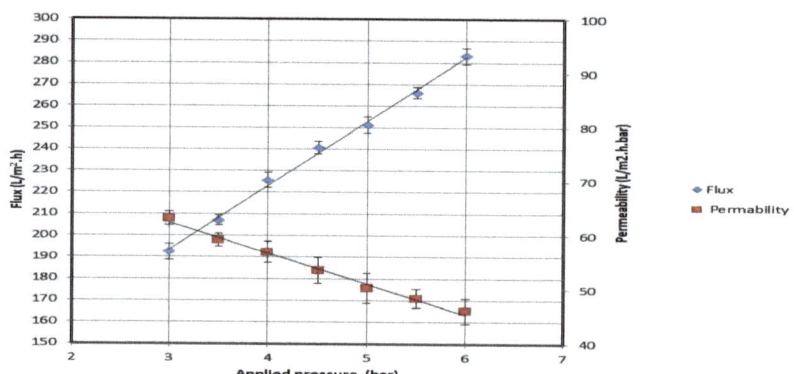

**Figure 8.** Water flux rate and permeability of fabricated membrane at different pressures.

As shown in Figure 8, the flux changed from 191.38 to 284.99 L/m$^2$·h on increasing the applied pressure from 3 to 6 bar. The data in Figure 9 were fitted with Darcy's law [23] and the membrane permeabilities were obtained. High water permeability was observed for the fabricated membrane at low operating pressures. The obtained results were comparable to those from the RO process [24]. Based on Figure 8, the standard deviation (5.36) is lower than the mean (179.84), indicating that the data is reliable. In addition, a 90% confidence level is 185.10–174.58, i.e., we are 90% certain that the mean lies between 185.10 and 174.58 with a small margin of error.

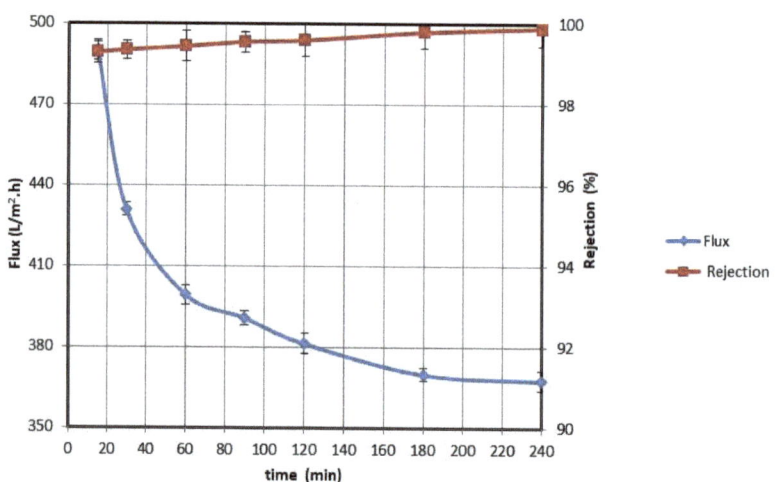

**Figure 9.** Water/synthetic oil flux rate through membrane and percentage rejection by a fabricated membrane at different times.

The oil emulsion was tested at 5 bar (500 kPa) to measure the flux over time using a feed concentration of 64 NTU. Based on the standards of the industry, UF can be conducted in the range 4–7 bar (400–700 kPa) [25]. Hence, to determine a perfect water flux, the process was run at 5 bar (500 kPa). The water flux rate was calculated from equation 1 and the rejection was determined from equation 2 (Figure 9).

Figure 9 shows that the water flux rate decreased over 4 h from 490.28 to 367.32 L/m²·h. This behavior could be due to a considerable quantity of oil deposited on the membrane over the course of testing, resulting in a decreased flux through membrane fouling. Nandi et al. [26] observed similar results warranting an efficient cleaning procedure to remove foulants from the membrane.

The standard deviation (3.65) is lower than the mean (404.25), indicating that the data is reliable. In addition, the 90% confidence level is 404.26–404.24, i.e., we are 90% certain that the mean is between 404.26 and 404.24 with a small margin of error.

3.2.2. Evaluation with Aramco Oil-Contaminated Water

Contaminated water obtained from Aramco was tested at 5 bar (500 kPa) to measure the flux over time using a feed concentration of 80 NTU. The water flux rate was calculated from Equation (1) and the rejection was determined from Equation (2) (Figure 10).

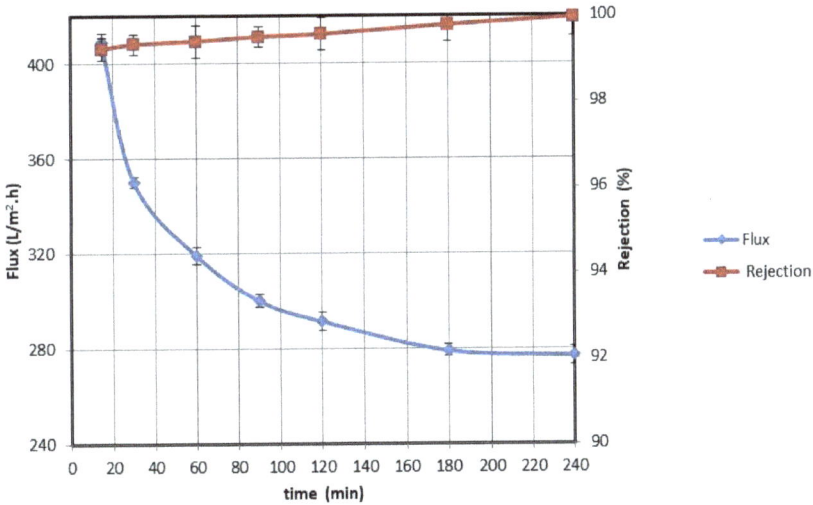

**Figure 10.** Flux rate of a real oil-contaminated water sample through the fabricated membrane and percentage rejection at different times.

It was observed that the water flux decreased because of oil deposition on the membrane surface causing membrane fouling, while the oil rejection changed from 99.23 to 99.96%. The decrease in the water flux was related to oil precipitation on the membrane. Based on Figure 10, the standard deviation (3.64) is lower than the mean (318.19) indicating that the data is reliable; the 90% confidence level is 318.19–318.15.

*3.3. Cleaning Mechanism of a Fouled Membrane and Cyclic Filtering Test*

Membrane fouling has detrimental effects on membrane performance. During oil-water separation, fouling occurs due to the interaction between the membrane and oil droplets in the wastewater; the cohesion between the foulant and membrane surface depends on membrane surface properties, such as its zeta potential and hydrophilicity [23]. Here, the fouling problem on the membrane was investigated via a cyclic filtrating test performed for 1 h, and the flux rate was determined. Next, backwashing was performed for 10 min to clean foulants from the membrane by pushing water mixed with air into the membrane. Then, the water flux was recalculated, and a total of seven experimental cycles were performed (Figure 11).

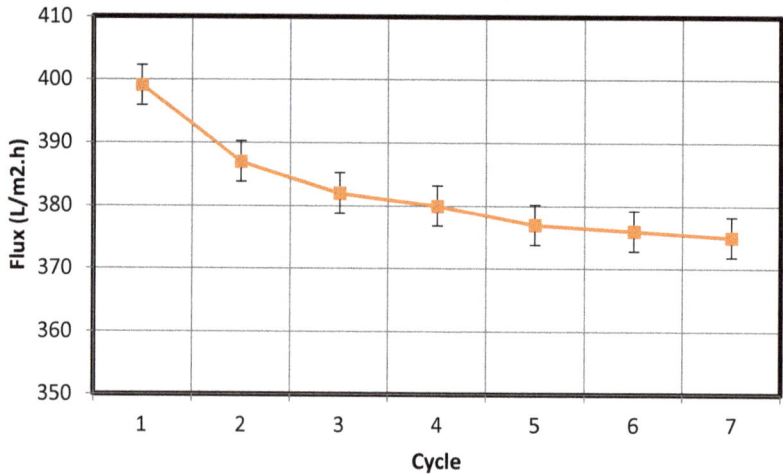

**Figure 11.** The water flux of seven experimental cycles filtration for oil separation by fabricated membrane.

Performing the filtration periodically can prevent the tendency of fouling and maintain a nearly constant flux rate value owing to the repeated backwash purification of the membrane. The plot of the water flux rate as a function of the number of cycles (Figure 11) revealed that the membrane retained a significant negative surface charge, as shown in the zeta potential plot of the fabricated membrane (Figure 3), resulting in an almost constant water flux. This also allowed for the membrane to be efficiently cleaned. It is possible to purify and reform the negatively charged membrane during backwashing because the oil droplets are also negatively charged (Figure 3), yielding suitable water flux with sufficient cycling. The backwashing potential arising from the repulsion between the negatively charged oil droplets and the membrane surface renders the developed membrane suitable for use in oil-water separation.

Figure 11 shows that the standard deviation (6.24) is lower than the mean (287.89); therefore, the data is reliable. In addition, the 90% confidence level is 296.32–279.45.

The effectiveness of the fabricated membrane was similar to those of membranes described in the literature [9,27–30], and some comparisons are presented in Table 2. It is evident that the fabricated membrane presented favorable performance with red clay, which is locally available, and an inexpensive material. Table 2 also presents data for some expensive membrane materials, such as NaA zeolite deposited on $\alpha$-$Al_2O_3$. Additionally, our tubular ceramic membrane has a higher water flux (367.32 $L/m^2$ h) than the tubular PVDF-UF at the same operating pressure (5–6 bar, 309 $L/m^2$ h).

Table 2. Performance comparison of the fabricated membrane in this study with membranes investigated in the literature.

| Ref | Membrane Type | Solution | Pressure (bar) | Permeation Flux (L/m² h) |
|---|---|---|---|---|
| [27] | cellulose microfiltration membranes | Synthetic produced water | 3 | 200 |
| [28] | PAN nanofiber membrane | Synthetic produced water | 0.1 | 810 |
| [29] | a-$Al_2O_3$ ceramic membrane | Synthetic produced water | 1.37 | 66 |
| [30] | Magnesium bentonite hollow fiber ceramic membrane | Synthetic produced water | 1 | 224 |
| [31] | Tubular UF module equipped with polyvinylidene fluoride and inorganic nano-sized $Al_2O_3$ | Oilfield | 1 | 170 |
| [9] | NaA zeolite/α-$Al_2O_3$ tubular ceramic membrane | Synthetic produced water | 0.5 | 85 |
| This work | Red clay tubular ceramic membrane | Synthetic produced water | 5 | 367.32 |

The practical implication of the membrane fabricated in this study is the clean, oil-free water, which can be used in agriculture and cooling systems. The scientific contribution of this study is the use of new materials, such as $CaF_2$ and nano-activated carbon, to fabricate a ceramic membrane from a low-cost material like red clay. Our working hypotheses of using low-cost materials for ceramic membrane fabrication were confirmed by the obtained results. However, the membrane fabricated in this study is only limited to separating oil and water. To expand its for use in industry, its performance must be investigated regarding different types of industrial pollutants. In addition, irreversible fouling also requires further research.

## 4. Conclusions

An efficient new membrane was fabricated from red clay combined with $CaF_2$ as a binder and nano-activated carbon as a pore former. These materials were used to fabricate a porous membrane by the extrusion technique, and the membrane was applied for the purification of oil-contaminated water. $CaF_2$ was used to minimize the sintering temperature of the red clay membrane and increase the mechanical strength of the membrane as a nucleation promoter.

The fabricated membrane was tested using both, a synthetic oil-water emulsion and water produced from an oilfield from Aramco. The fabricated membrane had an average pore size of 95.46 nm; thus, it qualified as a UF membrane. It showed a good bending strength of 54.13 MPa and a contact angle of 67.3°, indicating hydrophilicity. The performance of the fabricated membrane complied with the standards of the national wastewater. The clean water also met the desired standards.

In the separation of oil from water, the flux was increased upon increasing the applied pressure. High water permeability was obtained for the fabricated membrane under low operating pressure, and this result was fitted with Darcy's law. The membrane flux decreased by oil deposition on the membrane surface; regardless, the maximum oil rejection was 99.96% at the oil concentration of 80 NTU and pressure of 5 bar (500 kPa). The prepared membrane showed high efficiency in removing foulants by the backwash technique because of the charge repulsion forces between the oil molecules and the negatively charged membrane. The fabricated membrane showed good potential applicability in oil-water separation treatments.

**Funding:** This research received no external funding.

**Acknowledgments:** The author appreciates KACST for their encouragement during this study.

**Conflicts of Interest:** The author declare no conflict of interest.

## References

1. Mercado, M.; Acuna, J.C.; Vasquez, J.E.; Caballero, C.; Soriano, J.E. Successful field application of a high temperature conformance polymer in Mexico. *SPE Int. Symp. Oilfield Chem. Soc. Pet. Eng.* **2009**. [CrossRef]
2. Brian, B.; Jianhua, Z.; Xing, W.; Zongli, X. A review on current development of membranes for oil removal from wastewaters. *Membranes* **2020**, *65*, 65. [CrossRef]
3. Padaki, M.; Surya Murali, R.S.; Abdullah, M.S.; Misdan, N.; Moslehyani, A.; Kassim, M.A.; Hilal, N.; Ismail, A.F. Membrane technology enhancement in oil–water separation. A review. *Desalination* **2015**, *357*, 197–207. [CrossRef]
4. 4-Hakami, M.W.; Alkhudhiri, A.; Al-Batty, S.; Zacharof, M.P.; Maddy, J.; Hilal, N. Ceramic microfiltration membranes in wastewater treatment: Filtration behavior, fouling and prevention. *Membranes* **2020**, *10*, 248. [CrossRef] [PubMed]
5. Li, M.; Zhao, Y.; Zhou, S.; Xing, W. Clarification of raw rice wine by ceramic microfiltration membranes and membrane fouling analysis. *Desalination* **2010**, *256*, 166–173. [CrossRef]
6. Hankins, N.P.; Lu, N.; Hilal, N. Enhanced removal of heavy metal ions bound to humic acid by polyelectrolyte flocculation. *Sep. Purif. Technol.* **2006**, *51*, 48–56. [CrossRef]
7. Judd, S.; Jefferson, B. *Membranes for Industrial Wastewater Recovery and Re-Use*; Elsevier Science: Amsterdam, The Netherlands, 2003.
8. Hsu, B.M.; Yeh, H.H. Removal of giardia and cryptosporidium in drinking water treatment: A pilot-scale study. *Water Res.* **2003**, *37*, 1111–1117. [CrossRef]
9. Cui, J.; Zhang, X.; Liu, H.; Liu, S.; Yeung, K.L. Preparation and application of zeolite/ceramic microfiltration membranes for treatment of oil contaminated water. *J. Membr. Sci.* **2008**, *325*, 420–426. [CrossRef]
10. Liu, W.; Yang, G.; Huang, M.; Liang, J.; Zeng, B.; Fu, C.; Wu, H. Ultrarobust and biomimetic hierarchically macroporous ceramic membrane for oil–water separation templated by emulsion-assisted self-assembly method. *Acs Appl. Mater. Interfaces* **2020**, *12*, 35555–35562. [CrossRef]
11. Dongwei, L.; Zhang, T.; Gutierrez, L.; Ma, J.; Croué, J.P. Influence of surface properties of filtration-layer metal oxide on ceramic membrane fouling during ultrafiltration of oil/water emulsion. *Env. Sci. Technol.* **2016**, *50*, 4668–4674.
12. Lobo, A.; Cambiella, Á.; Benito, J.M.; Pazos, C.; Coca, J. Ultrafiltration of oil-in-water emulsions with ceramic membranes: Influence of pH and crossflow velocity. *J. Membr. Sci.* **2016**, *278*, 328–334. [CrossRef]
13. Luo, L.; Gang, H.; Chung, T.; Weber, M.; Staudt, C.; Maletzko, C. Oil/water separation via ultrafiltration by novel triangle-shape tri-bore hollow fibre membranes from sulfonated polyphenylenesulfone. *J. Membr. Sci.* **2015**, *476*, 162–170. [CrossRef]
14. Atallah, C.; Mortazavi, S.; Tremblay, A.; Doiron, A. In-process steam cleaning of ceramic membranes used in the treatment of oil sands produced water. *Ind. Eng. Chem. Res.* **2019**, *33*, 15232–15243. [CrossRef]
15. Issaoui, M.; Lionel, L. Low-cost ceramic membranes: Synthesis, classifications, and applications. *Comptes Rendus Chim.* **2019**, *22*, 175–187. [CrossRef]
16. Mittal, P.; Jana, S.; Mohanty, K. Synthesis of low-cost hydrophilic ceramic–polymeric composite membrane for treatment of oily wastewater. *Desalination* **2011**, *282*, 54–62. [CrossRef]
17. Zhu, L.; Chen, M.; Dong, Y.; Tang, C.Y.; Huang, A.; Li, L. A low-cost mullite-titania composite ceramic hollow fiber microfiltration membrane for highly efficient separation of oil-in-water emulsion. *Water Res.* **2016**, *90*, 277–285. [CrossRef] [PubMed]
18. Liu, N.; McPherson, B.J.; Li, L.; Lee, R.L. Factors determining the reverse-osmosis performance of zeolite membranes on produced-water purification. *Int. Symp. Oil Field Chem. Soc. Pet. Eng.* **2007**. [CrossRef]
19. Nandi, B.K.; Moparthi, A.; Uppaluri, R.; Purkait, M.K. Treatment of oily wastewater using low cost ceramic membrane: Comparative assessment of pore blocking and artificial neural network models. *Chem. Eng. Res. Des.* **2010**, *88*, 881–892. [CrossRef]
20. Goswami, K.P.; Pugazhenthi, G. Credibility of polymeric and ceramic membrane filtration in the removal of bacteria and virus from water: A review. *J. Environ. Manag.* **2020**, *268*, 110583. [CrossRef]
21. Ben Amar, R.; Khemakhem, M.; Oun, A.; Cerneaux, S.; Cretin, M.; Khemakhem, S. Decolorization of dyeing effluent by novel ultrafiltration ceramic membrane from low cost natural material. *J. Membr. Sci. Res.* **2018**, *4*, 101–107.
22. Hubadillah, S.K.; Othman, M.H.D.; Matsuura, T.; Ismail, A.F.; Rahman, M.A.; Harun, Z.; Jaafar, J.; Nomura, M. Fabrications and applications of low cost ceramic membrane from kaolin: A comprehensive review. *Ceram. Int.* **2018**, *44*, 4538–4560. [CrossRef]
23. Jepsen, K.; Bram, M.; Pedersen, S.; Yang, Z. Membrane fouling for produced water treatment: A review study from a process control perspective. *Water* **2018**, *10*, 847. [CrossRef]

24. Franks, R.; Chilekar, S.; Bartels, C.R. The unexpected performance of highly permeable SWRO membranes at high temperatures. *IDA J. Desalin. Water Reuse* **2012**, *4*, 52–56. [CrossRef]
25. Goff, H. Dairy product processing equipment. *Handb. Farm Dairy Food Mach.* **2007**, 193–214. [CrossRef]
26. Nandi, B.K.; Das, B.; Uppaluri, R.; Purkait, M.K. Microfiltration of mosambi juice using low cost ceramic membrane. *J. Food Eng.* **2009**, *95*, 597–605. [CrossRef]
27. Meng, X.; HuHong, M.; NiuXian, L.; ChenHua, Z. Natural cellulose microfiltration membranes for oil/waternanoemulsions separation. *Colloids Surf. A Physicochem. Eng. Asp.* **2019**, *564*, 142–151.
28. Wang, Y.; Li, Y.; Yang, H.; Liang, Z. Super-wetting, photoactive $TiO_2$ coating on amino-silane modified PAN nanofiber membranes for high efficient oil-water emulsion separation application. *J. Membr. Sci.* **2019**, *580*, 40–48. [CrossRef]
29. He, Y.; Jiang, Z.W. Treating oilfield wastewater: Technology review. *Filtr. Sep.* **2008**, 14–16. [CrossRef]
30. Olabode, Y.; Othman, M.; Nordin, N.; Tai, Z.; Usman, J.; Mamah, S.; Ismail, A.; Rahman, M.; Jaafar, J. Fabrication of magnesium bentonite hollow fibre ceramic membrane for oil-water separation. *Arab. J. Chem.* **2020**, *13*, 5996–6008.
31. He, Z.; Miller, D.J.; Kasemset, S.; Wang, L.; Paul, D.R.; Freeman, B.D. Fouling propensity of a poly(vinylidene fluoride) microfiltration membrane to several model oil/water emulsions. *J. Membr. Sci.* **2016**, *514*, 659–670. [CrossRef]

*Article*

# Analysis of Microstructure and Mechanical Properties of Bi-Modal Nanoparticle-Reinforced Cu-Matrix

Fadel S. Hamid [1,*], Omayma A. Elkady [2], A. R. S. Essa [1,3], A. El-Nikhaily [1], Ayman Elsayed [2] and Ashraf K. Eessaa [4]

[1] Mechanical Department, Faculty of Technology and Education, Suez University, Suez 43519, Egypt; ahmed.eessa@suezuniv.edu.eg (A.R.S.E.); ahmedeassa1946@gmail.com (A.E.-N.)
[2] Powder Technology Division, Manufacturing Technology Department, Central Metallurgical R & D Institute, 1 Elfelezat St. Eltebeen, Cairo 11421, Egypt; o.alkady68@gmail.com (O.A.E.); ayman_elsayed_11@yahoo.com (A.E.)
[3] Mechanical Engineering Department, Egyptian Academy for Engineering & Advanced Technology, Affiliated to Ministry of Military Production, Cairo 3056, Egypt
[4] Nanotechnology Lab El Nozha, Electronic Research Institute (E.R.I.), Cairo 12622, Egypt; ashrafkamal888@hotmail.com
* Correspondence: fadel.shaban@suezuni.edu.eg

**Abstract:** Bi-modal particles are used as reinforcements for Cu-matrix. Nano TiC and/or $Al_2O_3$ were mechanically mixed with Cu particles for 24 h. The Cu-TiC/$Al_2O_3$ composites were successfully produced using spark plasma sintering (SPS). To investigate the effect of TiC and $Al_2O_3$ nanoparticles on the microstructure and mechanical properties of Cu-TiC/$Al_2O_3$ nanocomposites, they were added, whether individually or combined, to the copper (Cu) matrix at 3, 6, and 9 wt.%. The results showed that titanium carbide was homogeneously distributed in the copper matrix, whereas alumina nanoparticles showed some agglomeration at Cu grain boundaries. The crystallite size exhibited a clear reduction as a reaction to the increase of the reinforcement ratio. Furthermore, increasing the TiC and $Al_2O_3$ nanoparticle content in the Cu-TiC/$Al_2O_3$ composites reduced the relative density from 95% for Cu-1.5 wt.% TiC and 1.5 wt.% $Al_2O_3$ to 89% for Cu-4.5 wt.% TiC and 4.5 wt.% $Al_2O_3$. Cu-9 wt.% TiC achieved a maximum compressive strength of 851.99 $N/mm^2$. Hardness values increased with increasing ceramic content.

**Keywords:** copper; nanocomposites; metal-matrix composites (MMCs); mechanical properties; spark plasma sintering

## 1. Introduction

Copper strengthening is a current priority due to the pressing need to use it in various applications requiring a balance of properties [1–3]. Metal-matrix composites are most promising in achieving balanced mechanical properties between nano and microstructure materials [4–8]. Copper is used in many industries owing to its low cost, ease of manufacturing, and good corrosion resistance [9]. The main drawbacks of pure copper are its substantial low strength, high coefficient of thermal expansion (CTE), and generally poor mechanical properties [10]. One effective way to overcome these limitations is to reinforce copper with ceramic particles to obtain composites with superior properties. The effectiveness of dispersed particles in matrix strengthening depends primarily on particle characteristics: size, distribution, spacing, thermodynamic stability, and low solubility and diffusivity of its constituent elements in the matrix. Among ceramic particles, alumina nanoparticles have shown outstanding mechanical properties even at high temperatures, as well as low production costs [11,12]. In addition, TiC is an attractive candidate for metallic matrices such as copper (Cu), iron (Fe), aluminum (Al), titanium (Ti), and nickel (Ni) because of its high hardness, high melting point, and abrasion resistance with good electrical conductivity [12–14]. Due to the aforementioned factors, Cu reinforced with (TiC-Al$_2$O$_3$) composites led to a more viable material.

Numerous techniques have been used to fabricate reinforced copper matrix composites (CMCs), including molecular-level mixing (MLM) [15], in situ metallurgy [12,13], flake powder metallurgy [16,17], high-energy ball milling (HEBM) [7,18–20], friction stir processing [21–26], high-pressure torsion [26], and rolling [27–30]. Although these techniques enhanced the mechanical properties of processed composites, they resulted in an inhomogeneous distribution of particle reinforcements within the matrix. Additionally, they have the potential to cause morphological and structural damage, as demonstrated through carbon nanotubes (CNTs) within a copper matrix [31].

The spark plasma sintering (SPS) method, developed recently, is a new technique for synthesizing metal matrix composites. The SPS technique has piqued researchers' interest due to its advantages of sintering at relatively low temperatures, higher heating speeds, shorter processing times, and the absence of pre-compression as in conventional sintering. Thus, the SPS technique enables the fabrication of nanostructured composites without the high grain growth rate associated with traditional sintering methods. As a result, SPS composites exhibit exceptional mechanical properties at room temperature, even at elevated temperatures [32].

To the authors' knowledge, few papers discuss the solid-state spark plasma sintered Cu-Al$_2$O$_3$ [33] and Cu-TiC [34–36], respectively. However, no information on the synthesis and mechanical investigation of hybrid Cu-Al$_2$O$_3$-TiC through mechanical alloying and SPS techniques have been released. Thus, this work fabricated three separate nanocomposites of Cu-TiC, Cu-Al$_2$O$_3$, and hybrid Cu-TiC-Al$_2$O$_3$, using mechanical alloying and SPS processing. The influence of the TiC and Al$_2$O$_3$ nanoparticles content on the microstructure and mechanical properties of the prepared nanocomposites was also investigated.

## 2. Materials and Methods

### 2.1. Materials

Copper (Cu) powder with 99.9% purity (supplied by AlphaChemical, MA, USA) with an average particle size of 10 µm was used as a metal matrix. Alumina (Al$_2$O$_3$) nanopowder with 99.7% purity (supplied by Alpha Chemicals, MA, USA) with an average particle size of 50 nm and titanium carbide (TiC) nanopowder with 99.7% purity (supplied by Inframat Advanced Materials, L.L.C., CT, USA) with an average size of 100 nm were used. Both TiC and Al$_2$O$_3$ were used as individual/hybrid reinforcement. The Cu powder was mixed with 3, 6, and 9 wt.% of individual/hybrid reinforcement of TiC and Al$_2$O$_3$ using a ball milling technique for 24 h. The powders were mixed in a stainless-steel vial and protected from oxidation using highly pure argon gas using a 25:1 ball to powder ratio (BPR), 110 rpm, and a ball diameter of 5 mm. Stearic acid (1.5 wt.%) was used as a process controlling agent (PCA). Figure 1 and Table 1 show the composition of fabricated samples.

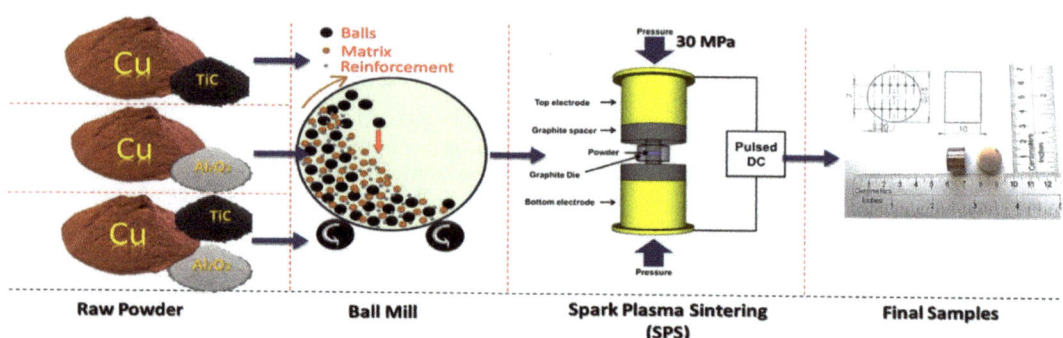

**Figure 1.** Schematic diagram of Cu with TiC and Al$_2$O$_3$ and (TiC + Al$_2$O$_3$) hybrid nano reinforcement composites.

**Table 1.** Composition of the prepared specimens and their contents in the Cu matrix.

|   | Materials after Sintering by (SPS) | Composition | | |
|---|---|---|---|---|
|   |   | Matrix | Reinforcement | |
|   |   | Cu [wt.%] | TiC [wt.%] | $Al_2O_3$ [wt.%] |
| 0 | Pure copper | 100 | — | — |
| I | Cu-3 wt.% TiC | 97 | 3 | — |
|   | Cu-6 wt.% TiC | 94 | 6 | — |
|   | Cu-9 wt.% TiC | 91 | 9 | — |
| II | Cu-3 wt.% $Al_2O_3$ | 97 | — | 3 |
|   | Cu-6 wt.% $Al_2O_3$ | 94 | — | 6 |
|   | Cu-9 wt.% $Al_2O_3$ | 91 | — | 9 |
| III | Cu-1.5 wt.% TiC and 1.5 wt.% $Al_2O_3$ | 97 | 1.5 | 1.5 |
|   | Cu-3 wt.% TiC and 3 wt.% $Al_2O_3$ | 94 | 3 | 3 |
|   | Cu-4.5 wt.% TiC and 4.5 wt.% $Al_2O_3$ | 91 | 4.5 | 4.5 |

*2.2. Spark Plasma Sintering (SPS)*

The sintering process was performed using a spark plasma sintering technique (DR. SINTER LAB Model: SPS-1030, Syntex, Osaka, Japan). In all experiments, the powder was loaded into a graphite die with an inner diameter of 15 mm with graphite foil and enclitic by 0.5 mm thick graphite cover to prevent the friction of the sample with the die during the compaction process and to minimize heat loss. Before sintering, the SPS chamber was evacuated to a pressure below 5 Pa. The samples were heated from room temperature up to 950 °C by pulsed D.C. current using the heating rate of 20 °C/min. The samples were then held at the maximum temperature for 45 min under a uniaxial pressure of 30 MPa applied since the first minute of heating. This processing route was used to fabricate the Cu-TiC/$Al_2O_3$ nanocomposites.

*2.3. Mechanical Properties*

The hardness was measured along the polished surface of the specimen using a Vickers hardness tester (HMV-2T Model SHIMADZU, Kyoto, Japan). The test was carried out under 100 g load for 15 s dwell time.

The microhardness values were evaluated for an average of twelve readings on the surface of each sample. The compression test for the investigated specimens was carried out using a universal testing machine. In the compression test, three samples were investigated, and average results were obtained. The dimensions of the specimens for compression tests were 6 mm in diameter and 15 mm in length. The applied crosshead speed was 0.05 mm/s, and the test was performed at room temperature.

### 3. Results

*3.1. XRD Analysis*

Figure 2 shows the XRD patterns of the prepared ten samples, pure Cu and 3, 6, 9 wt.% TiC/$Al_2O_3$ (individual and hybrid) nanocomposites. Only peaks corresponding to Cu, TiC, and $Al_2O_3$ appeared, whereas pattern-like Cu was observed in the case of 3 wt.% Cu/TiC/$Al_2O_3$ samples; this may be due to the lower percentage of both TiC and $Al_2O_3$ that are below the limits of the XRD device. This may be attributed to the controlled milling and sintering process in an argon atmosphere which shows that no other peaks for any new phases or intermetallic compounds were formed due to the rapid consolidation process (45 min) during the SPS technique.

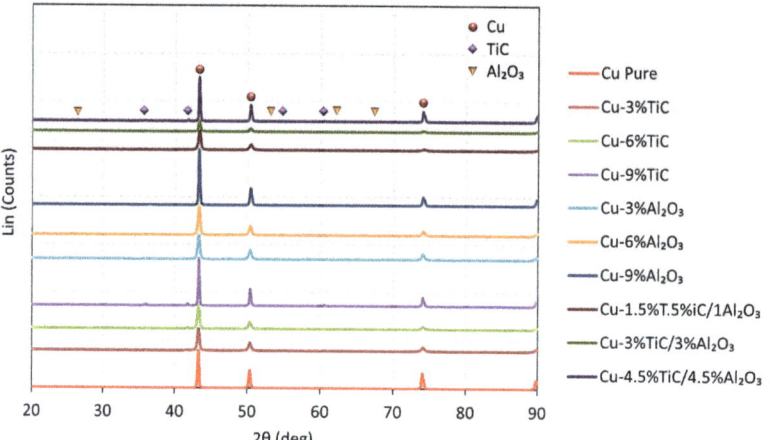

**Figure 2.** XRD patterns of composites after (SPS) process.

The crystallite size was assessed by the classical Williamson–Hall method (FWHM) from the broadening of XRD peaks and using the following formula [37,38]:

$$\frac{\beta \cos \theta}{\lambda} = \frac{k}{d} + 2\varepsilon \left( \frac{2\sin \theta}{\lambda} \right) \quad (1)$$

where β is the full width at half maximum height (FWHM), θ is the Bragg's angle of the peak, λ is the wavelength of X-ray (0.15406 nm), K is a dimensionless shape factor (0.9), which depends on the material, d is the crystallite size, and ε is the microstrain.

Figure 3 shows the effect of ceramic ratio on the crystallite size. A clear reduction of the crystallite size with increasing wt.% of ceramic additives was observed. Al$_2$O$_3$ and TiC are ceramic materials that act as internal balls that reduce the particle size [39,40]. In addition, the SPS technique achieved the consolidation process, which is a rapid method for the sintering in which no chance for the grain growth of the particles occurs [41,42]. The crystallite size of pure copper was ~105 nm, whereas the crystallite size for the produced composites was in the range of 5–25 nm.

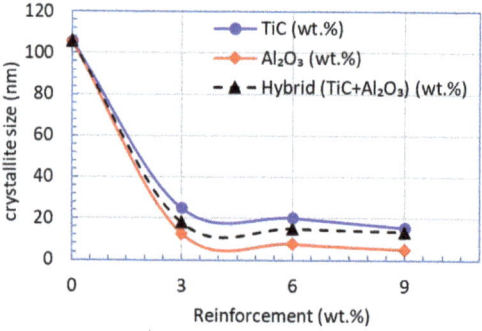

**Figure 3.** The effect of ceramic additions on the crystallite size.

### 3.2. Densification

The density of composite material is the most important parameter, which significantly affects both physical and mechanical properties. The relative density is calculated and plotted in Figure 4. Relative density is the ratio of the measured and theoretical density of the sample. Measured density was determined by the Archimedes method, and the theoretical density was calculated from the simple rule of mixtures. Each percent of pure Cu and 3, 6, 9 wt.% TiC/Al$_2$O$_3$ (individual and hybrid) nanocomposite were tested by three samples, and the average results were obtained. It was observed that the relative density decreased with increasing reinforcement content for all composites, as shown in Figure 4. The maximum relative density (~96%) was achieved by adding 3 wt.% TiC to copper, whereas the minimum relative density (89%) was obtained for 9 wt.% Al$_2$O$_3$/copper composite. TiC (4.91 g/cm$^3$) and Al$_2$O$_3$ (3.987 g/cm$^3$) also have lower densities than Cu [7,35]. So, the addition of a light material to a denser one decreased the overall density of the prepared composites. This may be attributed to the presence of hard ceramic material with a high melting point into a ductile metal such as Cu that may hinder the high densification and increase the porosity content accompanied by the high fraction of reinforcement [7,43].

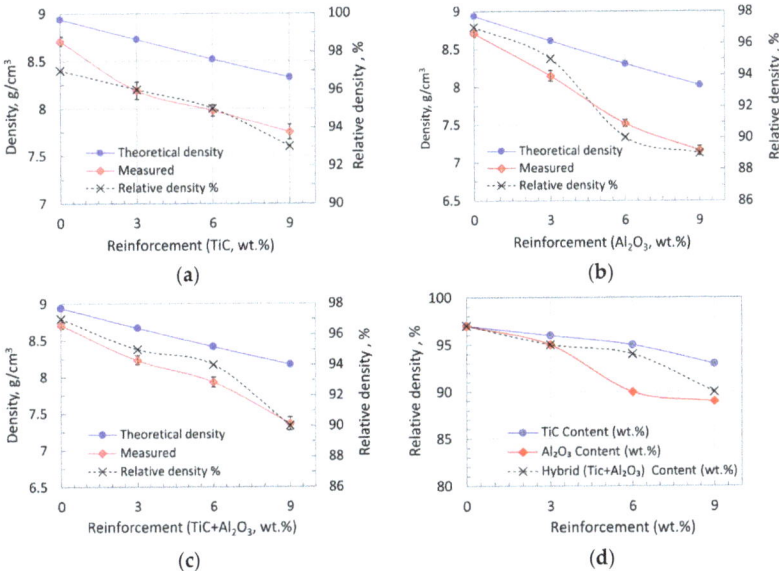

**Figure 4.** The effect of reinforcement fraction on the density of the produced composites. (**a**) Cu/TiC composites, (**b**) Cu/Al$_2$O$_2$ composites, (**c**) Cu/hybrid composite, and (**d**) Comparison of relative density between three series.

Ayman Elsayed et al. [11] studied experimental investigations for the synthesis of W–Cu nanocomposite through spark plasma sintering, and they concluded that using the SPS technique led to reaching a maximum of 90% relative density. On the other hand, a relative density of 98.1% was reached for Cu-Fe-Al$_2$O$_3$-MoS$_2$ composite sintered using the SPS route [33]. Moreover, Babapoor et al. [44] investigated the effects of spark plasma sintering temperature on the densification of TiC. They reached a relative density of 99.4% at 1900 °C for 7 min under 40 MPa using the TiC powder with a mean particle size of 7 μm. They also suggested that there is an optimum temperature for reaching the maximum density.

*3.3. Microstructure Analysis*

Figure 5 shows the FE-SEM micrograph of TiC-reinforced copper composite using 3%, 6%, and 9% TiC addition to Cu. Two phases are observed; the dark-gray phase represents the Cu matrix, and the black phase is the TiC particles. For 3 wt.% samples, TiC and Al$_2$O$_3$ particles are concentrated along the grain boundaries in a chain form, while 6 and 9 wt.% samples were homogeneously distributed all over the Cu matrix. This may be attributed to the suitable mechanical milling parameters and good SPS technique applied. The SPS technique leads to a finer structure compared with traditional routes [18,44].

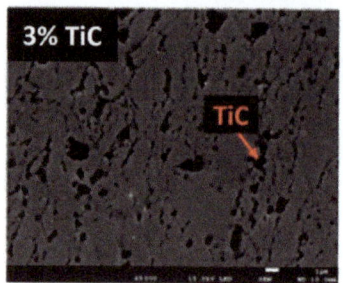 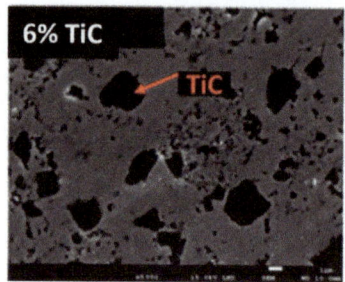 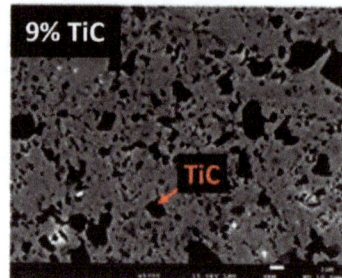

**Figure 5.** SEM of Cu/TiC nanocomposite with TiC percentage of 3, 6, and 9% prepared using spark plasma sintered route.

The SEM investigation of Cu/Al$_2$O$_3$ microstructure is shown in Figure 6. Two phases are observed; the dark-gray phase represents the Cu matrix while the white phase represents the Al$_2$O$_3$ particles. The dispersion of Al$_2$O$_3$ inside the copper matrix is observed for Cu/3% Al$_2$O$_3$ with a little agglomeration of Al$_2$O$_3$ reinforcement. On the other hand, white areas of agglomerated alumina reinforcement are revealed within the Cu/6% Al$_2$O$_3$ matrix grain boundaries, whereas very fine particles are dispersed within the grain interior. Moreover, the SEM of Cu/9% Al$_2$O$_3$ composite shows that most of the Al$_2$O$_3$ nanoparticles are agglomerated along grain boundaries, and a small percentage are dispersed with the grains. Some authors have also concluded that increasing agglomeration steadily occurs, along with increasing the weight percentage of reinforcement [9,15].

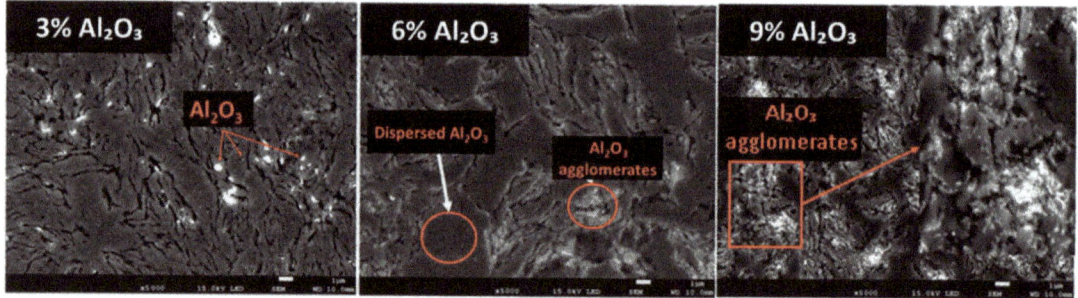

**Figure 6.** SEM of Cu/Al$_2$O$_3$ nanocomposite with Al$_2$O$_3$ percentage of 3, 6, and 9%.

The combination of both TiC and Al$_2$O$_3$ for reinforcing copper (Cu/hybrid nanocomposite) with point analysis EDS is shown in Figure 7a,b. The homogeneous distribution of TiC nanoparticles is predominant, while the agglomeration of some Al$_2$O$_3$ is observed along grain boundaries. More agglomeration of alumina particles along grain boundaries is observed with increasing hybrid percentage (TiC and Al$_2$O$_3$).

### 3.4. Mechanical Strength

Figure 8 represents the stress–strain curves for pure copper and TiC/Al$_2$O$_3$-reinforced copper matrix composites, while the key mechanical properties obtained from the compression test are plotted in Figure 9. The compressed samples are photographed in Figure 10. The addition of TiC enhanced the compression strength of copper and reached its maximum compression strength of ~852 N/mm$^2$ at 9 wt.% TiC, whereas Al$_2$O$_3$ additions exhibited a dramatic effect on the Cu strength. Increasing Al$_2$O$_3$ from 3 to 6 wt.% increased the Cu strength, but a clear failure of strength is noticed at 9 wt.% ratios at which a minimum value of compression strength 367.8 N/mm² resulted. After the compression test, the Cu-6% Al₂O₃ and 9% Al₂O₃ samples were destroyed (see Figure 10). clearpage

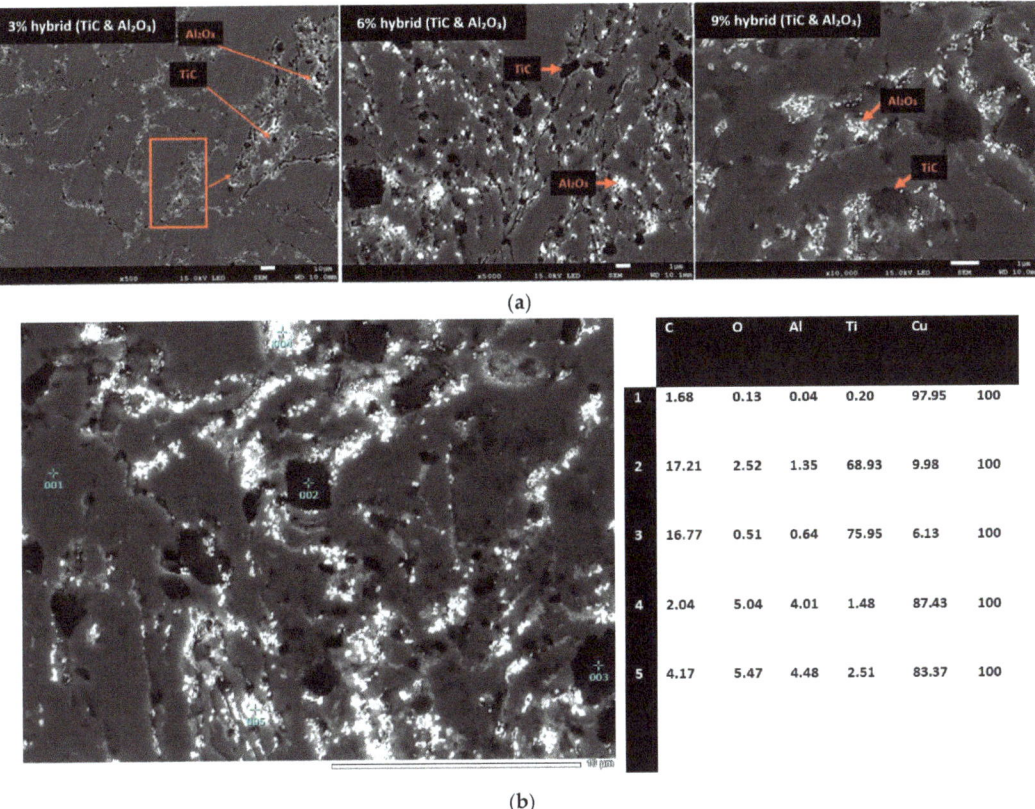

**Figure 7.** (a) SEM of Cu/hybrid nanocomposite with (TiC+Al₂O₃) weight percentage of 3, 6, and 9% prepared using the spark plasma sintered route; (b) point analysis for Cu-reinforced hybrid ceramic matrix composite containing 9 wt.% hybrid ratio.

Moreover, the strength of hybrid composites increased firstly with increasing the percentage of reinforcement up to 6% and then slightly decreased. The extreme drop in the compression strength of the Cu/9% Al₂O₃ composite may be attributed to particle-to-particle contact resulting from ceramic particle agglomeration (see Figure 6). The high compression strength of TiC-strengthened Cu prepared by the SPS route compared with Cu/Al₂O₃ composite with the same wt.% is attributed to a combined effect of ultrafine grain (UFG) structure by the Hall–Petch mechanism and the obstruction of dislocation movement by nanoscale ceramic particles in the grain interior by the Orowan mechanism [8,43].

The effect of ceramic additions on the hardness of copper is illustrated in Figure 11. The hardness steadily increases with increasing the wt.% of reinforcement for synthesized composites. Cu/9% Al₂O₃ obtained the maximum hardness (211 HV), whereas two composites that obtained the minimum hardness value of 112 HV are Cu/3% TiC and Cu/3% hybrid. Many reasons could explain this. The first is that adding high-hardness and high-strength ceramic materials such as TiC and Al₂O₃ on the ductile Cu matrix increases the overall hardness. The second is that the addition of nanomaterials with the incorporation of nanoparticles between the Cu particles improves the hardness as a grain reinforcement takes place accordingly.

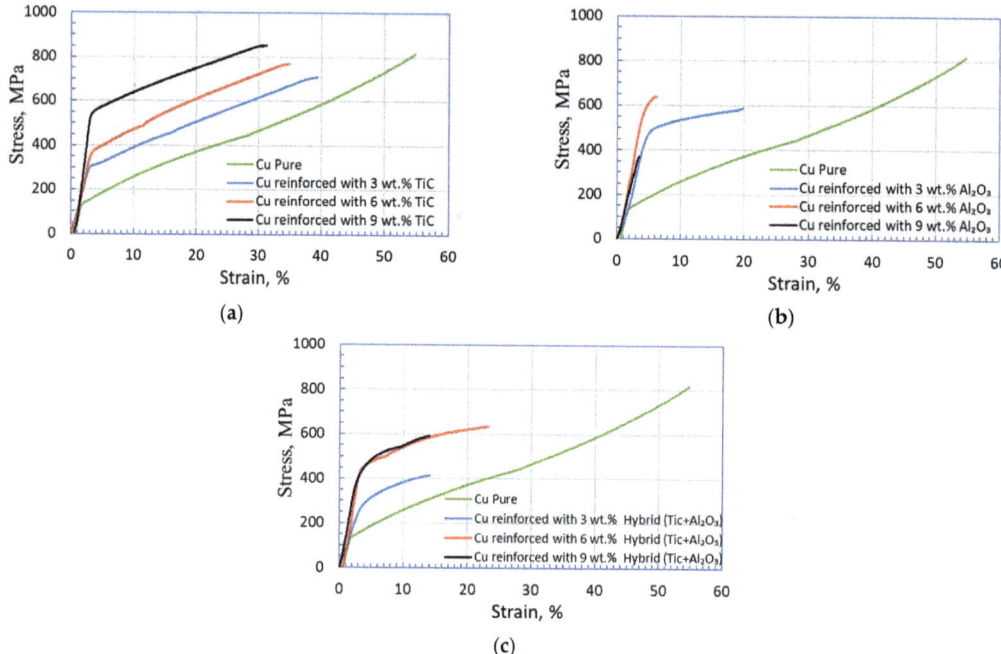

**Figure 8.** Compressive stress–strain curve for pure Cu and Cu-reinforced composites prepared by the spark plasma sintering route, (**a**) Cu/TiC composites, (**b**) Cu/Al$_2$O$_3$ composites, and (**c**) Cu/hybrid composite.

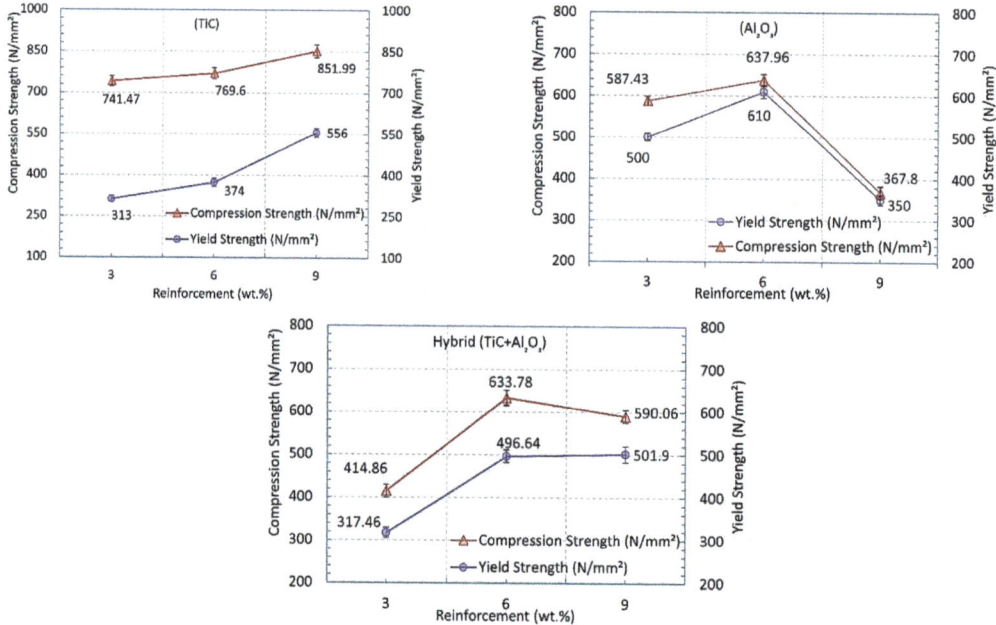

**Figure 9.** Effect of nanosized TiC and Al$_2$O$_3$ contents on the mechanical properties of the Cu-based composites prepared by spark plasma sintering.

**Figure 10.** Compressed test samples for pure Cu and Cu matrix nanocomposites.

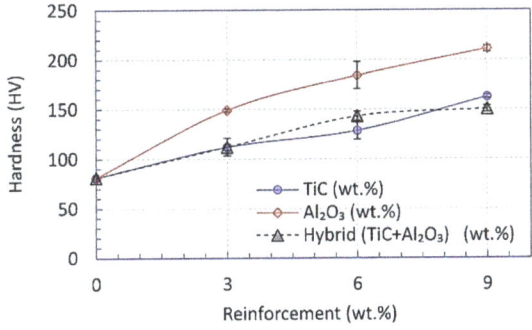

**Figure 11.** A diagram of the measured hardness for copper-based nanoceramics.

*3.5. Strengthening Criteria*

Increasing the strength of metallic materials is based on two competing factors. The first is work hardening, and the second is dynamic softening. Work hardening is caused by dislocations, multiplication, pileup, and tangle. Dynamic softening is caused by dislocations, rearrangement, and interactions. In the present work, the compression test and hardness measurement are carried out at room temperature, which is why the dynamic softening factor would not be probable, and the work hardening mechanism would affect the enhancement of strength and hardness. This is true for pure metal and alloys, unlike the composite materials where the contribution of ceramic additions to the matrix to enhance the properties should be considered.

Moreover, it is worth mentioning that some authors have considered the strengthening mechanisms in ceramic-reinforced composites [6,7,15,40,42–44]. The addition of TiC nanoparticles to the Cu matrix retained grain growth during sintering due to the peening effect of TiC for grain boundary movement and the strengthening effect of dispersed TiC in the Cu matrix grains where a mismatch of coefficient of thermal expansion is present [45] (see Figure 12). Furthermore, increasing the TiC fraction increased the strength and hardness of Cu-TiC composites in the present work. Another strengthening mechanism of TiC dispersion is the Orowan mechanism, especially at low fractions of TiC [6]. Compared to the other two cases, the composites with 3% and 6% $Al_2O_3$ gave the highest yield strength. This is a signal of increasing material strength with decreasing ductility. This may be attributed to the good adhesion between Cu matrix and TiC nanoparticles than between

Cu and Al$_2$O$_3$ [46]. An extreme drop in the Cu-Al$_2$O$_3$ composite strength is noticed at 9 wt.% of Al$_2$O$_3$. This unexpected behavior may be attributed to the agglomeration of some alumina particles in the Cu matrix, increasing the chance for particle-to-particle contact (see Figure 13). A balanced behavior was observed with the hybrid (TiC/Al$_2$O$_3$) additions to the Cu matrix in which the combined effects of both ceramics are clear.

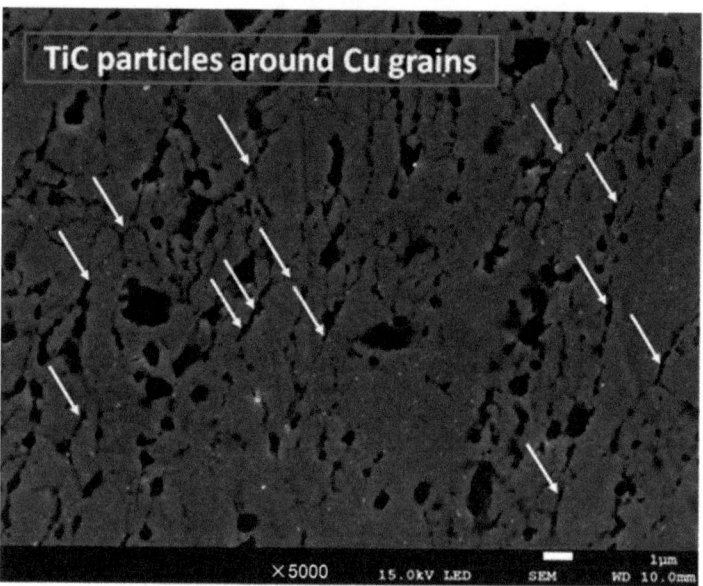

**Figure 12.** Spot white arrows indicate TiC around Cu grains.

**Figure 13.** SEM showing alumina distribution behavior in the Cu matrix.

Reinforcing the Cu matrix, which is ductile in nature with two types of ceramic materials—ceramic carbide (TiC) and ceramic oxide ($Al_2O_3$)—helps to improve the mechanical properties of the Cu matrix. Both TiC and $Al_2O_3$ are at the nanoscale; therefore, by high ball milling, they filled the interstitial voids between Cu particles. As a consequence, the strengthening effect of both of them is distributed all over the Cu matrix. The hardness estimation test increased as the nano-ceramic hard particles were increased. This can also be explained by the resistance of the hard ceramic particles to the indenter from greater depth in the Cu-composite surface.

Consequently, the hardness is enhanced [46,47]. For the compression test, the presence of the nano-ceramic particles dispersed formally in the Cu matrix prevents the dislocation of the particles. In addition, as these hybrid reinforcements are at the nanoscale, they fill the voids; consequently, the strength of samples is increased [47–49].

Table 2 shows a comparison between the present study and previous work used to fabricate copper composites reinforced with alumina and titanium carbide nanoparticles. In this work, different concentrations of nano alumina and/or nano titanium carbide particles were used as a reinforcement material to the Cu matrix manufactured by the SPS technique. This work is compared with the same composites prepared by traditional sintering, vacuum sintering, hot pressing, and hot extrusion. The table shows that the composites produced by the SPS technique have the best mechanical properties compared with the other consolidation techniques.

**Table 2.** Comparing the present study with literature data of previous investigations.

| Composite | Method | Density, (g/cm³) | Ultimate Stress, (MPa) | Yield Stress, (MPa) | Elongation, (%) | Hardness, (HV) | Ref. No |
|---|---|---|---|---|---|---|---|
| Pure copper | | 97 | N/A | 127.15 | 1.69 | 81 | |
| Cu-3 wt.% TiC | SPS at 950 °C | 96 | 741.47 | 313 | 3.97 | 111.9 | [Present study] |
| Cu-3 wt.% $Al_2O_3$ | | 95 | 587.43 | 500 | 6.17 | 149 | |
| Cu-1.5 wt.% TiC and 1.5 wt.% $Al_2O_3$ | | 95 | 414.86 | 317.46 | 5.25 | 112 | |
| Cu-5 wt.% TiC | Hot Press at 700 °C | 93.3 | N/A | N/A | N/A | 67.3 | [50] |
| Cu-5 vol.%TiC | Hot extrusion | N/A | N/A | N/A | N/A | 112 | [51] |
| Cu-5 vol.% TiC | SPS | N/A | 712 | 661 | N/A | 221 | [14] |
| Cu-77 vol.% TiC | Sintering in Vacuum Furnace at 900 °C | 93.4 | N/A | N/A | N/A | 544 | [52] |
| Cu-5.3 vol.%TiC | SPS | N/A | 602 | 572 | N/A | 194 | [53] |
| Cu-3 wt.% $Al_2O_3$ with Coating Ag | Sintered at 950 °C | 95.9 | N/A | N/A | N/A | 85 | [7] |
| Cu-10 vol.% $Al_2O_3$ | Sintered at 880 °C | 83.49 | N/A | N/A | N/A | 71 | [54] |
| Cu-3 vol.% $Al_2O_3$ | Sintered at 850 °C | 91.5 | 350 | N/A | 0.51 | 77 | [55] |
| Cu-5 vol.% $Al_2O_3$ | Sintered at 850 °C | 88 | 550 | N/A | 0.46 | 100 | |
| Cu-5 vol.% $Al_2O_3$ | Sintering H2 at 850 °C | | 530 | 450 | 2.5 | 155 | [56] |
| Cu-2.7 wt.% $Al_2O_3$ | Sintered at 950 °C | 92.53 | 460 | 350 | N/A | 54.83 | [57] |
| Cu-5 vol.% $Al_2O_3$ | Conventional Sintering N₂ | 84.3 | N/A | N/A | N/A | 49 | [58] |
| Cu-5 vol.% $Al_2O_3$ | Conventional Sintering Ar | 84.3 | N/A | N/A | N/A | 48 | |
| Cu-5 vol.% $Al_2O_3$ | Conventional Sintering H₂ | 94.4 | N/A | N/A | N/A | 79 | |
| Cu-5 vol.% $Al_2O_3$ | SPS at 700 °C | 92.2 | N/A | N/A | N/A | 125 | |
| Cu-2.75 wt.% $Al_2O_3$ | Pulsed Electric Current Sintered (PECS) | 99.6 | N/A | N/A | N/A | 94.83 | [59] |

## 4. Conclusions

Cu-(TiC and/or $Al_2O_3$) nanocomposites were synthesized successfully using mechanical milling followed by the spark plasma sintering (SPS) technique.

The density decreased with increasing percentages of the nano reinforcements (TiC and/or $Al_2O_3$). A maximum relative density of ~96% was achieved with the addition of 3 wt.% TiC to copper, whereas the minimum relative density (89%) was obtained by adding 9 wt.% $Al_2O_3$ to copper.

Agglomerated areas of $Al_2O_3$ nanoparticles around grain boundaries were observed, and increased with increasing $Al_2O_3$ fractions that, in turn, adversely affect the mechanical properties.

The compression strength of Cu/TiC increased with increasing the TiC fraction, and a maximum value of 851.99 N/mm$^2$ was obtained by Cu/9% TiC. A dramatic behavior was observed for Cu/$Al_2O_3$ composites that gave the minimum compressive strength of 367.8 N/mm$^2$ resulted at 9% $Al_2O_3$.

The maximum hardness of 211 HV was obtained by Cu/9% $Al_2O_3$, whereas two types of composites obtained the minimum hardness value of 112 HV: the Cu/3% TiC and Cu/3% hybrid.

**Author Contributions:** F.S.H.: investigation, writing—original draft and preparation; O.A.E.: Investigation, writing—original draft, review, and editing; A.R.S.E.: conceptualization and formal analysis; A.E.-N.: investigation and review; A.E.: methodology, review, and editing; A.K.E.: review and editing. All authors have read and agreed to the published version of the manuscript.

**Funding:** This research received no external funding.

**Institutional Review Board Statement:** Not applicable.

**Informed Consent Statement:** Not applicable.

**Data Availability Statement:** The data that support the findings of this study are included in the article. Any further requested information can be addressed to the corresponding author.

**Acknowledgments:** The authors acknowledge the kind support by Katsuyoshi Kondoh, Composite Materials Processing Lab., Osaka University, for providing the spark plasma sintering machine in his laboratory to carry out the consolidation of the composites and thank the researchers and technicians of the Central Metallurgical R & D Institute (CMRDI) in Cairo, Egypt for their collaboration.

**Conflicts of Interest:** The authors declare no potential conflict of interest concerning this article's research, authorship, and/or publication.

## References

1. Daghigh, R.; Oramipoor, H.; Shahidian, R. Improving the Performance and Economic Analysis of Photovoltaic Panel Using Copper Tubular-Rectangular Ducted Heat Exchanger. *Renew. Energy* **2020**, *156*, 1076–1088. [CrossRef]
2. Abyzov, A.M.; Kidalov, S.V.; Shakhov, F.M. High Thermal Conductivity Composite of Diamond Particles with Tungsten Coating in a Copper Matrix for Heat Sink Application. *Appl. Therm. Eng.* **2012**, *48*, 72–80. [CrossRef]
3. Xiao, Y.; Yao, P.; Zhou, H.; Zhang, Z.; Gong, T.; Zhao, L.; Deng, M. Investigation on Speed-Load Sensitivity to Tribological Properties of Copper Metal Matrix Composites for Braking Application. *Metals* **2020**, *10*, 889. [CrossRef]
4. Cao, G.; Konishi, H.; Li, X. Mechanical Properties and Microstructure of Mg/SiC Nanocomposites Fabricated by Ultrasonic Cavitation Based Nanomanufacturing. *J. Manuf. Sci. Eng. Trans. ASME* **2008**, *130*, 0311051–0311056. [CrossRef]
5. Kaftelen, H.; Ünlü, N.; Göller, G.; Lütfi Öveolu, M.; Henein, H. Comparative Processing-Structure-Property Studies of Al-Cu Matrix Composites Reinforced with TiC Particulates. *Compos. Part A Appl. Sci. Manuf.* **2011**, *42*, 812–824. [CrossRef]
6. Morris, D.G. The Origins of Strengthening in Nanostructured Metals and Alloys. *Rev. Metal.* **2010**, *46*, 173–186. [CrossRef]
7. Sadoun, A.M.; Mohammed, M.M.; Fathy, A.; El-Kady, O.A. Effect of Al2O3 Addition on Hardness and Wear Behavior of Cu-Al2O3 Electro-Less Coated Ag Nanocomposite. *J. Mater. Res. Technol.* **2020**, *9*, 5024–5033. [CrossRef]
8. Zhang, Z.; Chen, D.L. Contribution of Orowan Strengthening Effect in Particulate-Reinforced Metal Matrix Nanocomposites. *Mater. Sci. Eng. A* **2008**, *483–484*, 148–152. [CrossRef]
9. Chandrakanth, R.G.; Rajkumar, K.; Aravindan, S. Fabrication of Copper-TiC-Graphite Hybrid Metal Matrix Composites through Microwave Processing. *Int. J. Adv. Manuf. Technol.* **2010**, *48*, 645–653. [CrossRef]
10. Fathy, A.; El-Kady, O. Thermal Expansion and Thermal Conductivity Characteristics of Cu-Al2O3 Nanocomposites. *Mater. Des.* **2013**, *46*, 355–359. [CrossRef]
11. Elsayed, A.; Li, W.; El Kady, O.A.; Daoush, W.M.; Olevsky, E.A.; German, R.M. Experimental Investigations on the Synthesis of W-Cu Nanocomposite through Spark Plasma Sintering. *J. Alloys Compd.* **2015**, *639*, 373–380. [CrossRef]
12. Shyu, R.F.; Ho, C.T. In Situ Reacted Titanium Carbide-Reinforced Aluminum Alloys Composite. *J. Mater. Process. Technol.* **2006**, *171*, 411–416. [CrossRef]
13. Ni, J.; Li, J.; Luo, W.; Han, Q.; Yin, Y.; Jia, Z.; Huang, B.; Hu, C.; Xu, Z. Microstructure and Properties of In-Situ TiC Reinforced Copper Nanocomposites Fabricated via Long-Term Ball Milling and Hot Pressing. *J. Alloys Compd.* **2018**, *755*, 24–28. [CrossRef]
14. Wang, F.; Li, Y.; Wang, X.; Koizumi, Y.; Kenta, Y.; Chiba, A. In-Situ Fabrication and Characterization of Ultrafine Structured Cu-TiC Composites with High Strength and High Conductivity by Mechanical Milling. *J. Alloys Compd.* **2016**, *657*, 122–132. [CrossRef]

15. Zhao, Q.; Gan, X.; Zhou, K. Enhanced Properties of Carbon Nanotube-Graphite Hybrid-Reinforced Cu Matrix Composites via Optimization of the Preparation Technology and Interface Structure. *Powder Technol.* **2019**, *355*, 408–416. [CrossRef]
16. Akbarpour, M.R.; Mousa Mirabad, H.; Khalili Azar, M.; Kakaei, K.; Kim, H.S. Synergistic Role of Carbon Nanotube and SiCn Reinforcements on Mechanical Properties and Corrosion Behavior of Cu-Based Nanocomposite Developed by Flake Powder Metallurgy and Spark Plasma Sintering Process. *Mater. Sci. Eng. A* **2020**, *786*, 139395. [CrossRef]
17. Akbarpour, M.R.; Alipour, S.; Farvizi, M.; Kim, H.S. Mechanical, Tribological and Electrical Properties of Cu-CNT Composites Fabricated by Flake Powder Metallurgy Method. *Arch. Civ. Mech. Eng.* **2019**, *19*, 694–706. [CrossRef]
18. Kumar, R.; Chaubey, A.K.; Bathula, S.; Jha, B.B.; Dhar, A. Synthesis and Characterization of Al2O3-TiC Nano-Composite by Spark Plasma Sintering. *Int. J. Refract. Met. Hard Mater.* **2016**, *54*, 304–308. [CrossRef]
19. Mohammadzadeh, A.; Akbarpour, M.R.; Heidarzadeh, A. Production of Nanostructured Copper Powder: Microstructural Assessments and Modeling. *Mater. Res. Express* **2018**, *5*, 065050. [CrossRef]
20. Akbarpour, M.R.; Alipour, S. Wear and Friction Properties of Spark Plasma Sintered SiC/Cu Nanocomposites. *Ceram. Int.* **2017**, *43*, 13364–13370. [CrossRef]
21. Hosseini, S.A.; Ranjbar, K.; Dehmolaei, R.; Amirani, A.R. Fabrication of Al5083 Surface Composites Reinforced by CNTs and Cerium Oxide Nano Particles via Friction Stir Processing. *J. Alloys Compd.* **2015**, *622*, 725–733. [CrossRef]
22. Fono-Tamo, R.S.; Tien-Chien, J.; Akinlabi, E.T.; Sanusi, K.O. Surface Characteristics of Stainless Steel Powder in Magnesium Substrate: A Friction Stir Processed Composite. In Proceedings of the IEEE 10th International Conference on Mechanical and Intelligent Manufacturing Technologies (ICMIMT 2019), Cape Town, South Africa, 15–17 February 2019; pp. 10–14. [CrossRef]
23. Mahmoud, E.R.I.; Ikeuchi, K.; Takahashi, M. Fabrication of SiC Particle Reinforced Composite on Aluminium Surface by Friction Stir Processing. *Sci. Technol. Weld. Join.* **2008**, *13*, 607–618. [CrossRef]
24. Tinubu, O.O.; Das, S.; Dutt, A.; Mogonye, J.E.; Ageh, V.; Xu, R.; Forsdike, J.; Mishra, R.S.; Scharf, T.W. Friction Stir Processing of A-286 Stainless Steel: Microstructural Evolution during Wear. *Wear* **2016**, *356–357*, 94–100. [CrossRef]
25. Escobar, J.D.; Velásquez, E.; Santos, T.F.A.; Ramirez, A.J.; López, D. Improvement of Cavitation Erosion Resistance of a Duplex Stainless Steel through Friction Stir Processing (FSP). *Wear* **2013**, *297*, 998–1005. [CrossRef]
26. Rathee, S.; Maheshwari, S.; Siddiquee, A.N.; Srivastava, M. Distribution of Reinforcement Particles in Surface Composite Fabrication via Friction Stir Processing: Suitable Strategy. *Mater. Manuf. Process.* **2018**, *33*, 262–269. [CrossRef]
27. Wen., H. *Processing, Microstructure, Mechanical Behavior and Deformation Mechanisms of Bulk Nanostructured Copper and Copper Alloys*; University of California: Davis, CA, USA, 2012.
28. Deng, H.; Yi, J.; Xia, C.; Yi, Y. Improving the Mechanical Properties of Carbon Nanotube-Reinforced Pure Copper Matrix Composites by Spark Plasma Sintering and Hot Rolling. *Mater. Lett.* **2018**, *210*, 177–181. [CrossRef]
29. Kim, K.T.; Cha, S., II; Hong, S.H.; Hong, S.H. Microstructures and Tensile Behavior of Carbon Nanotube Reinforced Cu Matrix Nanocomposites. *Mater. Sci. Eng. A* **2006**, *430*, 27–33. [CrossRef]
30. Asgharzadeh, H.; Eslami, S. Effect of Reduced Graphene Oxide Nanoplatelets Content on the Mechanical and Electrical Properties of Copper Matrix Composite. *J. Alloys Compd.* **2019**, *806*, 553–565. [CrossRef]
31. Deng, H.; Yi, J.; Xia, C.; Yi, Y. Mechanical Properties and Microstructure Characterization of Well-Dispersed Carbon Nanotubes Reinforced Copper Matrix Composites. *J. Alloys Compd.* **2017**, *727*, 260–268. [CrossRef]
32. German, R.M. *Particulate Composites*; Springer International Publishing: Cham, Switzerland, 2016; ISBN 978-3-319-29915-0.
33. Nautiyal, H.; Srivastava, P.; Khatri, O.P.; Mohan, S.; Tyagi, R. Wear and Friction Behavior of Copper Based Nano Hybrid Composites Fabricated by Spark Plasma Sintering. *Mater. Res. Express* **2019**, *6*, 0850h2. [CrossRef]
34. Oanh, N.T.H.; Viet, N.H.; Kim, J.C.; Kim, J.S. Synthesis and Characterization of Cu–TiC Nanocomposites by Ball Milling and Spark Plasma Sintering. *Mater. Sci. Forum* **2014**, *804*, 173–176. [CrossRef]
35. Oanh, N.T.H.; Viet, N.H.; Kim, J.S.; Dudina, D.V. Structural Investigations of TiC–Cu Nanocomposites Prepared by Ball Milling and Spark Plasma Sintering. *Metals* **2017**, *7*, 123.
36. Farías, I.; Olmos, L.; Jiménez, O.; Flores, M.; Braem, A.; Vleugels, J. Wear Modes in Open Porosity Titanium Matrix Composites with TiC Addition Processed by Spark Plasma Sintering. *Trans. Nonferrous Met. Soc. China* **2019**, *29*, 1653–1664. [CrossRef]
37. Abdel-Aziem, W.; Hamada, A.; Makino, T.; Hassan, M. Microstructural Evolution during Extrusion of Equal Channel Angular-Pressed AA1070 Alloy in Micro/Mesoscale. *Mater. Sci. Technol.* **2020**, *36*, 1169–1177. [CrossRef]
38. Zak, A.K.; Majid, W.A.; Abrishami, M.E.; Yousefi, R. X-Ray Analysis of ZnO Nanoparticles by Williamson–Hall and Size–Strain Plot Methods. *Solid State Sci.* **2011**, *13*, 251–256. [CrossRef]
39. Fathy, A.; Elkady, O.; Abu-Oqail, A. Production and Properties of Cu-ZrO2 Nanocomposites. *J. Compos. Mater.* **2018**, *52*, 1519–1529. [CrossRef]
40. Sohag, M.A.Z.; Gupta, P.; Kondal, N.; Kumar, D.; Singh, N.; Jamwal, A. Effect of Ceramic Reinforcement on the Microstructural, Mechanical and Tribological Behavior of Al-Cu Alloy Metal Matrix Composite. *Mater. Today Proc.* **2020**, *21*, 1407–1411. [CrossRef]
41. Cavaliere, P. Spark Plasma Sintering of Materials: Advances in Processing and Applications. *Spark Plasma Sinter. Mater. Adv. Process. Appl.* **2019**, 1–781.
42. Pan, Y.; Xiao, S.Q.; Lu, X.; Zhou, C.; Li, Y.; Liu, Z.W.; Liu, B.W.; Xu, W.; Jia, C.C.; Qu, X.H. Fabrication, Mechanical Properties and Electrical Conductivity of Al$_2$O$_3$ Reinforced Cu/CNTs Composites. *J. Alloys Compd.* **2019**, *782*, 1015–1023. [CrossRef]
43. Tu, J.P.; Wang, N.Y.; Yang, Y.Z.; Qi, W.X.; Liu, F.; Zhang, X.B.; Lu, H.M.; Liu, M.S. Preparation and Properties of TiB2 Nanoparticle Reinforced Copper Matrix Composites by in Situ Processing. *Mater. Lett.* **2002**, *52*, 448–452. [CrossRef]

44. Babapoor, A.; Asl, M.S.; Ahmadi, Z.; Namini, A.S. Effects of Spark Plasma Sintering Temperature on Densification, Hardness and Thermal Conductivity of Titanium Carbide. *Ceram. Int.* **2018**, *44*, 14541–14746. [CrossRef]
45. Yin, Z.; Huang, C.; Zou, B.; Liu, H.; Zhu, H.; Wang, J. Study of the Mechanical Properties, Strengthening and Toughening Mechanisms of Al2O3/TiC Micro-Nano-Composite Ceramic Tool Material. *Mater. Sci. Eng. A* **2013**, *577*, 9–15. [CrossRef]
46. Peng, T.; Yan, Q.; Zhang, X.; Zhuang, Y. Role of Titanium Carbide and Alumina on the Friction Increment for Cu-Based Metallic Brake Pads under Different Initial Braking Speeds. *Friction* **2021**, *9*, 1543–1557. [CrossRef]
47. Venkateswarlu, M.; Kumar, M.A.; Reddy, K.H.C. Thermal Behavior of Spark Plasma Sintered Ceramic Matrix-Based Nanocomposites. *J. Bio-Tribo-Corrosion* **2020**, *6*, 6013–6028. [CrossRef]
48. Wagih, A.; Abu-Oqail, A.; Fathy, A. Effect of GNPs Content on Thermal and Mechanical Properties of a Novel Hybrid Cu-Al2O3/GNPs Coated Ag Nanocomposite. *Ceram. Int.* **2019**, *45*, 1115–1124. [CrossRef]
49. Kumar, E.G.; Ahasan, M.; Venkatesh, K.; Sastry, K.S.B.S.V.S. Design, Fabrication of Powder Compaction Die and Sintered Behavior of Copper Matrix Hybrid Composite. *Int. Res. J. Eng. Technol.* **2018**, *5*, 876–882.
50. Akkaş, M.; Islak, S.; Özorak, C. Corrosion and Wear Properties of Cu-TiC Composites Produced by Hot Pressing Technique. *Celal Bayar Üniversitesi Fen Bilim. Derg.* **2018**, *14*, 465–469. [CrossRef]
51. Palma, R.H.; Sepúlveda, A.H.; Espinoza, R.A.; Montiglio, R.C. Performance of Cu-TiC Alloy Electrodes Developed by Reaction Milling for Electrical-Resistance Welding. *J. Mater. Process. Technol.* **2005**, *169*, 62–66. [CrossRef]
52. Akhtar, F.; Askari, S.J.; Shah, K.A.; Du, X.; Guo, S. Microstructure, Mechanical Properties, Electrical Conductivity and Wear Behavior of High Volume TiC Reinforced Cu-Matrix Composites. *Mater. Charact.* **2009**, *60*, 327–336. [CrossRef]
53. Wang, F.; Li, Y.; Yamanaka, K.; Wakon, K.; Harata, K.; Chiba, A. Influence of Two-Step Ball-Milling Condition on Electrical and Mechanical Properties of TiC-Dispersion-Strengthened Cu Alloys. *Mater. Des.* **2014**, *64*, 441–449. [CrossRef]
54. Öksüz, K.E.; Şahin, Y. Microstructure and Hardness Characteristics of Al2O3-B4C Particle-Reinforced Cu Matrix Composites. *Acta Phys. Pol. A* **2016**, *129*, 650–652. [CrossRef]
55. Panda, S.; Dash, K.; Ray, B.C. Processing and Properties of Cu Based Micro- and Nano-Composites. *Bull. Mater. Sci.* **2014**, *37*, 227–238. [CrossRef]
56. Orolínová, M.; Ďurišin, J.; Ďurišinová, K.; Danková, Z.; Ďurišin, M. Effect of Microstructure on Properties of Cu-Al$_2$O$_3$ Nanocomposite. *Chem. Mater. Eng.* **2013**, *1*, 60–67. [CrossRef]
57. Fathy, A.; Shehata, F.; Abdelhameed, M.; Elmahdy, M. Compressive and Wear Resistance of Nanometric Alumina Reinforced Copper Matrix Composites. *Mater. Des.* **2012**, *36*, 100–107. [CrossRef]
58. Dash, K.; Ray, B.C.; Chaira, D. Synthesis and Characterization of Copper-Alumina Metal Matrix Composite by Conventional and Spark Plasma Sintering. *J. Alloys Compd.* **2012**, *516*, 78–84. [CrossRef]
59. Ritasalo, R.; Liu, X.W.; Söderberg, O.; Keski-Honkola, A.; Pitkänen, V.; Hannula, S.P. The Microstructural Effects on the Mechanical and Thermal Properties of Pulsed Electric Current Sintered Cu-Al2O3 Composites. *Procedia Eng.* **2011**, *10*, 124–129. [CrossRef]

Article

# Effect of Copper Addition on the AlCoCrFeNi High Entropy Alloys Properties via the Electroless Plating and Powder Metallurgy Technique

Mohamed Ali Hassan [1], Hossam M. Yehia [2], Ahmed S. A. Mohamed [1,3], Ahmed Essa El-Nikhaily [4] and Omayma A. Elkady [5,*]

1. Mechanical Department, Faculty of Technology and Education, Sohag University, Sohag 82524, Egypt; Mohammed_Ali@techedu.sohag.edu.eg (M.A.H.); Ahmedquse2000@yahoo.com (A.S.A.M.)
2. Mechanical Department, Faculty of Technology and Education, Helwan University, Cairo 11795, Egypt; Hossamelkeber@techedu.helwan.edu.eg
3. High Institute for Engineering and Technology, Sohag 82524, Egypt
4. Mechanical Department, Faculty of Technology and Education, Suez University, Suez 41522, Egypt; Ahmedeassa1946@gmail.com
5. Powder Technology Department, Central Metallurgical R & D Institute, P.O. Box 87 Helwan, Cairo 11421, Egypt
* Correspondence: o.alkady68@gmail.com

**Citation:** Hassan, M.A.; Yehia, H.M.; Mohamed, A.S.A.; El-Nikhaily, A.E.; Elkady, O.A. Effect of Copper Addition on the AlCoCrFeNi High Entropy Alloys Properties via the Electroless Plating and Powder Metallurgy Technique. *Crystals* **2021**, *11*, 540. https://doi.org/10.3390/cryst11050540

Academic Editors: Sergio Brutti and Fawad Inam

Received: 23 February 2021
Accepted: 26 March 2021
Published: 12 May 2021

**Publisher's Note:** MDPI stays neutral with regard to jurisdictional claims in published maps and institutional affiliations.

**Copyright:** © 2021 by the authors. Licensee MDPI, Basel, Switzerland. This article is an open access article distributed under the terms and conditions of the Creative Commons Attribution (CC BY) license (https://creativecommons.org/licenses/by/4.0/).

**Abstract:** To improve the AlCoCrFeNi high entropy alloys' (HEAs') toughness, it was coated with different amounts of Cu then fabricated by the powder metallurgy technique. Mechanical alloying of equiatomic AlCoCrFeNi HEAs for 25 h preceded the coating process. The established powder samples were sintered at different temperatures in a vacuum furnace. The HEAs samples sintered at 950 °C exhibit the highest relative density. The AlCoCrFeNi HEAs model sample was not successfully produced by the applied method due to the low melting point of aluminum. The Al element's problem disappeared due to encapsulating it with a copper layer during the coating process. Because the atomic radius of the copper metal (0.1278 nm) is less than the atomic radius of the aluminum metal (0.1431 nm) and nearly equal to the rest of the other elements (Co, Cr, Fe, and Ni), the crystal size powder and fabricated samples decreased by increasing the content of the Cu wt%. On the other hand, the lattice strain increased. The microstructure revealed that the complete diffusion between the different elements to form high entropy alloy material was not achieved. A dramatic decrease in the produced samples' hardness was observed where it decreased from 403 HV at 5 wt% Cu to 191 HV at 20 wt% Cu. On the contrary, the compressive strength increased from 400.034 MPa at 5 wt% Cu to 599.527 MPa at 15 wt% Cu with a 49.86% increment. This increment in the compressive strength may be due to precipitating the copper metal on the particles' surface in the nano-size, reducing the dislocations' motion, increasing the stiffness of produced materials. The formability and toughness of the fabricated materials improved by increasing the copper's content. The thermal expansion has increased gradually by increasing the Cu wt%.

**Keywords:** high entropy alloys; electroless copper plating; thermal expansion; hardness; compressive strength

## 1. Introduction

The high entropy alloys (HEAs) are a novel class of materials that are different from conventional alloys, which contains only one or two base elements. HEAs usually consist of more than four main elements. It has already been coined because the entropy is substantially higher under two conditions. One is when there are more significant elements in the mix, and the other is when their proportions are nearly equal. HEAs are characterized by high strength and hardness, high thermal stability, and excellent corrosion resistance. However, it suffers from high brittleness [1–9]. The lattice crystal structures of most HEAs are either body-centered cubic (BCC) [10] or face-centered cubic (FCC) [11]. HEAs with BCC structure exhibit low plasticity yet high strength, while FCC-structured HEAs have increased mobility with low strength [12]. The choice of elements for HEAs' design is essential when taking the specific applications and economical alloy into consideration [13,14]. A variety of alloys focusing on the influence of different elements on HEAs, such as Al [15], Ni [16], Cu [17], Ti [18], and Sn [19], have been widely published.

Recently, many HEA systems have been exploited, and most research focused on the microstructure and properties of the Al$x$CoCrFeNi HEAs system [20–22]. The addition of copper (Cu) to a HEA matrix has been widely investigated [23]; this is probably because the atomic radii of elements are close to one another and because there is a tendency to form a reliable solution according to the phase rule and phase diagram between them [24]. The main difference between AlxCoCrFeNi and AlCoCrFeNiCux is the presence of the Cu-rich phases at last. This is because Cu has positive enthalpy mixtures with most other elements and has thus repelled to the inter-dendrite (IR) region [25]. Moreover, the remnant Cu in the dendritic area also clusters and forms various sediments [26]. The investigators also found that the Cu element can enhance the FCC phase formation [27]. Furthermore, the increase in copper content decreases the oxidation ratio for AlCoCrCu$_x$FeNi HEAs [28].

The AlCoCrFeNiCu is preferentially produced via conventional metallurgy, for example, arc melting or induction melting [29]. These methods result predominantly in a dendritic microstructure. Additionally, high interfacial energy between the molten metals and Cu reinforcement particles reduces the wettability, which causes the formation of pores that profoundly affect electrical conductivity. Consequently, it renders the casting technique applications for the copper matrix composites [30]. Furthermore, the machining needed for the final parts could be challenging due to these alloys' high hardness. Therefore, powder metallurgy (PM) could be a suitable alternative technique for the manufacturing process. The process has been described in the literature, where it is performed in three main steps: mixing, cold forming, and then sintering at a suitable temperature [31–34]. The PM process is characterized by no need for further machining, minimized scrap loss of raw materials, and facilitated alloying elements. To establish good adhesion and distribution for the copper element with the different alloying elements of the HEAs, and consequently, achieving high strength, the electroless precipitation process is recommended. Additionally, this technique was suitable for the production of nano-sized copper metal, which improves the strength of material according to the Hall–Petch equation [35,36].

Hence, the purpose of the current study was to investigate the effect of sintering temperatures on the density, microstructure, physical, and mechanical properties of the $Cu_x/(AlCoCrFeNi)_{1-x}$ ($x = 5$, 10, 15, and 20 wt%) fabricated by the powder metallurgy (PM) technique. Three various sintering temperatures were considered for all the prepared samples, 900, 950, and 1000 °C, to optimize the suitable sintering temperature for the preparation of $Cu_x/(AlCoCrFeNi)_{1-x}$ HEA.

## 2. Experimental Procedure

### 2.1. Material

In the present study, elemental powders with high purity (>99.5%) of Al, Cr, Co, Ni, and Fe with an average particle size (<45 µm) supplied from Nore industrial and laboratory chemicals (Nilcco)—Cairo, Egypt were used as the start materials. The electroless plating of copper on the AlCoCrFeNi alloy powder particles was performed using a bath containing copper (II) sulfate pentahydrate, sodium hydroxide, potassium sodium tartrate, and formaldehyde. All chemicals were supplied by El-Alsharq Al'awsat Company for chemicals—Cairo, Egypt.

### 2.2. HEA Preparation

Al, Co, Cr, Ni, and Fe with an equimolar ratio were mixed in a stainless-steel container for 25 h using a stainless-steel balls with a 10-mm diameter at 120 rpm rotational speed. The process proceeded under argon atmosphere to avoid any oxidation. The ball to powder ratio (BPR) was 5:1. The mixing process was followed by the electroless copper coating from the bath, as shown in Table 1, to prepare four powder samples $Al_{19}Co_{19}Cr_{19}Fe_{19}Ni_{19}Cu_5$, $Al_{18}Co_{18}Cr_{18}Fe_{18}Ni_{18}Cu_{10}$, $Al_{17}Co_{17}Cr_{17}Fe_{17}Ni_{17}Cu_{15}$, and $Al_{16}Co_{16}Cr_{16}Fe_{16}Ni_{16}Cu_{20}$ of high entropy alloys. The coating process was established by dissolving copper sulfate, adjusting the pH ~ 12 by sodium hydroxide, and adding $Al_{20}Co_{20}Cr_{20}Ni_{20}Fe_{20}$ mixed powders. Finally, formaldehyde as a reducing agent was added to start the reaction. The coated HEAs powder samples were filtered, then dried in an electric furnace at 90 °C for 1 h. After the coating process was completed, the AlCoCrFeNi/Cu powders were filtered and washed with distilled water, and dried in an electric furnace at 90 °C for 2 h. The weight percentage of Cu was adjusted by weighing each powder sample after the coating process.

**Table 1.** Electroless chemical composition bath of copper precipitation.

| Materials | Weight |
|---|---|
| Copper (II) sulfate, ($CuSO_4$) | 70 g/L |
| Potassium sodium tartrate, ($KNaC_4H_4O_6 \cdot 5H_2O$) | 170 g/L |
| Sodium hydroxide, (NaOH) | 50 g/L |
| Formaldehyde | 200 mL/L |
| Temperature | 60 °C |

The nano copper was precipitated in cuprous oxide (CuO) due to Cu's oxidation during the electroless deposition process. Accordingly, in order to reduce CuO to pure Cu metal, the prepared HEAs powders were heated at 450 °C for 1 h under hydrogen atmosphere.

Cold compaction at 800 MPa using a universal hydraulic press was applied to consolidate the prepared $Cu_x/(AlCoCrFeNi)_{1-x}$ powder samples. All the obtained green materials were sintered in a vacuum furnace at different heating temperatures of 900, 950, and 1000 °C for 90 min by the heating cycle shown in Figure 1. Three holding temperatures during the sintering process were used. The first was the de-waxing step at 250 °C to remove paraffin wax, the second was at 850 °C to achieve high diffusivity for cobalt and other elements, and the last one was at 950 °C to complete the sintering process. Figure 2 illustrates the procedures of the preparation and fabrication of $(AlCoCrFeNi)_{1-x}/Cu_x$ HEAs materials.

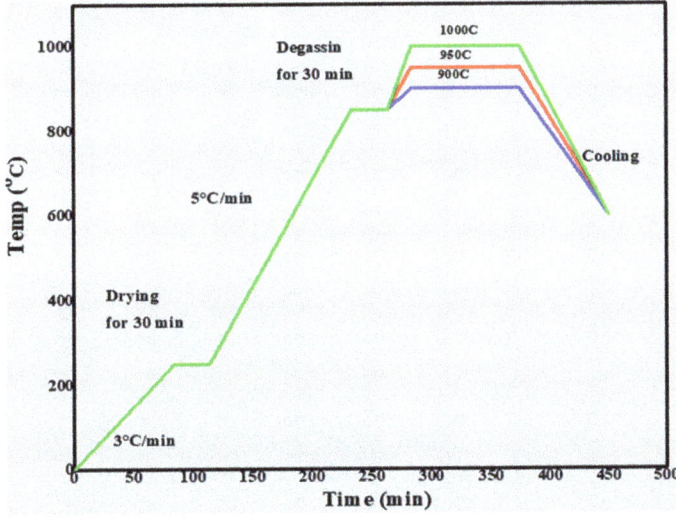

**Figure 1.** Heating cycle of the sintering process.

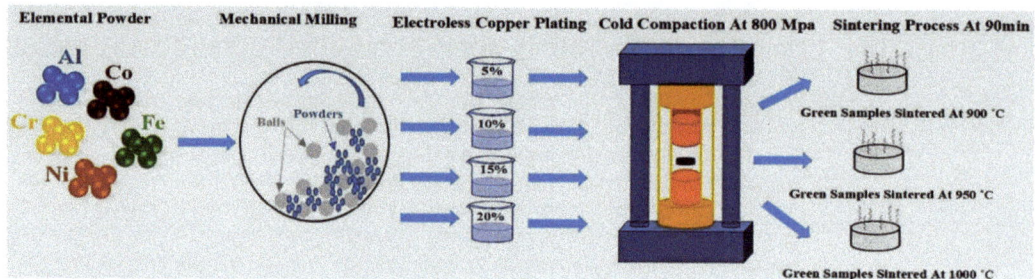

**Figure 2.** Flowchart of $Cu_x/(AlCoCrFeNi)_{1-x}$ HEAs powder samples' preparation and fabrication.

2.3. Characterization and Analysis

The bulk density of the sintered alloys was measured using Archimedes' rule according to the standard (ASTM B962-14) [37]. The sintered samples were weighed at room temperature in air and in distilled water as floating liquid.

$$\rho_{Arch} = \frac{W_{air}}{W_{air} - W_{water}} \quad (1)$$

The relative density ($R_d$) was determined using the following equation:

$$R_d = \frac{\rho_{Arch}}{\rho_{th}} \rho_{liquid} \quad (2)$$

where $\rho_{Arch}$ is the Archimedes density, $\rho_{th.}$ is the theoretical density, and $\rho_{liquid}$ is the liquid density.

The theoretical density $\rho_{th}$ of the samples was calculated by using the following equation:

$$\rho_{theo} = \rho_{Cu} wt\%_{Cu} + \rho_{Al} wt\%_{Al} + \rho_{Co} wt\%_{Co} + \rho_{Cr} wt\%_{Cr} + \rho_{Fe} wt\%_{Fe} + \rho_{Ni} wt\%_{Ni} \quad (3)$$

where ($\rho_{Cu}$) and ($wt\%_{Cu}$) are the density and the weight percent of the Cu element, ($\rho_{Al}$) and ($wt\%_{Al}$) are the density and the weight percent of the Al element, ($\rho_{Co}$) and ($wt\%_{Co}$) are the density and the weight percent of the Co element, and ($\rho_{Cr}$) and ($wt\%_{Cr}$) are the density and weight percent of the Cr element, respectively.

In order to ensure the repeatability of the test, we recommend performing it for three times as a range. For microstructure observation, sample surfaces were prepared using silicon carbide papers (SiC) with 600, 800, 1000, and 1200 grades, respectively.

The crystal and phase structure of the powder and consolidated samples were investigated using (X'pert PRO PANalytical) a PANalytical X'Pert Pro device (Panalytical, Almelo, The Netherlands) X-ray diffraction with Cu kα radiation (λ = 0.15406 nm). The 0.02° scan rate in a range of 20° to 100° was used. The Scherrer equation [38] was used to calculate the crystallite size of the coated powders using the XRD peak broadening equation as follows:

$$D = \frac{0.9\lambda}{B\cos\theta} \quad (4)$$

While the lattice strain ε was determined based on Equation (5) introduced by Danilchenko et al., [39] as follows:

$$\varepsilon = \frac{B}{4\tan\theta} \quad (5)$$

where $D$ is the crystallite size, $B$ is the full width at half maximum (FWHM), λ is the wavelength, and θ is the peak position.

Scanning electron microscopy (SEM, FEI model 'Philips XL30') FE-SEM (Philips XL 30, Royal Dutch Philips Electronics Ltd., Amsterdam, The Netherlands) equipped with back scattered-electron (BSE) mode and an energy dispersive X-ray spectroscopy (EDX) microanalysis system (operated at accelerating voltage 20 kV, 1.2 nA beam current, and 10 microsecond dwell time for minimizing SEM image noise) was used to characterize the morphology and microregion composition of sintered samples.

The Vickers hardness (micro and macro-hardness) of fabricated HEAs alloys using a Leitz Durimet microhardness tester (Leitz, Oberkochen, Germany) and Vickers hardness tester model 5030 SKV England (Indentec 5030 SKV, Stourbridge, West Midlands, UK) was measured. The micro-hardness test was performed by a Vickers indenter under a static load of 25 g and a dwell time of 5 s according to the (ASTM E384-11) at the temperature of 25 °C $\pm$ 3 °C [40]. On the other hand, Vicker's macro hardness was measured at 10 kg for 15 s. The average of five readings for each hardness value was calculated. A digital indicator with 0.001-mm precision and an electrical furnace were used to investigate the thermal strain of the $Cu_x/(AlCoCrFeNi)_{1-x}$ alloys, and consequently, the coefficient of thermal expansion (CTE). The 3 °C/min heating and cooling rates in an argon atmosphere were used. The measurements of the thermal expansion were performed in the range of 100 to 350°C and repeated for three times.

The compressive strength test of the samples was measured using a uniaxial SHIMADZU universal testing machine (Shimatzu, Kyoto, Japan) (UH-F500KN). The ratio between the length and width of the used samples was 1. The applied cross-head speed of the used universal test machine was 3 mm/min. The test was performed at room temperature (25 °C).

## 3. Results and Discussion

### 3.1. Density Measurement

The effect of the Cu proportion on the relative density of the $Cu_x/(AlCoCrFeNi)_{1-x}$ alloys at 900, 950, and 1000 °C are shown in Figure 3. In general, the sintered samples' relative density increased at all sintering temperatures by increasing the Cu ratio. Results revealed that 950 °C is the best suitable sintering temperature for achieving the $Cu_x/(AlCoCrFeNi)_{1-x}$ HEAs' highest densification. Experimentally, the AlCoCrFeNi alloy was not established in these sintering conditions, which is why its density was excluded from the present results. The addition of copper to the AlCoCrFeNi HEA facilitates the sintering process and a 98% relative density at $Al_{16}Co_{16}Cr_{16}Fe_{16}Ni_{16}Cu_{20}$ was achieved. Copper precipitated in the nano size, which helps to reduce the heating temperature of sintering. Additionally, its presence on the grain boundaries of the AlCoCrFeNi facilitates the interfacial bonding and then impedes the formation of micropores or voids during the sintering process. On the other hand, the electroless process' reduction in crystallite size helped increase the surface area, which enhances the diffusion phenomenon. Accordingly, the relative density gets closer to the theoretical one [41,42]. Sintering at 900 °C records the lowest densification for the prepared samples. This reduction in densification may be due to the particles' low diffusion temperature of the different elements. So, little densification takes place and, consequently, low density is achieved. On the other hand, the reduction in density values at 1000 °C may be due to copper swelling during the sintering process [43].

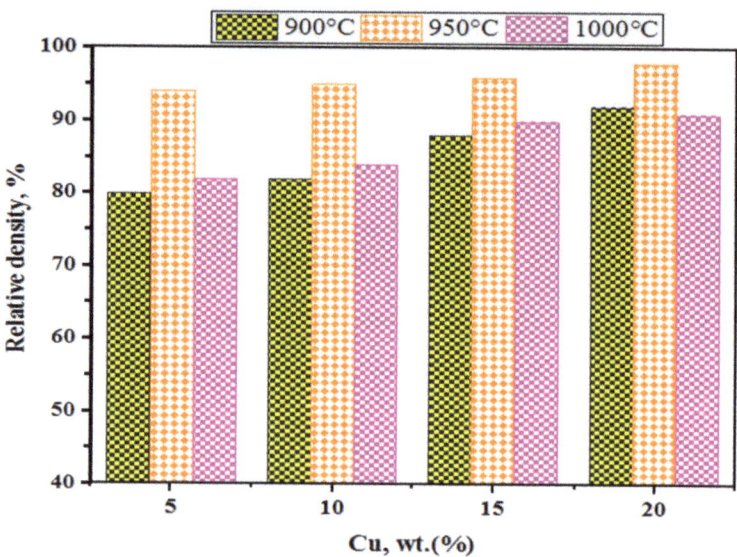

**Figure 3.** Relative density for $Cu_x/(AlCoCrFeNi)_{1-x}$ HEAs sintered at 900, 950, and 1000 °C.

*3.2. Microstructure Investigation*

The prepared $Cu_x/(AlCoCrFeNi)_{1-x}$ alloy powders were characterized by FE-SEM (BSE), as shown in Figure 4. High homogeneity between all the constituents of the differently prepared powder alloys was observed. As a result of milling AlCoCrFeNi equiatomic HEAs for 25 h, cold welding between the powders occurred, and a flake powder shape was noted. Such preliminaries combined leads one to predict the formation of alloying between the used elements that will in turn form solid solution and consequently, the high entropy. Additionally, the reduction in the particle sizes facilitates the accumulation between the powders. The microstructure also shows a good coating of the nano-copper particle on AlCoCrFeNi particles' surface due to the high quality of the electroless Cu deposition process. An image for the precipitated nano copper after reduction by hydrogen is shown in the last picture to confirm its nano size.

The mapping analysis of the 20 wt% $Cu_x/(AlCoCrFeNi)_{1-x}$ powder alloy is shown in Figure 5. The mapping illustrates a high homogeneous distribution of the prepared element powder alloys. It confirmed the presence of all the used raw materials. No oxides were detected, which means an excellent mixing control atmosphere and a reasonable reduction of CuO.

Figure 6 illustrates the Cu content's effect on the microstructure of the AlCoCrFeNi HEAs fabricated at 950 °C. The microstructures of the sintered samples a, b, c, and d show a homogeneous distribution of all the constituent elements that are well-bonded and connected. Three distinct regions are observed from the microstructure: the first is the dark area, which is believed to be the Cr-rich BCC; the second is the bright one, which refers to the Cu rich FCC; and the third is the AlCoCrFeNi matrix FCC. Agglomerations of Cr appeared due to the milling process. As shown, Al, Co, Fe, and Ni entirely merged with each other, and an FCC solid solution formed. The solid solution formed in the principal return of the prior mechanical milling and establishes the alloying process between the different elements. On the other hand, the Cu and Cr elements did not dissolve to participate in the formation of high entropy alloys. Based on this result of the microstructure, the completion of high entropy alloys formation was not achieved with the powder metallurgy technique under the current conditions. As the Cu element increases, the dominant phases of the $Cu_x/(AlCoCrFeNi)_{1-x}$ alloy switches to the FCC (matrix, Cu rich) phase, and the amount of the Cr-rich BCC phase decreases. Cr versus Cu's reduction resulting from negative vacancy-formation energies indicates a thermodynamic drive for Cr to segregate [44]. Coating AlCoCrFeNi HEA by copper reduces the surface energy between the surfaces of the elements and consequently improves the wettability, and excellent adhesion between all elements of the alloys is achieved. No pores were noticed, which means high densification for all fabricated alloys.

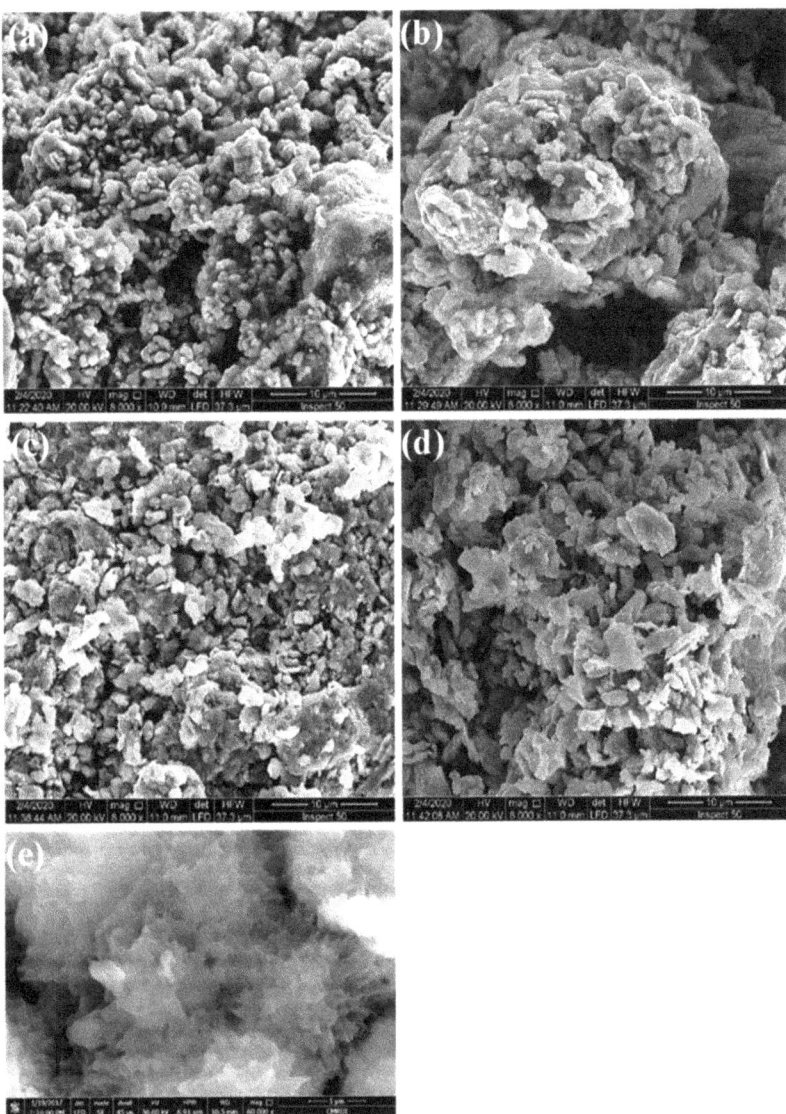

**Figure 4.** SEM (BSE) micrographs of the fabricated (**a**) $Al_{19}Co_{19}Cr_{19}Fe_{19}Ni_{19}Cu_5$, (**b**) $Al_{18}Co_{18}Cr_{18}Fe_{18}Ni_{18}Cu_{10}$, (**c**) $Al_{17}Co_{17}Cr_{17}Fe_{17}Ni_{17}Cu_{15}$, (**d**) $Al_{16}Co_{16}Cr_{16}Fe_{16}Ni_{16}Cu_{20}$ and (**e**) Pure precipitated Cu powder alloys.

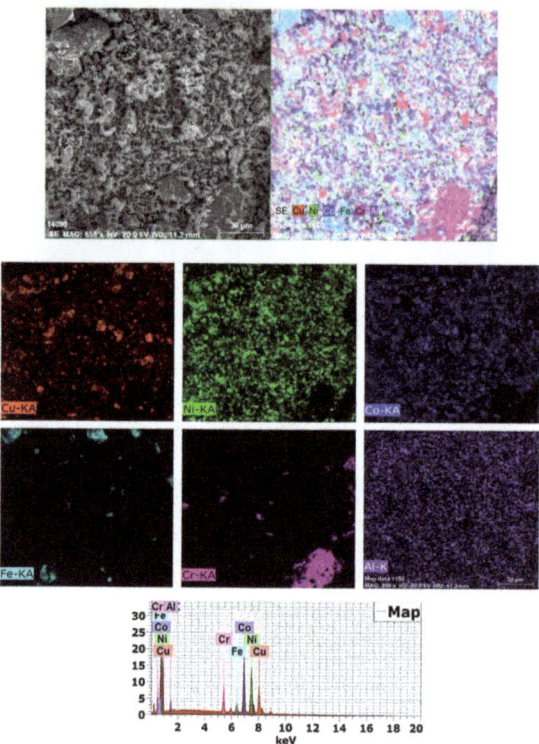

**Figure 5.** Mapping of the $Al_{16}Co_{16}Cr_{16}Fe_{16}Ni_{16}Cu_{20}$ powder sample.

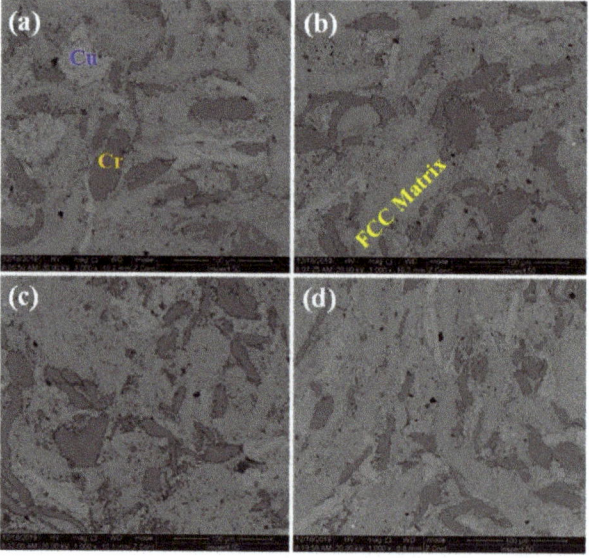

**Figure 6.** SEM-BSE micrographs of the fabricated (**a**) $Al_{19}Co_{19}Cr_{19}Fe_{19}Ni_{19}Cu_5$, (**b**) $Al_{18}Co_{18}Cr_{18}Fe_{18}Ni_{18}Cu_{10}$, (**c**) $Al_{17}Co_{17}Cr_{17}Fe_{17}Ni_{17}Cu_{15}$, and (**d**) $Al_{16}Co_{16}Cr_{16}Fe_{16}Ni_{16}Cu_{20}$ samples.

The SEM-EDX elemental mapping of the four fabricated HEAs samples is shown in Figure 7. The mapping represents the distribution of each element in the formed piece with a different color. Six principal components, including Al (red), Co (rose), Cr (green), Cu (yellow), Fe (blue), and Ni (magenta), were observed. Clear agglomeration of the Cr element, which prevents the formation of the high entropy alloys, is evident in all samples. As shown, the Cu element tends to dissolve with the FCC phase that represents the matrix of all the fabricated pieces.

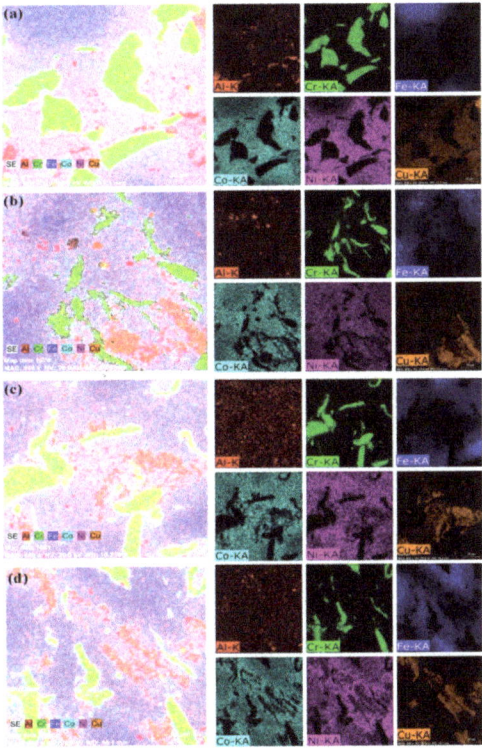

**Figure 7.** Mapping chart of the fabricated (**a**) $Al_{19}Co_{19}Cr_{19}Fe_{19}Ni_{19}Cu_5$, (**b**) $Al_{18}Co_{18}Cr_{18}Fe_{18}Ni_{18}Cu_{10}$, (**c**) $Al_{17}Co_{17}Cr_{17}Fe_{17}Ni_{17}Cu_{15}$, and (**d**) $Al_{16}Co_{16}Cr_{16}Fe_{16}Ni_{16}Cu_{20}$ samples at 950 °C.

### 3.3. XRD of the Powder and Sintered HEAs

The X-ray diffraction details of the $Cu_x/(AlCoCrFeNi)_{1-x}$ HEA powders at room temperature are shown in Figure 8. The figure reflects the impact of the Cu content on the chemical composition and crystal structure of the AlCoCrFeNi. It was observed that the peak intensity of Cu became shinier by increasing its percentage. No new intermetallics were formed as a result of milling or electroless plating processes. The first peak of Al and Co completely disappeared, which is consistent with Shivam et al., [45]. Shivam et al. investigated the influence of milling times at 0, 5, 10, 15, 20, 25, and 30 h on the crystal structure and chemical composition of AlCoCrFeNi HEA powders. The results revealed that the first peak of Al dissipated at 10, which is attributed to the lower melting temperature of Al, indicating the possibility of dissolving in the host lattice of Fe/Cr. They observed that the Fe structure's solid solution is formed by increasing the milling time up to 20 h. Further milling proceeded up to 30 h to become more homogenized solid solution with more refined grains. Suryanarayana C. et al., [46] revealed that forming a solid solution phase might mainly be attributed to the high entropy and stronger bonding among constituent elements, atomic size difference, and electro-negativities of the atoms.

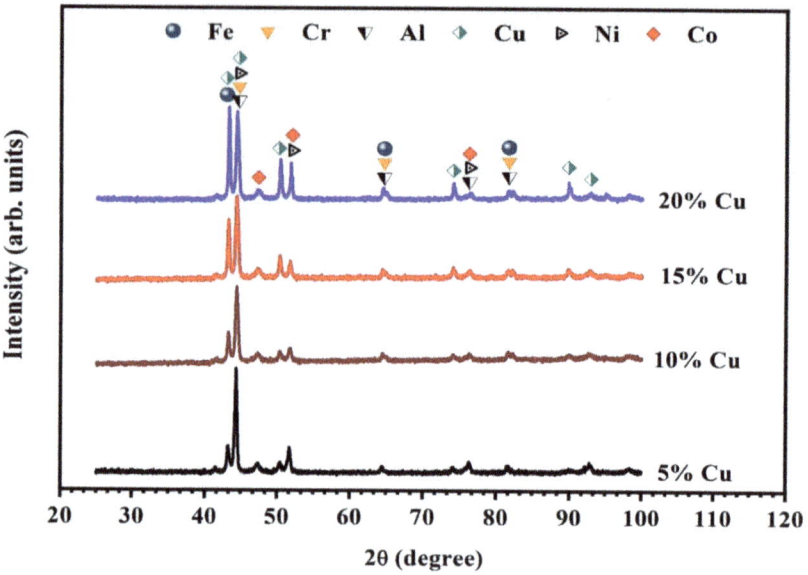

**Figure 8.** XRD patterns of the $Cu_x/(AlCoCrFeNi)_{1-x}$ powder alloys after milling and Cu coating.

As Shang et al., [47] investigated the effect of different milling times on CoCrFeNiW HEA powders' physical properties, they found that the used element's spectrum was still visible and evident up to 40 h. This study also demonstrated that the elements' peaks do not disappear simultaneously; for example, Cr and Fe's crystalline peaks go thoroughly after 40 h, while Co also dissolves after 5 h of milling. From those mentioned above, one can notice the elements dissolving and transforming into FCC or BCC phases. It is not related to a specific milling time and does not indicate the ideal milling conditions. Copper oxide phases (CuO and $Cu_2O$) were not found at the XRD spectrum, proving that the deposited copper has high purity with all molecule compositions in the electroless bath [48].

The crystal size and lattice strain of the $Cu_x/(AlCoCrFeNi)_{1-x}$ powder alloys were calculated from the X-ray peak broadening using Scherrer's formula and are presented in Table 2. The crystallite size was significantly refined as the weight percentage of the increased Cu. The refining in the particle size by increasing the Cu wt% was attributed to the precipitation of copper in nanoscale size. Coating the mixed Al, Co, Cr, Fe, and Ni with a layer of nano Cu particles prevents the green growth between the constituent particles. So, grain refining takes place. On the other hand, the lattice strain results revealed that the lattice string increases by increasing the Cu wt% deposited layer. Generally, the reasons for the increment in the lattice strain can be attributed to the size mismatch effect between the elements, increasing the grain boundary fraction, and high mechanical deformation [49].

**Table 2.** Crystal size and lattice strain of the HEAs powder samples.

| Wt% Cu | Crystal Size (nm) | Lattice Strain (%) | Full Width at Half Maximum (FWHM) |
|---|---|---|---|
| 5 | 29.78 | 0.081 | 0.32 |
| 10 | 23.17 | 0.104 | 0.411 |
| 15 | 22.26 | 0.109 | 0.428 |
| 20 | 21.98 | 0.111 | 0.434 |

The XRD patterns of the HEAs sintered samples are shown in Figure 9. The peaks' broadening, height, and number changed after the sintering process, where the detected peaks' extension became more comprehensive, and their intensities smaller. Some peaks disappeared during the sintering process. These changes may be attributed to two main factors, i.e., high lattice strain and refined crystal size. The formation of a solid solution phase may mainly have participated in the high entropy and stronger bonding among constituent elements, atomic size difference, and electro-negativities of the atoms [50]. The X-ray reflects three crystalline structures (Cu-rich FCC 1, Cr-rich BCC, and matrix FCC 2). The total number of phases in the prepared alloys is well below the maximum number of equilibrium phases allowed by the Gibbs phase principles [51]. As seen from the XRD patterns, the Cr fertile BCC phase intensity is much higher than that of both FCC phases. This may be attributed to Al and Cr's presence in equal or higher concentrations than the other elements that stabilize the BCC structure [52]. The Cr fertile BCC phase's relative intensity decreased by increasing the Cu content, which suggests that the weight fraction of the FCC phase increases with Cu reinforcement against the BCC phase [53]. The relative intensity of both FCC phases becomes more durable and more reliable by increasing the Cu content up to 20%. This is because the presence of Cu with higher concentrations supports the formation of the FCC structure [54]. It is primarily due to the purpose of BCC or FCC lattice crystallography investigated by the average periodic table group number or average valence electron concentration (VEC). FCC phases are motivated to be more stable at higher values of VEC ($\geq 8$) while BCC phases maintain stability at lower VEC ($\leq 6.87$). The addition of aluminum particles reduces the VEC value [55]. Thus, the alloy system spontaneously promotes crystal structure transition from FCC to BCC to minimize the lattice distortion energy.

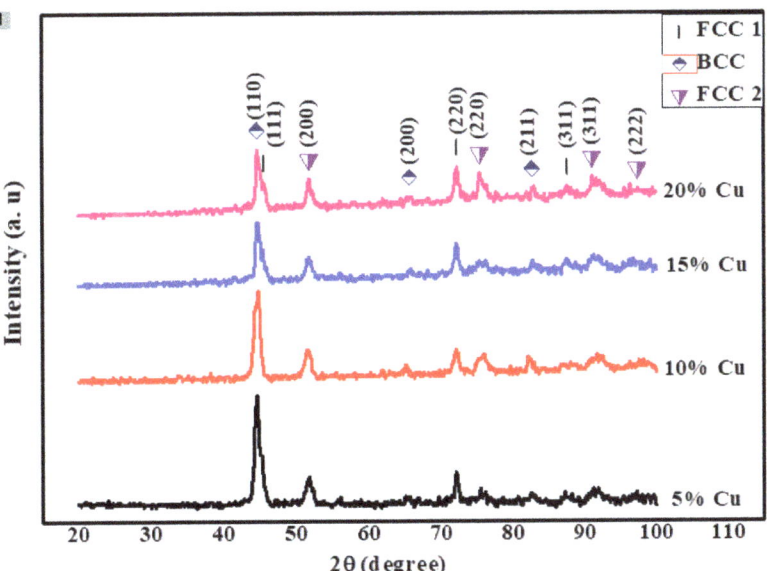

**Figure 9.** XRD patterns of the $Cu_x/(AlCoCrFeNi)_{1-x}$ sintered at 950 °C.

The crystal size and lattice strain parameters of the fabricated $Cu_x/(AlCoCrFeNi)_{1-x}$ alloys are arranged in Table 3. As shown in the table, the crystal size of fabricated HEAs decreased by increasing the precipitated Cu content, which is similar to the result of powder alloys. Additionally, the lattice strain increases by increasing the Cu content. The coating process of the milled AlCoCrFeNi probably affects the atoms' orders and, consequently, plans of the new material during the sintering process. So, the lattice strain distortion increased by increasing the content of the precipitated Cu layer on the surfaces of the AlCoCrFeNi HEA particles.

Table 3. Crystal size and lattice strain of the fabricated HEAs samples.

| Wt% Cu | Crystal Size (nm) | Lattice Strain (%) | Full Width at Half Maximum (FWHM) |
|---|---|---|---|
| 5 | 96.64 | 2.09 | 1.45 |
| 10 | 90.41 | 2.23 | 1.32 |
| 15 | 91.05 | 2.22 | 1.33 |
| 20 | 73.24 | 0.53 | 5.95 |

Figure 10 compares the crystallite sizes between the BCC and both FCC phases by the Scherrer formula. It is significantly noted that the crystallite size of the Cr fertile BCC phase was increased with an increase in the Cu electroless concentration. The crystallite sizes of both FCC phases were decreased with an increase in the Cu electroless concentration. This may be attributed to the increasing in the FCC weight fraction. The effect of Al addition on the BCC phase of the $Al_xCoCrCuFeNi$ HEAs using the Scherrer formula was investigated [56]. The calculations show that the increase in the Al content participates in increasing the BBC phase fraction and decreasing the crystallite size.

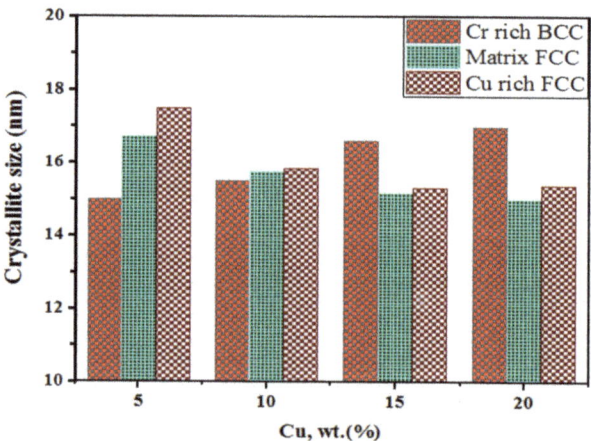

Figure 10. Crystallite sizes of the BCC and both FCC phases for the sintered samples.

Table 4 represents the chemical composition of the three phases obtained by SEM-EDX analysis of the fabricated $Cu_x/(AlCoCrFeNi)_{1-x}$ HEAs samples. It was observed that the Al element has the lowest value in the chemical composition of all manufactured alloys. This may be attributed to the evaporation of Al at the sintering temperature, where it has the lowest melting temperature [30,57]. It also may be related to its high diffusivity in the FCC matrix. Moreover, apparently out of Table 4, both Al and Ni have the same content at Cr-rich BCC due to the segregation attributed to the different mixed enthalpy between Al, Ni, and Cr [58]. The melting point of elements and mixing enthalpy are two main influencing factors affecting the alloying rates [57]. In the solid state, the item with a low melting point has a higher intrinsic diffusion coefficient [59]. At lower temperatures (<300 °C), Cu and Cr are widely immiscible in the solid state and have large positive enthalpies of mixing [60,61].

Table 4. Chemical composition of the FCC and BCC phases obtained by SEM-EDX analysis from fabricated samples.

| Alloy | Region | Mole Fraction/at% | | | | | |
|---|---|---|---|---|---|---|---|
| | | Al | Co | Cr | Fe | Ni | Cu |
| $Al_{19}Co_{19}Cr_{19}Fe_{19}Ni_{19}Cu_5$ | Nominal | | | $(AlCoCrFeNi)_{95}$ | | | 5 |
| | Cr rich BCC | 0.55 | 3.65 | 88.93 | 2.12 | 3.03 | 1.71 |
| | matrix FCC | 5.01 | 30.12 | 14.41 | 5.74 | 26.31 | 18.40 |
| | Cu rich FCC | 1.46 | 10.13 | 5.80 | 2.60 | 15.52 | 64.49 |
| $Al_{18}Co_{18}Cr_{18}Fe_{18}Ni_{18}Cu_{10}$ | Nominal | | | $(AlCoCrFeNi)_{90}$ | | | 10 |
| | Cr rich BCC | 0.31 | 3.78 | 88.25 | 3.31 | 3.27 | 1.09 |
| | matrix FCC | 1.74 | 38.93 | 6.04 | 2.60 | 33.40 | 17.29 |
| | Cu rich FCC | 15.94 | 20.39 | 3.89 | 6.05 | 25.83 | 27.89 |
| $Al_{17}Co_{17}Cr_{17}Fe_{17}Ni_{17}Cu_{15}$ | Nominal | | | $(AlCoCrFeNi)_{85}$ | | | 15 |
| | Cr rich BCC | 0.04 | 3.40 | 87.89 | 3.39 | 2.91 | 2.37 |
| | matrix FCC | 1.51 | 20.42 | 21.45 | 23.59 | 16.90 | 16.14 |
| | Cu rich FCC | 0.85 | 10.13 | 5.60 | 5.72 | 12.01 | 65.68 |
| $Al_{16}Co_{16}Cr_{16}Fe_{16}Ni_{16}Cu_{20}$ | Nominal | | | $(AlCoCrFeNi)_{80}$ | | | 20 |
| | Cr rich BCC | 0.14 | 3.71 | 87.45 | 3.88 | 2.42 | 2.39 |
| | matrix FCC | 2.90 | 21.87 | 9.63 | 33.77 | 16.99 | 14.84 |
| | Cu rich FCC | 0.93 | 6.30 | 4.67 | 6.03 | 9.47 | 72.60 |

### 3.4. Hardness

Figure 11 illustrates the copper content effect on the micro and macro-hardness of the (AlCoCrFeNi) HEA sintered at 950 °C. Although improving the grain boundary of the FCC phases versus the BCC phase, as shown in Figure 9, the results revealed that the (AlCoCrFeNi) HEA's hardness is decreased gradually by increasing the Cu content. It is evident that the micro and macro hardness has the same trend and are nearly the same, which means high homogeneity of the new HEA. This gradual reduction in hardness can be easily described in three points: The first is incorporating more energetic binding elements, and high melting point elements like Cr in a composite increase Young's modulus and the slip resistance [62]. The second is a drop in the BCC phase with increases in the copper content FCC phase that encourages deform flexibility [27,63]. The third is attributed to the solid solution strengthening effect and the high work hardening capability for this composite.

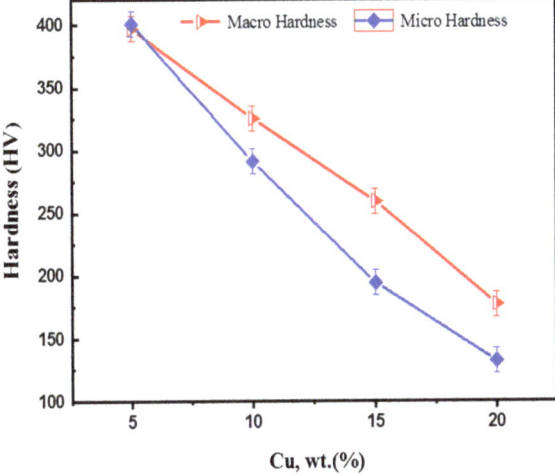

Figure 11. Vickers macro and micro hardness of HEAs fabricated samples at 950 °C.

## 3.5. Compression Strength

The deformation behavior of the cylindrical fabricated samples at 950 °C under a compression force in the form of stress–strain curves is shown in Figure 12. The compressive strength $\sigma_{max}$ and total strain $\varepsilon$ of all samples are listed in Table 5 to make the comparison easier. The results show that the compressive strength increases gradually by increasing the nano-copper content up to 15 wt% then decreased. Moreover, the total strain increases by increasing the amount of Cu, enhancing the alloys' flexibility and, consequently, the toughness. Increasing the strength of the fabricated HEAs samples may be due to the nano size of the precipitated copper. According to the Hall–Petch equation, there is an inversely proportional relationship between the material's strength and grain-size, where the reduction in grain size leads to an increase in material strength [64,65]. As shown, the AlCoCrFeNi HEAs strength has increased by increasing the percentage of the nano-copper particles that precipitated using the electroless coating process up to 15 wt% Cu then decreased. The increase in the rate of Cu nanoparticles to 20 wt% increases the agglomerations, and consequently, the grain particles that participate in dropping the strength again. On the other hand, the strain managed to grow because copper's high ductility improved its formability. The high strength and high strain of the fabricated materials mean that they have high absorption capacity (consumes a large amount of energy to deform until fracture). Additionally, increasing the strength may be due to the superb distribution and excellent adhesion among the composites' different constituents, as shown in the microstructure in Figure 6. The $Cu_x/(AlCoCrFeNi)_{1-x}$ HEAs' high strength is also relevant to the solid solution and the durable bonding influence among the composite metallic elements [66]. Incorporating Cr with high melting point elements in the BCC phase improves the Young's modulus and slip resistance [67]. On the other hand, the good elasticity and large work hardening capacity can be described by the $Cu_x/(AlCoCrFeNi)_{1-x}$ composite phase composition. According to the principal structural factor, a structure with a more slippery system leads to lower lattice friction during dislocation motion and increases the samples' elasticity [68]. The FCC structure has 48 slip systems against the BCC structure with 12 slip systems [69]. In this study, the composite contains two FCC phase residues and a single BCC phase. The FCC phases become dominant and hence increases the elasticity and work hardening effect. Improving the toughness of fabricated alloys by increasing the Cu's content is related to the copper's properties, where it is characterized by high toughness compared with the other used elements.

Table 5. Compressive strength and strain of the fabricated HEAs samples.

| Wt% Cu | Compressive Stress (MPa) | Strain % |
|---|---|---|
| 5 | 400.03 | 11.53 |
| 10 | 416.34 | 20.71 |
| 15 | 599.53 | 30.82 |
| 20 | 535.51 | 31.29 |

## 3.6. Coefficient of Thermal Expansion (CTE)

The CTE of all specimens was estimated from room temperature up to 350 °C with a 3 °/min heating rate. The effect of the Cu content and heating temperature on the CTE of the AlCoCrFeNi for all sintered samples is shown in Figure 13. It is increased gradually by increasing the heating temperature. As the temperature increases, the kinetic energy of atoms increases, and so the CTE of the heated material does consequently. Another phenomenon from the results could be observed, which is increasing the CTE of the sintered samples with increasing Cu content. The $Cu_{0.05}/(AlCoCrFeNi)_{0.95}$ recorded the lowest value of CTE $10.37 \times 10^{-6}$ °$C^{-1}$ at 150 °C, whereas the $Cu_{0.2}/(AlCoCrFeNi)_{0.8}$ recorded the highest one $10.95 \times 10^{-6}$ °$C^{-1}$ at the same temperature. Additionally, $Cu_{0.05}/(AlCoCrFeNi)_{0.95}$ recorded the lowest value of CTE $10.71 \times 10^{-6}$ °$C^{-1}$ at 350 °C, whereas the $Cu_{0.2}/(AlCoCrFeNi)_{0.8}$ recorded the highest value of $11.46 \times 10^{-6}$ °$C^{-1}$ at the same temperature. Increasing the CTE of the AlCoCrFeNi HEA as a result of its coating with Cu may be attributed to the high CTE of the Cu ($17.6 \times 10^{-6}$) when compared to other elements, such as Cr ($8.0 \times 10^{-6}$), Co ($12.0 \times 10^{-6}$), Ni ($13.0 \times 10^{-6}$), and Fe ($12.0 \times 10^{-6}$). This fact means that the $Cu_x/(AlCoCrFeNi)_{1-x}$ can be connected with steel items or deposited as coatings on their surfaces to increase hardening and wear resistance [70,71].

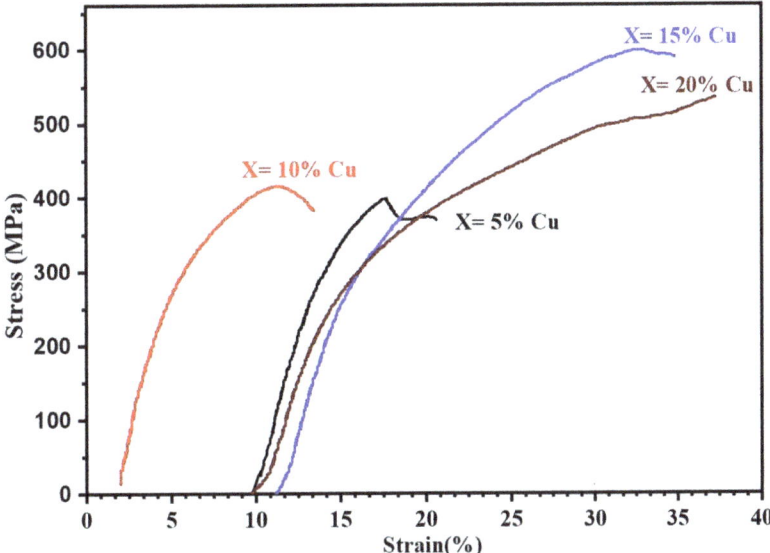

**Figure 12.** Compressive stress–strain curves of $Cu_x/(AlCoCrFeNi)_{1-x}$ HEAs sintered at 950 °C.

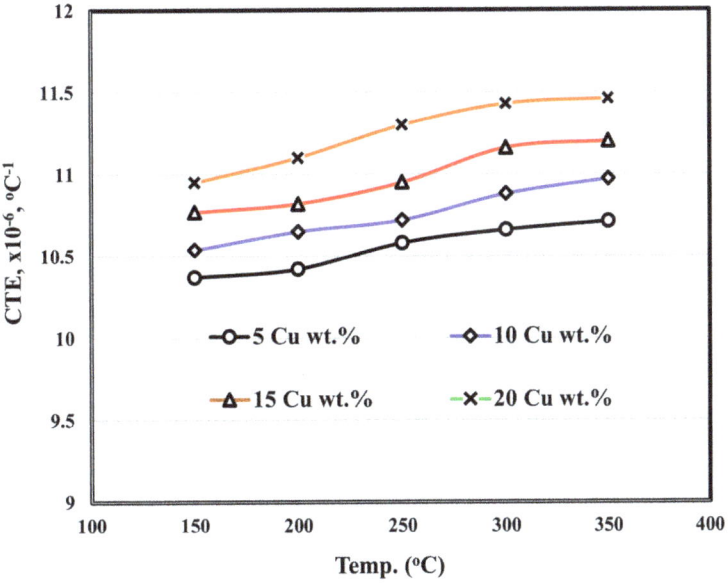

**Figure 13.** The coefficient of thermal expansion (CTE) of fabricated HEAs samples at different temperatures.

So, by increasing the nano Cu percentage, the quality of Cu coating of the different elements particles increases. This leads to an increase in the overall thermal expansion of the prepared alloys. Additionally, increasing the Cu percentage from 5 up to 20 wt% increases the densification of the samples. This means that no pores are detected, which causes an increase in the dimensions of the samples by heat.

Lowering the CTE of the PM sheets makes it suitable for many thermal applications that require thermal stability of materials to keep the shape of the samples without any deformation affected by the thermal stresses during the working process. This finally leads to an increase in the lifetime of materials, which is pretty much valuable from an economic point of view [72].

## 4. Conclusions

Four compositions ($Al_{19}Co_{19}Cr_{19}Fe_{19}Ni_{19}Cu_5$, $Al_{18}Co_{18}Cr_{18}Fe_{18}Ni_{18}Cu_{10}$, $Al_{17}Co_{17}Cr_{17}Fe_{17}Ni_{17}Cu_{15}$, and $Al_{16}Co_{16}Cr_{16}Fe_{16}Ni_{16}Cu_{20}$) of high entropy alloys were prepared by the powder metallurgy process at 900, 950, and 1000 °C for 90 min. Based on the results and their discussion, copper content's effect can be summarized in the following points.

- The $(AlCoCrFeNi)_{1-x}/Cu_x$ HEAs samples that sintered at 950 °C achieved the highest densification.
- The AlCoCrFeNi sample was not established with the powder metallurgy technique.
- Adding the copper element to the AlCoCrFeNi HEAs enhances its formability by the powder metallurgy technique.
- The microstructure revealed that the mechanical alloying that preceded the sintering process achieved a high homogeneity between the different elements of the AlCoCrFeNi (HEAs), which made the elements AlCoFeNi compose the FCC matrix. An agglomeration for the Cr was established and prevented the complete forming of the HEAs.
- The mapping showed that the Cu element tends to dissolve with the FCC matrix.
- The crystallite size of the HEA powders was significantly refined as the weight percentage of Cu increased. On the other hand, the results of the lattice strain revealed that the lattice strain increased by increasing the Cu wt%.
- The crystal size of fabricated HEAs was decreased by increasing the content of the precipitated Cu, which is similar to the result of powder alloys. Additionally, the lattice strain increased by increasing the Cu content.
- The hardness of the manufactured AlCoCrFeNi HEAs decreased gradually by increasing the Cu content.
- The results showed that the compressive strength gradually increased by increasing the nano-copper content up to 15 wt% then decreased.
- According to the area under the stress–strain curves, the toughness was enhanced with the copper content.
- Precipitation of the copper in the nano size by the electroless coating process enhances the strength of the produced HEAs material according to the Hall–Petch equation.
- The CTE gradually increased by increasing the heating temperature and the content of the Cu wt%.

**Author Contributions:** M.A.H.: Conceptualization, data analysis, writing original draft and preparation, H.M.Y.: Investigation, writing original draft, review and editing, A.S.A.M.: review and editing, A.E.E.-N.: Investigation and review, O.A.E.: Methodology, review and editing. All authors have read and agreed to the published version of the manuscript.

**Funding:** This research received no external funding.

**Informed Consent Statement:** Not applicable.

**Data Availability Statement:** Not applicable.

**Acknowledgments:** The authors would like to thank Aya Emad Ewais for her help in the proofreading of this paper.

**Conflicts of Interest:** The authors declare no conflict of interest.

## References

1. Yeh, J.W.; Chen, S.K.; Lin, S.J.; Gan, J.Y.; Chin, T.S.; Shun, T.T.; Tsau, C.H.; Chang, S.Y. Nanostructured high-entropy alloys with multiple principal elements: novel alloy design concepts and outcomes. *Adv. Eng. Mater.* **2004**, *6*, 299–303. [CrossRef]
2. Cantor, B.; Chang, I.; Knight, P.; Vincent, A. Microstructural development in equiatomic multicomponent alloys. *Mater. Sci. Eng.* **2004**, *375*, 213–218. [CrossRef]
3. Zhang, Y.; Zuo, T.T.; Tang, Z.; Gao, M.C.; Dahmen, K.A.; Liaw, P.K.; Lu, Z.P. Microstructures and properties of high-entropy alloys. *Mater. Sci.* **2014**, *61*, 1–93. [CrossRef]
4. Ye, Y.; Wang, Q.; Lu, J.; Liu, C.; Yang, Y. High-entropy alloy: challenges and prospects. *Mater. Today* **2016**, *19*, 349–362. [CrossRef]
5. Gorr, B.; Mueller, F.; Christ, H.-J.; Mueller, T.; Chen, H.; Kauffmann, A.; Heilmaier, M. High temperature oxidation behavior of an equimolar refractory metal-based alloy 20Nb20Mo20Cr20Ti20Al with and without Si addition. *J. Alloys Compd.* **2016**, *688*, 468–477. [CrossRef]
6. Gludovatz, B.; Hohenwarter, A.; Catoor, D.; Chang, E.H.; George, E.P.; Ritchie, R.O. A fracture-resistant high-entropy alloy for cryogenic applications. *Science* **2014**, *345*, 1153–1158. [CrossRef] [PubMed]
7. Zhao, J.; Ji, X.; Shan, Y.; Fu, Y.; Yao, Z. On the microstructure and erosion–corrosion resistance of AlCrFeCoNiCu high-entropy alloy via annealing treatment. *Mater. Sci. Technol.* **2016**, *32*, 1271–1275. [CrossRef]
8. Hsu, C.-Y.; Sheu, T.-S.; Yeh, J.-W.; Chen, S.-K. Effect of iron content on wear behavior of AlCoCrFe$_x$Mo$_{0.5}$Ni high-entropy alloys. *Wear* **2010**, *268*, 653–659. [CrossRef]
9. Amar, A.; Li, J.; Xiang, S.; Liu, X.; Zhou, Y.; Le, G.; Wang, X.; Qu, F.; Ma, S.; Dong, W. Additive manufacturing of high-strength CrMnFeCoNi-based High Entropy Alloys with TiC addition. *Intermetallics* **2019**, *109*, 162–166. [CrossRef]
10. Hao, J.; Ma, Y.; Wang, Q.; Zhang, C.; Li, C.; Dong, C.; Song, Q.; Liaw, P.K. Formation of cuboidal B2 nanoprecipitates and microstructural evolution in the body-centered-cubic Al$_{0.7}$NiCoFe$_{1.5}$Cr$_{1.5}$ high-entropy alloy. *J. Alloys Compd.* **2019**, *780*, 408–421. [CrossRef]
11. Wei, D.; Li, X.; Schönecker, S.; Jiang, J.; Choi, W.-M.; Lee, B.-J.; Kim, H.S.; Chiba, A.; Kato, H. Development of strong and ductile metastable face-centered cubic single-phase high-entropy alloys. *Acta Mater.* **2019**, *181*, 318–330. [CrossRef]
12. Kao, Y.-F.; Chen, T.-J.; Chen, S.-K.; Yeh, J.-W. Microstructure and mechanical property of as-cast, -homogenized, and -deformed Al$_x$CoCrFeNi ($0 \leq x \leq 2$) high-entropy alloys. *J. Alloys Compd.* **2009**, *488*, 57–64. [CrossRef]
13. Fu, X.; Schuh, C.A.; Olivetti, E.A. Materials selection considerations for high entropy alloys. *Scripta Mater.* **2017**, *138*, 145–150. [CrossRef]
14. Kukshal, V.; Patnaik, A.; Bhat, I. Effect of cobalt on microstructure and properties of AlCr$_{1.5}$CuFeNi$_2$Co$_x$ high-entropy alloys. *Mater. Res. Express.* **2018**, *5*, 046514. [CrossRef]
15. Stepanov, N.; Yurchenko, N.Y.; Shaysultanov, D.; Salishchev, G.; Tikhonovsky, M. Effect of Al on structure and mechanical properties of Al$_x$NbTiVZr (x = 0, 0.5, 1, 1.5) high entropy alloys. *Mater. Sci. Technol.* **2015**, *31*, 1184–1193. [CrossRef]
16. Wang, X.-R.; He, P.; Lin, T.-S.; Wang, Z.-Q. Microstructure, thermodynamics and compressive properties of AlCrCuNi$_x$Ti (x = 0, 1) high entropy alloys. *Mater. Sci. Technol.* **2015**, *31*, 1842–1849. [CrossRef]
17. Qin, G.; Zhang, Y.; Chen, R.; Zheng, H.; Wang, L.; Su, Y.; Ding, H.; Guo, J.; Fu, H. Microstructures and mechanical properties of (AlCoCrFeMn)$_{100-x}$Cu$_x$ high-entropy alloys. *Mater. Sci. Technol.* **2019**. [CrossRef]
18. Du, W.; Liu, N.; Peng, Z.; Zhou, P.; Xiang, H.; Wang, X. Technology. The effect of Ti addition on phase selection of CoCrCu$_{0.5}$FeNi high-entropy alloys. *Mater. Sci. Technol.* **2018**, *34*, 473–479. [CrossRef]
19. Zheng, Z.; Li, X.; Zhang, C.; Li, J. Microstructure and corrosion behaviour of FeCoNiCuSn$_x$ high entropy alloys. *Mater. Sci. Technol.* **2015**, *31*, 1148–1152. [CrossRef]
20. Zhang, C.; Zhu, J.; Zheng, H.; Li, H.; Liu, S.; Cheng, G. A review on microstructures and properties of high entropy alloys manufactured by selective laser melting. *Int. J. Extreme. Manuf.* **2020**, *2*, 032003. [CrossRef]
21. Yang, Y.; Luo, X.; Ma, T.; Wen, L.; Hu, L.; Hu, M. Effect of Al on characterization and properties of Al$_x$CoCrFeNi high entropy alloy prepared via electro-deoxidization of the metal oxides and vacuum hot pressing sintering process. *J. Alloys Compd.* **2021**, *864*, 158717. [CrossRef]
22. Wang, W.-R.; Wang, W.-L.; Wang, S.-C.; Tsai, Y.-C.; Lai, C.-H.; Yeh, J.-W. Effects of Al addition on the microstructure and mechanical property of Al$_x$CoCrFeNi high-entropy alloys. *Intermetallics* **2012**, *26*, 44–51. [CrossRef]
23. Tsai, M.-H.; Yeh, J.-W. High-entropy alloys: A critical review. *Mater. Res. Lett.* **2014**, *2*, 107–123. [CrossRef]
24. Xie, L.; Brault, P.; Thomann, A.-L.; Yang, X.; Zhang, Y.; Shang, G. Molecular dynamics simulation of Al–Co–Cr–Cu–Fe–Ni high entropy alloy thin film growth. *Intermetallics* **2016**, *68*, 78–86. [CrossRef]
25. Santodonato, L.J.; Zhang, Y.; Feygenson, M.; Parish, C.M.; Gao, M.C.; Weber, R.J.; Neuefeind, J.C.; Tang, Z.; Liaw, P.K. Deviation from high-entropy configurations in the atomic distributions of a multi-principal-element alloy. *Nat. Commun.* **2015**, *6*, 1–13. [CrossRef] [PubMed]
26. Ogata, S.; Umeno, Y.; Kohyama, M. Engineering. First-principles approaches to intrinsic strength and deformation of materials: perfect crystals, nano-structures, surfaces and interfaces. *Mater. Sci. Eng.* **2008**, *17*, 013001.
27. Tung, C.-C.; Yeh, J.-W.; Shun, T.-T.; Chen, S.-K.; Huang, Y.-S.; Chen, H.-C. On the elemental effect of AlCoCrCuFeNi high-entropy alloy system. *Mater. Lett.* **2007**, *61*, 1–5. [CrossRef]
28. Dąbrowa, J.; Cieślak, G.; Stygar, M.; Mroczka, K.; Berent, K.; Kulik, T.; Danielewski, M. Influence of Cu content on high temperature oxidation behavior of AlCoCrCu$_x$FeNi high entropy alloys (x = 0; 0.5; 1). *Intermetallics* **2017**, *84*, 52–61. [CrossRef]

29. Yu, Y.; Shi, P.; Feng, K.; Liu, J.; Cheng, J.; Qiao, Z.; Yang, J.; Li, J.; Liu, W. Effects of Ti and Cu on the microstructure evolution of AlCoCrFeNi high-entropy alloy during heat treatment. *Acta Metallurgica Sin.* **2020**, *33*, 1–14. [CrossRef]
30. Jones, N.; Izzo, R.; Mignanelli, P.; Christofidou, K.; Stone, H. Phase evolution in an $Al_{0.5}CrFeCoNiCu$ high entropy alloy. *Intermetallics* **2016**, *71*, 43–50. [CrossRef]
31. Yehia, H.M.; Daoush, W.M.; Mouez, F.A.; El-Sayed, M.H.; El-Nikhaily, A.E. Microstructure, Hardness, Wear, and Magnetic Properties of (Tantalum, Niobium) Carbide–Nickel–Sintered Composites Fabricated from Blended and Coated Particles. *Mater. Perform. Charact.* **2020**, *4*, 543–555. [CrossRef]
32. Yehia, H.M.; Allam, S. Allam Hot Pressing of Al-10 wt% Cu-10 wt% Ni/x ($Al_2O_3$–Ag) Nanocomposites at Different Heating Temperatures. *Metal Mater. Int.* **2020**. [CrossRef]
33. El-Tantawy, A.; El Kady, O.A.; Yehia, H.M.; Ghayad, I.M. Effect of Nano $ZrO_2$ Additions on the Mechanical Properties of Ti-12Mo Composite by Powder Metallurgy Route. *Key Eng. Mater.* **2020**, *835*, 367–373. [CrossRef]
34. Eißmann, N.; Klöden, B.; Weißgärber, T.; Kieback, B. High-entropy alloy CoCrFeMnNi produced by powder metallurgy. *Powder Metall.* **2017**, *60*, 184–197. [CrossRef]
35. Yehia, H.M.; Abu-Oqail, A.; Elmaghraby, M.A.; Elkady, O.A. Microstructure, hardness, and tribology properties of the $(Cu/MoS_2)$/graphene nanocomposite via the electroless deposition and powder metallurgy technique. *J. Compos. Mater.* **2020**, 1–12. [CrossRef]
36. El-Kady, O.; Yehia, H.M.; Nouh, F.; Materials, H. Preparation and characterization of Cu/(WC-TiC-Co)/graphene nano-composites as a suitable material for heat sink by powder metallurgy method. *Int. J. Refract. Met. Hard Mater.* **2019**, *79*, 108–114. [CrossRef]
37. ASTM. *Standard Test Methods for Density of Compacted or Sintered Powder Metallurgy (PM) Products Using Archimedes' Principle*; ASTM: West Conshohocken, PA, USA, 2014; p. 7.
38. Scherrer, P. Determination of the size and internal structure of colloidal particles using X-rays. *Nachr. Ges. Wiss. Göttingen* **1918**, *2*, 98–100.
39. Danilchenko, S.; Kukharenko, O.; Moseke, C.; Protsenko, I.Y.; Sukhodub, L.; Sulkio-Cleff, B.J.C.R.; Experimental, T.J.O.; Crystallography, I. Determination of the bone mineral crystallite size and lattice strain from diffraction line broadening. *J. Exper. Ind. Crystallogr.* **2002**, *37*, 1234–1240. [CrossRef]
40. ASTM Standard E384; 19428-2959; ASTM International: West Conshohocken, PA, USA, 2011; Available online: https://www.astm.org/Standards/E384 (accessed on 30 March 2021).
41. Jianhong, L.; Yun, J.; Haiping, L.; Jiguo, T.; Shu-qiang, J. Influence of EDTA/THPED Dual-Ligand on Copper Electroless Deposition. *Int. J. Electrochem.* **2018**, *13*, 6015–6026. [CrossRef]
42. Yuan, Y.; Gan, X.; Lai, Y.; Zhao, Q.; Zhou, K. Microstructure and properties of graphite/copper composites fabricated with Cu-Ni double-layer coated graphite powders. *Comp. Interf.* **2020**, *27*, 449–463. [CrossRef]
43. Yusoff, M.; Hussain, Z. Manufacturing. Effect of sintering parameters on microstructure and properties of mechanically alloyed copper-tungsten carbide composite. *Int. J. Mater. Mech. Manuf.* **2013**, *1*, 283–286.
44. An, Z.; Jia, H.; Wu, Y.; Rack, P.D.; Patchen, A.D.; Liu, Y.; Ren, Y.; Li, N.; Liaw, P.K. Solid-solution CrCoCuFeNi high-entropy alloy thin films synthesized by sputter deposition. *Mater. Res. Lett.* **2015**, *3*, 203–209. [CrossRef]
45. Shivam, V.; Basu, J.; Pandey, V.K.; Shadangi, Y.; Mukhopadhyay, N. Alloying behaviour, thermal stability and phase evolution in quinary AlCoCrFeNi high entropy alloy. *Adv. Powder Technol.* **2018**, *29*, 2221–2230. [CrossRef]
46. Suryanarayana, C.; Ivanov, E.; Boldyrev, V.A. The science and technology of mechanical alloying. *Mater. Sci. Eng.* **2001**, *304*, 151–158. [CrossRef]
47. Shang, C.; Axinte, E.; Sun, J.; Li, X.; Li, P.; Du, J.; Qiao, P.; Wang, Y. CoCrFeNi ($W_{1-x}Mo_x$) high-entropy alloy coatings with excellent mechanical properties and corrosion resistance prepared by mechanical alloying and hot pressing sintering. *Mater. Des.* **2017**, *117*, 193–202. [CrossRef]
48. Bonache, V.; Salvador, M.; Fernández, A.; Borrell, A.; Materials, H. Fabrication of full density near-nanostructured cemented carbides by combination of $VC/Cr_3C_2$ addition and consolidation by SPS and HIP technologies. *Int. J. Refract. Met. Hard Mater.* **2011**, *29*, 202–208. [CrossRef]
49. Fang, S.; Chen, W.; Fu, Z. Microstructure and mechanical properties of twinned $Al_{0.5}CrFeNiCo_{0.3}C_{0.2}$ high entropy alloy processed by mechanical alloying and spark plasma sintering. *Mater. Des.* **2014**, *54*, 973–979. [CrossRef]
50. Yeh, J.-W.; Lin, S.-J.; Chin, T.-S.; Gan, J.-Y.; Chen, S.-K.; Shun, T.-T.; Tsau, C.-H.; Chou, S.-Y. Formation of simple crystal structures in Cu-Co-Ni-Cr-Al-Fe-Ti-V alloys with multiprincipal metallic elements. *Metall. Mater. Trans.* **2004**, *35*, 2533–2536. [CrossRef]
51. Ibrahim, A.; Abdallah, M.; Mostafa, S.; Hegazy, A.A. An experimental investigation on the W–Cu composites. *Mater. Des.* **2009**, *30*, 1398–1403. [CrossRef]
52. Huang, P.K.; Yeh, J.W.; Shun, T.T.; Chen, S.K. Multi-principal-element alloys with improved oxidation and wear resistance for thermal spray coating. *Adv. Eng. Mater.* **2004**, *6*, 74–78. [CrossRef]
53. Ogura, M.; Fukushima, T.; Zeller, R.; Dederichs, P.H. Structure of the high-entropy alloy $Al_xCrFeCoNi$: fcc versus bcc. *J. Alloys Compd.* **2017**, *715*, 454–459. [CrossRef]
54. Zhu, J.-M.; Meng, J.-L.; Liang, J.-L. Microstructure and mechanical properties of multi-principal component $AlCoCrFeNiCu_x$ alloy. *Rare Met.* **2016**, *35*, 385–389. [CrossRef]

55. Kim, D.G.; Jo, Y.H.; Park, J.M.; Choi, W.-M.; Kim, H.S.; Lee, B.-J.; Sohn, S.S.; Lee, S. Effects of annealing temperature on microstructures and tensile properties of a single FCC phase CoCuMnNi high-entropy alloy. *J. Alloys Compd.* **2020**, *812*, 152111. [CrossRef]
56. Sriharitha, R.; Murty, B.; Kottada, R.S. Phase formation in mechanically alloyed Al$_x$CoCrCuFeNi (x = 0.45, 1, 2.5, 5 mol) high entropy alloys. *Intermetallics* **2013**, *32*, 119–126. [CrossRef]
57. Chen, Y.-L.; Hu, Y.-H.; Hsieh, C.-A.; Yeh, J.-W.; Chen, S.-K. Competition between elements during mechanical alloying in an octonary multi-principal-element alloy system. *J. Alloys Compd.* **2009**, *481*, 768–775. [CrossRef]
58. Liu, X.-T.; Lei, W.-B.; Li, J.; Ma, Y.; Wang, W.-M.; Zhang, B.-H.; Liu, C.-S.; Cui, J.-Z.J.R.M. Laser cladding of high-entropy alloy on H13 steel. *Rare Met.* **2014**, *33*, 727–730. [CrossRef]
59. Porter, D.; Easterling, E.; Sherif, M. *Crystal Interfaces and Microstructure. Phase Transformations in Metals and Alloys*; CRC Press: New York, NY, USA, 1992.
60. Takeuchi, A.; Inoue, A. Classification of bulk metallic glasses by atomic size difference, heat of mixing and period of constituent elements and its application to characterization of the main alloying element. *Metall. Mater. Trans.* **2005**, *46*, 2817–2829. [CrossRef]
61. Gwalani, B.; Choudhuri, D.; Soni, V.; Ren, Y.; Styles, M.; Hwang, J.; Nam, S.; Ryu, H.; Hong, S.H.; Banerjee, R. Cu assisted stabilization and nucleation of L12 precipitates in Al$_{0.3}$CuFeCrNi$_2$ fcc-based high entropy alloy. *Acta Mater.* **2017**, *129*, 170–182. [CrossRef]
62. Wu, J.-M.; Lin, S.-J.; Yeh, J.-W.; Chen, S.-K.; Huang, Y.-S.; Chen, H.-C. Adhesive wear behavior of Al$_x$CoCrCuFeNi high-entropy alloys as a function of aluminum content. *Wear* **2006**, *261*, 513–519. [CrossRef]
63. Reed-Hill, R.E.; Abbaschian, R. *Physical Metallurgy Principles*, 3rd ed.; PWS-KENT Publishing Company: Boston, MA, USA, 1994; pp. 140–146.
64. Qiu, X.-W. Microstructure and properties of AlCrFeNiCoCu high entropy alloy prepared by powder metallurgy. *J. Alloys Compd.* **2013**, *555*, 246–249. [CrossRef]
65. Nyanor, P.; El-Kady, O.; Yehia, H.M.; Hamada, A.S.; Hassan, M.A. Effect of Bimodal-Sized Hybrid TiC–CNT Reinforcement on the Mechanical Properties and Coefficient of Thermal Expansion of Aluminium Matrix Composites. *Met. Mater. Int.* **2020**, *27*, 753–766. [CrossRef]
66. Harwood, J. *Strengthening Mechanisms in Solids*; ASM Seminar; ASM International: Materials Park, OH, USA, 1960.
67. Wen, L.; Kou, H.; Li, J.; Chang, H.; Xue, X.; Zhou, L. Effect of aging temperature on microstructure and properties of AlCoCrCuFeNi high-entropy alloy. *Intermetallics* **2009**, *17*, 266–269. [CrossRef]
68. Dieter, G.E.; Bacon, D.J. *Mechanical Metallurgy*; McGraw-Hill: New York, NY, USA, 1986; Volume 3.
69. Zhang, K.; Fu, Z.; Zhang, J.; Shi, J.; Wang, W.; Wang, H.; Wang, Y.; Zhang, Q. Annealing on the structure and properties evolution of the CoCrFeNiCuAl high-entropy alloy. *J. Alloys Compd.* **2010**, *502*, 295–299. [CrossRef]
70. Barakat, W.S.; Elkady, O.; Abu-Oqail, A.; Yehya, H.M.; EL-Nikhaily, A. Effect of Al$_2$O$_3$ Coated Cu Nanoparticles on Properties of Al/Al$_2$O$_3$ Composites. *J. Petroleum Min. Eng.* **2020**, *22*. [CrossRef]
71. Nadutov, V.; Makarenko, S.Y.; Svystunov, Y.O. Effect of Al content on magnetic properties and thermal expansion of as-cast high-entropy alloys Al$_x$FeCoNiCuCr. *aoa o oo* **2015**, *37*, 987–1000.
72. Yehia, H.M.; Elkady, O.A.; Reda, Y.; Ashraf, K.M. Electrochemical surface modification of aluminum sheets prepared by powder metallurgy and casting techniques for printed circuit applications. *Trans. Indian Inst. Met.* **2019**, *72*, 85–92. [CrossRef]

Article

# A New Approach to Direct Friction Stir Processing for Fabricating Surface Composites

Abdulla I. Almazrouee [1,*], Khaled J. Al-Fadhalah [2] and Saleh N. Alhajeri [1]

[1] Department of Manufacturing Engineering Technology, College of Technological Studies, P.A.A.E.T., P.O. Box 42325, Shuwaikh 70654, Kuwait; sn.alhajeri@paaet.edu.kw

[2] Department of Mechanical Engineering, College of Engineering & Petroleum, Kuwait University, P.O. Box 5969, Safat 13060, Kuwait; khaled.alfadhalah@ku.edu.kw

* Correspondence: ai.almazrouee@paaet.edu.kw

**Abstract:** Friction stir processing (FSP) is a green fabrication technique that has been effectively adopted in various engineering applications. One of the promising advantages of FSP is its applicability in the development of surface composites. In the current work, a new approach for direct friction stir processing is considered for the surface fabrication of aluminum-based composites reinforced with micro-sized silicon carbide particles (SiC), eliminating the prolonged preprocessing stages of preparing the sample and filling the holes of grooves. The proposed design of the FSP tool consists of two parts: an inner-threaded hollow cylindrical body; and a pin-less hollow shoulder. The design is examined with respect to three important tool processing parameters: the tilt angle of the tool, the tool's dispersing hole, and the tool's plunge depth. The current study shows that the use of a dispersing hole with a diameter of 6 mm of and a plunge depth of 0.6 mm, in combination with a tilting angle of 7°, results in sufficient mixing of the enforcement particles in the aluminum matrix, while still maintaining uniformity in the thickness of the composite layer. Metallographic examination of the Al/SiC surface composite demonstrates a uniform distribution of the Si particles and excellent adherence to the aluminum substrate. Microhardness measurements also show a remarkable increase, from 38.5 Hv at the base metal to a maximum value of 78 Hv in the processed matrix in the surface composites layer. The effect of the processing parameters was also studied, and its consequences with respect to the surface composites are discussed.

**Keywords:** direct friction stir processing; in situ composites; surface composites

## 1. Introduction

Wrought aluminum alloys are considered one of the most significant metallic materials in today's fabrication and manufacturing production, especially in the transportation industries [1,2]. In general, many aluminum alloys offer an excellent combination of properties such as high strength-to-weight ratio, corrosion resistance, good formability, and weldability. However, in specific applications where dynamic loading is imposed, aluminum alloys' fatigue performance is considered a significant drawback. Other properties such as wear resistance are also of concern. Surface treatments such as coating, shot peening and induction friction stir processing (FSP) have been developed as solid-state processes capable of enhancing the surface properties of aluminum alloys [3]. The FSP process was developed based on the principles of friction stir welding (FSW) [3–5]. It involves the use of a relatively high-speed rotating non-consumable tool, which consists mainly of a shoulder with/without a pin to process the sheet/plate surface of the metal. During FSP, a localized heat produced by friction is generated at the interface of the rotating tool and the workpiece, resulting in metal softening and plasticization; the rotating pin, on the other hand, allows significant stirring and mixing, leading to severe plastic deformation and thus producing microstructure refinement in the stirred zone (SZ) [6–10]. The high-speed rotating tool then traverses along a specified path of interest to process and modify the material's matrix.

The microstructural alterations are characterized by equiaxed and ultra-fine grains for many metals and alloys [4,11]. Such grain refinement by FSP has been shown to enhance mechanical properties in several Al alloys, including tensile strength, ductility, and creep and fatigue strength [4,7,12–16]. The processing of other high-temperature materials such as steels [17] and Ti alloys [18] with FSP has also been reported.

The application of FSP for the development of surface composite matrixes was first introduced by Mishra et al. [4]. An in situ surface composite was produced by smearing the aluminum surface with reinforcement SiC particles and methanol and applying FSP to produce a surface composite layer. Since then, different methods for adding reinforcement particles have been reported [4,12,19–24]. For example, several studies proposed a new FSP method for making in situ composites, known as the groove method, by filling the reinforcement particles into a machined groove in the base material's top surface using FSP [25–28]. The groove method can be divided into three different steps: (1) machining the groove with the desired dimensions, (2) processing with a pin-less tool to pack the reinforcement particles, and (3) processing the entrap reinforcement particles by a tool with a pin in the desired processing parameters. The last two steps can be substituted by placing a thin sheet on the groove of the same materials and then applying the process using a tool with a pin to process the reinforcement particles entrapped in the groove by the thin sheet. Another approach for adding reinforcement particles, which eliminates the need for pin-less tool processing, is drilling several holes in a line or lines and then packing them with reinforcement particles, followed by processing with a tool with a pin to develop the surface composites [27,29]. The dimensions and number of grooves and holes play a vital role in acquiring the second phase's desired volume fraction [7,9]. All the previous methods have used a non-consumable tool. However, a consumable rod can be used [27] with drilled holes placed at different positions along a radial line filled with reinforcement particles to provide excellent results. The aforementioned methods require rigorous preparations of the surface or the consumable rod. More recently, Huang et al. [30] fabricated a surface composite by direct friction stir processing (DFSP) using a hollow tool without preprocessing to add the reinforcement particles. This design allows the in situ implementation of reinforcement particles.

Furthermore, the tool's design in FSW/P plays a critical and decisive role in the welding and processing, as well as the fabrication, of materials and surface composites. Different tool designs have been reported in previous studies [4,9,13,31–33]. Types of tools used for FSW/P are generally categorized into three types, namely, fixed, adjustable, and self-reacting [10]. The first type is the fixed-pin tool, where the tool is a single piece that includes the pin and the shoulder. The second type is the flexible tool with an adjustable pin, which can also be made from another material. The third type is the self-reacting tool, which is similar to the second type but with the addition of a bottom shoulder, and which acts as a backing anvil for the processed piece during the process. These three tool types have been used in FSW/P to weld and process several metallic materials.

The work of Huang et al. [30] proposed an efficient tool design for making in situ surface composites. The tool design consists of a pin-less hollow shoulder that is tapered at its lower end. This design was shown to allow the efficient spread and mixing of the reinforcement particles into the matrix surface during FSP. The shoulder is tapered from the center of the tool to minimize unnecessary frictional contact between the shoulder and the metal surface. In this study, a modified design of the FSP tool is proposed for making aluminum-based surface composites reinforced by SiC particles, based on the concept of the tool design reported in [30]. The modification includes several alterations to the tool design, including a two-part design, several hole sizes, and shoulder shape. A new approach for the fabrication of in situ surface composites by FSP has also been proposed in this study. It aims to reduce the probabilities of clogging the hole in the shoulder or any back extrusion during the processing. The role of the new design and technique on the microstructure and microhardness of the matrix surface are studied.

## 2. Materials and Methods

### 2.1. Tool Design and Processing Approach

In this study, a new design for the tool and a new processing approach are proposed to carry out the friction stir processing to develop surface composites. A two-part tool is fabricated where the two parts are the top part of a hollow cylindrical body and the lower part of a tool head, as schematically illustrated in Figure 1a,b.

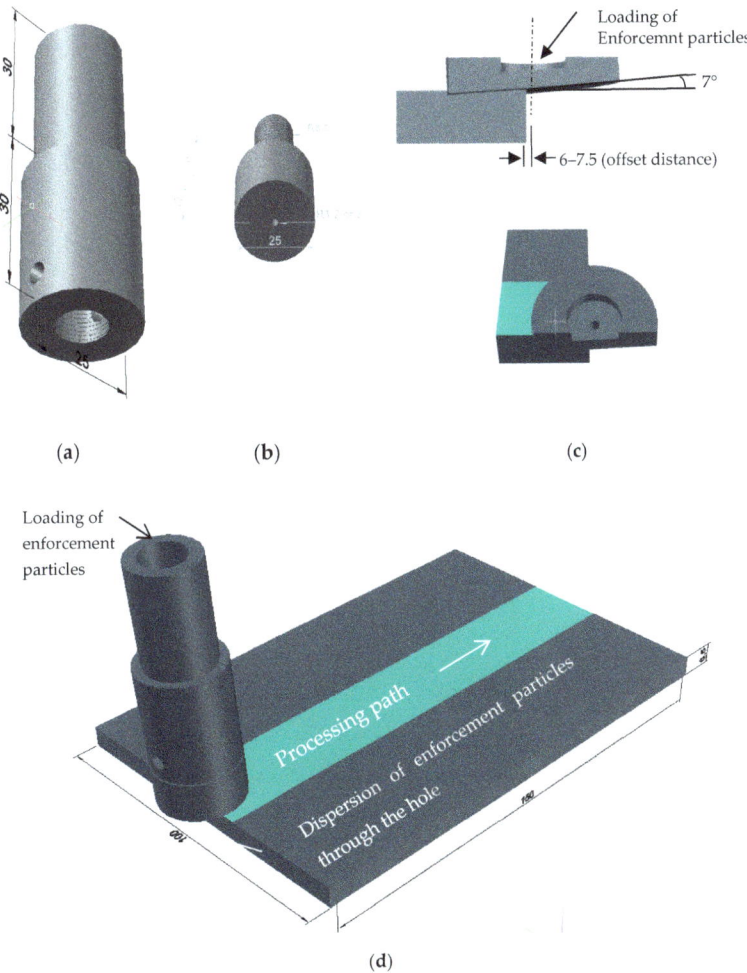

**Figure 1.** Schematic illustration of the FSP tool and design and processing procedures: (**a**) upper part, and (**b**) lower part of the FSP tool; (**c**) sectioned views of the tilting of tool head with respect to the processed aluminum plate; and (**d**) processing setup. All dimensions are in mm.

The top part consists of a body made from H13 steel with an outer diameter of 25 mm and a 13.5 mm hollow body with 15 mm of threading at the 30 mm end. The height of the top portion of the tool is 60 mm; half of the distance is prepared for clamping, and the other half remaining is used as a casing for the lower part. The lower portion is the head, containing a shoulder with a hole in its middle instead of a pin. Three different sizes of holes were used in experiments 2, 4, and 6 mm. The head also contains threading that fits into the top part of the tool's body. The upper portion of the lower part consists of a

15-mm thread with a diameter of 11 mm, and then the head body is 25 mm in diameter and 12 mm in thickness for the lower portion of the lower part, as depicted in Figure 1b. The total length of the whole body, assembled from the two parts, is 72 mm. Two bolts are also used to add more stability to the tool during processing by secularly tightening the lower part's threaded part and the top part. The tool can be fabricated as one part; however, the two-part design is easier to manufacture and allows for more extension of the tool; thus, more powder can be loaded inside the tool from the top part, as illustrated in Figure 1c. The two-part tool also has many benefits with respect to the flexibility of the tool's use, as well as advantages such as the ability to use different materials and designs for the head, such as the different hole sizes used in this experiment. The new design also facilitates head/tool body reuse, reduces the materials needed, and reduces the associated costs.

Commercial silicon carbide (SiC) particles with an average size of 20 µm were used as the reinforcement particles, as shown by the secondary electron image in Figure 2. The SiC particles were poured into the cavity of the lower part of the tool. The two-part tool was assembled and clamped into a vertical milling machine to conduct FSP. The processing was carried out with a constant tool rotating rate of 3000 rpm, using a clockwise rotation and a 20 mm/min travel speed. The reinforcement SiC particles were fed into the matrix via gravitational force. A ball bearing of 11 mm in diameter was placed on the top of the powder inside the hollow tool to facilitate the dispersion of the reinforcement particles into the metal surface to produce surface composites at the top of the metal matrix. To allow the use of different hole sizes, the tool was tilted by 7°, as illustrated in Figure 1c. By doing so, it was possible for a part of the tool, less than half of shoulder surface, to be in contact with the workpiece. Offset distances of 6 and 7.5 mm were used between the point of contact of tool with the workpiece and the center of the tool (Figure 1c). Two main processing parameters, namely the offset distance and the tilting angle, were evaluated to provide the best possible combination of plunge depth required to enhance the particle dispersion processes with minimum back extrusion effect or tool hole blockage by the processed material. The use of the two offsets of 6 and 7.5 mm, in combination with a tilting angle of 7°, was shown to result in plunge depths of approximately 0.9 and 0.6 mm, respectively.

**Figure 2.** Secondary electron images of the as-received commercial silicon carbide particles in aggregate state.

*2.2. Material and Experimental Procedures*

T6 tempered aluminum alloy 1100 plates were purchased in hot extrusion condition. The chemical composition of the alloy is given in Table 1. The aluminum plates were used as the base for fabricating the surface composite. The plates were sectioned perpendicular to the extrusion direction. The dimensions of the FSP plates were $100 \times 150 \times 6.5$ mm, as schematically illustrated in Figure 1d. Coupons were cut transversely from the middle of the processed plate for metallographic investigations. A standard metallographic procedure was used to prepare the sample for examination. The cross-section of the coupons was examined by a metallurgical microscope (AxioImager A1M, ZEISS, Oberkochen, Germany). Additionally, a field emission scanning electron microscope (FESEM) (JEOL, model: JSM-7001F, Tokyo, Japan), equipped with an Energy Dispersive X-Ray Spectroscopy (EDS) detector (Aztec-Energy, Oxford, High Wycombe, UK) was used to examine surface composite layer development in the FSP zone and to identify the elemental distribution across the surface composite/matrix interface. The EDS analysis was carried out at an accelerating voltage of 20 kV and a working distance of 11.5 mm to allow for a sufficient depth of penetration (1–2 µm). The surface composite's elemental analysis was acquired from the top surface to the alloy matrix toward the plate's bottom, and horizontally at the center of the processed zone.

**Table 1.** Chemical composition of the 1100 aluminum plate used in the study.

|         | Composition (wt.%) |       |       |       |       |       |       |      |
|---------|------|-------|-------|-------|-------|-------|-------|------|
|         | Si   | Fe    | Mn    | Cu    | Ti    | Cr    | Zn    | Al   |
| Al 1100 | 0.140 | 0.250 | 0.001 | 0.051 | 0.019 | 0.001 | 0.002 | Bal. |

Vickers microhardness measurements of the polished samples were conducted. Indentations were carried out using an Innovatest Falcon 500 Hardness Tester (Innovatest, Maastricht, The Netherlands) with a load of 100 gf and a dwell time of 20 seconds. Microhardness measurements were taken along two lines from the top surface toward the bottom of the samples. The first line is vertical in the middle of the SZ from the upper surface toward the bottom, and the indentations were carried out with increments of 0.1 mm in the processed area and 0.25 mm in the alloy matrix for a total distance of 3 mm. The other line is a horizontal line on the processed area at a distance 150 µm away from the top surface of the SZ, with increments of 0.5 mm, as schematically illustrated in Figure 3.

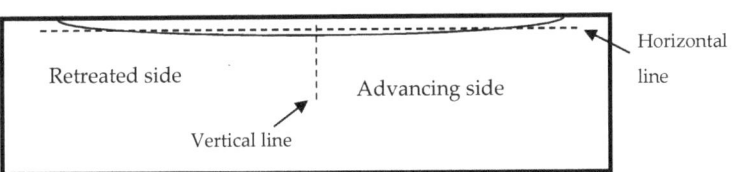

**Figure 3.** Schematic illustration of the microhardness measurements vertical and horizontal lines.

## 3. Results and Discussion

### 3.1. Processing of the Surface Composite

As a result of the two-part tool's flexibility, different hole sizes of the lower part of the tool (2, 4, and 6 mm) were used to examine particle dispersion efficiency. Visual examination showed that there was a noticeable change in the dispersion characteristics with increasing the hole size. Apparently, there was a flow discontinuity of the enforcement particles into the workpiece when processing with a 2-mm hole size since there was a blockage of the tool hole by a large amount of the SiC particles, as shown in Figure 4b. This discontinuity can most likely be attributed to the small size of the hole used and the enforcement particles adhesion characteristics. The blockage might be assisted by the high rotational speed, which might enhance the material flow due to the presence of significant centrifugal forces. Therefore, the workpiece processed by the tool with a 2-mm hole size showed no signs of reinforcement particles for plunge depths of either 0.6 or 0.9 mm. In addition, the use of a tool with a 4-mm hole size resulted in flow irregularity of the enforcement particles into the workpiece. This led to the non-uniform distribution of the enforcement particles in the upper region of the matrix. On the other hand, processing with a 6-mm hole size was shown to promote a continuous flow of the enforcement particles into the workpiece. This resulted in uninterrupted feeding of SiC particles into the aluminum matrix. However, there were back-extruded pieces of aluminum formed in the hole when using a tool with hole sizes of 4 and 6 mm, as shown in Figure 5. This was particularly evident with an offset distance of 6 mm, i.e., a plunge depth of 0.9 mm. The occurrence of back extrusion is believed to occur due to the buildup of friction-stirred material into the hole by the forging action of the tool. The easy access to the hole due to the short distance between the processed material and the hole opening aids in back extrusion, as do the softness and the quantity of the materials, as a result of tool rotation and plunge depth.

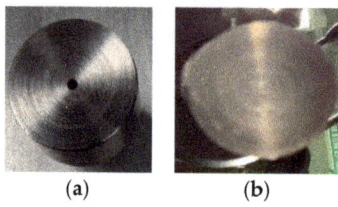

**Figure 4.** Photographs of (**a**) the tool before processing, and (**b**) blockage of 2 mm hole after processing using 2 mm hole size.

**Figure 5.** Photographs of the back-extruded aluminum during processing when the offset between the hole and the processing material is low during processing using 4- and 6-mm hole sizes.

In general, the processing and tool design parameters, such as the dispersion mechanism, the tool hole's size, the tilting angle, and the offset distance, were shown to significantly contribute in the blockage of tool hole and back extrusion buildups. Besides that, the characteristics and size of the SiC enforcement particles might have an effect on the flow and dispersion of enforcement particles. However, only one enforcement particle type and size were used, and consequently, the current study was focused on the critical processing parameters that significantly control the flow and dispersion of the SiC particles, i.e., hole size and plunge depth; both parameters were shown to affect the SiC particle blockage in the tool hole and the buildup of the back-extruded material during processing. Therefore, to avoid excessive amounts of back-extruded aluminum, a balance is required between the processed area and the offset distance between the tool's hole and the processed workpiece surface. This was achieved by adopting a tool design with a 6-mm hole size, using an offset distance of 7.5 mm, resulting in a plunge depth of 0.6 mm. This was shown to provide the best processing parameters for the development of a surface composite layer in the current study.

Figure 6a shows the aluminum matrix's processed area reinforced with SiC particles as a surface composite at the upper surface using a tool with a 6-mm hole size and a 0.6-mm plunge depth. A well-distinguished layer of surface composites of Al/SiC was shown to develop, with no evidence of porosity in the processed zone, illustrating a good mixture and adherence to the aluminum substrate. However, as previously mentioned, using a tool with a hole size of 4 or 6 mm combined with a plunge depth of 0.9 mm resulted in limited dispersion of the Si particles into the aluminum matrix and non-uniform surface composite layer, as illustrated in Figure 6b. This is due to the enforcement particle flow's irregularity, and more aluminum material is back extruded rather than being mixed with the SiC particles. The high rotational speed and the tilting degree hence plunge depth, aids in developing the required forging and frictional forces that produce the heat required for plasticization and the implementation of reinforcement particles into the matrix's upper surface, especially when enough enforcement particles are available. The reinforcement particles are stirred and bonded to the upper part of the SZ during FSP. Although the presence of SiC is expected to cause wear to the H13 tool, the H13 tool was visually inspected, and there is no evidence of tool wear. This can be attributed to the soft nature of the processed material examined in the current study, i.e., aluminum alloy 1100. This is also supported by the fact that the FSP tool used in this study had no pin, and thus the friction stirring action occurred at a shallow depth, resulting in a maximum thickness of 300 µm at the center of the processed zone.

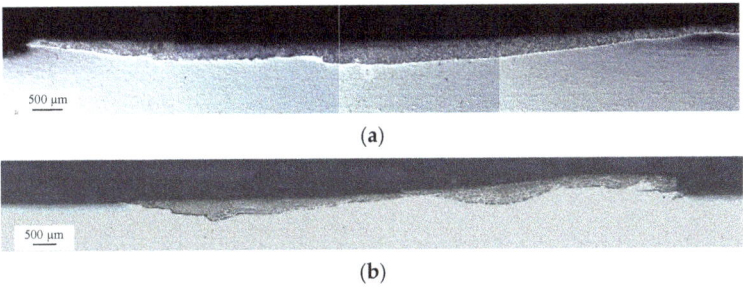

**Figure 6.** Panoramic image of the cross-section of the processed samples showing the composite surface using: (**a**) a tool with a hole size of 6 mm and a plunge depth of 0.6 mm; and (**b**) a tool with a hole size of 6 mm and a plunge depth of 0.9 mm.

In addition, the results show that some variation in the thickness of the developed surface composite layer had occurred. Although the surface composite approximately covered the main width of the shoulder, which is around 21 mm, the thickness of the processed layer varied slightly, reaching a maximum thickness of 300 µm at the center of the processed zone, i.e., near the hole, and minimum thickness of approximately 50 µm near the edge of the tool shoulder. The maximum thickness at the center of the processed zone was most likely achieved due to the high quantity of enforcement particles dispersed at the center during the processing. Some layer depth uniformity can be noticed around the center of the tool, which can be attributed to the high rotational speed of 3000 rpm and the moderate forging force generated from the tool's shoulder with a tilting angle of 7° and plunge depth of 0.6 mm. On the other hand, the workpiece processed with a plunge depth of 0.9 mm exhibited a large variation in thickness along the width of the processed zone. Unlike for the 0.6-mm plunge depth, the center of the processed zone (near the hole) exhibited the minimum thickness. This is a direct indication that back extrusion adversely inhibits the dispersion process at this center region. Additionally, the surface irregularities (inclinations) of the composite are evident, which suggests that the increase in the plunge depth was unnecessary to process the soft aluminum matrix.

The dispersion uniformity of the reinforcement particles generally depends on several factors, such as the vertical pressure on the base metal, number of passes, tool rotational direction, travel speed of the rotating tool, and the traverse tool speed. The vertical pressure on the base metal is reported [34] to develop a better forging and improve material flow and particles dispersion. Additionally, the increase in the number of passes plays a role in developing a better uniformity of the distributed second phase particles in addition to eliminating the porosity and developing refined microstructure [17,21,34,35]. Furthermore, significant homogeneity in the dispersion of SiC particles into the composite matrix has been reported [36] to occur as a result of changing the tool rotational direction between passes. Other factors such as the travel speed of the rotating tool and the traverse tool speed have various effects depending on the heating cycles developed during processing, as a result of deformation processes similar to extrusion and forging [20,35,37].

*3.2. Elemental Analysis of the Surface Composite*

Only the samples developed using the tool with a hole size of 6 mm, a tilting angle of 7° and a plunge depth of 0.6 mm were studied. Figure 7 presents the EDS elemental mapping of the surface composite region, illustrating the distribution of major elements in the processed zone (Al, Si, C, and O). No noticeable Fe content was detected in the cross-section of the surface composite. The processed area was mainly a mixture of aluminum matrix and dispersed microsized SiC particles, as well as some other oxides. The dispersed microsized SiC particles can be seen to be uniformly distributed in the aluminum matrix. The presence of oxygen is evidence for the possible formation of silicon or carbon oxides during FSP. Small pockets of oxygen are also present. Moreover, the concentration of silicon seems to be well distributed throughout the processed zone, indicating a sufficient dispersion of SiC particles in the matrix and a lack of clustering or agglomerations of SiC particles.

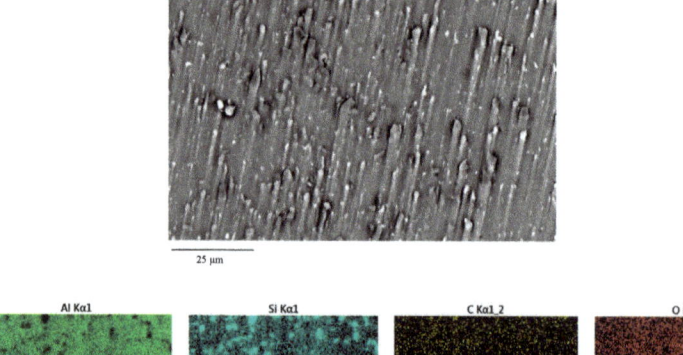

**Figure 7.** EDS elemental mapping of the surface composite.

The through-thickness distribution of the silicon content is further examined in Figure 8. The spatial elemental concentration, presented in terms of counts per second, indicates that the aluminum content is uniform, and is rich within the base metal through to the composite/matrix interface. In the processed zone, the SiC content increases with lower aluminum content, which indicates good dispersion of SiC particles. On the other hand, the Si content is almost zero in the base metal, and increases toward the composite's outer layer. Other elements, such as oxygen and carbon, show a negligible change in content through the thickness.

The variation in silicon throughout the width of the processed zone was also examined, as presented in Figure 9. To eliminate the presence of foreign particles, the EDS measurements were carried out roughly 100 μm beneath the top layer of the processed zone. The results show that silicon and aluminum are randomly present. The EDS spectrum generally indicates a larger number of counts for aluminum compared to silicon content for all points along the line of measurements. However, at a measurement distance of approximately 110 μm, there is a large peak of silicon content presenting higher counts than those for aluminum. The higher silicon content might be a result of the clustering of

SiC particles. Other Si peaks are also shown in the spectrum, indicating small regions of SiC clusters throughout the processed zone. The presence of silicon is evident from the line of measurements, which indicates sufficient dispersion of SiC particles.

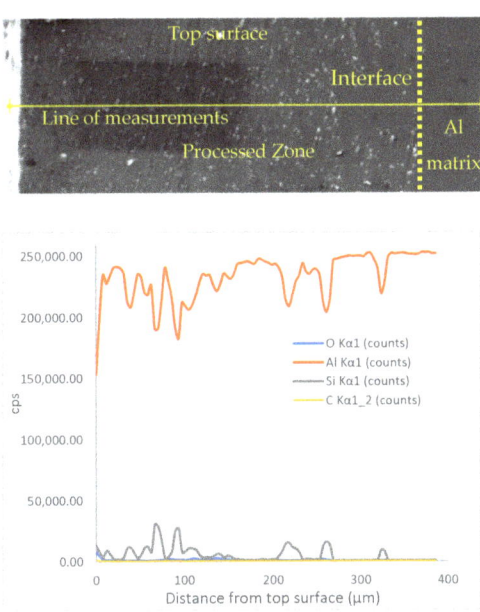

**Figure 8.** Vertical line EDS spectrum at the center of the surface composite starting from the top surface of the composite layer to the composite/matrix interface.

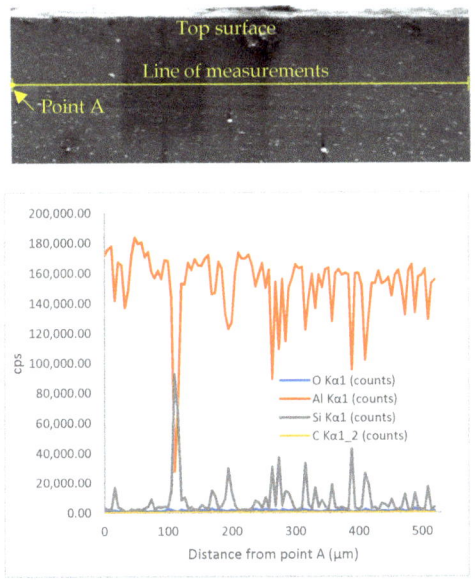

**Figure 9.** Horizontal line EDS spectrum of surface composite measured at 100 µm below top surface.

## 3.3. Microhardness

The widthwise microhardness profile of the processed zone, along with the average microhardness of the as-received 1100 aluminum, are shown in Figure 10 for three processing conditions. The microhardness indentations were measured at 150 μm below the top surface of the SZ. The average microhardness of the as-received 1100 aluminum was 38.5 Hv. In the case of FSP with negligible SiC dispersion, the results show a reduction in the hardness below the base metal value for all points measured in the processed zone. This indicates that the as-received aluminum, which was initially in cold-worked condition, was exposed to high frictional heat during FSP, resulting in a drop in hardness in the heat-treatable 1100 aluminum alloy. For the workpiece processed using a plunge depth of 0.6 mm, the surface composite's microhardness was, remarkably, higher than the base metal due to the dispersion-strengthening mechanism attained as a result of the presence of the reinforcement SiC particles in the processed zone [4,35,38,39]. High hardness values, approaching 76 Hv, were recorded near the center of the processed zone, which is consistent with the OM and EDS results, supporting the production of excellent SiC dispersion in the thick composite layer at the center of the processed zone. The increase of hardness extends roughly 5 mm from the center of the processed zone. Beyond 5 mm, a reduction in the hardness was shown to occur, reaching a minimum of approximately 24 Hv at the composite/matrix interface, which is lower than the hardness recorded for the as-received aluminum (38.5 Hv). The reduction in hardness demonstrates that the dispersion of SiC particles into the aluminum matrix became less effective as the distance from the tool center increases. Over such large distances, the hardness is strongly governed by the aluminum matrix.

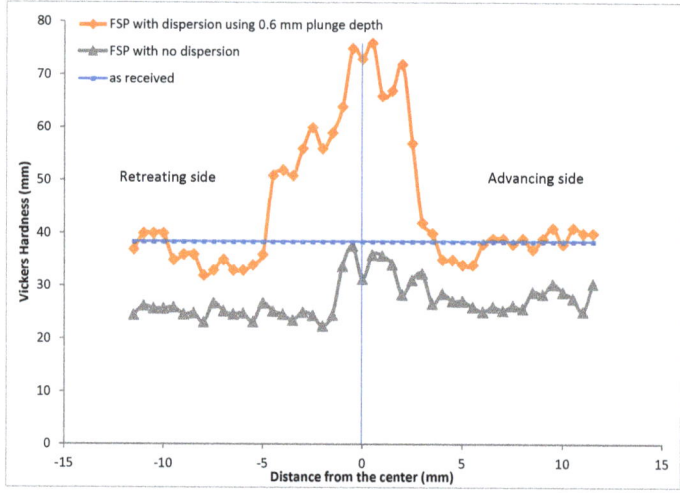

**Figure 10.** Vickers microhardness profile of a horizontal line at the upper layer of the sample.

In addition, the hardness profile of the workpiece processed using a plunge depth of 0.6 mm was not symmetrical. The hardness was slightly higher on the retreating side. Additionally, the decrease in hardness below the base metal value was less present on the retreating side. Such non-symmetrical behavior can mostly be attributed to the temperature increase in the advancing side in comparison to the retreating side during processing, as reported by [40,41]. The hardness profile recorded for the workpiece processed with a plunge depth of 0.6 mm was reported by Sharma et al. [26], indicating steep fluctuations in the hardness values from the center towards the composite/matrix interface. The fluctuations in hardness are strongly related to the SiC agglomeration and formation of clusters/bands of and/or the number of dispersed particles. The volume fraction of SiC particles is expected to influence the microhardness in the processed zone. As the volume percent of the SiC particles increases, the microhardness eventually increases. This was reported by Mishra et al. [4], where the microhardness of the surface composites increased by more than 50 Hv when the vol.% of SiC particles was increased from 13% to 27%. The measured microhardness values for the workpiece processed using a plunge depth of 0.6 mm were generally in good agreement with the results of the semi-quantitative elemental mapping presented in Figures 8 and 9.

Figure 11 presents the through-thickness hardness profile at the center of the three cases examined in Figure 10. For the workpiece processed by FSP without particle dispersion, there is a decline in the hardness below the as-received value, reaching a minimum of 28 Hv at approximately 0.4 mm below the top surface. The hardness values gradually increase at increasing depth, reaching the hardness of the as-received material at a depth of 3 mm. For the workpiece by a plunge depth of 0.6 mm, the dispersion of SiC particles at the top surface resulted in an increase in hardness to 76 Hv and a decrease at greater depths, dropping to approximately 45 Hv at the composite/matrix interface (0.5 mm from the top surface). The remarkable increase in the microhardness at the outer processed surface is most likely attributable to the presence of the reinforcement SiC particles, as demonstrated by the ESD mapping presented in Figures 8 and 9. In addition, a further reduction in hardness to 35 Hv was recorded at depths greater than the composite/matrix interface (between 0.5 and 1 mm). However, the hardness gradually increased to the hardness of the as-received material at a depth of 1 mm. Compared to the workpiece processed using FSP without SiC dispersion, the decrease in hardness in the aluminum matrix region next to the composite/matrix interface was less severe for the workpiece processed using SiC dispersion via FSP. This indicates that the amount of frictional heat absorbed by the aluminum matrix was greater for the workpiece processed by FSP without SiC dispersion, and thus the reduction in hardness was not only stronger, but also occurred at a greater depth below the processed surface. It can also be deduced that dispersion of SiC on the top surface of the workpiece acted as a lubricant, reducing the transition of frictional heat in the aluminum matrix due to FSP, and thus producing less loss in hardness in the base metal.

The plunge depth was shown to strongly affect the development of uniform thickness of the composite layer, as well as the uniformity of SiC particle dispersion and hardness development. For a given angle of tilt (°7), the increase in plunge depth resulted in an increase in the contact area between the shoulder and the workpiece. This eventually promoted greater depth in the composite layer, considering that the back-extrusion force was not sufficiently high to cause matrix plastic flow into the tool hole. However, the increase in plunge depth could have a deteriorating effect on the matrix microhardness due to the high amount of heat generated during processing, as previously reported by Rathee et al. [42]. However, the reduction in matrix microhardness was shown to be small when the plunge depth was carefully chosen at a value of 0.6 mm, as also demonstrated in Figure 11.

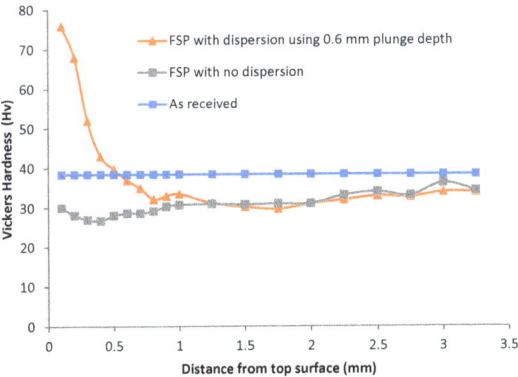

**Figure 11.** Vickers microhardness through-thickness profile of the FSP aluminum alloy.

## 4. Conclusions

A new approach and tool design for direct friction stir processing to eliminate the step of preplacing reinforcement particles into the base metal was proposed for developing a surface composite layer on 1100 aluminum alloy. The FSP tool consists of a two-part hollow body, i.e., a hollow shank and a shoulder. The new processing approach requires less than half of the tool shoulder to contact the surface, allowing the dispersion of reinforcement particles onto the workpiece's surface through the hole and mitigating the blockage of the hole during processing. Three different sizes of hole were used, but the 6-mm hole was the optimum. The high rotational speed rate, a tilting angle of 7° for the tool, and a moderate plunge depth produced the heat required for plasticization, leading to the application of reinforcement particles. These tool design and processing approaches were shown to effectively disperse the microsized SiC particles into the aluminum matrix during FSP, reaching a uniform distribution of SiC particles in the SZ at the top surface of the base metal with no porosity. The Al/SiC surface composite layer has a thickness ranging from 50 µm, near the edge of the processed area, to about 300 µm, at the center of the processed area. The microhardness increased remarkably from 38.5 Hv at the base to about 80 Hv at the composite's top layer.

**Author Contributions:** Conceptualization, A.I.A.; methodology, A.I.A., K.J.A.-F. and S.N.A.; software, A.I.A., K.J.A.-F. and S.N.A.; validation, A.I.A., K.J.A.-F. and S.N.A.; formal analysis, A.I.A., and K.J.A.-F.; investigation, A.I.A., K.J.A.-F. and S.N.A.; resources, A.I.A., K.J.A.-F. and S.N.A.; data curation, A.I.A., K.J.A.-F. and S.N.A.; writing—original draft preparation, A.I.A., K.J.A.-F. and S.N.A.; writing—review and editing, A.I.A., K.J.A.-F. and S.N.A.; visualization, A.I.A., K.J.A.-F. and S.N.A.; supervision, A.I.A., K.J.A.-F. and S.N.A.; project administration, A.I.A. All authors have read and agreed to the published version of the manuscript.

**Funding:** The authors acknowledge the support provided by the Public Authority for Applied Education and Training (PAAET) Grants No. TS-18-12. The authors also acknowledge the support provided by Kuwait University General Facility (Grant No. GE 01/07) for sample preparation, OM, EDS analysis, and microhardness measurements.

**Acknowledgments:** The authors acknowledge the help of engineers Ahmad Shehata and Shaji Michael in setting the experiments and preparing samples.

**Conflicts of Interest:** The authors declare no conflict of interest.

## References

1. Jawalkar, C.; Kant, S. A Review on Use of Aluminium Alloys in Aircraft Components. *i-Manag. J. Mater. Sci.* **2015**, *3*, 33.
2. Fridlyander, I.N.; Sister, V.G.; Grushko, O.E.; Berstenev, V.V.; Sheveleva, L.M.; Ivanova, L.A. Aluminum Alloys: Promising Materials in the Automotive Industry. *Metal Sci. Heat Treat.* **2002**, *44*, 365–370. [CrossRef]
3. Karthikeyan, P.; Mahadevan, K.; Thomas, W.; Mishra, R.; Ma, Z.; Elangovan, K.; Balasubramanian, V.; Elangovan, K.; Balasubramanian, V.; Karthikeyan, P. Influences of tool pin profile and welding speed on the formation of friction stir processing zone in AA2219 aluminium alloy. *J. Appl. Sci.* **1991**, *12*, 1–78.
4. Mishra, R.S.; Ma, Z.Y.; Charit, I. Friction stir processing: A novel technique for fabrication of surface composite. *Mater. Sci. Eng. A* **2003**, *341*, 307–310. [CrossRef]
5. Mishra, R.S.; Mahoney, M.W.; McFadden, S.X.; Mara, N.A.; Mukherjee, A.K. High strain rate superplasticity in a friction stir processed 7075 Al alloy. *Scr. Mater.* **1999**, *42*, 163–168. [CrossRef]
6. Padhy, G.K.; Wu, C.S.; Gao, S. Friction stir based welding and processing technologies—Processes, parameters, microstructures and applications: A review. *J. Mater. Sci. Technol.* **2018**, *34*, 1–38. [CrossRef]
7. Rathee, S.; Maheshwari, S.; Siddiquee, A.N. Issues and strategies in composite fabrication via friction stir processing: A review. *Mater. Manuf. Process.* **2018**, *33*, 239–261. [CrossRef]
8. Sudhakar, M.; Srinivasa Rao, C.H.; Saheb, K.M. Production of Surface Composites by Friction Stir Processing-A Review. *Mater. Today: Proc.* **2018**, *5*, 929–935. [CrossRef]
9. Sharma, V.; Prakash, U.; Kumar, B.V.M. Surface composites by friction stir processing: A review. *J. Mater. Process. Technol.* **2015**, *224*, 117–134. [CrossRef]
10. Zhang, Y.N.; Cao, X.; Larose, S.; Wanjara, P. Review of tools for friction stir welding and processing. *Can. Metall. Q.* **2012**, *51*, 250–261. [CrossRef]
11. Mishra, R.S.; Ma, Z.Y. Friction stir welding and processing. *Mater. Sci. Eng. R Rep.* **2005**, *50*, 1–78. [CrossRef]

12. Sharma, V.; Prakash, U.; Kumar, B.V.M. Challenges in Fabrication of Surface Composites by Friction Stir Processing Route. In *Advanced Composites for Aerospace, Marine, and Land Applications II*; Sano, T., Srivatsan, T.S., Eds.; Springer International Publishing: Cham, Switzerland, 2016; pp. 93–100.
13. Shojaeefard, M.H.; Akbari, M.; Khalkhali, A.; Asadi, P. Effect of tool pin profile on distribution of reinforcement particles during friction stir processing of B4C/aluminum composites. *Proc. Inst. Mech. Eng. Part L J. Mater. Des. Appl.* **2016**, *232*, 637–651. [CrossRef]
14. Chaudhary, A.; Kumar Dev, A.; Goel, A.; Butola, R.; Ranganath, M.S. The Mechanical Properties of Different alloys in friction stir processing: A Review. *Mater. Today Proc.* **2018**, *5*, 5553–5562. [CrossRef]
15. Węglowski, M.S. Friction stir processing—State of the art. *Arch. Civ. Mech. Eng.* **2018**, *18*, 114–129. [CrossRef]
16. Al-Fadhalah, K.J.; Almazrouee, A.I.; Aloraier, A.S. Microstructure and mechanical properties of multi-pass friction stir processed aluminum alloy 6063. *Mater. Des.* **2014**, *53*, 550–560. [CrossRef]
17. Singh, S.; Kaur, M.; Saravanan, I. Enhanced microstructure and mechanical properties of boiler steel via Friction Stir Processing. *Mater. Today Proc.* **2020**, *22*, 482–486. [CrossRef]
18. Ding, Z.; Fan, Q.; Wang, L. A Review on Friction Stir Processing of Titanium Alloy: Characterization, Method, Microstructure, Properties. *Metall. Mater. Trans. B* **2019**, *50*, 2134–2162. [CrossRef]
19. Zayed, E.M.; El-Tayeb, N.S.M.; Ahmed, M.M.Z.; Rashad, R.M. Development and Characterization of AA5083 Reinforced with SiC and Al2O3 Particles by Friction Stir Processing. In *Engineering Design Applications*; Öchsner, A., Altenbach, H., Eds.; Springer International Publishing: Cham, Switzerland, 2019; pp. 11–26.
20. Buradagunta, R.S. Different strategies of secondary phase incorporation into metallic sheets by friction stir processing in developing surface composites. *Int. J. Mech. Mater. Eng.* **2016**, *11*. [CrossRef]
21. Guo, J.F.; Liu, J.; Sun, C.N.; Maleksaeedi, S.; Bi, G.; Tan, M.J.; Wei, J. Effect of Nano-Al$_2$O$_3$ Particle Addition on Grain Structure Evolution and Mechanical Behavior of Friction-Stir-Processed Al. *Mater. Sci. Eng. A* **2014**, *602*, 143. [CrossRef]
22. Min, Y.; Chengying, X.; Chuansong, W. Fabrication of AA6061/Al$_2$O$_3$ Nano Ceramic Particle Reinforced Composite Coating by using Friction Stir Processing. *J. Mater. Sci.* **2010**, *45*, 4431.
23. Lorenzo-Martin, M.C.; Ajayi, O.O. Surface Layer Modification of 6061 Al Alloy by Friction Stir Processing and Second Phase Hard Particles for Improved Friction and Wear Performance. *J. Tribol.* **2014**, *136*, 044501. [CrossRef]
24. Muthukumar, P.; Jerome, S. Surface coating (Al/Cu &Al/SiC) fabricated by direct particle injection tool for friction stir processing: Evolution of phases, microstructure and mechanical properties. *Surf. Coat. Technol.* **2019**, *366*, 190–198.
25. Sathiskumar, R.; Murugan, N.; Dinaharan, I.; Vijay, S.J. Characterization of boron carbide particulate reinforced in situ copper surface composites synthesized using friction stir processing. *Mater. Charact.* **2013**, *84*, 16–27. [CrossRef]
26. Sharma, V.; Gupta, Y.; Kumar, B.V.M.; Prakash, U. Friction Stir Processing Strategies for Uniform Distribution of Reinforcement in a Surface Composite. *Mater. Manuf. Process.* **2016**, *31*, 1384–1392. [CrossRef]
27. Miranda, R.M.; Santos, T.G.; Gandra, J.; Lopes, N.; Silva, R.J.C. Reinforcement strategies for producing functionally graded materials by friction stir processing in aluminium alloys. *J. Mater. Process. Technol.* **2013**, *213*, 1609–1615. [CrossRef]
28. Lee, C.J.; Huang, J.C.; Hsieh, P.J. Mg based nano-composites fabricated by friction stir processing. *Scr. Mater.* **2006**, *54*, 1415–1420. [CrossRef]
29. Iwaszko, J.; Kudła, K.; Fila, K. Friction stir processing of the AZ91 magnesium alloy with SiC particles. *Arch. Mater. Sci. Eng.* **2016**, *77*, 85–92. [CrossRef]
30. Huang, Y.; Wang, T.; Guo, W.; Wan, L.; Lv, S. Microstructure and surface mechanical property of AZ31 Mg/SiCp surface composite fabricated by Direct Friction Stir Processing. *Mater. Des.* **2014**, *59*, 274–278. [CrossRef]
31. Elangovan, K.; Balasubramanian, V.; Valliappan, M. Effect of Tool Pin Profile and Rotational speed on Mechanical Properties of friction stir welded in AA6061 aluminium alloy. *Mater. Manuf. Process.* **2008**, *23*, 251. [CrossRef]
32. Mahesh, V.P.; Arora, A. Effect of Tool Shoulder Diameter on the Surface Hardness of Aluminum-Molybdenum Surface Composites Developed by Single and Double Groove Friction Stir Processing. *Metall. Mater. Trans. A* **2019**, *50*, 5373–5383. [CrossRef]
33. Elangovan, K.; Balasubramanian, V.; Valliappan, M. Influence of tool pin profile and axial force on the formation of friction stir processing zone in AA6061 aluminium alloy. *Int. J. Adv. Manuf. Technol.* **2008**, *38*, 285. [CrossRef]
34. Asadi, P.; Faraji, G.; Besharati, M.K. Producing of AZ91/SiC composite by friction stir processing (FSP). *Int. J. Adv. Manuf. Technol.* **2010**, *51*, 247–260. [CrossRef]
35. Arora, H.S.; Singh, H.; Dhindaw, B.K. Composite fabrication using friction stir processing—A review. *Int. J. Adv. Manuf. Technol.* **2012**, *61*, 1043–1055. [CrossRef]
36. Dolatkhah, A.; Golbabaei, P.; Besharati Givi, M.K.; Molaiekiya, F. Investigating effects of process parameters on microstructural and mechanical properties of Al5052/SiC metal matrix composite fabricated via friction stir processing. *Mater. Des.* **2012**, *37*, 458–464. [CrossRef]
37. Arbegast, W.J. A flow-partitioned deformation zone model for defect formation during friction stir welding. *Scr. Mater.* **2008**, *58*, 372–376. [CrossRef]
38. Wang, W.; Shi, Q.-y.; Liu, P.; Li, H.-k.; Li, T. A novel way to produce bulk SiCp reinforced aluminum metal matrix composites by friction stir processing. *J. Mater. Process. Technol.* **2009**, *209*, 2099–2103. [CrossRef]
39. Choi, D.-H.; Kim, Y.-I.; Kim, D.A.E.U.; Jung, S.-B. Effect of SiC particles on microstructure and mechanical property of friction stir processed AA6061-T4. *Trans. Nonferrous Met. Soc. China* **2012**, *22*, s614–s618. [CrossRef]

40. Mohammed, M.H.; Subhi, A.D. Exploring the influence of process parameters on the properties of SiC/A380 Al alloy surface composite fabricated by friction stir processing. *Eng. Sci. Technol. Int. J.* **2021**. [CrossRef]
41. Hamilton, C.; Kopyściański, M.; Senkov, O.; Dymek, S. A Coupled Thermal/Material Flow Model of Friction Stir Welding Applied to Sc-Modified Aluminum Alloys. *Metall. Mater. Trans. A* **2013**, *44*, 1730–1740. [CrossRef]
42. Rathee, S.; Maheshwari, S.; Siddiquee, A.N.; Srivastava, M. Effect of tool plunge depth on reinforcement particles distribution in surface composite fabrication via friction stir processing. *Def. Technol.* **2017**, *13*, 86–91. [CrossRef]

MDPI
St. Alban-Anlage 66
4052 Basel
Switzerland
Tel. +41 61 683 77 34
Fax +41 61 302 89 18
www.mdpi.com

Crystals Editorial Office
E-mail: crystals@mdpi.com
www.mdpi.com/journal/crystals

www.ingramcontent.com/pod-product-compliance
Lightning Source LLC
LaVergne TN
LVHW070143100526
838202LV00015B/1884